Orchid Biology: Reviews and Perspectives, X

Orchid Biology: Reviews and Perspectives, X

Edited by

Tiiu Kull
Department of Botany,
Institute of Agricultural and Environmental Sciences
Estonian University of Life Sciences, Tartu, Estonia

Joseph Arditti
Department of Developmental and Cell Biology
University of California, Irivne, CA, USA

Sek Man Wong
Department of Biological Sciences,
National University of Singapore, Singapore

Springer

Editors
Tiiu Kull
Department of Botany
Institute of Agricultural and
Environmental Sciences
Estonian University of Life Sciences
Tartu, Estonia

Joseph Arditti
Department of Developmental
and Cell Biology
University of California
Irivne, CA
USA

Sek Man Wong
Department of Biological Sciences
National University of Singapore
Singapore

ISBN: 978-1-4020-8801-8 e-ISBN: 978-1-4020-8802-5

Library of Congress Control Number: 2008932566

Printed on acid-free paper

9 8 7 6 5 4 3 2 1

springer.com

Dedicated by the editors to

Kee Yoeup Paek
Norris Williams
Edward Yeung

Editors

Founding Editor

Associate Editors

Board of Editors

In addition to the editors and associate editors the following individuals took part in the production of this volume by providing advice, comments and reviews.

Manuscripts and Publication Policy

Orchid Biology, Reviews and Perspectives (*OB*) publishes only reviews, not original research papers. It is also not the policy of *OB* to publish letters to the editor. Although most *OB* chapters are invited, submitted manuscripts are also considered. Authors of such manuscripts are advised, but not required, to contact one of the editors well in advance before starting to write. Manuscripts in a language other than English and/or not in strict adherence with the format used by recent *OB* volumes will be rejected outright. Authors of such manuscripts will be notified that their manuscripts have been rejected, but only if they provided an e-mail address. The manuscripts will not be returned unless accompanied by self-addressed postage prepaid envelopes. All submitted manuscripts which meet the conditions set above will be reviewed by the editors and at least two additional reviewers. Acceptance or rejection will be based on reviewer recommendations. In cases of split opinions acceptance or rejections will be based on the recommendations of the majority of the reviewers.

Persons to Whom this Volume is Dedicated

Kee-yoeup Paek was born on 24 February 1951 in Korea. He received his Ph.D. from the department of horticulture at Kyungbuk National University in 1984. After that he was a postdoctoral fellow at the Plant Physiology Research Group at the University of Calgary from 1985 until 1986. He joined the Dcpartment of Horticultural Science, Chungbuk National University in 1982 and became director of the Research Center for the Development of Advanced Horticultural Technology in March, 1996.

Other appointments and activities include:

Invited Professor, Institute of Agricultural Biochemistry, Tsukuba University, Japan, 1989–1990

Project Leader, Orchid Research Group (supported by the Korea Science and Engineering Foundation), 1993

Invited Professor, Department of Horticulture, Chiba University, Japan, 1994

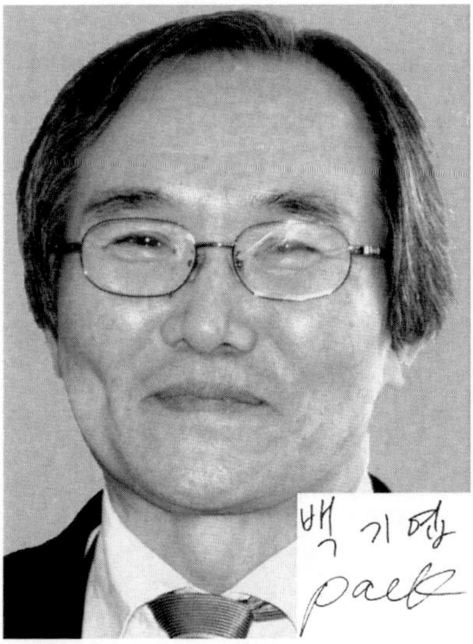

Member, Scientific and Judging Committees, Joint Federation of Oriental Orchids, 1995
Member, Scientific and Judging Committees, Asia Pacific Orchid Conference, 1995–to date
Member, Technology Support Group for Consulting on Export of Agricultural Products Export and Korea Agro Fisheries Trade Cooperation, 2005
Invited Professor, Institute of Plant Physiology, Russian Academy of Science, Moscow, 2006
Chairman of Scientific Committee, 9th Asia Pacific Orchid Conference, 2007

Honors received by Prof. Paek include:

Appreciation Plaque from the Association of Korean Orchid Growers, 2002
Certificate of Appreciation, 8th Asia-Pacific Orchid Conference, Taiwan, 2004
Judginge, Korea Orchid Exhibition Committee, 2005
Member, Technology Support Group for Consulting on Export of Agricultural Products Export and Korea Agro Fisheries Trade Cooperation, 2005.
Professor Paek has published widely in scientific journals, symposium volumes, monograph series and magazines – *Eun-Joo Hahn, Research Professor, Research Center for the Development of Advanced Horticultural Technology, Chungbuk National University Cheongju, 361–763, South Korea, ejhahn@chungbuk.ac.kr*

Norris Williams was born in Anniston, Alabama on 31 March 1943. He received his B.S. and M.S. degrees in Biology from the University of Alabama, and his Ph.D. in Biology at the University of Miami under the direction of Calaway Dodson in 1971. Following several postdoctoral positions at the University of Miami, the Smithsonian Institution, and the Fairchild Tropical Garden, he accepted a position as Assistant Professor of Biology at Florida State University in Tallahassee in 1973.

In 1981, he moved to Gainesville, Florida to assume his present position as Keeper of the Herbarium in the Florida Museum of Natural History and Affiliate Professor of Botany at the University of Florida. He and his wife Nancy have two sons who have inherited their father's love of antique MG cars.

As a Ph.D. student under Dodson, Norris travelled extensively in Central America and Ecuador. During that time he absorbed information on orchid biology from Dodson and Robert Dressler, and was often accompanied by fellow graduate students Kiat Tan (formerly Director of Parks in Singapore, now retired), Harold Hills, Ralph Adams, Kathy Gregg and Hans Wiehler.

Norris's dissertation was a revision of *Brassia* (Oncidiinae), but his research interests quickly shifted to the pollination biology of orchids pollinated by fragrance-collecting male euglossine bees (a topic first studied by Dodson, Dressler and their students). He expanded the work of the Dodson group, pioneering the use of gas chromatography-mass spectrometry to analyze the floral fragrances of orchids and the glandular secretions of male euglossine bees and laying the foundation for our current understanding of the orchid/bee interaction.

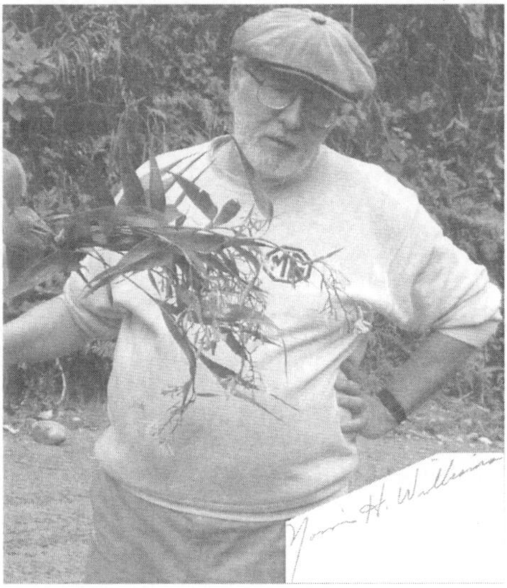

Dr. Williams's other early research interests included orchid anatomy and palynology. In the 1990s, orchid systematics was increasingly influenced by data from DNA sequencing, and Norris's research focus soon shifted back to Oncidiinae, this time utilizing sequence data and phylogenetic analyses. Together with Mark Chase and coworkers, his studies have revolutionized our concepts of evolutionary relationships in the Oncidiinae, one of the largest and most horticulturally important groups of neotropical orchids.

Perhaps Norris's most enduring contribution to orchidology will be his graduate students who have also become leading orchid researchers and teachers. They include John Atwood (formerly of Selby Botanical Gardens), James Ackerman (University of Puerto Rico), Alec Pridgeon (Royal Botanic Gardens, Kew) and Mark Whitten (Florida Museum of Natural History). His research continues to focus on molecular systematics of neotropical orchids, and his active laboratory currently includes Ph.D. students from Costa Rica, Ecuador, Colombia, and the United States – *Mark Whitten, University of Florida, whitten@flmnh.ufl.edu*

Edward C. Yeung was born in Hong Kong on 21 October 1947. He received his B.Sc. (1973) from Guelph University (graduating with distinction), and Ph.D. (1977) from Yale University. After spending one year as a postdoctoral fellow at the University of Ottawa, he joined the Department of Biological Sciences at University of Calgary in July 1978. He was promoted to Associate Professor in 1984 and became a Full Professor six years later. In addition, he also served on the Natural Sciences and Engineering Research Council of Canada (NSERC) committee. The University of Calgary recognized his teaching excellence with the Faculty of Science Teaching Excellence Award in 2002. His primary research interests have been reproduction biology of higher plants, especially the structural and physiological aspects of embryo development. In his early research he used *Epidendrum ibaguense*

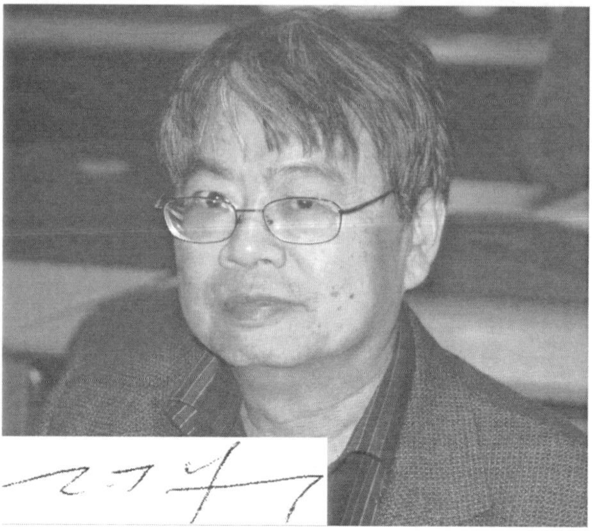

as his major plant material. He also used some native Canadian orchid species, such as *Calypso bulbosa* and *Cypripedium passerinum* to study the ovule and embryo development. Later he collaborated with Dr. Zee Sze-Yong (University of Hong Kong) and Dr. Y. Xiu-Lin (South China Institute of Botany) focusing on the embryology of *Cymbidium sinense* and *Phaius tanervilliae*. Recently, he studied seed germination and embryo development on *Cypripedium formosanum*, *Paphiopedilum delenatii*, *Calanthe tricarinata* and *Phalaenopsis amabilis* var. *formosa*. He has had three Ph.D students and five Master's students so far. Dr. Yeung enjoys cooking for friends, students and family, eating, Sake, playing Spider Solitaire and singing Karaoke – *Yung Lee*, *leeyung@hotmail.com*

Preface

As stated many times before the purpose of *Orchid Biology, Reviews and Perspectives* (*OB*) is to present reviews on all aspects of orchids. The aim is not to balance every volume, but to make a balanced and wide ranging presentation of orchids in the series as a whole. The chapters in this, the last volume of the series, range over a number of topics which were not covered before.

Singapore is justly famed for its orchids. They can be seen on arrival (or departure) in its modern, highly efficient and comfortable Changi Airport and on the way from it to town. *Vanda* Miss Joaquim, the first hybrid to come from Singapore became its National Flower. This natural hybrid can be seen on its currency, stamps, and public and private decorations. Many excellent breeders, starting with Prof. Eric Holttum who bred the first man made hybrid (*Spathoglottis* Primrose), produced numerous magnificent hybrids and won countless awards in Singapore and elsewhere. These hybrids served to enrich the country's orchid mystique. In the opening chapter of this volume Dr. Teoh Eng Soon (Western style: Eng Soon Teoh), himself a prize winning orchid breeder, grower and author writes about some of the breeders who contributed to the Singapore orchid fame.

Prof. Hans Fitting was one of the best known plant physiologists of his time. As a young man he studied the effects of pollen on orchid flowers. His studies led him to become the first plant scientist to propose that plants produce hormones. Fitting assumed that pollen exerts its effects on flowers through a hormone which he named *Pollenhormon*. Some suggested that *Pollenhormon* was auxin. Fitting never accept this suggestion. Now it seems that the extract Fitting called *Pollenhormon* was probably a mixture of auxin and other substances. Dr. T. W. Yam and his coauthors describe Fitting's work and draw conclusions which are based on modern plant physiology.

The bible is an important part of the lives of many people. It is regarded as absolute truth by the devout and studied by both believers and non believers. Orchids are not directly mentioned in the Bible, but several biblical plants have been described as being Orchidaceous. Professor Emeritus A. S. Dunn, an animal biologist who has studied both the Bible and the Talmud as an avocation and Professor Emeritus J. Arditti, who asserts repeatedly that he only knows orchids, examine the question of orchids in the bible in their chapter.

Food hairs play an important role in the biology of orchids. In chapter four Dr. K. L. Davis surveys their form and the effects they have on orchid diversification.

Pollen dispersal of orchids has been researched extensively and its consequences have been discussed widely. Prof. E. Pacini's chapter is the first review on the subject.

Orchids have found their way into the arts, letters and even into music. K. E. Quinn discusses orchids in art in Chapter 6.

Vanilla is the only orchid grown as a plantation crop for human consumption. Its cultivation became possible only after the discovery of a hand pollination method. The question of who discovered the method is the subject of Chapter 7 by Professors J. Arditti, A. N. Rao and H. Nair.

Viruses which attack orchids are a major problem because they cause significant financial losses. Effective control of these viruses requires an understanding of their nature. Dr. Ajjikuttira and Prof. Wong provide such understanding in Chapter 8.

Appendixes in this series have always been practical in nature and intended to provide information which can be used directly by readers. The appendix in this volume is a list of more than 2000 books on orchids in several languages.

* * *

The founding editor, Joseph Arditti who guided all ten volumes conceived the idea of this series in during the late 1960s, started to collect and write chapters in the early 1970s and managed to get the first volume published in 1977. For a complete list of chapters and appendices please consult the closing pages of this volume). Since its inception the statistics for the series are:

volumes, ten (including this one),
pages, approximately 4,500,
founding editor, Joseph Arditti,
co editors, Tiiu Kull (vols. 8, 9, 10), Alec M. Pridgeon (vol. 7), Kenneth M. Cameron (vol. 9), Sek Man Wong (vol. 10),
chapters, 71,
appendixes, 10,
authors and contributors, *ca* 88 from 26 countries,
publishers, 5.

The founding editor retired from the University of California, Irvine, in 2001 and became Professor Emeritus, but continued his involvement with *OB*. He stated in the preface to volume IX that his involvement with the series will end with the present volume. As it turns out this, the tenth volume in the series will also be the last one. This series will be terminated with the present volume. The decision to terminate the series is firm but not irreversible. Anyone interested in continuing the series should contact Prof. Arditti with a detailed plan. All plans will be considered. We thank all those (too many to list) who helped us make this series a success.

Tiiu Kull, Tartu, Estonia
Joseph Arditti, Irvine, California, USA
Sek Man Wong, Singapore

Contents

Editors vii

Manuscripts and Publication Policy ix

Persons to Whom this Volume is Dedicated xi

Preface xv

Contributors xix

Biographies

1 Unsung Heroes of the Singapore-Malaysian Orchid World (1951–2000) 1
Eng Soon Teoh

History-Physiology

2 Pollination Effects on Orchid Flowers and the First Suggestion by Professor Hans Fitting (1877–1970) that Plants Produce Hormones 37
Tim Wing Yam, Yong Neng Chow, Popuri Nageswara Avadhani, Choy Sin Hew, Joseph Arditti, and Hubert Kurzweil

Biblical and Talmudic Literature

3 Are Orchids Mentioned in the Bible? 141
Arnold S. Dunn and Joseph Arditti

Morphology

4 Food-Hair Form and Diversification in Orchids 159
Kevin. L. Davies

Pollination Biology

5 Orchids Pollen Dispersal Units and Reproductive Consequences 185
Ettore Pacini

Art

6 A Rare Beauty: The Orchid in Western Art 219
Karen E. Quinn

History-Pollination

7 Hand-Pollination of *Vanilla*: How Many Discoverers? 233
Joseph Arditti, Adisheshappa Nagaraja Rao, and Helen Nair

Virology

8 Molecular Biology of Two Orchid Infecting Viruses: Cymbidium Mosaic Potexvirus and Odontoglossum Ringspot Tobamovirus 251
Prabha Ajjikuttira and Sek-Man Wong

Appendix

A List of Orchid Books 279
Tim Wing Yam, Benjamin Singer, Choy Sin Hew, Tiiu Kull, Irina Tatarenko, and Joseph Arditti

Index of Persons 425

Index of Organism Names 435

Subject Index 441

Contents of Volumes I–IX 447

Contributors

Prabha Ajjikuttira received her B.Sc. degree in Chemistry and Biology in1987 and her M.Sc. in Botany, specializing in Microbiology, in 1989, both from Bangalore University, India. She earned an M.Phil in Biotechnology in 1991, studying oleogenesis in sunflower seeds at the Central Food Technological Research Institute, India. After a short teaching stint, she worked as a research scientist in commercial plant tissue culture laboratories. These opportunities developed her interest in research. As a graduate student in the laboratory of Prof. Sek-Man Wong at the National University of Singapore, her Ph.D. thesis focused on orchid-infecting viruses. Graduating in 2004, she worked as a postdoctoral fellow at the Institute of Bioengineering and Nanotechnology, Singapore and the Stem Cell Bank, Singapore Stem Cell Consortium. She recently moved to Canada, where she hopes to continue her research interests. Her e-mail address is: prabhatalks@hotmail.com.

Joseph Arditti was born in Bulgaria in May 1932 and immigrated to the USA in 1954. He received his Ph.D. from the University of Southern California (USC) in Los Angeles, spent his entire career at the University of California, Irvine and retired in 2001 after 35 years of service. He is now Professor Emeritus. His scientific interests center on orchids and he spent many of his summers and sabbatical leaves doing research on these plants at the National University of Singapore (with Professors Avadhani, Hew and Rao), Botanical Gardens in Bogor, Indonesia (with the late Saleh Idris and he late Dr. Djunaedi 'Adjun' Gandawijaja) and the University of Malaya (with Prof. Helen Nair). Professor Arditti founded Orchid Biology, Reviews and Perspectives in 1974. The first volume was published in 1977. Now, after 30 years of editing the series he has decided to retire from it. This is the last volume he will edit. Prof. Arditti raised his son, Jonathan (now 24) from the age of 6 years as a single father. Jonathan has followed in his father's footsteps and also graduated from the University of Southern California. He received his B.A. in psychology in May 2008. In Southern California tradition and parlance Jonathan is a second generation USC Trojan, a fact which pleases both father and son enormously. When not at USC Jonathan lives with his father in Irvine, California. Prof. Arditti's e-mail address is jarditti@uci.edu. Jonathan's is arditti@usc.edu.

Popuri Nageswara Avadhani was born in December 1932 in Andhra Pradesh, India. He obtained his Bachelor's (Honors) and Master's degrees from Andhra University in 1952 and 1953 respectively, majoring in Botany. After that, in 1953, he proceeded to study at Durham University (Kings' college) where he was awarded the Ph.D degree. Following postdoctoral study at McGill University in Canada for 2 years, he joined the University of Malaya in Singapore (now the National University of Singapore), where he taught Plant Physiology for 32 years before retiring. Professor Avadhani also worked very closely with eminent scientists at several laboratories in different parts of the world. His involvement with Botany continues in retirement as a consultant and advisor. His e-mail address is avadhani@pacific.net.sg.

Yong Neng Chow received his doctorate from Queen's University in Belfast for work on tissue culture of Narcissus. He carried out post-doctoral research on orchids at the Department of Botany, National University of Singapore before accepting a position as a research scientist in a commercial plant tissue culture establishment in Singapore where he worked for 4 years. Dr Chow was Director of Research and Development at EcoFirst Laboratories, a subsidiary of EcoFirst Consolidated Bhd., in Malaysia where he spearheaded the company's efforts in entering the agro-biotechnology sector in that country. His e-mail address is everboleh@gmail.com.

Kevin. L. Davies is primarily a plant anatomist with interests in pteridology, palaeobotany, zinc toxicity in grasses and in particular, orchids. He graduated in 1978 with an Honors degree in botany from Swansea University (Wales, UK) before embarking on his Master's degree, during which time he was in frequent correspondence with the late Professor Eric Holttum, then associated with the Royal Botanic Gardens, Kew. For a number of years, he was employed at the botany department of the National Museum and Gallery of Wales, Cardiff and soon thereafter, commenced his doctorate at Cardiff University. As a postdoctoral researcher at that same institution, his research involved ultrastructural studies of Silurian and Lower Devonian spores and the anatomy of Eocene wood. In 1992, he and his supervisor, Professor Dianne Edwards, published a paper confirming the vascular status of *Cooksonia*, finally establishing it as the archetypal, vascular land plant (Nature 357: 683–685). Since then, his research has been almost exclusively related to orchids (mainly Maxillariinae, labellar micromorphology and anatomy as well as floral food rewards). In collaboration with Dr. Malgorzata Stpiczyska (Lublin, Poland), the research has been extended to include the ultrastructure of floral nectaries and elaiophores. Recently, Dr Davies established an orchid group at the newly-built National Botanic Garden of Wales and much of his spare time is spent tending his own living orchid collection. He is at the School of Earth, Ocean and Planetary Sciences, Cardiff University, Main Building, Park Place, Cardiff CF10 3YE, UK. His e-mail address is: kevinldavies@btinternet.com,

Arnold S. Dunn received his B.S. from George Washington University in 1950 and doctorate from the University of Pennsylvania in1955. In1962 he joined the faculty at the University of Southern California (USC). He retired in 2001, but is still teaching as Professor Emeritus. Dr. Dunn received an award for teaching excellence in 1969, an award for research excellence in 1972, the USC Raubenheimer Award for balanced contributions to teaching, research and administration in 1981, the

Mortarboard National Honor Society Faculty Excellence Award in 1988 and 2001, the Gamma Sigma Alpha National Academic Society Professor of the Year award in 2004, and the USC Lifetime Achievement Award in 2005. A secular biblical and talmudic student as an avocation, Dr. Dunn was awarded an honorary doctorate, Doctor of Humane Letters (Honoris Causa) by Hebrew Union College in 1995. During his research career Prof. Dunn specialized in hormonal mechanisms which regulate metabolism. Despite working with animals Dr. Dunn served as a mentor (1962–1965) for a graduate student in plant physiology at USC by the name of Joseph Arditti. Dr. Dunn's e-mail address is arnolddu@usc.edu.

Choy Sin Hew received his PhD at Queen's University, Canada and was a postdoctoral fellow with Professor Martin Gibbs at Brandeis University in the USA. After that he served as Professor of Botany at the National University of Singapore until his retirement in 2003, specializing in the physiology of tropical orchids. He was awarded the Singapore National Science Award in 1977 for his valuable contributions to orchid research. Professor Hew serves as consultant to several public and private orchid organizations in Singapore and Malaysia. He is also an advisor to the Research Centre for Chinese orchids in Guangdong, China. His e-mail address is choysinhew@yahoo.com.sg.

Tiiu Kull received her Ph.D. in botany from the University of Tartu, Estonia in 1997 with a dissertation on population dynamics of *Cypripedium calceolus*. She has been associated with the Institute of Zoology and Botany (now called Institute of Agricultural and Environmental Sciences) during her entire career. Since 1998 she has been head of the department of botany there. In addition to orchid population biology her research interests include reproductive biology of vascular plants and the protection of biodiversity. Most of her 40 scientific papers deal with orchids. Prof. Kull has also participated in the compilation of the *Flora of the Baltic Countries, Red Data* books for Baltic region and the *Key-Book* of Estonian plants. She also led projects on Estonian Biodiversity Strategy and Action Plan and Atlas of Vascular Plants. Dr. Kull is a member of a steering committee of European Platform for Biodiversity Research Strategy and she is participating in several Europe wide research consortia. Prof. Kull was the president of the Estonian Orchid Protection Club for 16 years. Her e-mail address is *tiiu.kull@emu.ee*.

Hubert Kurzweil is a herbarium taxonomist at the Singapore Botanic Gardens and currently involved in a taxonomic study of the genus *Calanthe* (in collaboration with botanists in the UK) as well as in the drafting of the orchid volume of the Flora of Thailand. Originally from Vienna (Austria), where he also grew up and completed his university studies, Dr. Kurzweil spent 20 years in Cape Town (South Africa). While there he studied the floral morphology and ontogeny the sub-Saharan orchids. In the course of this research on Southern African orchids he published extensively on various aspects of the indigenous orchid flora. He also made several contributions to a phylogenetic study of the orchid tribe Diseae. These studies were carried out in collaboration with Prof. P. Linder (currently in Zurich). The work culminated in a joint comprehensive and well-illustrated treatment of the orchids of South Africa. Dr. Kurweil always had a strong interest in Asian orchids, and undertook several botanical trips to Nepal, India, Sri Lanka, Malaysia, Thailand and Myanmar. His recent move to Singapore at the end of 2005 enables him to pursue this research interest. His e-mail address is Hubert_KURZWEIL@nparks.gov.sg.

Helen Nair was Professor and Head of the Department of Biotechnology, at the Asian Institute of Medicine, Science and Technology, the first private university in northern Malaysia to offer a range of science and technology-based degree programs. She accepted the post in 2001 after serving for 27 years on the Botany Department Faculty at the University of Malaya in Kuala Lumpur, where as Professor she held the Chair of Plant Physiology for six years. Her long-standing interest in orchid biology covers physiology, molecular biology and postharvest handling of cut-flowers. Currently, she is also working on the cryopreservation of orchids as part of a comprehensive orchid biotechnology program. Her husband, James Bonney is a Chartered Accountant. They have two children, Glenn a physician who resides in the UK and Sharm, a lawyer married to a physicist who lives in the Netherlands. Her e-mail addresses are helen.nair@gmail.com and jtb@pc.jaring.my.

Ettore Pacini graduated from the University of Siena, Faculty of Biological and Natural Sciences in November 1967 where he has been working since then. Since January 1984 he has been s "Professore Ordinario." In November 2006 he became a corresponding member of the Accademia Nazionale dei Lincei. He is author of research papers and reviews on anther tapetum, pollen development, pollination, allergenic pollen, and gametophytic incompatibility. Prof. Pacini has published more than 130 papers and edited 6 books. He also contributed 11 book chapters. His e-mail address is: pacini@unisi.it.

Karen E. Quinn is Kristin and Roger Servison Curator of Paintings, Art of the Americas, Muscum of Fine Arts, Boston. She received her Bachelor's degree at McGill University and her Master's degree at the University of Pennsylvania. Since joining the MFA in 1987, she has contributed to the exhibitions and catalogues for Weston's Westons: Portraits and Nudes (1989), The Lure of Italy (1992), and John Singleton Copley in America (1995). She has organized or co-organized the exhibitions Ansel Adams: The Early Years (1991), Weston's Westons: California and the West (1994), Martin Johnson Heade (1999), Edward Weston: Photography and Modernism (2000), The Lane Collection and American Modernism: Georgia O'Keeffe and her Time (2004–2009), and Rockwell and the Red Sox (2005), and written for their accompanying publications. Ms Quinn also manages the Martin Johnson Heade catalogue raisonné (2nd edition published in 2000). She is an adjunct professor at Middlesex Community College. Most recently she has been researching the working methods of Fitz Henry Lane with colleagues in paintings conservation. Her e-mail address is: kquinn@mfa.or.

Adisheshappa Nagaraja Rao was Born in Channapatna, Karnataka, India, on 12 November 1925. He received his B.Sc. Honors and M.Sc. degrees from Mysore University in India and Ph.D. from Iowa State University in the USA. in 1959. Dr. Rao was a postdoctoral fellow at the University of Manchester in the U. K.

He joined the University of Malaya in Singapore as Lecturer in 1960, rose to the rank of Professor and was Head of the Botany Department from 1967 until his retirement in 1992. After retiring Professor Rao joined the International Plant Genetics Research Institute (IPGRI) and remained there until 2000. At present he is the editor of the Journal of Tropical Medicinal Plants. Prof. Rao is a member of many organizations including Honorary Fellow of the Singapore Institute of Biology; member of Indian Society of Plant Morphologists; Vice President of International Association of Plant Tissue Culturists (IAPTC) for 9 years; Chairman, Governing Boards of SEARCA in the Philippines and Biotropical Institute for South East Asia, Bogor, Indonesia (BIOTROP) Bogor, Indonesia; consultant and committee member, Global Environmental Facility (GEF) of the World Bank; United Nations Environmental Programme (UNEP); United Nations Development Programme (UNDP); United Nations Environmental, Scientific and Cultural Organization (UNESCO) in Paris; International Development Research Council (IDRC) in Canada; International Bamboo and Rattan Organisation (INBAR) and others. He has been the Secretary of the Asian Network of Biological Sciences for 12 years. Professor Roa has published more than 200 research papers, edited and or authored more than 25 books, published in UK, India, Philippines, Indonesia, Malaysia, Singapore, Germany and other countries. He does not have an e-mail address.

Ben Singer was born in 1943 in the Dutch city of Enkhuizen. He became interested in orchids at the age of twelve after receiving a postcard with a picture of a *Cattleya* from his mother. Some years later he left school to pursue the beginning of what would become a successful career in the orchid business. He moved to the city of Aalsmeer to work as an apprentice to G.A. de Jong, the father of the modern Dutch orchid industry. He worked there for 2 years, and at the age of 17 set up his own nursery with the help of his father. He began with one thousand *Cymbidium* plants, and after discovering tissue culture in 1963 he started producing miniature cymbidiums on a large scale. When his competitors caught up with him, he turned his attention to *Paphiopedilum*, in particular *Paphiopedilum lawrenceanum* and *P. ciliolare*. In addition to growing orchids, Singer has put together an impressive collection of orchid books and antiques over the years. In 1979, after growing weary of the orchid business, he sold his nursery. Since then he has remained active within the orchid world, traveling to see orchids in their natural habitats, and attending conferences. Ben and his Mexican wife Eugenia have a daughter named Laelia. His e-mail address is singerorchids@hotmail.com.

Irina Vasilyevna Tatarenko graduated from Moscow State University in 1986. She received her Ph.D. from the same university in 1991 for work on the orchids of Far Eastern Russia. Between 1991 and 1994 Dr. Tatarenko worked on plants and fungi for the Russian Red Data Book. Since 1994 Dr. Tatarenko has held a position as senior researcher at the Moscow Pedagogical State University. In 2000–2001 she was an invited visiting researcher at Hiroshima University in Japan and carried on cooperation projects on Japanese and Russian floras in 2002–2006. Her Doctor of Science dissertation – “Biomorphology of Russian and Japanese orchids” was completed in 2007. The systematic, distribution, ontogenesis, seasonal development and mycorrhiza analyses were based on morphology of vegetative and generative organs. Dr. Tatarenko has published a monograph and more than 60 research papers on morphology, demography and mycorrhiza of terrestrial orchids. Her e-mail address is tulotis@yandex.ru.

Eng Soon Teoh (Chinese style Teoh Eng Soon) is an obstetrician-gynecologist who has been in practice for over 40 years. A graduate of the University of Singapore he undertook postdoctoral research in London, Sweden and Miami, Florida. He has published extensively in refereed international medical journals and is the recipient of several awards from the Royal College of Obstetricians and Gynecologists in London for his work on trophoblastic cancer and fetal monitoring. Dr. Teoh is a Fellow of the Royal College of Obstetricians and a Fellow of the American College of Surgeons. Starting to take a serious interest in orchids in 1965, Dr. Teoh read Rebecca Northern's *Home Orchid Growing* for relaxation while he was in London preparing for a higher degree examination. In Miami he studied Williams's *The Orchid Grower's Manual* on the advice of Roy Fields. His interest in *Phalaenopsis* started in Miami. Dr. Teoh did not start growing orchids in earnest until 1970 when he returned to Singapore. He was active in the Orchid Society of Southeast Asia during the 1970s and 1980s, serving as Editor of the *Malayan Orchid Review* (1978–1982), on the Committee, on the Panel of Judges, and as President in 1981/82. In addition Dr. Teoh edited a volume on orchids to Commemorate the Golden Anniversary of OSSEA in 1978. He is the author of *Orchids of Asia* published by Marshall Cavendish (this book has seen three editions 1980, 1989, 2005), and *A Joy Forever (Vanda Miss Joaquim; Singapore's National Flower)* published by Times Editions (1982; revised and reprinted, 1998). Dr. Teoh has been married to Phaik Khuan Teoh since 1963. They have two children, John Tee Teong who works for a property investment firm in Singapore, and Kristine Leok Kheng, a cardiothoracic surgeon in London. Dr. Teoh's e-mail address is teohengsoon@pacific.net.sg.

Wong Sek Man (Western style: Sek Man Wong) received his B.Sc. in Biology from Nanyang University in Singapore in 1980. He continued to study for his Honors degree program and participated as a part-time tutor at the Botany Department in the University of Singapore for one semester in 1980. Prof. Wong earned his M.Sc. and Ph.D. degrees from the Department of Plant Pathology at West Virginia University, and Cornell University, USA, respectively. He joined the National University of Singapore (NUS) in 1987. At present he is a Professor in the Department of Biological Sciences and a Vice Dean (Special Duties) in the Faculty of Science. At NUS, he has received several teaching awards and the Outstanding University Researcher Award in 1999. He has supervised more than 10 Ph.D., 20 M.Sc. and 30 Honors degree students in his Molecular Plant Virology Laboratory. Currently Prof. Wong is the President of the Plant Protection Society (Singapore) and was the President of the Asian Association of Societies of Plant Pathology from 2005 until 2007. His e-mail address is: dbswsm@nus.edu.sg.

Tim Wing Yam (Chinese style: Yam Tim Wing) was born in Hong Kong. He received his doctorate in botany from the University of Hong Kong and spent nearly 3 years as a postdoctoral fellow in Professor Joseph Arditti's laboratory before assuming his current position as a Senior Researcher in Orchidology at the Singapore Botanic Gardens. He is responsible for the orchid breeding and conservation programs at that institution. Dr. Yam has produced many outstanding orchid hybrids, lectures world wide and is the author of many papers in international peer reviewed journals. His e-mail address is Yam_Tim_Wing@NPARKS.gov.sg.

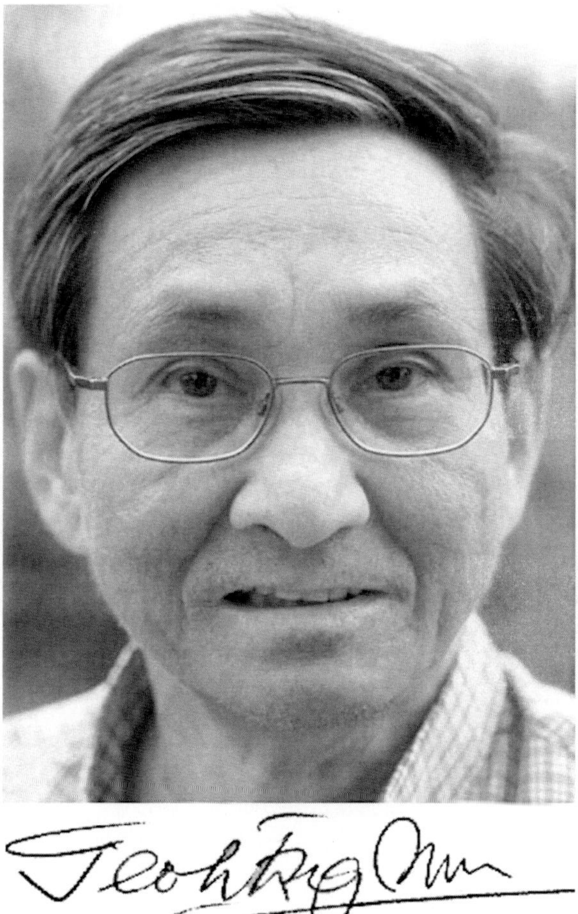

BIOGRAPHIES

1

Unsung Heroes of the Singapore-Malaysian Orchid World (1951–2000)

ENG SOON TEOH

Contents

Introduction 2
Whampoa 2
Ridley 3
Holttum 3
Lee Kim Hong 5
Tan Hoon Siang 7
Yeoh Bok Choon 8
Quek Kiah Huat (1923–1983) 9
George Alphonso 11
Sum Lai Wah 13
John Ede 13
Syed Yusof Alsagoff 14
Koh Keng Hoe 16
Wong Leong Fatt 17
Chong Chok Chye 18
How Yee Peng 18
Soh Kim Kang 19
Mak Chin On 20
Mok Choi Yew 22
David Lim 23
Phoon Yoon Seng 24
Au Yong Nang Yip 25
Anthony Lamb 26
Ooi Leng San 27
Cheah Kheng Cheong 27
Lum Hon 27
The Ladies 27
Gracia Lewis 28
Wendy Scott 29
Rosalind Lee 31
Amy Ede 32
Ong Siew Hong 32
Madam Yong 33
Glossary 33
Literature Cited 34

Kull, Arditti and Wong (eds.), *Orchid Biology: Reviews and Perspectives*, *X*,

Introduction

Enough has been said about the scientists, explorers and the pioneering orchid enthusiasts who started collecting orchids in the 19th century, and due respect has been given to the people who started the Malayan Orchid Society which promoted orchids both as a hobby and as an export trade. I am full of admiration for these people and do not begrudge the honors conferred on them (Teoh, 1978). It follows that there should also be better information about the leaders, growers, collectors, breeders and adventurers who sustained the orchid interest in Singapore and Malaysia during the second half of the 20th century. Not everyone appreciates their role and it saddens mc to observe how very soon they are forgotten. Without them we would not have the beautiful orchids that now grace our gardens, the thriving orchid societies in the region, and the orchid flower industry in the area. This is a contentious topic because one cannot be absolutely comprehensive, and I admit that my selection in this chapter is personal.

Whampoa

Whampoa (Fig. 1-1), a Cantonese ship-chandler, the wealthiest Singaporean Chinaman of the 19th century, was possibly the first person to grow orchids on a large scale in Singapore. He was a chosen representative of the Chinese residents and a member of the Agro-Horticultural Society. When the British government wanted to establish a new Botanic Garden he was persuaded to exchange the beautiful piece of undulating real estate which is now the present Singapore Botanic Gardens for a piece of low lying flat land at Serangoon Road. Here, Whampoa built a grand home which contained a magnificent Chinese garden. He was familiar with the senior officers in the Royal Navy and a friend of the merchantmen, so it was no surprise that he managed to collect all sorts of plants for his garden. He had *Victoria regia* growing in the pond, a gift from the King of Thailand (Teoh, 2006). Sadly, Whampoa's Chinese garden has fallen victim to progress and there is no record that any of his orchids have managed to survive. Indeed the only orchid from the 19th century to remain in cultivation is a hybrid, Singapore's very own *Vanda* Miss Joaquim, now its national flower. A number of books and articles have been published on this orchid and Miss Agnes Joaquim is well celebrated (Alphonso, 1981; Phoon, 1981; Teoh, 1981, 1982, 2004; Hew, Yam, and Arditti, 2002, 2004; Dalpethado, 2004; Johnson, 2004; Arditti and Hew, 2007).

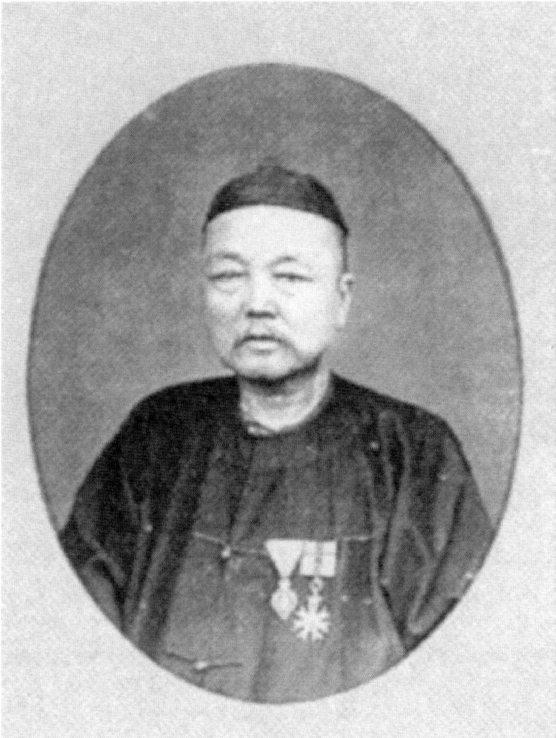

Fig. 1-1. Whampoa.

Ridley

Henry Ridley, then director of the Botanic Gardens registered this hybrid for Miss Joaquim. The director made several collecting trips for orchid species and published a volume on the Malayan species. He was also a great innovator. He campaigned strenuously for the cultivation of *Hevea braziliensis*, a proposal taken up by Tan Chay Yan who is remembered by the famous *Vanda* hybrid named after him. Ridley is remembered as the "Father of Natural Rubber." With the rising oil prices natural rubber is enriching the planters who have not replaced their rubber estates with other crops. Thus it is surprising that Henry Ridley never embarked on orchid hybridization when he was shown the hybrid by Agnes Joaquim in 1893.

Holttum

Hybrids constitute the backbone of orchid collections worldwide, and Singapore and Malaysia are no exceptions. Hybridization of orchids became a practical reality in Singapore when Hans Burgeff showed Eric Holttum, director of the Singapore

Botanic Gardens and later Professor of Botany at the university, how to achieve asymbiotic orchid seed germination in 1928 (Holttum, 1978). At that time there were few free flowering plants in the local gardens. Knowing that orchid flowers were long lasting on the plant, and inspired by the floriferousness of the remarkable *Vanda* Miss Joaquim, Holttum's aim was to produce other orchid hybrids that would thrive and flower equally freely in the tropical lowlands. He achieved his first success with *Spathoglottis*. He named the hybrid *Spathoglottis* Primrose *(plicata × aurea)*. From pollination to flowering took only 2½ years, in part assisted by the rapid ripening of the seed capsule at Penang Hill. Its F1 hybrid *Spathoglottis* Chrysops flowered in 17 months. However, crosses involving other genera which he made in Singapore took much longer to flower. But there were many lucky outcomes which spurred him on. The first of the spider orchids, *Aranda* Deborah flowered 7 years after the seeds were sown. It proved to be a remarkably free flowering hybrid, often carrying up to a dozen sprays with hundreds of flowers on a plant less than a meter tall. Over the course of the next two decades, by trial and error, outstanding stud plants were identified and others eliminated. This information proved to be extremely useful to the breeders who commenced breeding orchids after World War II.

During the Japanese Occupation (1942–1945), the Singapore Botanic Gardens was renamed Syonan-to Botanic Gardens and Professor Hidezo Tanakadate was appointed as Administrative Head. Professor Kwan Koriba and Marquis Tokugawa also played a role in the administration of the Gardens. They were scholars and must be remembered for their indirect contribution to the Singapore orchids. They allowed Holttum and E.H.J. Corner to remain in the Gardens to continue their botanical research. During this period, orchid hybridization continued under the charge of S.P. Livingston who continued to sow seeds by substituting local seaweed for agar. The monthly reports of the Gardens noted the numbers of orchid seedlings unflasked; the number of seedlings moved from the glass house to the main orchid nursery; the number of pollinations undertaken and the number of fruits produced; and the number that contained viable seeds; the nature of the new crosses; and the new orchid hybrids which had flowered (Syunin and Holttum 1945). Among these new hybrids was a cross between *Vanda dearei* and *Vanda sanderiana* which flowered in February 1944: but the War was on, and the Gardens lost out to the famous Hawaiian breeder John Noa who called it *Vanda* Ellen Noa (Yeoh, 1978). Many of the crosses germinated have not been registered and one wonders what happened to the plants. Among these were an interesting cross between *Spathoglottis plicata* and *Calanthe triplicata* (it had been the long cherished hope of Holttum to bring these two genera together) and numerous *Aranthera*. The records state that in September 1944 (Japanese Occupation Year 2604), some unflowered orchid seedlings were discarded: Presumably these were siblings of hybrid plants that flowered sometime ago, a case in point being the hybrid, *Arachnis hookeriana* × *Arachnis flos-aeries* which yielded free flowering progeny only with *Arach. hookeriana* var *luteola* and not with the standard *Arach. hookeriana* var. *alba*. George Alphonso who joined the Gardens in 1940 and Bajuri bin Sappan who came on board in 1937 continued to work there, but they did not seem to have been involved in the orchid program during the Japanese Occupation.

When Holttum was appointed to the chair of botany at the newly founded University of Malaya (located in Singapore) in 1949 the running of the Malayan Orchid Society fell on the local growers who rose to the responsibility. Nevertheless, it should not be forgotten that H.M. Burkill who succeeded as director of the Singapore Botanic Gardens gave his whole hearted support to the shows of the Malayan Orchid Society during his tenure in office and this support was continued by George Alphonso and subsequent directors. Singapore was not so crowded in those days and life still proceeded at a leisurely pace. Orchids were highly prized, and every effort was made to grow them into specimen plants that could be displayed at the flower shows. Beautiful orchid gardens were the pride of the people who could afford them and who had the commitment, skill and patience to bring the plants to flower. Among them were the following:

Lee Kim Hong

Lee Kim Hong (western style: Kim Hong Lee; Fig. 1-2) was a pioneer in the Singapore orchid trade. He dealt in gold bullion and orchids were his hobby. However, he was astute enough to realize that business and orchids could combine in the Singapore of the 1950s. He encouraged the local farmers to strive for export quality so that their flowers would fetch a better price. He was a prolific breeder himself after World War II: Dr. Yeoh Bok Choon (western style: Bok Choon Yeoh) who compiled the *List of Singapore-Malayan Orchid Hybrids up to 1963* said that Lee Kim Hong made more hybrids than anyone else. His name appears in the Hybrid Registry as "breeder' up to 1960, after which it was replaced by Singapore Orchids, the company he co-founded with John Laycock. Together they and their wives established the beautiful Mandai Orchid Garden located at the heart of Singapore. The site was a gentle hillside, located within the water catchments area, near to a reservoir and it had the highest rainfall in Singapore. Amy Ede recalled that Lee Kim Hong did all the work: "Daddy (John Laycock) learnt a lot from him" (Elliott et al., 2005). Lee Kim Hong was a warm, generous and helpful person and up to the 1960s, one could always go up to the Mandai Orchid Gardens to consult him about growing orchids. Among those whom he taught was Madam Ong Siew Hong (western style Siew Hong Ong, Dr. Tan Wee Kiat's [Kiat Wee Tan] mother) who also mastered the art of hybridization and raising orchids from seed. Dr. Kiat Tan contributed a warm reminiscence of his encounters with Lee Kim Hong while still a youngster in John Elliott's *Orchid Hybrids of Singapore, 1893–2003*. According to Kiat, Kim Hong was mainly interested in primary hybrids – what characteristics were transmitted when two different looking flowers mix their genes: he was more the curious scientist rather than the gardener. Nevertheless he

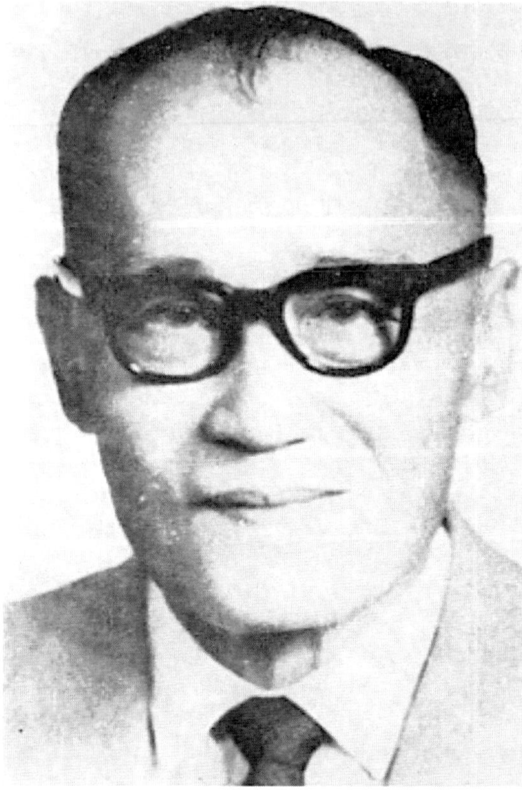

Fig. 1-2. Lee Kim Hong (western style: Kim Hong Lee).

grew a variety of garden plants and constructed, in his lifetime, the most beautiful orchid garden in Singapore.

In order to have total control of his breeding program, Lee Kim Hong mastered the process of raising orchids from seed. Singapore Orchids followed their own unique program and seldom ventured to follow the fashions of the day. Among Kim Hong's unique contributions are the delicate *Vandaenopsis* (now renamed *Paravanda*) hybrids, in particular the delicate yellow *Vandaenopsis* Prosperitas (*V.* dearei × *Pps. denevei*), *Vdnps.* Patience (*Pps. denevei* × *V.* Rothschildiana.). His name is commemorated in the miltigeneric hybrids *Leeara* Lissom Lucy (*Aranda* Lucy Laycock × *Vdps. lissochiloides*) and *Laycockara* Lee Kim Hong (*Arnps.* Lee Siew Chin × *Vdps. lissochiloides*). Dr. Lee Siew Chin (Siew Chin Lee) is the daughter of Lee Kim Hong. All these crosses were originated and registered by Singapore Orchids.

Tan Hoon Siang

Orchid growing among Singapore's populace took a tremendous setback during the World War II, and it was several years before the Society got back on its feet. To urge the members on, and provide the necessary leadership there was Tan Hoon Siang (Hoon Siang Tan; Fig. 1-3) who was elected president in 1956 (MOR, 1991). He was an ardent grower and breeder and insisted in growing his plants to perfection. His *Aranda lowii* earned one of the few Cultural Commendation Certificates ever granted by the MOS. The family tradition continues and his son, John Tan Jiew Hoe (Jiew Hoe John Tan) is president of the Singapore Gardening Society. John is an expert on begonias, and a staunch supporter of OSSEA.

Tan Hoon Siang's principal claim to fame lies in having bred the famous *Vanda* Tan Chay Yan (*V*. JVB × *V. dearei*) which won a First Class Certificate (FCC) in 1954 from the Royal Horticultural Society and set a trend for other strap leaf *Vanda* to be bred to the tetraploid *V*. Josephine van Brero (popularly referred to as JVB). Subsequently various clones of *V*. Tan Chay Yan received an Award of Merit (AM)

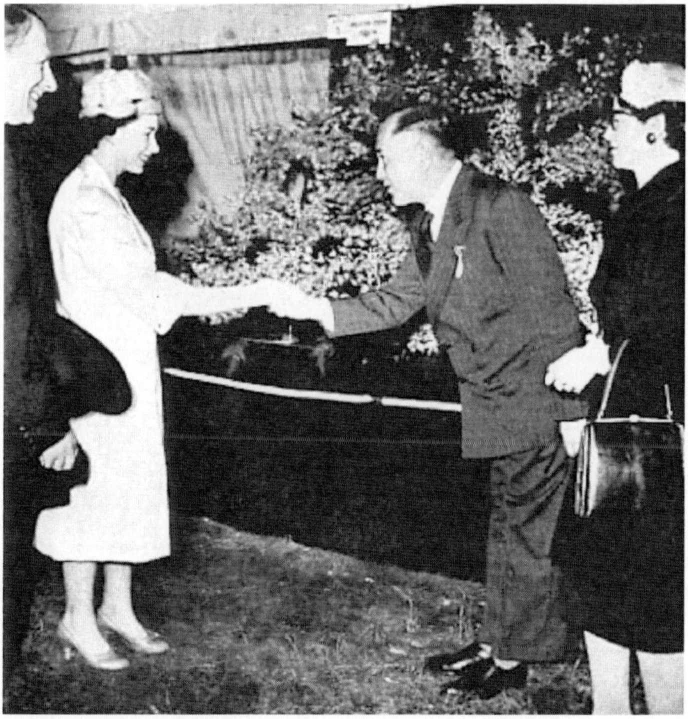

Fig. 1-3. Tan Hoon Siang being presented to H.M. Queen Elizabeth at the Chelsea Flower Show,1963.

from the Malayan Orchid Society. A rubber planter, Tan Hoon Siang was president of the Malayan Orchid Society when it made its successful bid at Chelsea to host the 4th World Orchid Conference in Singapore. *Vanda* Tan Hoon Siang (*V.* JVB × *V.* Somsri Pink) is a new spectacular bluish pink JVB hybrid named by Hoon Siang's son, John after his father.

Yeoh Bok Choon

Dr. Yeoh Bok Choon (Bok Choon Yeoh; Fig. 1-4) was the State Surgeon for the southern Malaysian state of Johor which is the immediate neighbor of Singapore. The proximity of Johor allowed Bok Choon to attend almost all monthly meetings of the MOS for a period of some 20 years. He was assigned the job of commenting on the plants brought by members for display and a friendly monthly competition. The job suited him perfectly because Bok Choon was an avid orchid grower and promoter. He believed that orchids benefited from occasionally

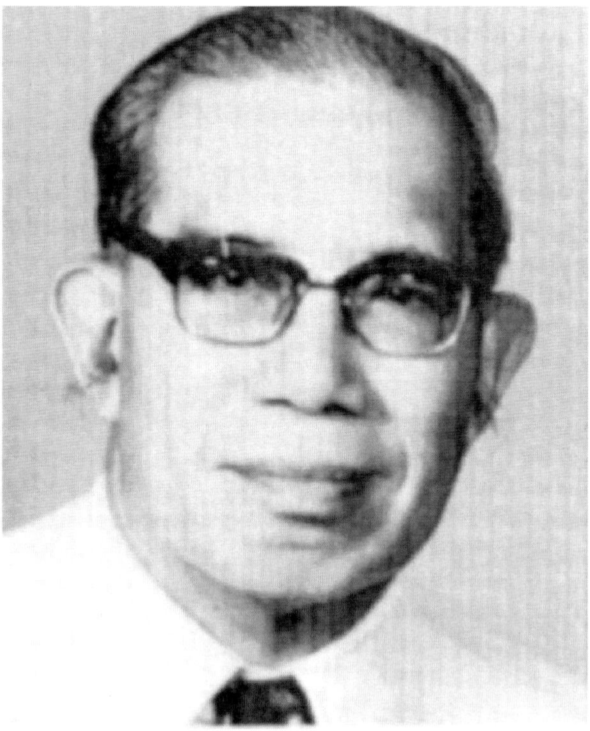

Fig. 1-4. Dr. Yeoh Bok Choon (Bok Choon Yeoh).

imbibing some beer and he never ceased to talk about it (Yeoh, 1962). He has also written several articles on orchids and compiled a *Handbook on the Orchid Hybrids of Singapore and Malaysia up to 1963* (Yeoh, 1962, 1963, 1978, 1981, 1982). A highly respected judge of orchids, Bok Choon was known to be generous with points if he really liked a plant, but nonetheless, the other judges thwarted his effort to give several plants the coveted First Class Certificate during his 35 year tenure as a judge.

Nevertheless, he was an indomitable and the longest serving member of OSSEA. Yeoh Bok Choon was elected president of the MOS in 1962. He continued attending monthly meetings of the Society where he was generally the principal speaker until his final days.

He is remembered in the quadrigeneric genus, *Bokchoonara* which is constituted by the combination of *Arachnis*, with *Paraphalaenopsis, Ascocentrum* and *Vanda*. The hybrid *Bokchoonara* Khaw Bian Huat (registered in 1977) is outstanding and has an Award of Merit from OSSEA. Bok Choon also made a number of hybrids. He was fondest of his white *(alba) Vanda* Miss Joaquim, to which he gave the varietal name John Laycock. It gained an Award of Special Mention at the 3rd WOC at Chelsea, England in 1960 (Yeoh, 1981).

Quek Kiah Huat (1923–1983)

Quek Kiah Huat (Kiah Huat Quek; Fig. 1-5) was one of the prime movers of Singapore's orchid circle during its heydays from the 1950s to the 1970s. Attending the WOC at Chelsea as Vice-president of the MOS when the Society launched its successful bid for the 4th WOC, Kiah Huat took over the reins of the Society the following year, serving both as its president and Show Chairman. He was a born leader: friendly, affable, decisive, open to new ideas and generous. A second generation banker and rubber planter, he got along well with people from all walks of life, and got things done.

Organizing the large orchid shows to which Singapore has now become accustomed was no small matter. It requires good team work, intensive planning, endless meetings, innumerable visits to orchid farms and amateur growers, lots of persuasion, days of landscaping work on site, registration, promotion, and generous funding. The unspoken understanding was that if things came to a head, Kiah Huat would underwrite the entire show. However, the Conference made a small profit allowing the MOS to make a donation to the Defense Fund during *Confrontasi* (Alsagoff, personal communication 1980). Professor Rapee Sagarik, long serving president of the Orchid Society of Thailand said that Kiah Huat played a large role in gaining world recognition for the orchids of Southeast Asia and he helped Thailand to move into the world orchid scene (Alphonso, 1983)

The 4th WOC marked a high point in the Society's history. The massive orchid show at the Singapore Turf Club set a new standard for orchid exhibits. It showcased the result of a 35 year hybridization effort to put free-flowering, colorful,

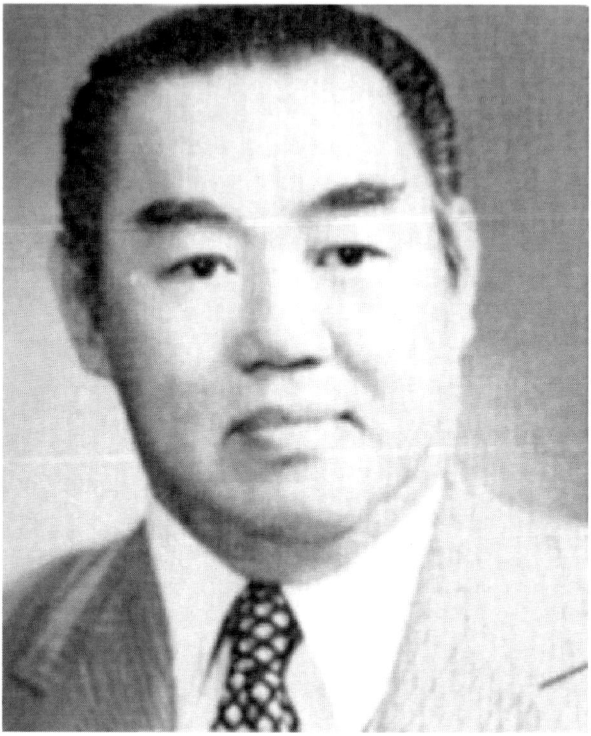

Fig. 1-5. Quek Kiah Huat (Kiah Huat Quek).

long-lasting orchids into the tropical garden. The main attractions were the wide range of scorpion orchids, sparkling semi-terete and quarter terete *Vanda*, and *Dendrobium* all suited for export as cut flowers. The show inspired an interest in commercial orchid cultivation throughout the tropics, particularly in Malaysia and Thailand which became major suppliers of orchids to the cut flower industry. Kiah Huat took a back seat after the 4th WOC but the loss of many major supporters through repatriation and age necessitated his recall. Kiah Huat assumed the presidency of the OSSEA, a post he retained until poor health required him to relinquish the post in1981- During this second term, OSSEA celebrated its 50th Anniversary with a spectacular show at the National Stadium and the publication of a commemorative volume on *Orchids* that featured its history. There was now a new range of orchids on show. The flowers were larger, substance much improved and the colors were deeper. There was also much evidence of selective clonal propagation spurred by the cut flower trade, although shoot tip cultures were yet to make their influence felt in the region. I had the privilege of working with Kiah Huat during this period and was much impressed by the way he held a team of very disparate personalities together.

George Alphonso

In the Malayan Orchid Society, George Alphonso (Fig. 1-6) was the leading expert on orchid habitats. He was the protégé of Professor Eric Holttum and had collected widely in peninsular Malaysia, Singapore (including its offshore islands), Thailand, Borneo, Indonesia and Papua New Guinea. Cheang Kok Choy (Kok Choy Cheang), superintendent of the Penang Botanic Gardens who was an expert on Malaysian species often accompanied him on such expeditions, and once in 1960 he traveled in the company of Major R. Brewer and the troops of the Singapore Guard Regiment who were on a jungle exercise in Kedah. George was an invited speaker on orchid species of the region at several WOC but this gave no indication of his vast knowledge of their breeding potential. In fact, George was responsible for most of the hybrids registered under Singapore Botanic Gardens during the 1950s and 1960s. Bajuri Bin Sappan who was the laboratory assistant from 1960–1969 assisted with the hybridization work during the 1960s.

In the lead chapter of volume IV of this series which he wrote on the *History of Orchidology in Southeast Asia*, George laments the disappearance of numerous

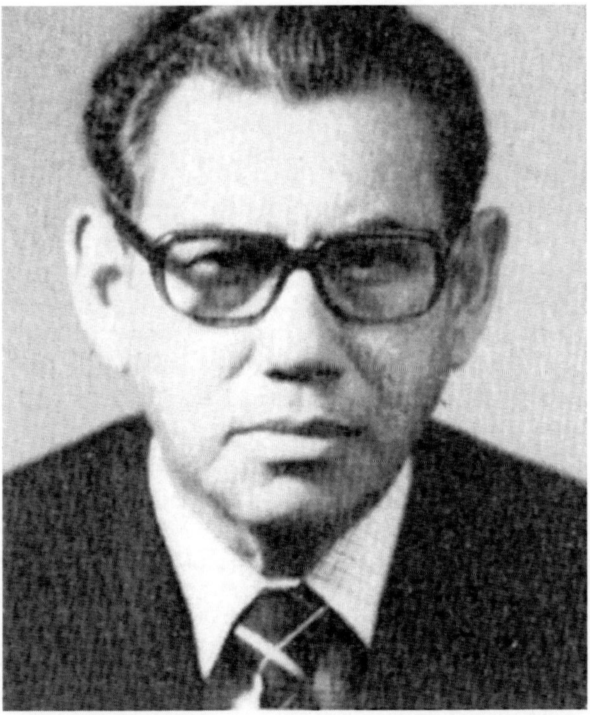

Fig. 1-6. George Alphonso.

beautiful species which he had once witnessed growing in abundance in their natural habitats: *Paphiopedilum niveum* in the Langkawi islands, *Habenaria carnea* in the limestone hills of Perlis, *Ascocentrum miniatum* on roadside trees and *Rhyncostylis retusa* on *Lagerstroemia* trees in Perlis and Kedah, *Vandopsis gigantea* which grew on rocky outcrops, *Paphiopedilum barbatum* on the granite hills of Penang. Wild *Arachnis hookeriana* is now extinct in Singapore, while *Dimorphis lowii, Phalaenopsis bellina* and *Vanda dearei* are threatened with extinction in Sarawak, and *Phalaenopsis gigantea* in Sabah. Already in 1966, at the 5th World Orchid Conference at Long Beach, California, and again in 1980 at the 9th WOC in Bangkok, he had stressed the importance of conserving such species in an invited paper:

George Alphonso was born in Kuala Lipis, in his own words, "by the railway tracks" to Norbert Alphonso and Sanada Kuma, his Sri Lankan father and Japanese mother, whose names are remembered in the early orchid hybrids. His father worked for the Malayan Railway and was transferred to Singapore in 1930. After finishing school George joined the Botanic Gardens which immediately sent him to Serdang, near Kuala Lumpur, for training. He worked for the Gardens during the Japanese Occupation where he was put in charge of plant propagation. With him were R.E. Holttum and E.J.H. Corner. In 1954, Holttum sent him to study horticulture at the Kew Botanic Gardens where he stayed for 2 years, graduating with a diploma in horticulture. During the War three *Bougainvillea* hybrids flowered in the Gardens, and there was some buzz about one that looked promising. George brought back *Bougainvillea* var. *poultonii* from South Africa in 1956. For several decades it was the No. 1 *Bougainvillea* in Singapore. Today, the bougainvilleas are still an important garden and roadside plant in Singapore.

George was appointed Curator at the Botanic Gardens in 1960 and put in charge of orchids and rose to the position of Acting Director Gardens in 1970. During the tenure of Director H.M. Burkhill at the Gardens, the Singapore Botanic Gardens contributed extensively in exhibition material and assistance by staff to the annual shows of the Singapore Gardening Society and the Malayan Orchid Society. George Alphonso was the Show Chairman for the massive 4th World Orchid Conference Show at the Singapore Turf Club in 1963, and afterwards he continued to serve as Show Chairman until 1968. Under his supervision, Singapore Botanic Garden's exhibits at the Singapore Orchid Shows and during the various Orchid Conferences in Singapore were spectacular. He has played an enormous role in promoting orchid cultivation in Singapore and in showcasing Singapore as a premier orchid nation.

George was editor of the Malayan Orchid Review (Alphonso, 1981), a past president and Fellow of the Orchid Society of Southeast Asia, and a past president of the Singapore Gardening Society.

Syed Yusof Alsagoff named the quadrigeneric *hybrid (Arachnis × Ascocentrum × Vanda × Vandopsis) Alphonsoara* Gus for George Alphonso, Gus being the nickname of George Alphonso.

Sum Lai Wah

Sum Lai Wah (Lai Wah Sum) was among the earliest commercial orchid cut flower growers. A past president of the Malayan Orchid Society, he was a gentleman, much admired by the local and foreign orchid enthusiasts, many of whom have fond memories of their encounters with this friendly, jovial grower. Lai Wah's mother tongue was Cantonese and sometimes he had to struggle to appreciate the fine points of the English language; but he always tried. Thomas Aana, one of our old Hawaiian friends, told me that when Sum Lai Wah was showing her group around Malaya after the 4th World Orchid Conference in 1964, there was a member who was fond of cracking jokes. Now and then, perhaps half an hour after the joke was uttered, Lai Wah would suddenly burst out in laughter. He was a notable orchid figure in Singapore from 1955 to 1970.

Mak Chin On tells me that Lai Wah had a sharp eye for spotting potential cut flower types. The day before he died, when Mak visited him, Lai Wah held on to Mak's hand and told him to concentrate on planting *Oncidium* Goldiana var. Golden Shower because he reckoned that there would always be a demand for its cut flower from Japan. To this day, almost four decades later, *Oncidium* Golden Shower is still a major export to Japan. Lai Wah's enormous Sun Kee Nursery in Serangoon focused early on cut flowers – *Dendrobium* Louisa Dark, *Den.* Pacific Island, *Aranda* Peter Ewart, *Aranda* Hilda Galistan, *Aranda* Tay Swee Eng, *Aranthera* James Storie, *Aranthera* Anne Black, *Vanda* Ruby, and the famous *Oncidium* Golden Shower. When the land was acquired by the government, Lai Wah's sons moved the nursery to Lim Chu Kang. However, the family's interest in orchids died with Lai Wah and Sun Kee soon faded from the scene.

John Ede

John Anthony Moore Ede (Fig. 1-7) was born in England and lived in India for 11 years before coming to Singapore in 1946 at the age of 34. He worked for the Cathay Organization for 16 years, was a member of the Legislative Assembly in the 1950s, and served as a member of the Presidential Council for Minority Rights from 1973 to 1983.

Marriage to Amy Laycock, the only daughter of the famous John Laycock in 1953 led him to the wonderful world of orchids with which he became deeply involved. When Laycock died in 1960, John Ede left the Cathay Organization to devote all his time to the Mandai Orchid Garden which he managed with Amy and Rosalind Lee. For decades, under their management Mandai was the best orchid garden this side of the Pacific. It was also a display and teaching garden for tropical plants. John Ede also continued Singapore Orchid's selective breeding program started by Lee Kim Hong and John Laycock, breeding several awarded plants for the company. *Vanda* John Ede (*V.* Josephine van Brero × *V.* Adrienne) was a remarkably

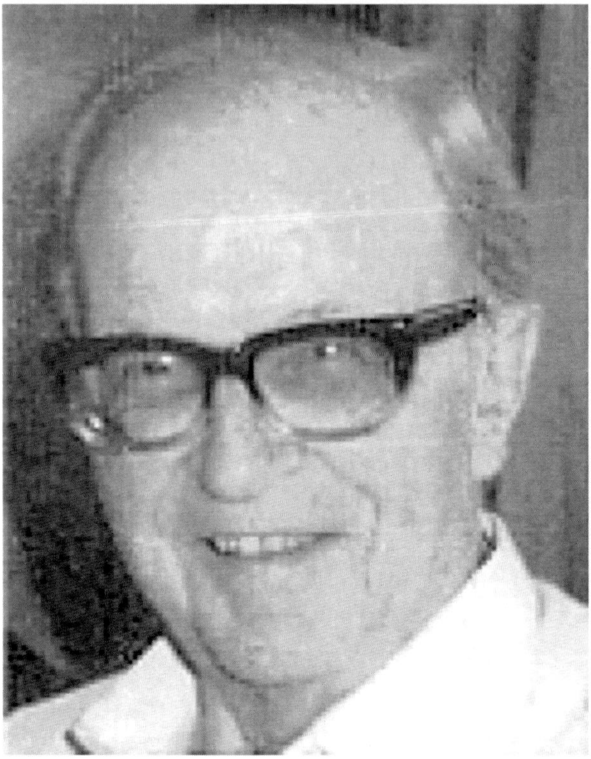

Fig. 1-7. John Anthony Moore Ede.

attractive JVB hybrid and won several best in show at the Singapore Orchid Shows. His garden was a major participant at the Singapore orchid shows. When Singapore Orchids started exporting orchid flowers, Ede studied the control of flowering in *Arachnis* Maggie Oei and discovered that exactly 14 days after all inflorescences were removed, new stalks would burst through the leaf axils, taking 70 days to mature. He lost no time in sharing this information with other growers through an article in the Malayan Orchid Review. By outlining his methodology, he also taught the farmers how to time their flowering for the best sales returns or for an event.

A quadrigeneric hybrid registered by Singapore Orchids commemorates his name: *Edeara* has *Arachnis, Paraphalaenopsis, Renanthera* and *Vandopsis*.

Syed Yusof Alsagoff

His is one of the most illustrious names in Singapore orchids in the last 50 years. A linear descendent of the Prophet, Syed Yusof Alsagoff (Fig. 1-8) is also the scion of a family which had settled in Singapore before its founding by Sir Stamford

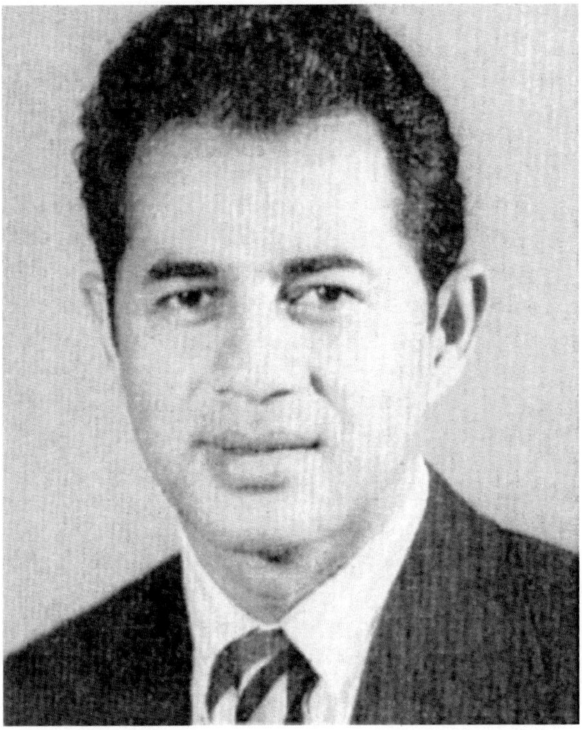

Fig. 1-8. Syed Yusof Alsagoff.

Raffles in 1819 (Elliott et al., 2005). The latter fact is not unexpected, given that Arab sailors and traders sailed between the Middle East and China for 2,000 years.

Yusof, as he is known to his friends, is the leading hybridizer of orchids in Singapore. He is the most prolific, with nearly 200 hybrids: he bred *Mokara* Zaleha Alsagoff which has the distinction of earning the only First Class Certificate ever awarded by the Orchid Society of Southeast Asia since it began granting awards in 1958. Many Awards of Merit and Highly Commended Certificates were awarded to his other hybrids. Today, Yusof's *Mokara* Zaleha Alsagoff is the most outstanding yellow scorpion orchid in the Singapore garden, *Aranda* Noorah Alsagoff is the most outstanding mauve scorpion orchid, and *Renanthera* Kalsom is the favorite, floriferous red orchid (Teoh, 2005). Yusof bred orchids not only to please himself and the judges. He played an important role in the development of the cut flower market, and he is a pioneer in the production of compact orchids favored by flat dwellers.

These achievements were not wrought overnight. Yusof has been breeding orchids for more than 50 years. He registered his first hybrid in 1953. In 1967 he succeeded Bajuri bin Sappan as secretary to the Awards Committee of the Orchid Society of Southeast Asia and later he became its chairman. He had a sharp eye for quality, and was an extremely demanding judge – qualities which enabled him to select the best parents for his hybrids (Alsagoff, 1968, 1988; Lim and Alsagoff, 1989; Whang and Alsagoff, 1991).

Koh Keng Hoe

Koh Keng Hoe (Keng Hoe Koh) is as colorful as his orchids. He was fond of saying that when no one was interested in buying his orchids, they were only as valuable as grass. But then Keng Hoe had the knack of turning grass into a fistful of dollars. Much of his success can be attributed to his warm personality, friendliness, sincerity, superb story-telling skills, and an ability to point out the good points of any particular orchid.

Mischievous and not particularly inclined to rote learning, Keng Hoe was nevertheless a very intelligent man with a high emotional IQ. He had an inborn ability to entertain and make friends easily. He did not earn a university degree but his marketing skills would be the envy of any MBA. He had an uncanny eye for good plants, remembered their names, parentage and progeny, and their history. In the mid-1950s Keng Hoe spent 6 months living with and learning from the Hawaiian orchid growers. At that time few Singaporeans had been to Hawaii. It was difficult to obtain foreign exchange for purchases in US dollars. Keng Hoe was quite unique in managing to import orchids from the leading Hawaiian hybridizers. Keng Hoe told Phoon Yoon Seng (Yoon Seng Phoon) that whenever he let out news of a new shipment he received enquiries throughout the day. There was lively bidding for these plants even before anyone had seen them. Collectors flocked to his home. Buyers would arrive at 11 pm to await a 2 am arrival. The boxes emptied within the hour. In those days, good plants could fetch a thousand Singapore dollars (at a time when mansions in good locations fetched only $10,000!) Keng Hoe bought two plants of an awarded *Vanda* Nellie Morley from George Ing of Hawaii for a Singapore collector. He paid $1,500 for a top cutting, and $2,500 for the mother plant. By comparison, a meter tall *Vanda* Tan Chay Yan was sold for $5,000.

The Korean War was heaping fortunes upon the rubber planters and merchants by sending rubber prices from the previous low of 99 cents per kg to $6.60 per kg. Their happiness was doubled by the ownership of the finest orchids which only the rich could enjoy, and made them the envy of their friends. Koh Keng Hoe played the important role of concert master by introducing quality plants into Singapore and encouraging the interest in orchids. Although he profited from the transactions, Keng Hoe treated everyone equally, whether or not they could afford his plants, and whether or not they bought anything from him.

This attitude paid off handsomely in the early 1980s when the orchid craze had died down in Singapore, in parallel with the stock market collapse. One day an African woman walked into his nursery. As usual, Keng Hoe bid her welcome and offered her a drink. Then he started chatting with her. He thought that she was a new resident, but it turned out that she was the First Lady in an African nation who wanted to introduce orchid cultivation into her country. Keng Hoe had no problem getting her the best cut flower orchid plants from Singapore because everyone was desperate to sell at that time. He spent 6 weeks in the African nation delivering the technology transfer with the plants. He was unprepared for the awesome sight of the presidential gardens which was said to be as large as the Singapore Botanic Gardens.

After the 1970s, Keng Hoe's sales have been based mainly on his own hybrids. An early success was *Aranthera* Beatrice Ng which was a breakthrough when it yielded a yellow clone which earned an AM and a Silver Medal at the Fourth World Orchid Conference in Singapore in 1963. Another, red clone also earned an Award of Merit.

Several of his outstanding *Dendrobium* crosses play important roles as cut flowers and as stud plants for other cut flower hybrids.

Wong Leong Fatt

Wong Leong Fatt (Leong Fatt Wong) is the only local breeder who has won the Eric Holttum Gold Medal (awarded to the breeder of an FCC or AM plant scoring the highest number of points for the year) twice, in 2 consecutive years; in 1967 for *Aranda* Hee Nui var. Khoo AM/OSSEA *(Arach.* Ishbel × *V.* Eisenhower) (Teoh, 2005: 316) and in 1968 for *Aranda* Chrisitine var. Pang Gan (*Arach. hookeriana* × *V.* Hilo Blue) (MOR, 1969: 80–81). *Aranda* Christine var. Singapore received an AM from the American Orchid Society in 1980.

Leong Fatt is probably best remembered as the breeder of the fabulous *Aranda* Christine whose several clones were so sought after during the Christine craze of 1972. In actual fact, only the Christine #1 form met the criteria of a good cut flower although Christine #80, which should be called var. Prachaub after its given varietal name when it received an Award of Merit from OSSEA, was the most beautiful (MOR 9, 1969). Its flowers were large, with broad petals, well arranged on a long upright rachis, and were of an even purple. It is not free flowering, but it is amphidiploid and has produced almost 60 fine hybrids well into the 21st century, including the second generation *Aranda* Baby Teoh and *Aranda* Multico Deluxe and the award winning *Mokara* Mak Chin On var. Maryland AM/OSSEA and *Mkra* Chark Kuan var. Yellow Bird AM/AOS.. Second generation progeny from *Aranda* Christine appeared from 1988 onwards.

The several clones of *Aranda* Christine (#1, #9, #27, Prachaub, #999) were all very different to the extent that one might think they were of dissimilar parentage. Christine #1 carried smaller flowers that were pink with dark purple spots, very pleasing either outdoors in the sun or under fluorescent light. It was grown on an extensive scale by many farms to meet demand right up to the late 1980s.

Leong Fatt was active in the orchid scene of the 1960s and 1970s, working through Seng Heng Nursery which was located at Thomson Road. Realizing the potential of *Aranda* as orchid cut flowers, he bought up what was available of *Aranda* Wendy Scott around 1962 and took them to a hillside farm in his hometown of Penang where he began to propagate them using chicken manure and other stimulants. The plants grew rapidly. Multiplication was rapid. I visited him in 1963 and was quite impressed by his growing technique. Unfortunately, a genetic change occurred in the plants causing their flowers to lose the soft blue color. Another problem was Penang's distinct dry season from November to February making it unsuitable as a place to produce cut flowers since demand is greatest from December to February.

Chong Chok Chye

Chok Chye Chong was a hobbyist grower but he was a breeder of considerable repute during the 1960s. Owning an unusual plant of *Aranda* Lily Chong which won an AM/MOS in 1962, he bred the famous *Aranda* Tay Swee Eng (*Aranda* Lily Chong × *V.* Piha Moon) which garnered an AM/OSSEA and two HCC/OSSEA (MOR, 1969, 9: 80–81). The improvement in form without loss of substance in this second generation *Aranda* ushered in the new trend in breeding forward with *Aranda*.

How Yee Peng

How Yee Peng (Yee Peng How; Fig. 1-9) hails from a farming family. He started growing orchids at Kings Road and when I came to know him in the mid-1960s he was busy propagating *Aranda* Wendy Scott, the first reliable, blue, popular, cut flower *Aranda*. He was a modest man and has remained so through all these years despite his numerous, remarkable achievements. Yee Peng is systematic, neat, clear in his mind about his goals when making a hybrid and is skilled in culture techniques. His opportunity came in the mid-1960s when he obtained a plant of the tetraploid *Vanda* Dawn Nishimura as a gift from Soh Kim Kang. It had been brought over from Bangkok by Devan, Kim Kang's top landscaper. The *Vanda* was bred to all the proven scorpion orchid parents and produced numerous outstanding hybrids for How Yee Peng for more than 20 years. Among them are the following:

Aranda How Yee Peng (*Aranda* Tay Theng Suan × Dawn) with three Awards of Merit, best hybrid in 1975, earning Yee Peng the Eric Holttum Gold Medal: *Aranda* Neo Hoe Kiat (*Arachnis* Maggie Oei × Dawn); *Aranda* Etsuko Nishio (*Arachnis* Ishbel × Dawn) earned an HCC/OSSEA in 1986; 1981 *Aranda* Multico Deluxe(*Aranda* Christine × Dawn): *Rntda* Soh Kim Kang (*Ren. storiei* × Dawn); *Aranda* Yee Peng Dawn (*Aranda* Yee Peng × Dawn); *Lewisara* Blue Moon (*Lwsra* Chittivan × Dawn); *Vascostylis* Blue Queen; *Vasco* Tham Yuen Hae × Dawn; *Bokchoonara* Looi Eng San *(Bkch* Khaw Bian Huat × Dawn); *Mokara* Soo Chee ((Mkra Chark Kuan × Dawn); *V* Koh Chong Joo (Joaquim × Dawn); *Vascostylis* Looi Eng San (*Vasco.* Than Yuen Hae × *V.* Dawnchild); *V.* Atsuko Johmori (JVB × Dawnchild); *V.* Soeryadjaya (Dawn × Emma van Deventer), etc.

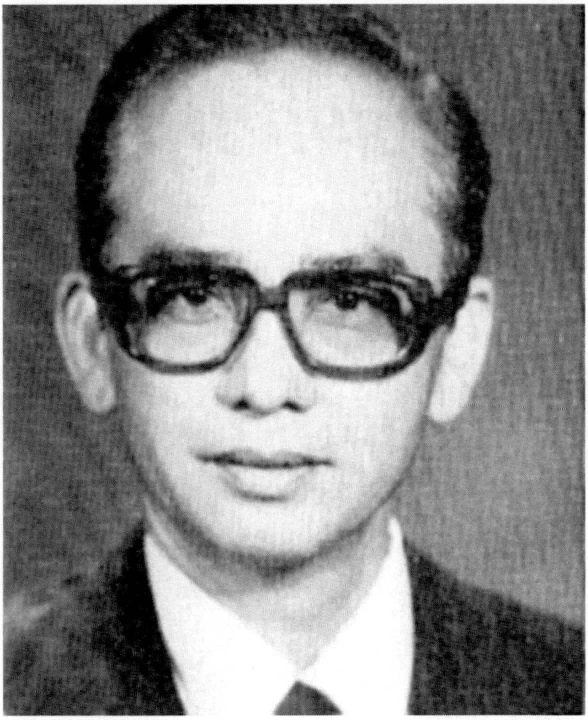

Fig. 1-9. How Yee Peng (Yee Peng How).

His son Wai Lai has continued using *V.* Dawn Nishimura to produce *Ascda* Children's World in 1999 further bred to *Aerides lawrenceae* to produce *Christierara* Nadeswari Desha.

Syed Yusof Alsagoff was also given some pollen of *V.* Dawn Nishimura by Kim Kang from which he bred the big flowered blue *Aranda* Noorah, *Aranda* Wong Bee Yeok (*Aranda* Lily Chong × Dawn), *Aranda* Ang Hee Seng. (*Arach.* Maroon Maggie × Dawn), *Trevorara* Ursula Holttum (*Arachnopsis* Eric Holttum × Dawn), and *Vandaenopsis* Yunus Alsagoff.

Soh Kim Kang

Kim Kang Soh was famous for his exhibits at the orchid shows of the late 1960s and early 1970s. To ensure that he would emerge as Best Exhibitor in the Show, he dispatched experts to source the best plants that were in bloom, not only in Singapore but in Malaysia and Thailand as well. Then he spared no expense in acquiring them. Coupled with the fact that the famous landscapist and orchidist, Deewan Raggardee came with the plants, Kim Kang's landscape exhibits were a sight to behold.

In 1969 he exhibited *Aranda* Christine "Prachaub" (bred by Wong Leong Fatt) and the second generation *Aranda* Tay Swee Eng (bred by Chong Chok Choy), both of which won an Award of Merit. Shortly thereafter, orchid growers in Singapore went crazy over these plants, and like Kim Kang, they too were willing to pay handsomely for top cuts or *keikis*.

Because of his willingness pay a good price for desirable plants, Kim Kang managed to acquire a few outstanding stud plants along with the attractive hybrids that were only good for the shows. *Vanda* Dawn Nishimura produced many outstanding hybrids for breeders who were friends of Soh Kim Kang. When his interest was at its height, Kim Kang undertook a difficult expedition into the mangrove swamps of Sarawak with Yap Kim Fatt and How Yee Peng to study the habitat of the newly discovered *Paphiopedilum linii* (Yap, Soh, and How, 1967).

To a large extent, Kim Kang's fascination with orchids grew out of his love for his Thai wife, Prachaub whose name is borne by the awarded clones of a dozen orchids. To the regret of the local orchid circle, Kim Kang lost interest in orchids after he lost his wife. However, his stud plants which were inherited by How Yee Peng continue to produce outstanding hybrids.

Mak Chin On

A master at plant propagation, Mak Chin On (Chin On Mak; Fig. 1-10) made multiple, significant contributions to the Singapore landscape, in addition to his role in the orchid cut flower industry. He is responsible for introducing the majestic *Bismarkia nobilis* palms that line the approach to Singapore's Changi Airport. His vast collection of 55 species of palm runs into the hundreds for many varieties. Mak (his friends generally do not call him Chin On) has propagated a deep liver-red clone of *Eugenia* to add color to the roadside. He is almost solely responsible for introducing the yellow rain tree (*Samanea samaan*) into the Singapore landscape. Although the first yellow rain tree was thriving at the entrance to the famous birth-quake, Kandang Kerbau Hospital during the 1960s, no one noticed its potential (perhaps because they were distracted by the 40,000 annual births at the hospital), until Mak started propagating the plant from seed in the 1990s. Today his rain trees produce spectacular stands of gold amidst the deep greens in the gardens, golf-courses and road-sides.

Mak Chin On has grown orchids since 1961, and by 1964 he had already perfected a way of obtaining 50 plantlets from a single plant of Aranda Wendy Scott. When Leong Fatt began mopping up all the *Aranda* Wendy Scott after the 4th World Orchid Conference in 1963 Mak sold off his plants and with the new found capital he gave up his teaching job and started a small nursery in Kok Nam Lane. He bought some plants of *Aranda* Christine from Leong Fatt and began propagating them in this new nursery. When the *Aranda* Christine craze hit Singapore in the late 1960s Mak already owned a large stock of the orchid and he could obtain more than a dozen plants from the single plant in a matter of months. Meanwhile, the buoyant

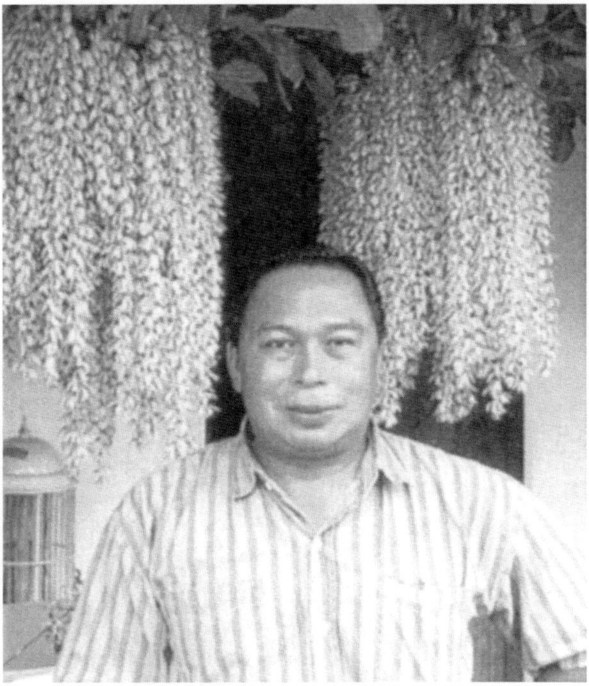

Fig. 1-10. Mak Chin On (Chin On Mak).

stock market lifted orchid prices. At the height of the Christine craze, a cutting of his Christine #80 fetched Mak more than a month's salary. The craze died when the stock market plummeted in 1973. By then, Mak had chosen a beautiful piece of sloping land in Ponggol to start his new orchid nursery. It faced the Straits of Johore and a strong breeze blew over the land every day.

Mak set himself two goals – first, to increase the supply of desirable hybrids for the cut flower industry; secondly to make new hybrids which met the criteria for a cut flower. To achieve his first goal, he bought two scorpion hybrids of *Vanda* Dawn Nishimura from their breeder, Syed Yusof Alsagoff. One he named after his mother, *Aranda* Wong Bee Yeok (Bee Yeok Wong). The other was named for his mother-in-law, *Aranda* Ang Hee Seng (Hee Seng Ang). He raised mericlones of awarded plants from both crosses and sold batches to various nurseries. The two *Aranda* hybrids have maintained their status as excellent garden or cut flower orchids.

Only about a dozen of Mak's own hybrids have been registered, but among them there are two outstanding *Dendrobium* and two *Mokara*. *Den*. Mary Mak (*Den* Theodore Takaguchi × *Den* May Neal) has yellow green flowers (it is named after his first wife Mary) of good form and substance and has been mericloned. *Dendrobium* Mak See (See Mak) which was named after his father has a contrasting deep purple lip with green tepals and is very striking. Mak won the Eric Holttum Gold Medal with *Mokara* Mak Chin On (MOR, 1981, 15:32), and at one time he had an acre of these plants in continuous bloom at Ponggol, but sadly they were wiped out when someone introduced a weed-killer into his watering system. They are featured in the endpaper of Teoh Eng Soon's *A Joy Forever* published by Times in 1982.

Mak had many friends among the Thai growers which gave him the opportunity to understand the hybrids that they favored. He imported and propagated three hybrids of *Cattleya* (the white *Lc.* Hawaiian Wedding Song, the mauve *Blc.* Pink Diamind and the yellow with red-lip. *Blc.* Alma Kee) which can be depended upon to thrive and bloom freely in Singapore. He grew many outstanding species plants. The spectacular flowering of *Coelogyne rochussenii* in his Seletar farm (10 pots bearing a total of some 25,000 flowers) was documented by Peter O'Byrne in the Vol. 25/1995 issue of the *Malayan Orchid Review*.

Mak played a big role in the Orchid Society of South East Asia during the decades when he served on its Committee from 1966–1985. He was a Vice-President for several years. He could be depended to contribute hundreds of assorted orchid cut flower whenever the Society needed to send a shipment for an Orchid Show somewhere in the world. Mak took care of everything, including packing and shipment, without any cost to the Society. He is well known to all the prominent orchidists in the region, both local and foreign.

When he was forced by Singapore's rapid urbanization to move his nursery from Ponggol to Seletar, and thence to Lim Chu Kang on the opposite side of the island in the short course of 20 years, Mak decided to stop growing orchids altogether, and to switch to foliage plants. Orchid prices tumbled during the last decade of the 20th century because the scarcity of land led to small private gardens or no garden at all, and there was also the onslaught of cheap mericlones initially from Thailand and subsequently from Taiwan. When a rogue company attempted to corner the market by dumping cut orchids below production cost, the prices of orchid cut flowers also declined. It never recovered. Mak stopped growing orchids in 1997. He stopped exporting them in 2000. Given the high cost of rent, labor and water, and the orchids demand for constant attention and frequent replanting, fiercely competitive pricing from Bangkok, the move that Mak took ensured his survival as a nurseryman. But it was a hard blow for orchids. Ironically some of his own orchid hybrids which he had given to Thai friends are now among the successful *Dendrobium* cut flowers from Bangkok.

Perhaps when he feels more secure, he might be persuaded to take up orchid growing again, even if only as a hobbyist. Mak and the founding editor of *Orchid Biology, Reviews and Perspectives,* Professor Joseph Arditti have been good friends for over 25 years. Because of that volume III of this series is dedicated to him.

Mok Choi Yew

Mok Choi Yew (Choi Yew Mok) was the first person to combine *Ascocentrum* with *Aranda,* not because he wanted to miniaturize but because he loved the free flowering input from the *Ascocent*rum. He attached his surname to the resultant hybrid and called it *Mokara*. Such trigeneric hybrids are now making the scene in the cut flower trade, while *Mokara* Zaleha Alsagoff won the first and only FCC ever conferred by OSSEA

C.Y. Mok has bred orchids for several decades and is locally famous for having bred the spectacular white terete *Vanda* Poepoe var. Diana, HCC/MOS and *Phalaenopsis* Mok Choi Yew which became a stepping stone for many outstanding *Phalaenopsis* hybrids.

David Lim

David Lim (Fig. 1-11) is very much a do-it-yourself man. As such he was more successful in producing *Phalaenopsis* hybrids than in turning a profit when he ran his own two hectare orchid nursery in Seletar. He has registered 55 *phalaenopsis* hybrids. Lim bred *Phal.* Amber Delite by crossing an outstanding *Phal.* Teoh Tee Teong with a fine form of *Phal. amboinensis*, which hybrid won Best Plant at the Penang Orchid Show and for David, the coveted OSSEA's Eric Holttum Gold Medal for 1987 (MOR, 1988). *Phal.* Teoh Tee Teong produced several fine hybrids

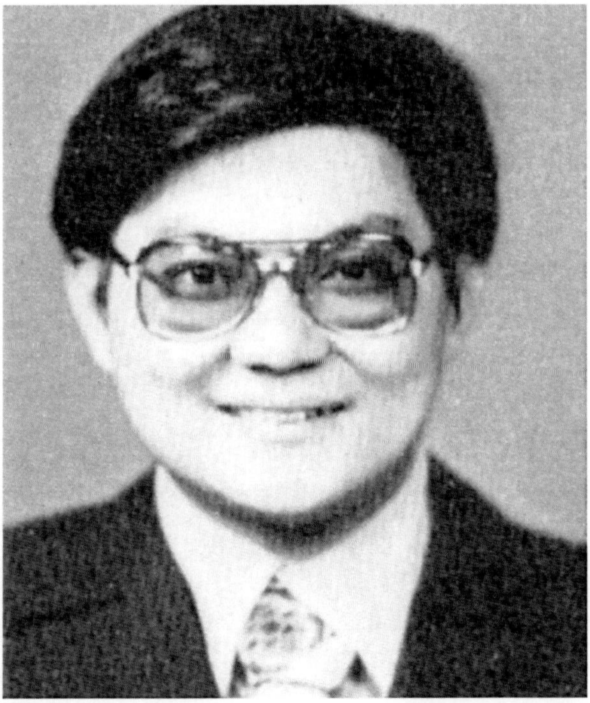

Fig. 1-11. David Lim.

for David, such as *Phal.* Singapore Swing (× *violacea*), *Phal.* Toto Han (× *Phal.* Golden Buddha), and *Phal.* How Yee Peng (× *Phal.* Ambonosa). He authored a paper on *Phal. amboinensis*, one of his favorite stud species and coauthored a paper on breeding with Yusof Alsagoff (Lim, 1982; Lim and Alsagoff, 1989.) David is an accomplished photographer and maintains an excellent photographic record of his studs and hybrids. David Lim served on the Committee and on the Awards Panel of OSSEA for many years during the 1970s and 1980s but has now taken a back seat.

Phoon Yoon Seng

Phoon Yoon Seng (Yoon Seng Phoon; Fig. 1-12) knows as much about the orchids of Singapore as any expert today. He has served on the Committee of the Orchid Society of South East Asia for nearly 40 years and was editor of its monthly bulletin and Chairman of the Panel of Judges for many years. Phoon has contributed numerous articles and photographs to the Malayan Orchid Review (Phoon, 1981, 1989, 1995, 1996, 1997, 1999a, b, 2002). His original love was *Paphiopedilum* but today he has a collection of outstanding specimens of numerous genera. A retired school

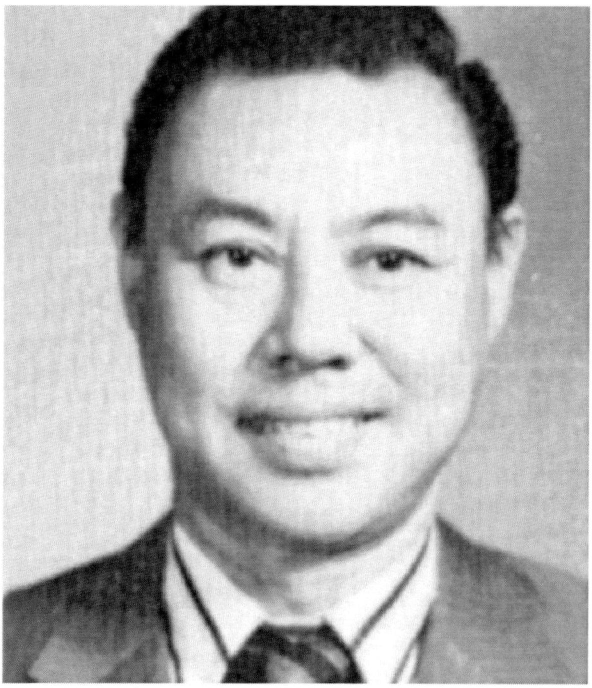

Fig. 1-12. Phoon Yoon Seng (Yoon Seng Phoon).

principal, Yoon Seng is nevertheless a very modest man and has never pushed himself forward for higher office.

Au Yong Nang Yip

Au Yong Nang Yip (or NangYip Au Yong) of Sarawak (Fig. 1-13) is already immortalized by Eric Hansens' sensational characterization of his outlook and personality. However, he is included here because of his important contributions to the appreciation of Bornean orchids rather than any fascination over his capabilities. His orchids are regularly featured at Orchid Shows in Singapore and the region over many decades. His most famous plant is an extraordinarily clear yellow *Vanda dearei* that is perfectly round with overlapping tepals of excellent shape, size and substance. However, his collection is vast and includes old hybrids that he acquired at great cost from Singapore and fine clones of *Phalaenopsis* and *Paphiopedilum* native to Sarawak. His knowledge of these plants is encyclopedic. He plays an important role in promoting awareness and love for the orchids of Sarawak and has contributed to the setting up of a civic nursery and public garden in Kuching where flowering orchids are on regular display.

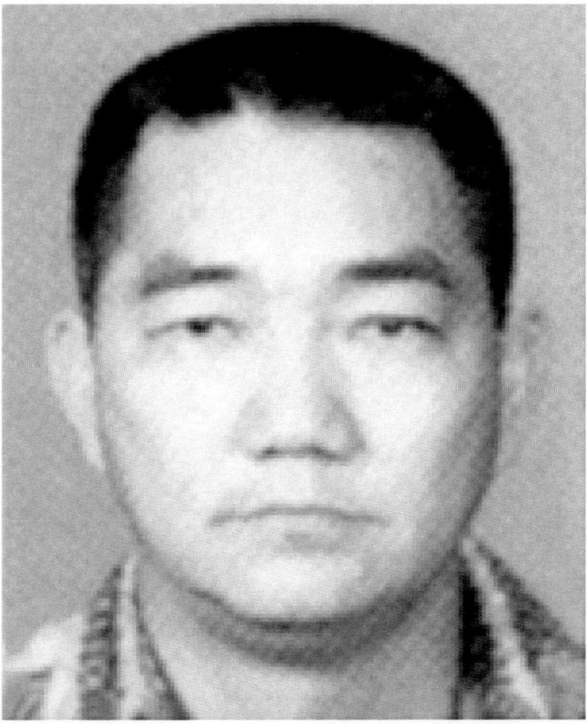

Fig. 1-13. Au Yong Nang Yip (or Nang Yip Au Yong) of Sarawak.

Anthony Lamb

Anthony Lamb (Fig. 1-14) is an authority on the orchids of Sabah. He set up the Tenom Orchid Centre which is a showcase for the orchids of the state. Following his studies at Cambridge and the University of the West Indies, he went to Sabah in 1962 to commence a career in agriculture in the Tawau area. His professional career involved oil palm and cocoa but his wider interest covered many aspects of agriculture as well as conservation. He made surveys for the World Wild Life Fund for Nature and the Sabah Parks and he is a member of the IUCN Orchid Specialist Group Species Survival Commission. He is celebrated for his knowledge of Sabah orchids but is also an expert on rhododendrons, hoya, gingers and pitcher plants. He has many publications on orchids and other subjects and participates in many international shows and seminars as judge and speaker (Lamb, 1978, 1985, 1993; Vermeulen and Lamb, 1988). He was elected an OSSEA Fellow at the 17th World Orchid Conference in Shah Alam, Malaysia in 2002.

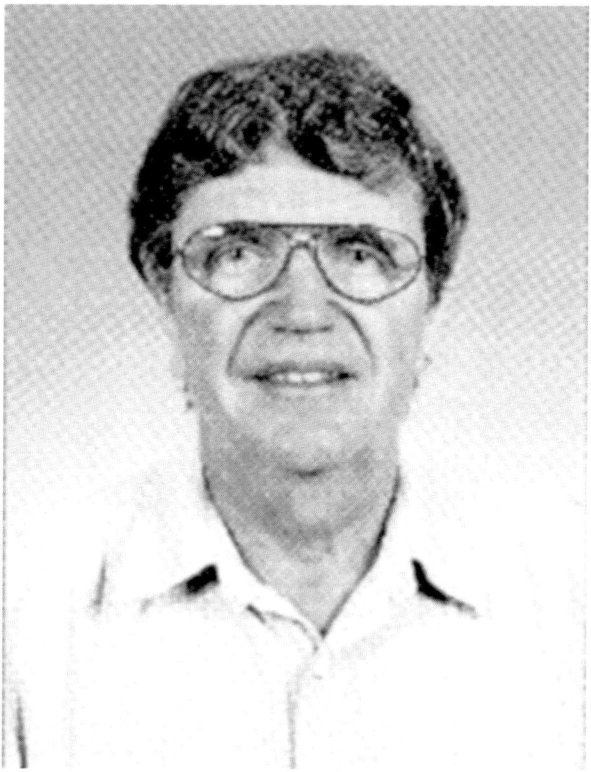

Fig. 1-14. Anthony Lamb.

Ooi Leng San

I came to know Ooi Leng San (Leng San Ooi) in 1964 when I was in charge of Butterworth Hospital. At that time he already had a large collection of cattleyas which were not the most free flowering orchids in the equatorial lowland but they certainly were the hardiest. Ooi Leng San also grew *Vanda* and *Ascocenda* which were among the most popular orchids at that time. He owned a huge piece of land and a kiln that made bricks and pottery, the latter a welcome tie-up with his orchid interest. Ooi Leng San's interest inspired his son, Michael to develop an interest in orchids, and it is Michael who eventually became a well respected orchid expert in Peninsular Malaysia. Father and son established a fine collection of popular local species and bred with them – *Phalaenopsis* (in particular *P. violacea*), *Paphiopedelum, Bulbophyllum*, and *Grammatophyllum*, The *Phal*. Penang series are line bred from the better strains of *P. violacea* which include the *alba* strain and they are being incorporated into some of the Taiwanese breeding programs. Michael's son will continue their family tradition of caring for the native orchids and their hybrids.

Cheah Kheng Cheong

Isolated at Malim Nawar near the city of Ipoh Cheah Keng Cheong (Keng Cheong Cheah) led in *Phalaenopsis* hybridization in Perak and his name appeared frequently in the hybrid register during the 1960s. The state is home to *Phalaenopsis violacea* and *Vanda hookeriana*. Kheng Cheong's pioneering work played an important role in promoting orchid cultivation in the state which remains a centre for *Phalaenopsis* breeding in Malaysia.

Lum Hon

Lum Hon (Hon Lum) is the sole proprietor of Lum Chin Orchid Nursery in Malaysia. He made numerous hybrids which are important to the cut flower industry and also bred the 1991 Eric Holttum Gold Medal Winner, *Christieara* Christie Low. The Chinese educated Lum Hon is shy and reclusive, but he is generous with his orchids.

The Ladies

Before 1980, Asian women in Singapore and Malaysia preferred to work behind the scene, and seldom sought to be recognized in their own right. The same is not true of their European sisters who gained suffrage a century earlier. Thus, the orchid circle was familiar with Gracia Lewis and Wendy Scott who both played major roles in the Malayan Orchid Society while the local women remained in the background although they did much of the work with their husbands' collections and in supporting and running the Orchid Society.

Gracia Lewis

Gracia Lewis (Fig. 1-15) was well known in her own right having spoken at the 3rd and 4th WOC: She is the breeder of many excellent early hybrids. In her case, her husband Max, is the one who takes a back seat although he was an active member of the MOS. He was a real gentleman who believed that his wife should have the limelight. Nevertheless, Max served on the Committee of the Malayan Orchid Society and took on the onerous task of Chairman of the Organizing Committee for the 4th WOC held in Singapore in 1963. During the 1950s and 1960s, the MOS, and later OSSEA, held their monthly meetings at the premises of Frazer & Neave, of which Max was then managing director. Being not a rich organization, the orchid

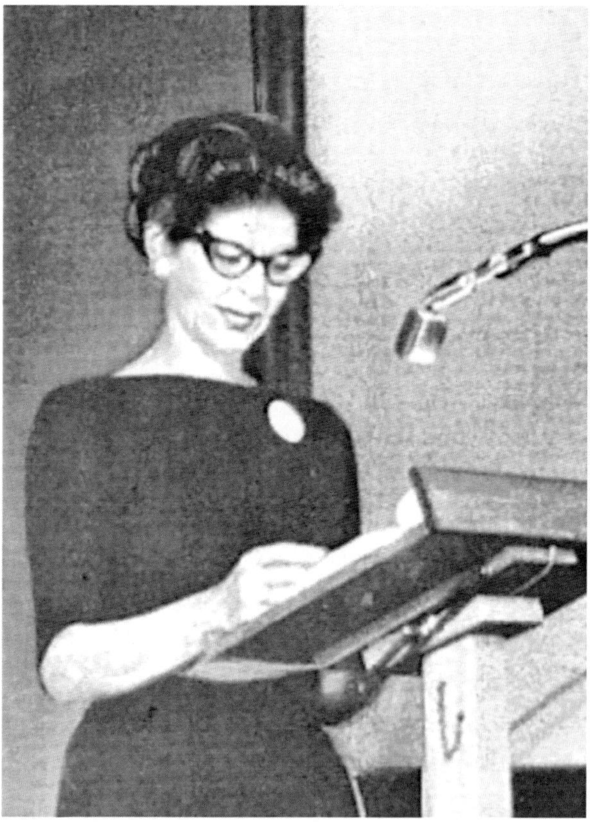

Fig. 1-15. Gracia Lewis.

society needed gracious organizations to allow the use of their premises for free to hold its monthly meetings.

Gracia Lewis maintained a large garden of fine orchids and won two FCC from the RHS at the 1954 Chelsea Flower Show for *Vanda* Tan Chay Yan, var. Pride of Malaya and *Vanda* Rothschildiana var Doo. Between 1959 and 1967, ten of her plants won an Award of Merit from MOS/OSSEA. These are truly remarkable achievements. The *Vdnps*. Patience var Francine was Lee Kim Hong's hybrid raised by Singapore Orchids. Gracia herself bred 23 hybrids, some of which were registered by Singapore Orchids Ltd., after Gracia returned to England (Elliot et al., 2005). Among the best known are *Aranthera* Gracia Lewis (*Arnth.* Mohamed Haniff × *Ren storiei*); the clone Burong Merah which was granted an Award of Merit by MOS in 1964 is perhaps the largest of the *Arantheras*: and *Aerides* Max Lewis *(A, odorata × jA. arkiana).*

Wendy Scott

Ronald Scott (Fig. 1-16) was the Managing Director and later Chairman of the Straits Times Press, a president of the Malayan Orchid Society and its staunch supporter. With Wendy as the editor of the Malayan Orchid Review, Ronald took

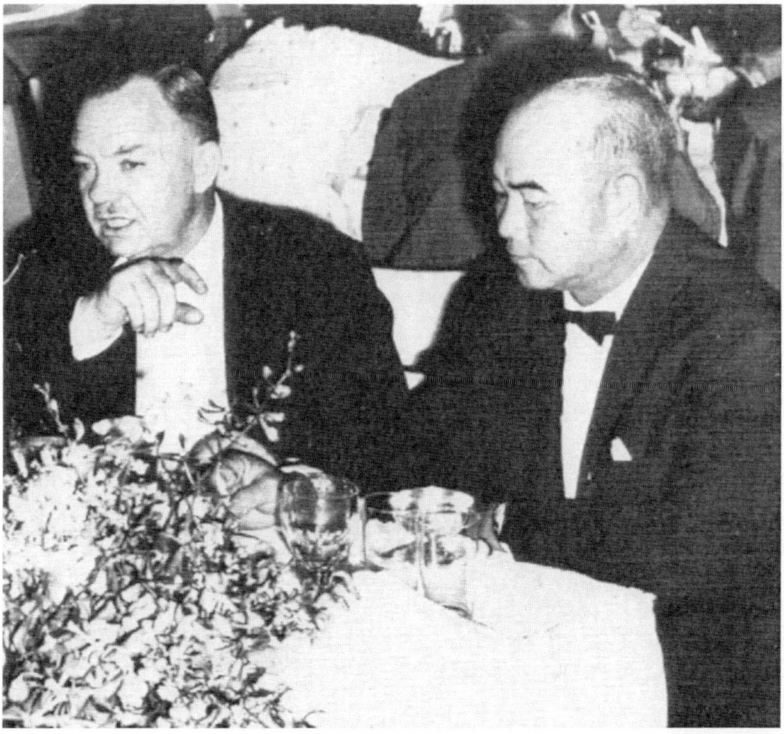

Fig. 1-16. Ronald Scott and Sum Lai Wah (right).

care that the magazine was well produced and in color as soon as color printing became available. The high quality of the *Malayan Orchid Review* played an important role in gaining worldwide recognition for the Malayan Orchid Society. Ronald and Wendy also helped the Society mount its exhibits at various World Orchid Conferences.

One of the earliest and most successful cut flower scorpion orchid is the lovely mauve *Aranda* Wendy Scott (*Arach. hookeriana* × *V.* Rothschildiana) (Fig. 1-17). It has an Award of Merit from the Royal Horticultural Society and the Malayan Orchid Society.

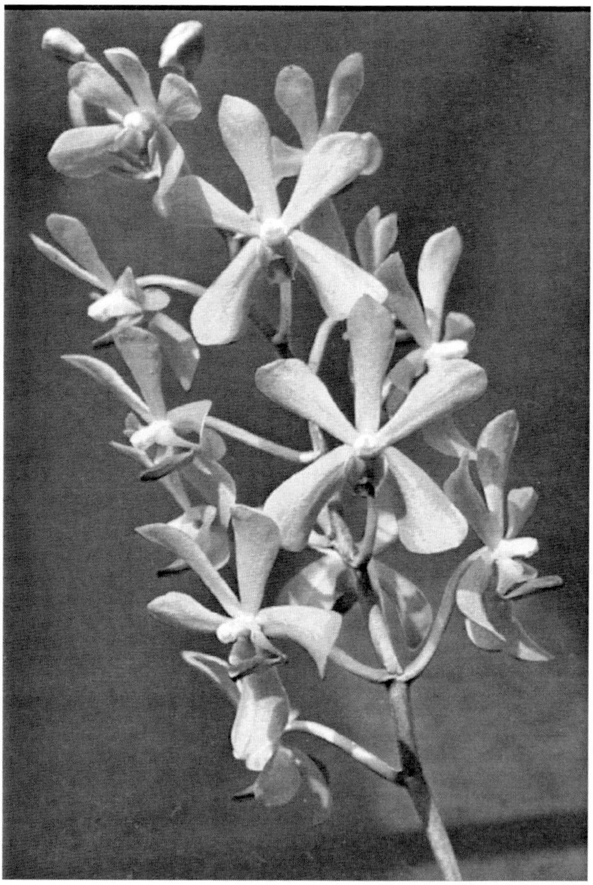

Fig. 1-17. *Aranda* Wendy Scott (*Arach. hookeriana* × *V.* Rothschildiana).

Rosalind Lee

Rosalind Lee is the wife of Lee Kim Hong who founded Singapore Orchids Ltd. and started the lovely Mandai Gardens of Singapore Orchids together with John Laycock. She and Amy Ede (Fig. 1-18) assisted their husbands in setting up and maintaining the beautiful orchid landscapes of Mandai which were a prime tourist attraction from the 1960s to the 1990s. The ladies were also responsible for collecting, packing and shipping the pioneering orchid exports of the 1960s. When the Malayan Orchid Society appealed for orchids to mount their display with 1,600 sprays at the 1960 Chelsea Flower Show to back their bid to host the 4th World Orchid Conference in Singapore, they ended up receiving 2,400 sprays from 45 well wishers in Singapore and Malaya. Rosalind and her husband Lee Kim Hong, confident after their successful trial in sending cut orchids around the world and finding the flowers still fresh when they arrived back in Singapore, undertook the packing which ended up in 35 cartons and 4 wooden boxes. At the London end, Amy and John Ede and the G.H. Addisons received, unpacked and treated the flowers. Only two sprays were discarded while 1,606 sprays were included in the exhibit. This MOS Exhibit of orchid cut flowers received a Gold Medal and the Queen complimented the MOS on their exhibit.

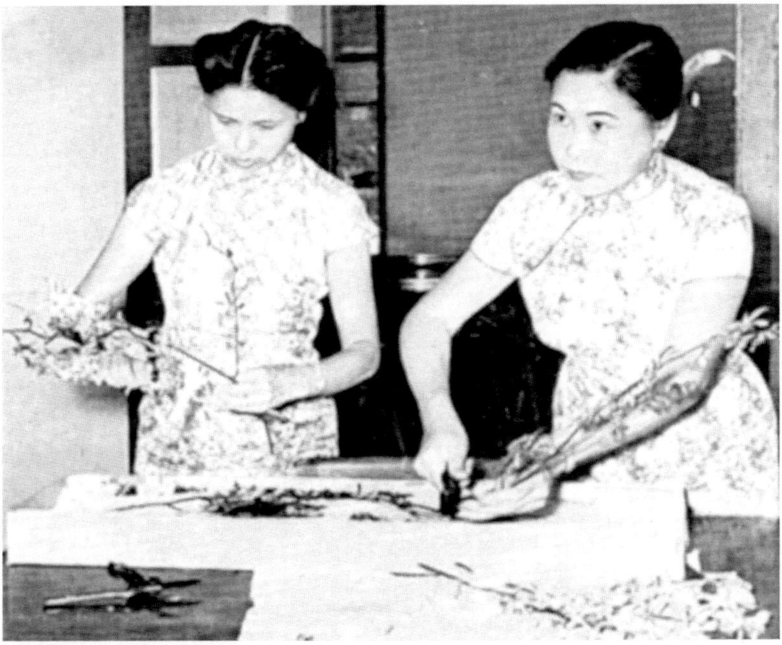

Fig. 1-18. Amy Ede and Rosalind Lee.

It consisted of 60 sprays of *Arachnis* Maggie Oei, 370 *Aranda* hybrids of various colours, 180 red *Aranthera* hybrids, 20 *Renanthera,* 390 terete, semi-teret and quarter terete *Vanda* that included *Vanda* Miss Joaquim and *Vanda* Tan Chay Yan, 70 strap leaf *Vanda* hybrids, 450 *Dendrobium*, 60 *Spathoglottis* and *Oncidium* Goldiana var. Golden Shower, and 6 *Rhynchostylis retusa* of which 3 were of the *alba* form. A beautiful photograph of this exhibit is featured in the *MOR* Vol 6/1, 1961.

Amy Ede

Amy Ede and Rosalind Lee (Fig. 1-18) worked almost as a pair assisting their husbands John and Kim Hong with the care of Mandai Orchid Gardens, its breeding program and the raising of their own hybrids, and the handling of cut orchids for export. Interest in Singapore orchids as cut flowers had emerged after the 1957 Chelsea Flower Show and with improvement in air travel, this was a window of opportunity for the local growers who hitherto had only been able to supply outlets in Singapore, Peninsular Malaya and Hongkong. Lee Kim Hong and John Ede urged local growers to strive for the quality required for export (MOR 6/1: 16–17, 1960). Singapore Orchids handled their produce. In the early days, gift parcels of exotic orchids to England, the USA, South Africa and Europe, peaking at Christmas, formed the mainstay of orchid exports from Singapore. Amy Ede took charge of Mandai Gardens for many years after the retirement of Lee Kim Hong and Rosalind and the demise of John Ede. Mandai Gardens was bought over by Heah Hock Heng in 2004.

Singapore Orchids had their own breeding program which stood apart from other Singapore breeders. They used their own stud plants and generally they do not repeat or line breed from crosses made by other breeders.

Ong Siew Hong

Madam Ong Siew Hong (Siew Hong Ong) was the wife of Tan Chee Tong (Chee Tong Tan), a distinguished orchid grower and breeder of *Vanda* and *Dendrobium* during the 1950s and 1960s. They raised many awarded plants which bear the varietal name, Siew Hong, and Dr. Yeoh Bok Choon ranked the Siew Hong variety of *V*. Tan Chay Yan as one of the loveliest. It was awarded the Best *Vanda* Cut Flower at the 2nd World Orchid Conference in Hawaii.

They contributed excellent cut flowers for the Society's exhibits at various internal shows. In their family the orchid tradition continues: Ong Siew Hong and Tan Chee Tong are the parents of Dr. Tan Wee Kiat.

Madam Yong

Madam Yong, the mother of Yong Kok Wah (Kok Wah Yong) was one of the most skillful commercial orchid growers during the 1960s and 1970s. Having undergone an operation to remove kidney stones, her husband was only permitted to entertain friends while she did all the work looking after a two hectare orchid farm and daily harvesting the orchid for export. Her green *Den.* Yong Kok Wah was the unchallenged green *Dendrobium* in the 1970s and it is the parent of many important commercial *Dendrobium* produced in Singapore and Thailand. Her secret was the constant use of dilute matured porcine manure and a modest amount of a slow release fertilizer.

It will be noticed that I have not included extensive discussions of Professor Eric Holttum, John Laycock, Emile Galistan and Dr. Tan Wee Kiat in my list because these four names are already world famous and they will forever be associated with the orchids of Singapore. I have also not mentioned Peter O'Byrne, Dr. Yam Tim Wing (Tim Wing Yam) Dr. John Elliott, Peggy Tan, Wendy Chew, Law Moi Hwa (Moi Hwa Law) and others who are active in the contemporary orchid scene because their contributions will be best summed up at a later date.

Glossary

AM. Award of Merit. This is awarded by a panel of experts of an orchid society to recognize the excellence of a spray of flowers of an orchid hybrid. Some societies insist that the awarded hybrid should excel both parents.

Amphidiploid. A doubling of the chromosome number in a heterogenous genome.

FCC. First Class Certificate. The highest floral quality award given by an orchid society. This ranks above the Award of Merit.

HCC. Highly Commended Certificate is an intermediate award given by an expert panel of an orchid society to a hybrid that is outstanding but jut falling short of the standard of an Award of Merit. On a different occasion this hybrid may or may not produce a flower spray of a higher standard.

Line breeding. A series of crosses or hybridizations that keeps reintroducing the genetic material of a desirable parent, either in part or in whole.

MOS. Malayan Orchid Society, founded in 1928 an headquartered in Singapore.

OSSEA. Orchid Society of Southeast Asia was the new name adopted by the Malayan Orchid Society when Singapore separated from Malaysia in 1965

Polyploid. A plant possessing more than the normal two sets of paired chromosomes.

Tetraploid. Possessing four sets of chromosomes.

WOC. World Orchid Conference.

Literature Cited

Alphonso, A.G. 1966. The need for conservation of Malaysian orchid species. In: L. de Garmo (ed.), *Proc. 5th World Orchid Conference*, Long Beach, CA, pp. 125–132.

Alphonso, A.G. 1980. Some colourful orchid species of Malaysia. In: M.R. Sukshom Kashemsanta (ed.), *Proc. 9th World Orchid Conf.*, Bangkok, Thailand, pp. 97–100.

Alphonso, A.G. 1981. Singapore's national flower (*Vanda* Miss Joaquim). *Malayan Orchid Review* 15: 9–14.

Alphonso, A.G. 1983. Obituary of Jimmy Quek Kiah Huat. *Malayan Orchid Review* 17: 19.

Alphonso, A.G. 1987. Orchidology in Southeast Asia: A History. In: J. Arditti (ed.), *Orchid Biology Reviews and Perspectives*, IV. Cornell Univ. Press, Ithaca, N.Y., pp. 21–31.

Alsagoff, S.Y. 1968. Awards of OSSEA. *Malayan Orchid Review* 9: 43–4.

Alsagoff, S.Y. 1988. Paraphalaenopsis and its hybrids. *Malayan Orchid Review* 22: 12–16.

Anonymous. 1962. M.O.S. Exhibit at Chelsea May 1960. Report on the Third World Orchid Conference. *Malayan Orchid Review* 6: 21–25, 41.

Anonymous. 1969. *Aranda* Christine, Prachaub, AM/OSSEA (color photo). *Aranda* Tay Swee Eng Var. Prachaub, AM/OSSEA (color photo). *Malayan Orchid Review* 9: 80–81.

Anonymous. 1969. *Phalaenopsis* Mok Choi Yew. *Malayan Orchid Review* 9: 63.

Anonymous. 1981. *Mokara* Mak Chin On, Maryland Nursery, AM/OSSEA (color photo). *Malayan Orchid Review* 14.

Anonymous. 1988. Phalaenopsis Amber Delite, AM.OSSEA (color photo), *Malayan Orchid Review* 22: 6.

Anonymous. 1991. Obituary: Tan Hoon Siang, 25: 91.

Arditti, J. and C.S. Hew. 2007. The origin of *Vanda* Miss Joaquim. In: K.M. Cameron, J. Arditti, and T. Kull (eds.), *Orchid biology, reviews and perspectives*, Vol IX. New York Botanical Garden Press, New York, pp. 261–309.

Dalpethado, M.E. 2004. Vanda Miss Joaquim. *Malayan Orchid Review* 38: 97–98.

Ede, J. 1963. Some commercial aspects of orchids in Singapore. Paper presented at the 4th World Orchid Conference in Singapore in 1963. Reprinted in *Malayan Orchid Review* 1993, 27: 77–80.

Elliott, J., P. Tan, S.Y. Alsagoff, and W. Chew. 2005. *Orchid hybrids of Singapore 1893–2003*. Orchid Society of Southeast Asia, Singapore.

Hansen, E. 2001. *Orchid fever*. London: Methuen.

Hew, C.S., T.W. Yam, and J. Arditti. 2002. *Biology of* Vanda *Miss Joaquim*. Singapore University Press, Singapore.

Holttum, R.E. 1978. Memories of early days. In: E.S. Teoh (ed.), *Orchids: a publication commemorating the Golden Anniversary of the Orchid Society of Southeast Asia (Founded 1928)*. Times Periodicals, Singapore.

Johnson, H. 2004. *Vanda* Miss Joaquim. *Malayan Orchid Review* 38: 99–107.

Lamb, A. 1978. The wild orchid species of Sabah. *Malayan Orchid Review* 12: 80–86.

Lamb, A. 1985. The exotic and strange *Paphiopedilum sanderianum* (Rchb,f) from Borneo. *Malayan Orchid Review* 19: 43–47

Lamb, A. 1993. Jewels in the moss. The helmet orchids of Sabah. *Malayan Orchid Review* 27: 36–41

Lee, K.H. and J. Ede. 1962. Cut orchids for export. *Malayan Orchid Review* 6: 16–17.

Lewis, G. 1962. Malaysian Vandas. *Malayan Orchid Review* 6: 26–28.

Lim, D. 1982. Phalaenopsis amboinensis. *Malayan Orchid Review* 16: 20–23.

Lim, D. and S.Y. Alsagoff. 1989. The selection of stud plants for hybridizing. *Malayan Orchid Review* 23: 98–102.

O'Byrne, P. 1995. *Coelogyne rochussenii* in full bloom – an incredible sight. *Malayan Orchid Review* 29: 60–62,

Phoon, Y.S. 1981. A Salute to *Vanda* Miss Joaquim. *Malayan Orchid Review* 15: 15–17.

Phoon, Y.S. 1989. Choosing and buying orchids. *Malayan Orchid Review* 23: 95–97.

Phoon, Y.S. 1995. A Hunting we will go! *Malayan Orchid Review* 29: 50–51.

Phoon, Y.S. 1996. OSSEA Orchid Show, 1996. *Malayan Orchid Review* 30: 13–15.
Phoon, Y.S. 1997. Singapore Orchid Show, 1997. *Malayan Orchid Review* 31: 15–17.
Phoon, Y.S. 1999a. Notes from an outstanding show. *Malayan Orchid Review* 33: 28–29.
Phoon, Y.S. 1999b. Chatting up Koh Keng Hoe. *Malayan Orchid Review* 33: 55–59
Phoon, Y.S. 2002. It's Exhilarating! *Malayan Orchid Review* 36: 18–19
Scott, W. 1962. The Society's exhibit at Chelsea, 1960. *Malayan Orchid Review* 6: 32–33
Syonan-to Botanical Gardens Monthly Report 2602–2605 (AD 1942–1945). Singapore Botanic Gardens, Singapore.
Teoh, E.S. 1978. Orchids. A Publication commemorating the Golden Anniversary of the Orchid Society of Southeast Asia (Founded 1928). Times Periodicals, Singapore
Teoh, E.S. 1981. Viewpoint: *Vanda* Miss Joaquim, National Flower of Singapore. *Malayan Orchid Review* 15: 5
Teoh, E.S. 1982. A joy forever. *Vanda* Miss Joaquim Singapore's National Flower. Times Editions, Singapore.
Teoh, E.S. 2004. On the origin of Vanda Miss Joaquim. *Malayan Orchid Review* 38: 96–97.
Teoh, E.S. 2005. *Orchids of Asia*, 3rd ed. Marshall Cavendish, Singapore.
Teoh, E.S. 2006. Orchids. In: T. Koh (ed.), Singapore: the Encyclopedia. Editions Didier Millet, Singapore, p. 398.
Vermeulen, J.J. and A. Lamb. 1988. *Bulbophyllum* – some interesting novelties from the Bornean jungle. *Malayan Orchid Review* 22: 44–47.
Whang, L.K. and S.Y. Alsagoff. 1991. *Vanda* Breeding plans at the Singapore Botanic Gardens. *Malayan Orchid Review* 25: 50–57.
Yam, T.W., J. Arditti, and C.S. Hiew. 2003. Several award-winning orchids and the women behind them. *Malayan Orchid Review* 37: 21–26.
Yam, T.W. 2004. The Origin of *Vanda* Miss Joaquim. How did *Vanda* Miss Joaquim really originate? *Malayan Orchid Review* 38: 86–95.
Yap, K.F., K.K. Soh, and Y.P. How. 1967. In search of *Paphiopedilum linii. Malayan Orchid Review* 9: 12–15.
Yeoh, B.C. 1962. Beer is best. *Malayan Orchid Review* 6: 104–106.
Yeoh, B.C. 1963 *A List of Singapore-Malayan Orchid Hybrids up to 1963*. Malayan Orchid Society, Singapore.
Yeoh, B.C. 1978. After the war. In: E.S. Teoh (ed.), *Orchids: a publication commemorating the Golden Anniversary of the Orchid Society of Southeast Asia (Founded 1928).* Times Periodicals, Singapore.
Yeoh, B.C. 1981. Our White Orchids. *Malayan Orchid Review* 14: 20–25.
Yeoh, B.C. 1982. Some Johor Orchids. *Malayan Orchid Review* 16: 17–19.

HISTORY-PHYSIOLOGY

2

Pollination Effects on Orchid Flowers and the First Suggestion by Professor Hans Fitting (1877–1970) that Plants Produce Hormones

TIM WING YAM, YONG NENG CHOW, POPURI NAGESWARA AVADHANI, CHOY SIN HEW, JOSEPH ARDITTI, AND HUBERT KURZWEIL

Contents

Introduction 38
Fritz Müller 38
Hans Fitting 41
Fitting's Research on Pollination and Developmental Physiology of Orchid Flowers 45
Pollination and Other Applications to Stigmas 45
Wounding 57
Killed Pollinia and Pollen Extract 57
Inter-Specific and Inter-Generic Pollination 83
Effects of Pollen of Non Orchidaceous Plants 83
Effects of the Location of Pollen Insertion 98
Ovary Swelling and Greening of Perianth 103
Confirming Experiments 104
Classification of Phenomena 104
Hormone 104
Kôichi Morita and Japanese Orchids 109
Relating Pollenhormon to Auxin 115
Conclusions 131
Dedication 131
Literature Cited 131
Appendix 1 134
Appendix 2 138
Plant Names Used in this Chapter 138
Orchids 138
Other Plants 140

Kull, Arditti and Wong (eds.), *Orchid Biology: Reviews and Perspectives*, *X*,

Introduction

A tropical orchid, *Vanilla* (Fig. 2-1D) was first brought to Europe in 1510 by the Spanish as a perfume (Lawler, 1984). Clusius (Fig. 2-1A) published what may be the first notice about it in his *Theatrum Botanicum* (Fig. 2-1A). The first plant may have reached England in 1739 (Lawler, 1984). *Bletia verecunda* (now a synonym of *Bletia purpurea*, Fig. 2-1F) was the first tropical orchid cultivated in England (Lawler, 1984). A plant received by Peter Colinson (Fig. 2-1B) from the Bahamas in 1731 was cultivated by Admiral Sir Charles Wager (Fig. 2-1G). Even before that a North American "Ladyes Slipper" (Fig. 2-1C) was reported to have been brought from North America and drawn by Sydney Parkinson (Parkinson, 1640; Lawler, 1984). A number of other orchids were brought to Europe between the 1640s and 1800. Still, tropical orchids remained a mystery and a source of fascination even as late as the 1860s when Darwin became interested in orchid pollination while vacationing at Torquay on the Devon coast with his daughter Henrietta. Once he became interested in orchids, Darwin was not satisfied with his own observations on British orchids. He read widely and corresponded extensively with a remarkable German botanist in Brazil, Fritz Müller (Fig. 2-2).

Fritz Müller

It is not uncommon for scientists to be progressive and free thinkers, disagree with established ideas, clash with autocratic regimes and run afoul of dogma. Galileo's (1564–1642) clash with the Church is probably the best known such conflict. In the Soviet Union those who disagreed with Trofim Lysenko (1898–1976), a charlatan who managed to pervert genetics under the banner of Marxism, ended in the Gulag and suffered or perished (Medvedev, 1971). The French revolution beheaded Antoine Lavosier (1743–1794) and the cultural revolution sent many "bourgeois" scientists to learn from the peasants. And, in the Germany of the 1852 it was not wise to disagree with Otto von Bismark (1815–1898), the Iron Chancellor.

Johann Friedrich Theodor 'Fritz' Müller (1821 or 1822, Thuringia, Germany-1897, Blumenau, Brazil; Fig. 2-2) began to question religion and became an atheist in 1846. He refused to take the medical oath in Germany at the time because it included the words "so help me God." Müller was also a supporter of the Prussian revolution in 1848 and held views which were too liberal for the regime. Therefore he had to leave the country (W. F. H. B, 1897). He left in 1852, settled in Brazil and reported on his numerous observations in journals and in an extensive correspondence with several scientists in Germany and Charles Darwin.

Müller became an early convert to the theory of evolution and in 1864 wrote his book *Für Darwin* (Fig. 2-2A) in which he used Brazilian crustaceans to support Darwin's views. He also corresponded extensively with Charles Darwin (Fig. 2-2C)

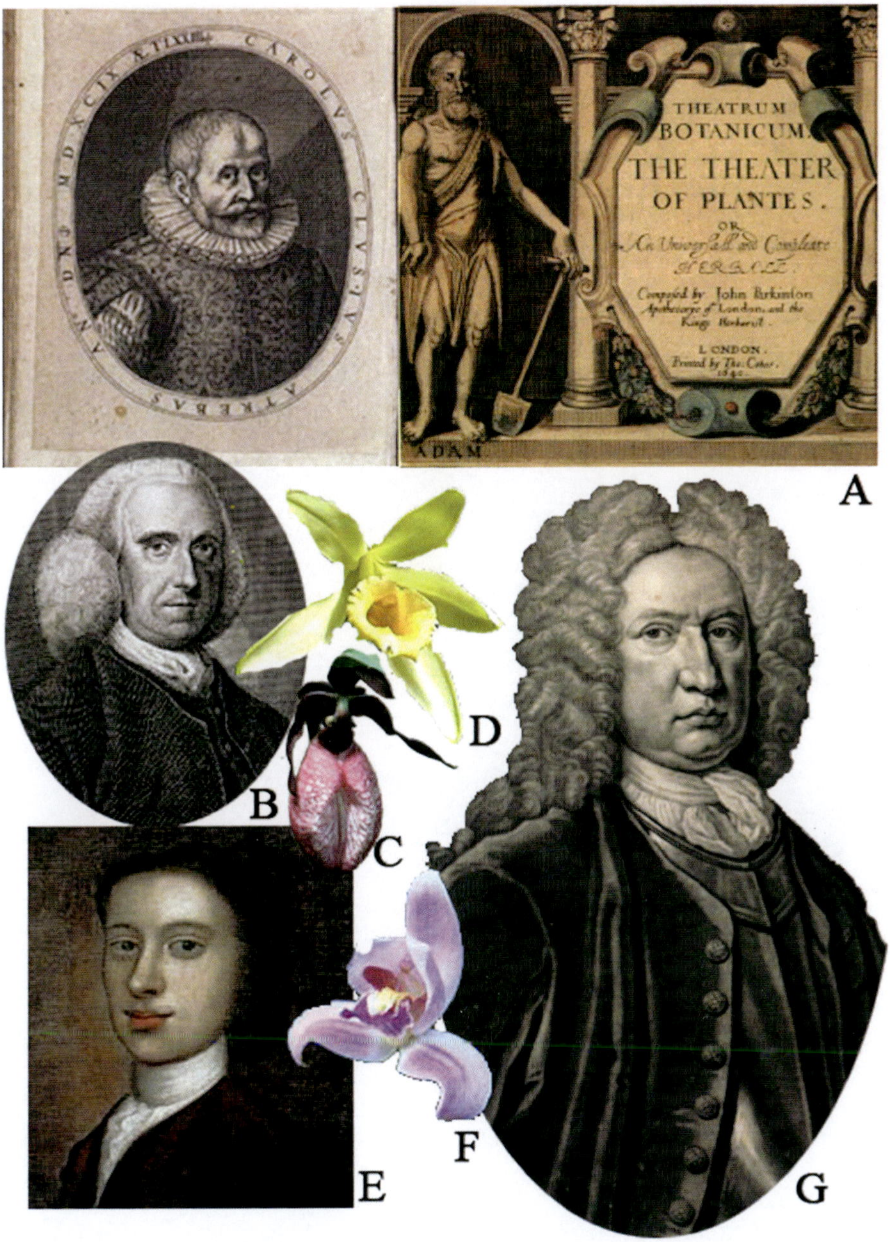

Fig. 2-1. The first orchids in Europe and persons associated with them. **A.** Charles l'Écluse, Carolus Clusius or Clusius (1526–1609) who published the first botanical notice of *Vanilla* and the cover of his *Theatrum Botanicum*. **B.** Peter Colinson (1694–1768). **C.** *Cypripedium acaule*, one possible "sort of our Ladyes Slipper" (Parkinson, 1640) from North America. **D.** *Vanilla* flower. **E.** Sydney Parkinson (*c*.1745–1771). **F.** *Bletia verecunda* flower. **G.** Sir Charles Wager [(1666–1743); (A, B, D, E and G, J. Arditti's collection; B, Wikipedia; C, courtesy Charles Cleland; D, J. Arditti; F, T. W. Yam).

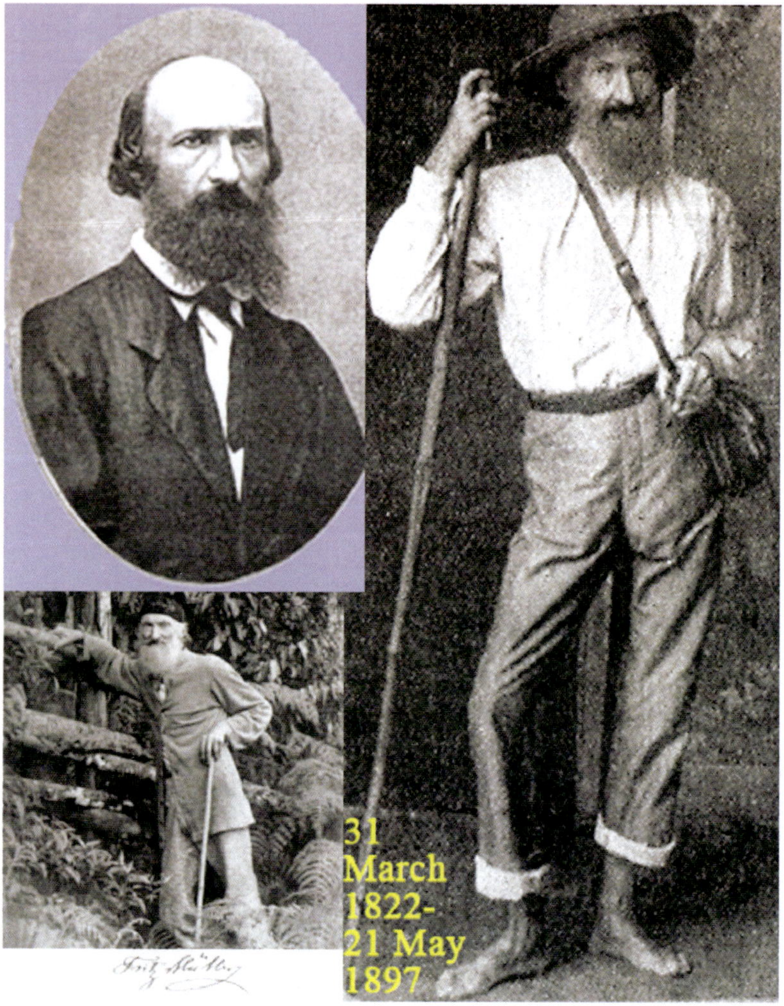

Fig. 2-2. Johann Friedrich Theodor 'Fritz' Müller (1821–1897) in Brazil during different stages of his life (Moeller, 1920, Wikipedia).

about orchids. Some of his letters and reports dealt with the effects of orchid pollen (Fig. 2-3B, 2-3D–2-3I) on flowers (Müller, 1868, 1886; Darwin, 1904; for a review see Avadhani et al., 1994) which Darwin described in *The Variation of Animals and Plants Under Domestication*, volume 2 as being "injurious and poisonous." Darwin seems to have accepted Müller's views and this lead to their wider acceptance. Müller's writings about orchids are very interesting and deserve a separate treatment. They will not be discussed in this chapter.

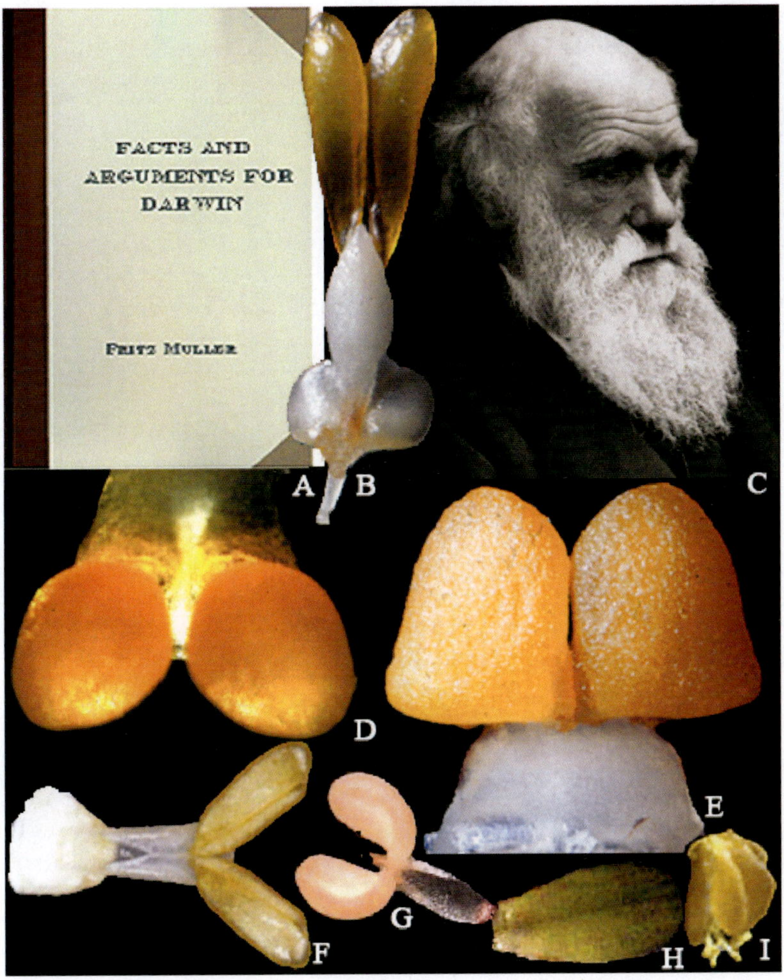

Fig. 2-3. "For Darwin," Darwin and orchid pollinia. **A.** Cover of an English translation of Fritz Müller's book *Für Darwin*. **B.** *Stanhopea* pollinarium. **C.** Charles Darwin. **D.** *Phalaenopsis* pollinia. **E.** *Cymbidium* pollinarium. **F.** *Catasetum* pollinarium. **G.** *Oncidium* pollinarium. **H.** *Masdevallia glandulosa* pollinia. **I.** *Cattleya* pollinia. (A, facsimile of Muller's 1869 *Facts and Arguments for Darwin*; B, D–F, I, J. Arditti; C, J. Arditti's collection; G, H, courtesy Lotte and Thomas).

Hans Fitting

The "theory that [was] very common in the older German literature on pollination biology, namely that the pollinia of many exotic orchids act like a poison during cross-pollination" attracted the attention of Hans Fitting (Fig. 2-4) in 1907 who also read a letter by Müller to Darwin dated 1 January 1867 which "elaborates on the toxicity of the orchid pollinia" [the statements in quotes are from a letter dated 10 November 1969 by the late Hans Fitting to Joseph Arditti (J.A.) translated by Hubert Kurzweil (H.K.); Appendix 1].[1] He obtained a travel grant and visited the Bogor Botanical Gardens in Indonesia to study the phenomenon.[2]

Johannes Theodor Gustav Ernst 'Hans' Fitting (Fig. 2-4) was born on 23 April 1877 in Halle a. d. Salle, Germany, the son of Herman Fitting a Professor of Law and his wife Clara (maiden name Merkel). At the age of nine young Hans declared that he will become a "*Blumenprofessor*" (Professor of Flowers). After attending gymnasium (1886–1895) and university in Halle (1895–1896) he served in the military (1896–1897). In January of 1900 Fitting became a doctoral student under H. Graf zu Solms-Laubach (1842–1915) and was granted a degree summa cum laude (Halbsguth, 1962, 1974). Due to lung disease Fitting had to take time off work and spend the first half of 1901 in the mountains.

After recovering Fitting worked with Wilhelm Pfeffer (1845–1920) in Leipzig[3] for a year (fall 1901–1902). He accepted a position under Professor Herman Vöchting (1847–1917) in Tübingen in October 1902 and remained there until the fall of 1907 when a *Deutschen Reichs* tropical stipend made it possible for him to travel until June 1908 and visit Sri Lanka (then Ceylon) and Java (specifically the Bogor Botanical Gardens). At Bogor Fitting carried out numerous experiments often as early as 06:00 and sometimes even earlier. Fitting seems to have enjoyed his stay in Bogor because he inquired about the gardens with interest and concern sixty years later in one of the letters[1] to J. A.

---

[1]In advanced age and suffering from heart problems when he wrote three letters to JA in longhand. They are dated 2 November 1969, 10 November 1969 and 10 December 1969. The letters include many details and citations and are very clear. Prof. Fitting clearly enjoyed being remembered and reminisced about work he did 61 years in the past with obvious pleasure.

Professor Kenneth V. Thimann recalled meeting Fitting at the 1935 Botanical Congress in Amsterdam and wrote that "he was in good form and gave a talk, not about pollination…he must have been 60 years old at the time, but was very lively" (letter by Prof. Kenneth V. Thimann to J. A. dated 27, January 1971).

The late professor Frits W. Went, the discoverer of auxin, wrote about Fitting that he "was…an indefatigable digger of facts which he published in incredible detail…but somehow, he always missed the boat." In "his work on the transmission of the phototropic stimulus…he failed to notice that most of his results were due to leakage across a wound gap…I was not impressed with his scientific reasoning…Every botanist had assumed that the stomata of desert plants would be closed, but Fitting…found that they were wide-open. His explanation was…'of course they should be open, otherwise the leaves could not photosynthesize. He did not even mention the opposite argument that they should be closed to conserve water" (letter dated 19 June 1974 by Prof. F. W. Went to J. A.).

Mrs. Sigrid Fitting provided photographs and details about Professor Fitting after his death.

All letters and photographs are part of the reprints collection J. A. donated to the Singapore Botanic Gardens.

[2]The names of orchids employed by Fitting in his experiments which are used in this chapter are those he included in his papers. Currently accepted names are listed in Appendix 2.

[3]There is an indirect connection to orchids in Fitting's association with Pfeffer. Prof. Lewis Knudson's used Pfeffer's solution in the research which lead to his first solution (Knudson B) for the asymbiotic germination of orchid seeds.

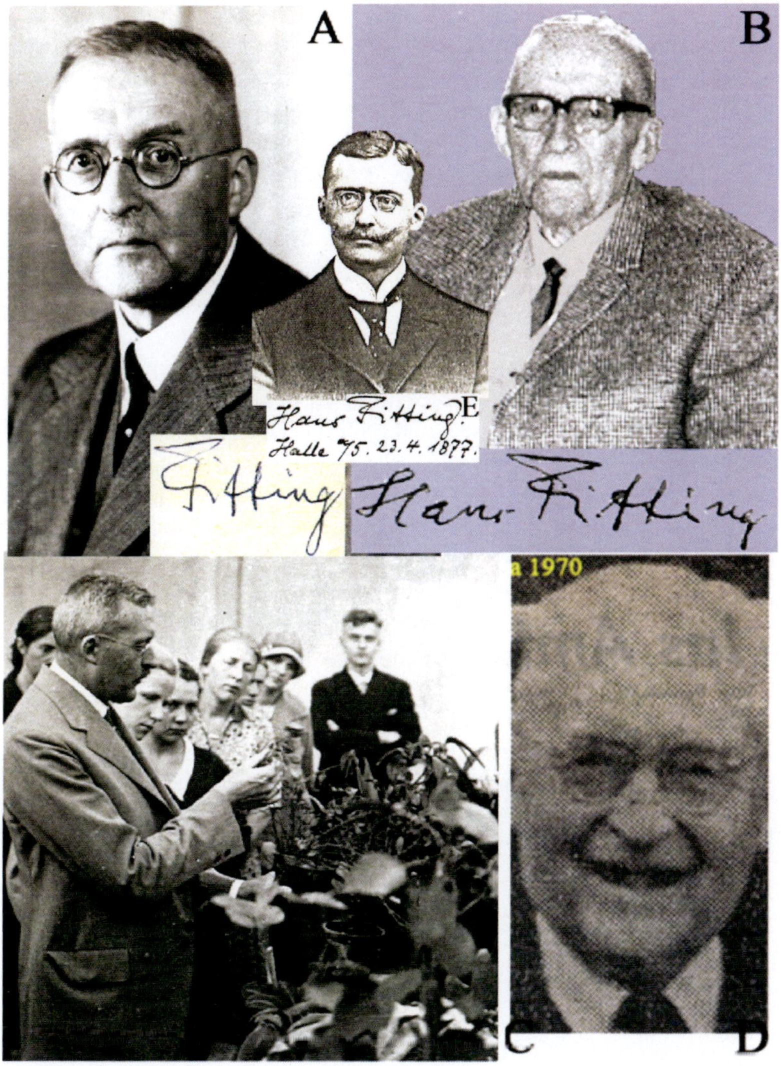

Fig. 2-4. Hans Fitting during various stages of his life. **A.** In mid life. **B.** In 1969. **C.** Teaching in the greenhouse, University of Bonn. **D.** In 1970 shortly before his death. **E.** Photograph from about the time he went to Bogor. Signature is from later with 1877 indicating year of birth.

On returning from Indonesia Fitting was appointed Associate (Extraordinarius) Professor at the University of Strasbourg.[4] He assumed the position in the summer of 1908 and carried out additional research on orchids there. In March–April 1910

[4]We thank Dr. Wolfgang Zierau, Institute of Solid State Theory, Westfälische Wilhelms Universität, Muenster, Germany for explaining the mysteries of German university titles.

Fitting left Strasburg for a 2.5 months study tour of Algerian Sahara. After that he accepted a position as Associate (Extraordinarius) Professor at the University of Halle and remained there until the fall of 1911. He left Halle to become Full Professor at the Colonial Institute in Hamburg and Director of the State Botanical Garden in the same city. While in Halle Fitting must have spent time in pursuits other than botany because he married Sigrid Meyer, the daughter of a University of Halle official in the summer of 1911. They had two daughters and three sons. All sons served in the military during World War II. The youngest, Rudolf, a lieutenant in an anti tank armored unit was killed between 15 and 16 of October 1943 in Volturno, Southern Italy.[5]

In 1912 Fitting was appointed to succeed Eduard Strasburger (1844–1912)[6] as Full Professor (Ordinarius) at the University of Bonn. This was one of the most prestigious botanical professorships in Germany and Fitting further increased its prestige. He remained at the University of Bonn for the rest of his life, became one of the best known plant physiologists in Germany at that time, retired on 1 October 1946 and was Professor Emeritus for the rest of his life.

World War II was not kind to Fitting and his wife. As already mentioned they lost one son in Italy. In addition, their house was destroyed during an air raid. The botanical institute was also damaged by a bomb. After the war, the University was reopened in the fall of 1945 with Fitting as Dekan (academic head) of the Mathematics and Natural History Faculty. His lack of the support for the Nazi regime and the saving of his Jewish colleague V. Simon served him in good stead during the postwar years (Halbsguth, 1974).

Fitting rebuilt his house in 1952 and remained in relatively good health until 1962. He had a heart attack then (his age was 85), but survived it. He died on 6 July 1970 at 93. When one of us (J. A.) initiated research on post pollination phenomena of orchid flowers in the late 1960s it did not seem reasonable to assume that Professor Fitting was still alive. As a result J. A. failed to visit Prof. Fitting on his trip to Germany in 1968. When the late Professor Erich L. Nuernbergk informed J. A. that Fitting was alive, Miss Brigitta H. Flick, J. A.'s technician at the time visited and photographed him in 1969 (Fig. 2-4B).

[5]Professor Erich Nuernbergk, an opponent of the Nazi regime in Germany, told one of us (J.A.) that Fitting also opposed the Nazis and helped Jewish colleagues (one of them being V. Simon) escape to safety. Fitting described himself as not being politically active, but a member of the Deutschen Volkspartei (according to Dr. Zierau this party which existed during the Weimar Republic in Germany was conservative and pro monarchy, but not pro Nazi).

[6]Eduard Strasburger was the first botanist to fix tissues and stain them and to describe meiosis and mitosis between 1877 and 1885. He coined the terms phototaxis, cytoplasm, nucleoplasm, prophase, metaphase, anaphase, plasmodesmata, haploid and diploid.

Fitting's Research on Pollination and Developmental Physiology of Orchid Flowers

The "Tropen (or Buitenzorg) Stipendium des Deutschen Reiches" (Halbsguth, 1962, 1974) made it possible for Fitting to engage in research at the Bogor (then Buitezorg) Botanical Gardens in Indonesia from September or November 1907 until June 1908 (Halbsguth, 1962, 1974). While in Bogor the 30 years old Fitting (Fig. 2-4E) carried out a total of 79 numbered experiments (some simple and consisting of one part and others multi part and elaborate), and many more that were mentioned but had no numbers (Fitting, 1909a, b). He described his experiments in great detail. At present his descriptions would be considered to be excessive, excruciatingly ponderous and tedious. Many of the details he provided are simply unnecessary. On the other hand the details about how he worked provide an insight into the mind of an influential plant physiologist and explain why Fitting became so well known during his life-time. In addition to being a resourceful and clever investigator Fitting was also a very hard worker. To carry that many experiments in the relatively short time he had, Fitting often started to work at 06:00 and worked 10–12 h a day 7 days a week.

Pollination and Other Applications to Stigmas

Despite being familiar with both Müller's and Darwin's writings Fitting established his own baselines (Table 2-1). He carried out his experiment with several orchid species (Figs. 2-5–2-9) probably for two reasons. One reason may have been his intent to determine whether the effects of pollen were universal or at least widespread in the Orchidaceae or limited to a few species. The second reason was practical: Even at the Bogor Botanical Gardens there was often a shortage of flowers of a single species (Fitting, 1909a, b). For this reason he also had to use a single flower for some experiments (i.e., there were no replications).

He pollinated flowers (self, cross, geitonogamous and xenogamous) and observed that in most cases pollination caused swelling of the ovary and gynostemium not long after the application of pollen to the stigma and shortened the life span of the flowers. Exceptions were flowers of: (1) *Liparis latifolia* (Fig. 2-8E) which did not wilt after six days (Table 2-1), (2) *Cymbidium sanguinolentum* (now *Cymbidium chloranthum)* which did not exhibit pronounced wilting, and (3) *Phalaenopsis violacea* (Fig. 2-8C) which yellowed, closed (Table 2-1) and underwent major changes (Fig. 2-6F) and, as shown by recent research (Tran et al., 1995), producing chlorophyll and probably becoming photosynthetic.

Fitting emasculated flowers of *Rhynchostylis retusa* (Table 2-1) and reported that changes were slower to occur than after pollination. This is in line with our current knowledge about the induction of post pollination phenomena and floral senescence in orchid flowers (Avadhani et al., 1994). He also made cuts in the gynostemium (Fig. 2-7H) and reported that the flower did not wilt even after 2 weeks

Table 2-1. Hans Fitting's experiments with orchid pollination at the Bogor Botanical gardens in 1908 (Fitting, 1909a).

Experiment number	Orchid[a]	Description	Results	Fitting's conclusions	Current explanation
1 (4–10 Jan 1908)	*Rhynchostylis retusa* (Fig. 2-7)	Placed sand in stigmas (11 flowers)	7 flowers wilted in 9h and all 11 wilted in 44h	Sand caused wilting but not swelling at 44h. Ovaries and gynostemia not swollen	Sand wounded stigmas causing shortened flower life. Ethylene evolution which induced senescence
		Six flowers pollinated (geitonogamous pollination)	All 6 flowers wilted in 96h. Ovaries and gynostemia swollen	Pollination caused both wilting and swelling. Emasculation did not shorten the life span of flowers	Emasculation should have done the same. Pollen effects are normal
		Unpollinated flowers	Flowers fresh until February		
2 (12–16 Jan 1908)	*Rhynchostylis retusa* (Fig. 2-7)	Ten flowers pollinated both geitonogamously and xenogamnously and self pollinated	All 10 flowers started to wilt after 24h; wilting increases after 48 and 72h	Sand shortened flower life span, and caused wilting, but not swelling of ovaries and gynostemia	Sand wounded stigmas causing ethylene evolution which induced wilting and senescence. Emasculation should have done the same
		Sand placed in stigmas of 21 intact flowers	No change after 24h, 3 flowers wilted after 48h and all 21 wilted after 72h	Pollination, but not emasculation caused all of these. Shortening of life spans does not require or depend on swelling of gynostemia and ovaries	Pollination effects are normal. Shortening of life span and swelling are not interdependent
		Sand placed in stigmas of 16 emasculated flowers	No change after 24h, 7 flowers wilted after 48h and all 16 wilted after 72h		
		Flowers emasculated	No change until start of February		
3 (8–25 Jan 1908) (10–21 Jan 1908)	*Rhynchostylis retusa* (Fig. 2-7)	Cut made in gynostemium below stigma, above the perianth point of origin	All 5 flowers unchanged and still fresh after 13 and 15 days	Wounding the gynostemium does not shorten the life span of flowers	If wounding the gynostemium induced ethylene evolution it should have shortened flower life span. Sand wounded the stigmas thereby inducing ethylene evolution which shortened the life span of flowers
		Gynostemium squeezed or crushed with forceps at level of or below stigma of 7 flowers	2 flowers, one after 5 days, the other after 9 days start to wilt and damaged areas turn black	The sand does not shorten the life span of flowers by damaging the gynostemium	

(continued)

(20–25 Jan 1908)		Gynostemium treated as above, but mostly crushed in 6 flowers	1 flower started to wilt after 2 days, damaged areas blackened in all flowers		
(31 Jan–12 Feb)		Cuts below stigmas made in gynostemia of 12 flowers	All flowers unchanged after 12 days		
5–20 Feb 1908)		Gynostemia removed	Flowers remained fresh until end of experiment		
4A (17–28 Jan 1908)	*Rhynchostylis retusa* (Fig. 2-7)	River sand placed in stigmas of 12 flowers	9 flowers started to to wilt after 5 days. All flowers wilted after 12 days	Sand does not have a mechanical effect which shortens the life of flowers	Either sea sand has no sharp edges and is not injurious or boiling it in acid eliminated sharp edges and rendered sea sand non injurious
		Acid washed sea sand placed in 12 stigmas	All 12 flowers not wilted after 12 days	Acid washing removed iron, aluminum, magnesium, manganese and cobalt from sea sand	
4B (20–29 Jan 1908)		River sand placed in stigma of 6 flowers	All 6 flowers started to wilt after 2 days 1 flower started		
		Acid washed sea sand placed in 6 stigmas	to wilt after 2 days, 5 not wilted after 10 days		
4 (22 Jan noon–24 Jan 7 a. M. 1908)		Pollinated (18 flowers)	All 18 flowers wilted after 43 h		
No number or date	Not listed	10% KNO_3 in water, aqueous solution of amylase, sea sand wetted with 1% iron chloride	None of these had any effect on the life span of flowers	None	Not necessary
5–16	The purpose of these experiments was to determine whether the observations with *Rhynchostylis retusa* were unique and specific to this one orchid. Unless indicated otherwise or is obvious from the experimental procedure the flowers used in the following experiments were not emasculated and the life span of the control blossoms was longer than that of the experimental ones				

(continued)

Table 2-1. (continued).

Experiment number	Orchid[a]	Description	Results	Fitting's conclusions	Current explanation
5 (12–18 Jan 1908)	*Liparis latifolia* (Fig. 2-8E)	Pollinated (2 flowers) River sand placed in stigma of 4 flowers	2 of 2 flowers not wilted after 6 days All 4 flowers not wilted after 6 days, but 2 abscised	The conclusions listed below in this column pertain to experiments 5–19	Current comments and explanations are opposite each conclusion below
6 (12–18 Jan 1908)	*Cymbidium sanguinolentum* (Fig. 2-8B)	Pollinated (2 flowers) River sand placed in stigma of 5 flowers	2 of 2 flowers do not exhibit pronounced wilting 5 of 5 flowers not exhibit pronounced wilting	1. River sand caused rapid closing and wilting only in flowers which are affected similarly by pollen	Interesting, but not very profound
7 (15–23 Jan 1908)	*Cymbidium finlaysonianum* (Fig. 2-8D)	Pollinated (single flower) River sand placed in stigmas of 3 flowers Diastase (amylase) in stigmas of 2 flowers	Flower slightly closed after 6 days, abscised after 7 days All 3 flowers slightly closed after 7–8 days and abscised Both flowers slightly closed after 5 days and abscised	2. River sand does not usually act more rapidly than pollen 3. River sand only shortens the lifespan of flowers 4. River sand does not bring about swelling of gynostemium and ovary	There is no reason why sand should act more rapidly Yes, it probably acts by inducing ethylene evolution Yes, because ethylene only brings about senescenc
8 (22 Jan–4 Feb 1908)	*Vanda tricolor* (Fig. 2-8H)	Pollinated (single flower) River sand placed in stigma A Acid washed sea sand placed in stigma A	Color changes of perianth started 2 days after pollination and proceeded slowly after that Flowers unchanged until 9th and 10th day	5. River sand had no effects on *Vanda tricolor* 6. Acid-washed sea sand and river sand usually have the same effect	Hard to explain why There is no reason why they should not
8a (13–24 Jan 1908)	*Oncidium incurvum* (Fig. 2-8G)	Pollinated (single flower) River sand placed in stigma	Flower wilted after ca 5 days Flowers wilted after ca. 6–7 days	7. Sand exerts its effects by wounding the stigma	Very astute observation for the time This is how sand probably acts

(continued)

9 (20–26 Jan 1908)	*Dendrobium antennatum* (Fig. 2-8A)	Pollinated (2 flowers) River sand places in stigmas of 2 flowers	All 4 flowers not wilted and nearly unchanged after 6 and 10 days	The conclusions lead Fitting to his next set of experiments which involved wounding of stigmas and gynostemia
10 (20–22 Jan 1908)	*Coelogyne pandurata* (Fig. 2-8F)	Pollinated (3 flowers) River sand placed in stigmas of 2 flowers Acid washed sea sand Placed in 4 stigmas	Treated flowers closed after 1 day. Untreated flowers remained open	On the whole Fitting's experiments and conclusions were very much ahead of their time
11 (17–23 Jan 1908)	*Coelogyne asperata* (Fig. 2-8I)	Pollinated (1 flower) River sand placed in stigmas of 2 flowers Acid washed sea sand Placed in 3 stigmas	Treated flowers closed after 1 day. Untreated flowers remained open 2–3 additional days	
12 (13–19 Jan 1908)	*Phalaenopsis esmeralda* (*P. regnieriana;* Fig. 2-8J)	Pollinated (2 flowers) River sand placed in stigmas of 2 flowers	All wilted after 3 days and were somewhat closed All wilted after 3–4 days and somewhat closed	The wounding experiments were numbered 17–40 by Fitting. They are summarized in Table 2-2
13A (18–22 Jan 1908)	*Phalaenopsis amabilis* (similar to Fig. 2-5, Fig. 2-13A)	Pollinated (2 flowers) River sand placed in stigmas of 2 flowers Acid washed sea sand placed in stigma of 1 flower	Both closed after 1½ days and wilted after that Both closed after 3 days and wilted after that Flower closed after 4 days and wilted after that	
13B (21–23 Jan 1908)		River sand placed in stigmas of 2 flowers	Both closed after 2 days and wilted after that	
13C (26–31 Jan 1908)		Pollinated (2 flowers) River sand places in stigmas of 2 flowers Acid washed sea sand placed in stigmas of 2 flowers	Both closed after 1½ days and wilted after that One flower closed after 3 days and wilted after that The second remained unchanged and closed next day Both remained unchanged	

(continued)

Table 2-1. (continued).

Experiment number	Orchid[a]	Description	Results	Fitting's conclusions	Current explanation
14A (18–23 Jan 1908)	*Phalaenopsis violacea* (Fig. 2-6F, 2-8C)	Pollinated (1 flower)	Flower closed and yellowed after 2 days		
		River sand placed in stigma of 1 flower	Flower closed and yellowed after 5 days		
14B (24–30 Jan 1908)		River sand placed in stigmas of 4 flowers	2 flowers closed and yellowed after 5 days; 2 did the same after 6 days		
15A (26 Jan–5 Feb 1908)	*Dendrobium superbum*	Pollinated (1 flower)	Closed and wilted after 1 day		
		River sand placed in stigmas of 2 flowers	Both flowers closed and wilted after 2 days		
15B (14–24 Feb 1908)		Pollinated (3 flowers) Rivers sand placed in stigmas of 3 flowers	All pollinated and sand-treated flowers closed and wilted after 1 day		
15C (16–22 Feb 1908)		River sand placed in stigmas of 2 flowers	1 flower closed and wilted after 2 days, the second closed and wilted after 3 days		
15D (16–20 Feb 1908)		Pollinated (2 flowers) River sand placed in stigmas of 2 flowers	All pollinated and sand-treated flowers closed and wilted after 1 day		
15E (24–29 Feb 1908)		River sand placed in stigmas of 2 flowers	Flower closed after 1 day, the other closed after 4 days		
16 (14–20 Feb 1908)	*Aerides odorata* (Fig. 2-9)	Pollinated (4 flowers) River sand placed in stigmas of 4 flowers	All pollinated and sand-treated flowers wilted after 4 days		

[a]The orchid names in this table are those used by Hans Fitting. Please see Appendix 2 for updated nomenclature.

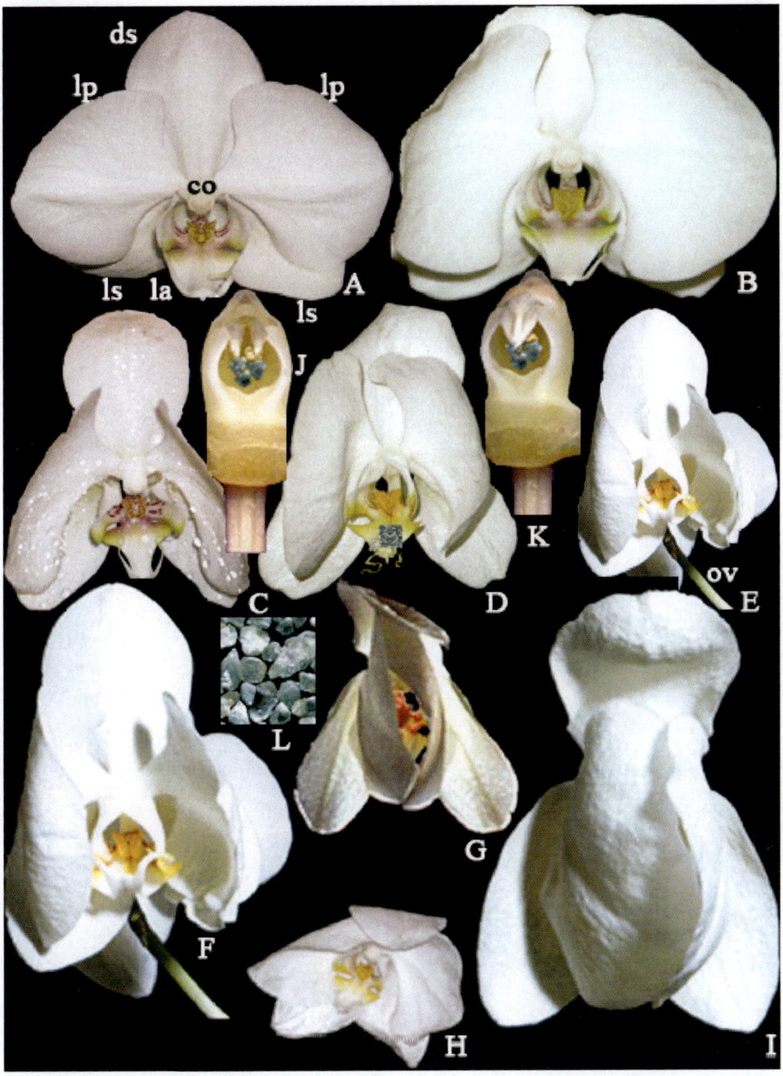

Fig. 2-5. Post-pollination hyponasty of lateral petals and floral aging in *Phalaenopsis amabilis*. **A.** Unpollinated flower. **B–I.** Post-pollination phenomena. **J.** Sand in stigma of emasculated gynostemia. **K.** Sand in stigma of an intact gynostemium. **L.** Sand. (J and K are computer-generated simulations). Explanation of symbols: co, column (gynostemium), ds, dorsal sepal; la, labellum (median petal); lp, lateral petal; ls, lateral sepal; ov, ovary (source: J. Arditti; some of the pictures were taken in Changi Airport, Singapore where one of the terminals was decorated with hundreds of *Phalaenopsis* plants).

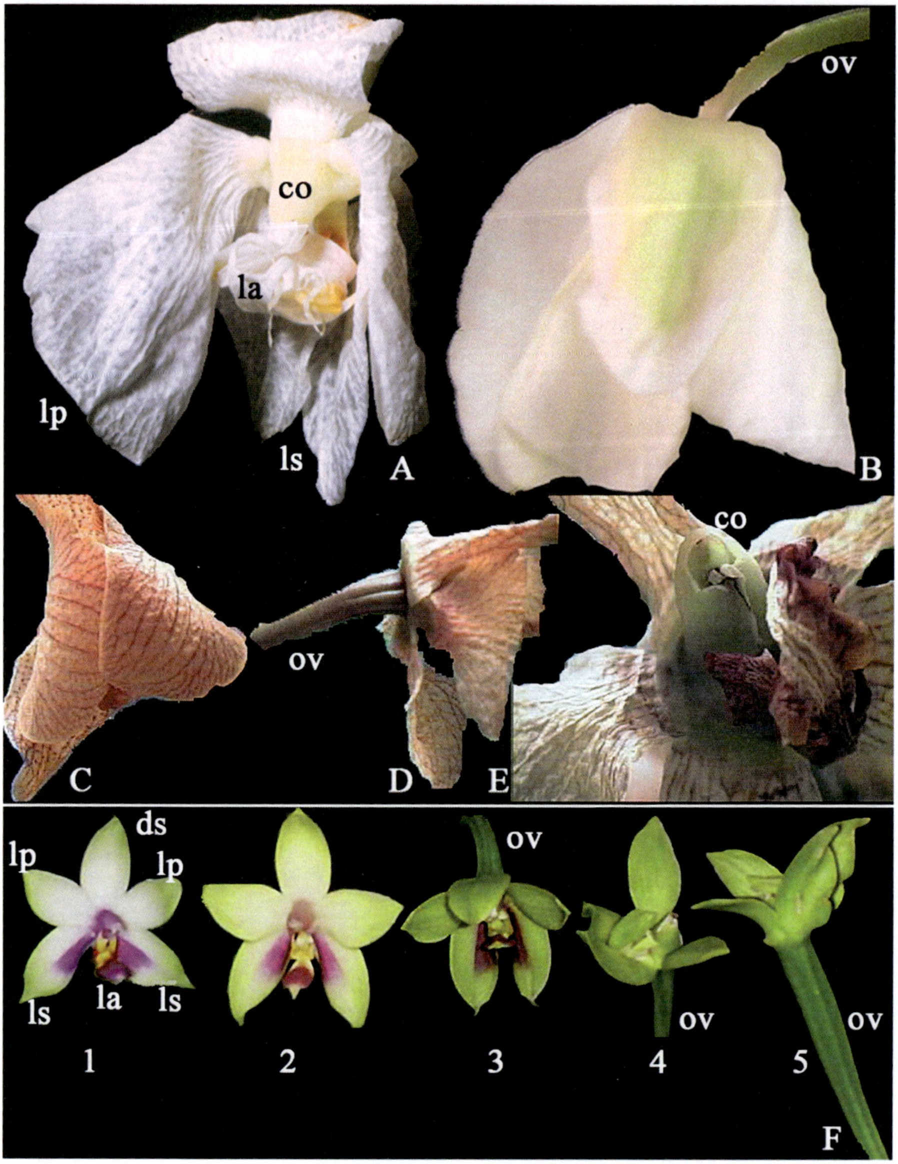

Fig. 2-6. Post-pollination phenomena in *Phalaenopsis* flowers. **A-E.** Senescence of perianth. **D.** Swelling of ovary. **E.** Greening and swelling of gynostemium. **F.** *Phalaenopsis violacea*. 1 unpollinated flower; 2–5 changes following pollination. Explanation of symbols: co, column (gynostemium), ds, dorsal sepal; la, labellum (median petal); lp, lateral petal; ls, lateral sepal; ov, ovary (Photographs and montage by J. Arditti).

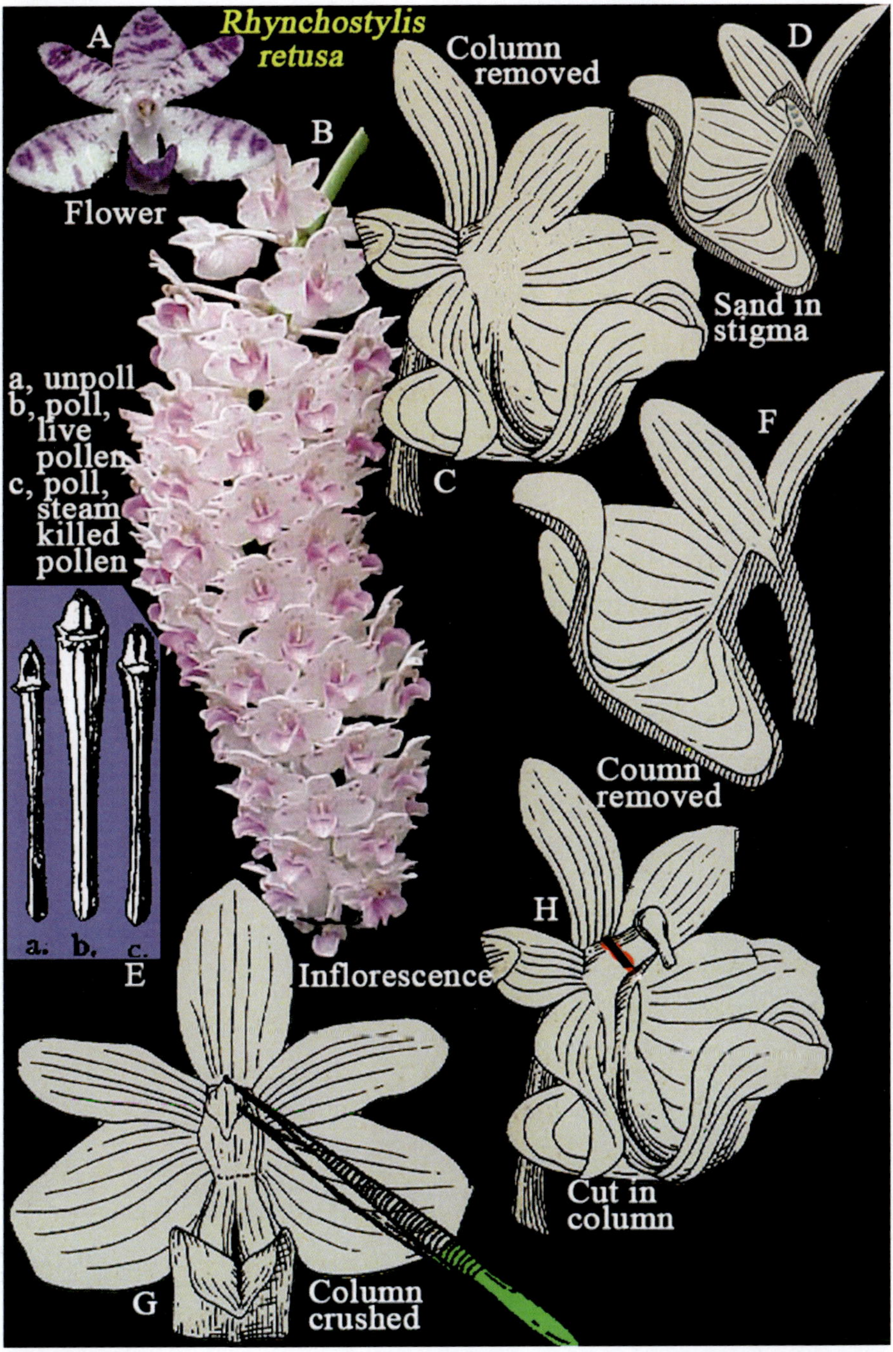

Fig. 2-7. *Rhynchostylis retusa*, a species used by Hans Fitting in his experiments. Explanations are on the figure itself (A, B, T. W. Yam, C–E, Fitting's papers).

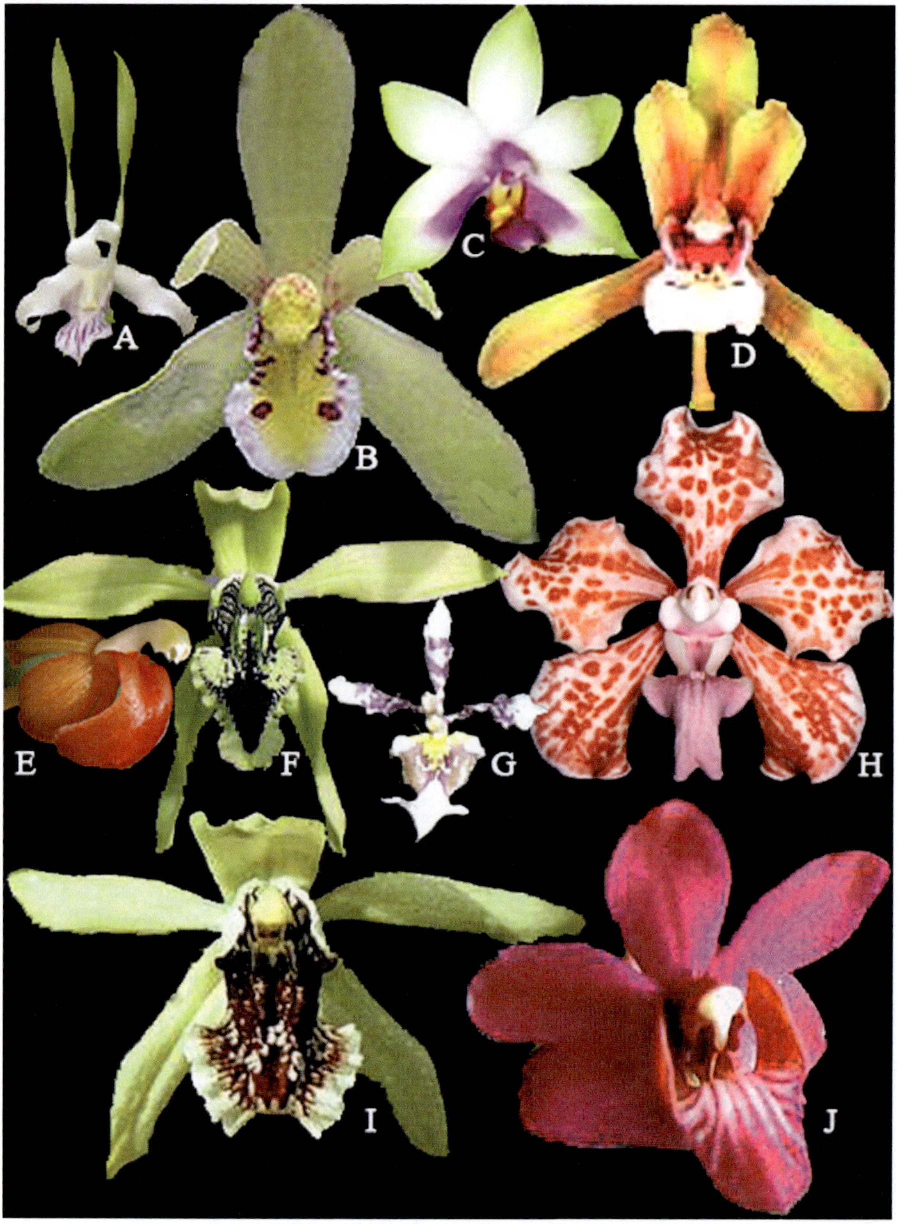

Fig. 2-8. Orchids used by Hans Fitting in his experiments. **A.** *Dendrobum antennatum.* **B.** *Cymbidium sanguinolentum.* **C.** *Phalaenopsis violacea.* **D.** *Cymbidium finlaysonianum.* **E.** *Liparis latifolia.* **F.** *Coelogyne pandurata.* **G.** *Oncidium incurvum.* **H.** *Vanda tricolor.* **I.** *Coelogyne asperata.* **J.** *Phalaenopsis esmeralda.* (A–C, F–J, T. W. Yam and J. Arditti; D, E, courtesy Eric Hunt).

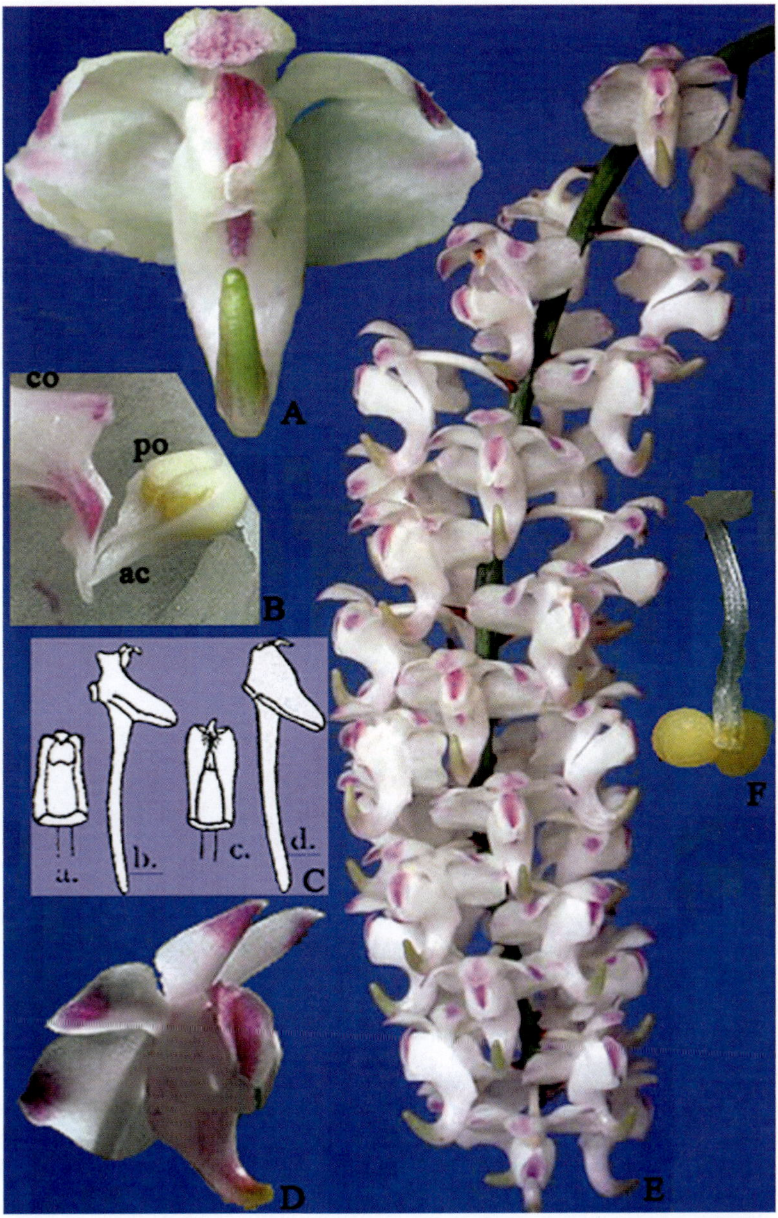

Fig. 2-9. *Aerides odorata* and *Aerides falcatum*. **A.** Front view of flower of *Aerides odorata*. **B.** Front of column of *Aerides odorata* showing detached pollinia inside anther cap. **C.** Column and ovary of *Aerides odorata*: a, front view of untreated gynostemium; b, side view of untreated ovary with labellum; c, front view of swollen gynostemium following application of pollen extract to stigma; d, side view of ovary with labellum following application of pollen extract to stigma. **D.** Side view of flower of *Aerides odorata*. **E.** Inflorescence of *Aerides odorata*. **F.** Pollinarium of *Aerides falcatum* (A, D, E, T. W. Yam and J. Arditti; B, F, courtesy Heinz Schneider; C, Fitting, 1909).

(Table 2-1). This is surprising since the cuts can be expected to evolve wound ethylene which would cause senescence. Removal of the gynostemium (Fig. 2-7C, F) also did not cause floral senescence (Table 2-1). It is possible that in both cases (Fig. 2-7C, F, H) the cuts were not severe enough to induce copious ethylene evolution. That this may be the case is suggested by the fact that crushing the gynostemium (Fig. 2-7G) did bring about senescence (Table 2-1).

For reasons which he did not explain, Fitting applied aqueous solutions of potassium nitrate (10% KNO_3) and amylase to stigmas of *Rhynchostylis retusa*. Both were without effect (Table 2-1). However when he applied an aqueous solution of "diastase [amylase]" to stigmas of two flowers of *Cymbdium finlaysonianum* both closed slightly after 5 days and abscised. Fitting did not explain the reasons for the different responses of the two species and they remain unclear. On the face of it there are no reasons why potassium nitrate should induce floral senescence and post pollination phenomena. The starch-hydrolyzing enzymes may have induced the slight stigmatic closure by breaking down starch reserves to glucose which raised the sugar levels inside cells. This in turn probably brought about water influx which caused the swelling that lead to stigmatic closure.

In a letter (to J. A.) about Hans Fitting, Prof. Frits W. Went[1] (Fig. 2-25C) wrote that "he was too much steeped in the 'stimulus' [the German word *Reizung* is a rough equivalent] concept." Fitting must have been concerned with *Reizung* even while in Bogor because he placed sand in stigmas of *Aerides odorata, Coelogyne asperata, Coelogyne pandurata, Cymbidium finlaysonianum, Cymbidium sanguinolentum* (now *Cymbidium chloranthum*), *Dendrobium antennatum, Dendrobium superbum* (now *Dendrobium anosmum*), *Liparis latifolia, Oncidium incurvum, Phalaenopsis amabilis* (Fig. 2-5D, 2-5J–L), *Phalaenopsis esmeralda* (now *Phalaenopsis pulcherrima*), *Phalaenopsis violacea, Rhynchostylis retusa* (Fig. 2-7) and *Vanda tricolor* (Table 2-1). He probably used what he described as "sand" without any additional details and also river sand to determine whether a purely physical stimulus would have any effects. The sand caused wilting in some flowers, but had no effects on *Cymbidium sanguinolentum* and *Liparis latifolia* which also did not wilt after pollination (Table 2-1). River sand caused wilting in *Rhynchostylis retusa*, but not if it was acid washed (Table 2-1). However, both washed and unwashed river sand caused wilting in both *Coelogyne* species and *Phalaenopsis amabilis*.

These findings suggest that *Cymbidium sanguinolentum* and *Liparis latifolia* flowers may be incapable of ethylene production, but this speculative suggestion requires experimental confirmation. The different effects of acid washed and unwashed sand on *Rhynchostylis retusa* on the one hand and on both *Coelogyne* species and *Phalaenopsis amabilis* on the other are more difficult to explain. If the acid somehow rounded off edges in the sand and blunted them thereby causing the sand not to injure the stigma and not to bring about ethylene production the effects should have been the same on both species. The same would be true if the acid removed injurious substances or heavy metals which could cause ethylene evolution. Differential sensitivity of the species seems to be the only remaining explanation.

Wounding

The possibility that sand exerted its effects by wounding the stigma lead to a series of experiments in which Fitting wounded gynostemia and stigmas in several places and in different ways (Fitting, 1909a, b). He also removed portions of gynostemia (Fig. 2-10: Table 2-2).

Wounding or damaging the stigma in any way brought about wilting of the perianth, senescence and shortened the life span of flowers (Fitting could not have known that the wounds induced ethylene production). The exceptions were:

- Wounds in the lower or middle part of the stigma (Table 2-2). This suggests that not all parts of the stigma can produce ethylene or produce enough of the gas to bring about senescence. The rostellum is known to be capable of producing large amounts of ethylene (for a review see Avadhani et al., 1994). Fitting's findings show that the upper parts of the stigma which are close to the rostellum have a similar capability.
- Wounded stigmas of *Aerides odorata* flowers (Fig. 2-9) which is surprising because pollination does bring about senescence of these blossoms (Table 2-2).
- Flowers of *Armodorum sulingi* (Fig. 2-11B), *Dendrobium macrophyllum* (Fig. 2-11A, 2-11D), *Renanthera maingayi* (now a synonym of *Arachnis* ×*maingayi*, Fig. 2-11C) and *Trichoglottis geminata* (Fig. 2-11E) all of which are fleshy. These flower also remained fresh and did not change following pollination. Not enough information is available at present regarding post pollination phenomena in fleshy orchid flowers to allow speculations regarding Fitting's findings with these species.

The effects of wounding are easy to explain at present: Wounding brought about the production of ethylene which induced senescence and other post-pollination phenomena (for a review see Avadhani et al., 1994). In those early days of plant physiology Fitting could not have known, let alone suggest the involvement of ethylene despite the fact that its effects on pea seedlings were already known (Neljubov, 1901). At a time when even solid substance, but soluble plant hormones were yet to be discovered the existence of a gaseous hormone would have been impossible to imagine. However, the facts are that Fitting observed both auxin and ethylene effects.

Killed Pollinia and Pollen Extract

Even if the wounding experiments were an aspect or an extension of Fitting's interest in "stimulus" or "*Reizung*" his interests were wider and very advanced for his time. He decided to determine the effects of dead pollinia killed by submerging them in chloroform or boiling water and subjecting them to steam (Figs. 2-13–2-15; Table 2-3; Fitting, 1909a, b).

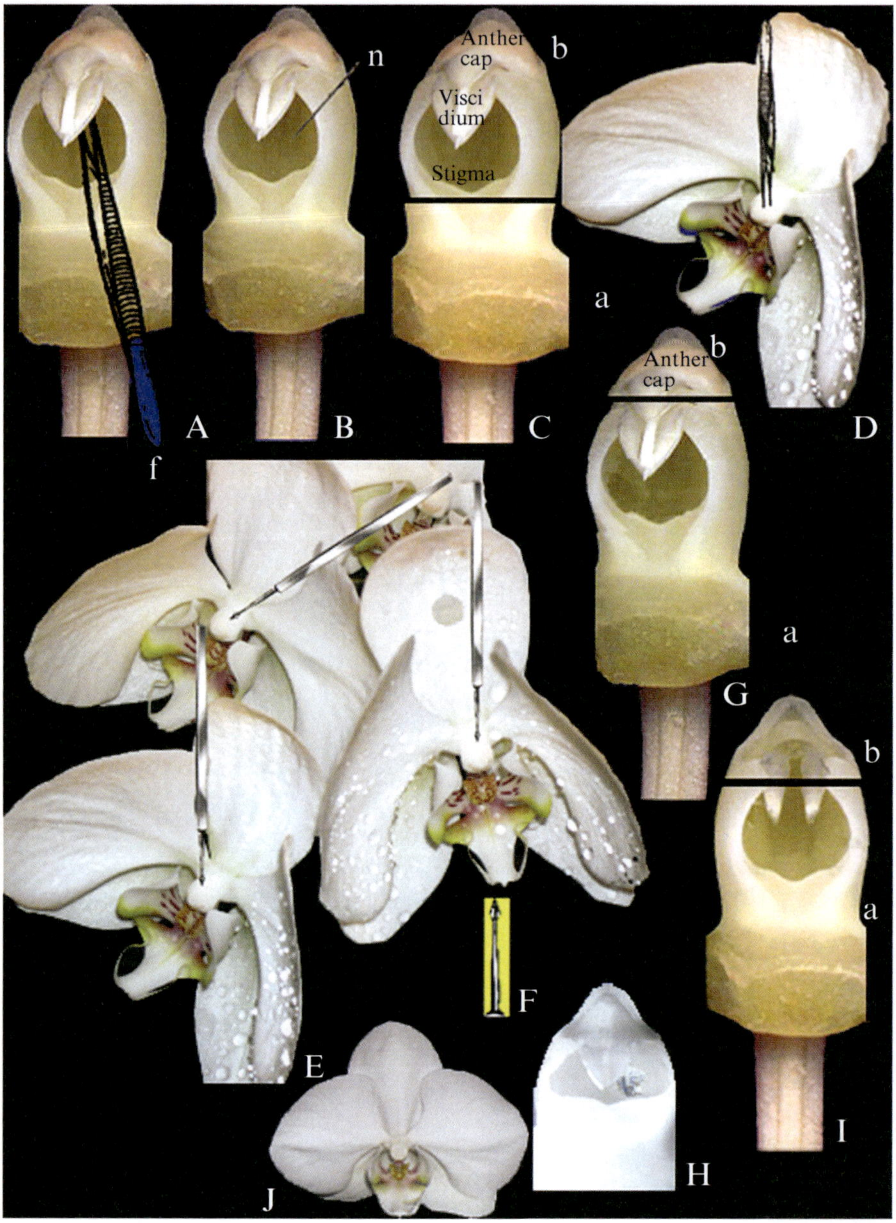

Fig. 2-10. Experimental treatments of *Phalaenopsis amabilis* gynostemia. **A.** Wounding or scratching stigma with forceps. **B.** Scratching or wounding stigma with needle. **C.** Removing all of gynostemium above the stigma (b) and leaving only the base (a) in place. **D.** Crushing gynostemium with forceps. **E.** Peeling epidermis from dorsal surface of gynostemium with "Starnadel." **F.** "Starnadel." **G.** Removing tip (b) from intact gynostemium above stigma leaving the rest (a) in place. **H.** Wad of paper or cotton in stigma. **I.** Removing tip (b) from gynostemium without anther cap and pollinia above stigma leaving the rest (a) in place. **J.** Gynostemium removed. Explanation of symbols: a, base or lower portion; b, top or tip; f, forceps; n, needle (J. Arditti and computer generated using a digital photograph).

Table 2-2. Hans Fitting's experiments involving the wounding of gynostemia and stigmas at the Bogor Botanical Gardens in 1908 (Fitting, 1909a)[a].

Experiment number	Orchid	Description	Results	Fitting's conclusions	Current explanation
17 (21 Jan–11 Feb 1908)	*Phalaenopsis amabilis* (Fig. 2-10A, 2-10B, 2-10; 2-13A)	Stigma of recently opened flower wounded with flamed forceps	Flower started to close after 24 hours, closed and started to wilt after 48 hours	Wounding of the stigma, not cutting or wounding of the gynostemium shortens the life span of flowers like pollination	Fitting's conclusion is consistent with current knowledge. Wounding of the stigma quickly induces copious ethylene evolution which brings about rapid senescence. Cutting or wounding the gynostemium induces much lower levels of ethylene evolution more slowly if at all
		Stigma of just opened flower wounded with flamed forceps	Flower started to close after 24 hours, closed and started to wilt after 48 hours		
		Stigma of newly opened flower wounded with needle	Flower closed after 24 hours		
		Flower pollinated	Flower closed after 24 hours		
		Top of gynostemium cut below stigma on 29 Jan	Flowers remained fresh until 1 March, then closed rapidly		
18A (dates not listed)	*Phalaenopsis amabilis* (Fig. 2-10A, 2-13A)	Stigmas of 3 flowers wounded by numerous scratches made with sterilized forceps	Flowers closed after 1-1½ days	Flowers close 1–1½ days after larger and 3 days after a smaller puncture wound with a needle	This is consistent with present day knowledge. Larger wounds induce more extensive ethylene evolution which bring about more rapid senescence and closing of flowers
18B (dates not listed)	*Phalaenopsis amabilis* (Fig. 2-10A, 2-13A)	Stigmas of 2 flowers wounded by 5 longitudinal scratches made with sterilized forceps	Flowers closed after 1-1½ days		
18C (dates not listed)	*Phalaenopsis amabilis* (Fig. 2-10A, 2-13A)	Stigma of 1 flower wounded by 4 longitudinal scratches made with sterilized forceps	Flowers closed after 1-1½ days		

(continued)

Table 2-2. (continued).

Experiment number	Orchid	Description	Results	Fitting's conclusions	Current explanation
18D (dates not listed)	*Phalaenopsis amabilis* (Fig. 2-10A, 2-13A	Stigmas of 2 flowers wounded by 3 longitudinal scratches made with sterilized forceps	1 flower closed after 1½ day, the other remains fresh		
18E (dates not listed)	*Phalaenopsis amabilis* (Fig. 2-10A, 2-13A)	Stigma of 1 flower wounded by 2 longitudinal scratches made with sterilized forceps	Flower closed after 1½ day		
18F (dates not listed)	*Phalaenopsis amabilis* (Fig. 2-10A, 2-13A)	Stigma of 1 flower wounded by longitudinal scratch made with sterilized forceps	Flower closed after 1 day		
18G (dates not listed)	*Phalaenopsis amabilis* (Fig. 2-10B, 2-13A)	Stigma of 1 flower punctured once with a needle	Flower remained fresh for 8 days		
		Multiple scratches made with a needle in stigma of 1 flower	Flower closed after 3 days		
19 (dates not listed)	*Phalaenopsis amabilis* Fig. 2-10C, 2-13A)	Gynostemia of 8 flowers cut just below the stigma	3 flowers remained fresh for a month, 1 for 19 days, 3 for 9 days and 1 for 8 days		
20A (dates as listed in 3rd and 4th columns)	*Phalaenopsis amabilis* (Fig. 2-10E, 2-10F, 2-13A)	7 scratches, 2–3 mm long and not very deep made at stigma levels on dorsal and lateral sides of gynostemium of 1 flower on 1 Feb 1908. A "Starnadel" was used	Flower remained fresh until 5 Feb 1908 when stigma was wounded lightly with a needle. Wound had no effect and flower was still fresh on 8 Feb 1908 when it was wounded more severely Flower wilted on 9 Feb 1908	Flower remained fresh when wounds on the gynostemium were not deep, extensive or severe	An explanation based on current knowledge would be that these wounds did not induce ethylene production which was not sufficient to bring about floral senescence

(continued)

20B (dates as listed in 3rd and 4th columns)	*Phalaenopsis amabilis* (Fig. 2-10E, 2-10F, 2-13A)	4 scratches, 2–3 mm long and not very deep made at stigma levels on dorsal and lateral sides of gynostemium of 1 flower on 31 Jan 1908. Starnadel used	Flower remained fresh until 7 Feb 1908 when stigma was wounded severely with a needle. The flower closed after 2 days
20C (dates as listed in 3rd and 4th columns)	*Phalaenopsis amabilis* (Fig. 2-10E, 2-10F, 2-13A)	4 scratches, 2–3 mm long and not very deep made at stigma levels on dorsal and lateral sides of gyno-stemium of 1 flower on 31 Jan 1908. Forceps used	Flower remained fresh until 8 Feb 1908 when stigma was wounded severely with a forceps. The flower closed after 2 days
20D (dates as listed in 3rd and 4th columns)	*Phalaenopsis amabilis* (Fig. 2-10E, 2-10F, 2-13A)	1 scratch, 2–3 mm long and not very deep made at stigma levels on dorsal side of gynostemium and 2 on its side (1 flower) on 31 Jan 1908. Needle used	Flower remained fresh until 7 Feb 1908 when stigma was wounded severely with a needle. The flower closed after 3 days
20E (dates as listed in 3rd and 4th columns)	*Phalaenopsis amabilis* (Fig. 2-10D, 2-13A)	Dorsal side of gynostemium crushed with forceps behind the anther on 4 Feb. Forceps used	Flower still fresh on 8 Feb when the upper part of the stigma was wounded severily with a needle. The flower closed after 1 day.
20F (dates as listed in 3rd and 4th columns)	*Phalaenopsis amabilis* (Fig. 2-10E, 2-10F, 2-13A)	The epidermis on the dorsal and lateral sides of the gynostemium above the stigma was peeled in 3 flowers on 9 Feb. "Starnadel" used	All flowers were still fresh on 19 Feb when the upper parts of the stigmas of 2 flowers were scratched. These flowers closed after 1½ days

(continued)

Table 2-2. (continued).

Experiment number	Orchid	Description	Results	Fitting's conclusions	Current explanation
21A (dates not listed)	*Phalaenopsis amabilis* (Fig. 2-10E, 2-10F, 2-13A)	Deep wounds were made on the dorsal surface and sides of the gynostemium of 1 flower with a "Starnadel"	Flower closed after 1½ days. Wound reached, but did not damage a vascular bundle	Deep wounds in the gynostemium have the same effects on flower longevity as wounding the stigma. Hence it seems that mechanical effects of wounding of the gynostemium are the same as sea sand in the stigma. The sand probably damages specific cells in the stigma	An explanation based on current knowledge is that the wounding and the sand damage to the stigma induce ethylene production levels which are sufficient to bring about closing of the flower
21B (dates not listed)		One deep wound was made near the stigma and two were deep enough to almost reach vascular bundles. Cuts were made with a "Starnadel" in 1 flower	Flower closed after 1–1½ days		
21C (dates not listed)	(Fig. 2-10G, 2-10I, 2-13A)	The tip of the of the gynostemium above the stigma was excised with a transverse cut which did not damage the stigma	Flower closed after 1½ days		
22A (dates not listed)	*Phalaenopsis amabilis* (Fig. 2-10H, 2-13A)	A cotton wad was inserted gently and carefully into the stigma of one flower The stigma was rubbed with the cotton wad 7 days after the initial insertion	Flower was still fresh 1½ days after the insertion The flower closed 1½ days after the rubbing	Superficial wounding of the dorsal surface and sides of the gynostemium did not shorten the life span of flowers. The same is true	An explanation based on current knowledge is that wounding and/or damage to parts of the gynostemium other than to the stigmatic surface do not

(continued)

22B (dates not listed)		Stigma of 1 flower was rubbed with a cotton wad	The flower closed 1½ days after the rubbing	for cuts at the base of the stigma. Only damage to the stigma (even if minor) and the wiping of stigmatic papillae and secretions reduce the life span flowers. However deep or sever wounds on the dorsal side of the gynostemium which come close to the stigmatic surface of vascular bundles do shorten the life span of flowers. span of flowers and bring about floral closure	induce ethylene production which was high enough to bring about closing of flowers. Deep or severe wounds do induce high enough levels of ethylene production to shorten the life
22C (dates not listed)		Stigma of 1 flower was rubbed with a cotton wad	The flower closed 2 days after the rubbing		
		A cotton wad was inserted gently and carefully into the stigma of one flower	Flower remained fresh		
		Uppermost part of stigma was wounded 7 days after insertion of the cotton	Flower closed 1½ days after the wounding		
22D (dates not listed)		Stigmas of 2 flowers were rubbed with cotton wads	1 flower closed 1½ days after the rubbing; the other remained fresh, but closed 1½ days after uppermost part of stigma was wounded		
22E (dates not listed)		Upper part of stigma was wiped with cotton wad (1 flower)	Flower closed after 1½ days		
		Lower part of stigma was wiped with a cotton wad (1 flower)	Flower closed after 1½ days		
		Fitting examined the closed flowers from this series of experiments and found that rubbing and wiping of stigmas removed stigmatic papillae and secretions but did not cause wounding other than at most damaging a single cell layer			

(continued)

Table 2-2. (continued).

Experiment number	Orchid	Description	Results	Fitting's conclusions	Current explanation
23 (dates not listed)	*Phalaenopsis amabilis* (Fig. 2-10E, 2-10F, 2-13A)	Uppermost parts of stigmas, below rostella were wounded in 15 flowers	12 flowers closed after 1½ days. Stigmas of the 3 flowers which remained open were wounded in the same spots. These flowers closed after 1½ days	Wounding of the upper part of the stigmatic surface brings about closing of the flowers. When only other parts of the stigma are wounded flowers do not close	This observation suggests that the site of ethylene production is in the upper part of the stigma near the rostellum. This seems logical since ethylene production requires energy and the rostellum contains numerous mitochondria
24 (dates not listed)	*Phalaenopsis amabilis* (Fig. 2-10E, 2-10F, 2-13A)	The lowermost part of the stigma just above the stylar canal and the cells of the stylar canal were wounded with a needle (9 flowers)	1 flower closed after 1½ days, 8 remained fresh after 5 days when the upper part of the stigma was wounded. Of these flowers 4 closed after 1½ days and 2 after 3 days, but 2 remained fresh after 4 days when they were wounded again. They closed after 3 additional days	Flowers also close if both the upper and another part are wounded	
		The middle part of the stigmatic surface was wounded in 2 flowers	Both were open after 4 days, but closed 1½ days after the upper parts of their stigmas were wounded		
25A (dates not listed)	*Phalaenopsis violacea* (Fig. 2-6F)	The stigmatic surfaces of 6 flowers were wounded	The flowers closed after 1–2 days	There are 6 conclusions:	These observations and conclusions can be explained in terms of ethylene evolution. If the wounding was severe enough to induce
25B (dates not listed)		Stigma of 1 flower was wounded minimally with a needle	The flower closed after 1½ days	1. With the possible exception of *Aerides odorata* wounding of the stigma	

(continued)

25C (dates not listed)		Stigma of 1 flower was wounded at insertion point of stylar canal	Flower is still fresh after 5 days, but close 2½ days after upper part of stigma is wounded	shortens the life span of flowers in all species in which river sand has the same effects 2. If pollination did not bring about changes in the perianth, river sand and and wounding of the stigma had no effects either 3. In two species (*Vanda tricolor* and *Calanthe veratrifolia*) pollination brings about changes in the perianth, but wounding of the stigma does not (experiments 33 and 35) 4. Wounding of the stigma had different effects on two *Vanda* species (experiments 32 and 33) 5. In all species in which wounding had any effects, wounding of the stigma was more effective than	ethylene evolution and if the affected area was capable of producing this gaseous hormone shortening of the life span of flowers was to be expected
25D (dates not listed)	(Fig. 2-10C is representative)	Gynostemia of 3 flowers cut below stigma	Flowers remained fresh for 6 Days, then closed		
25E (dates not listed)	(Fig. 2-10D is representative)	Dorsal surface and sides of the gynostemia of 2 flowers were crushed severely with forceps	Flowers closed after 2 days		
25F (dates not listed)		Several shallow scratches made with needle on dorsal surface and sides of gynostemium (1 flower)	Flowers remained fresh for an extended period		
25G (dates not listed)		Epidermis peeled from dorsal surface and sides of gynostemium (1 flower)	Flower closed after 1½ days		
25H (dates not listed)		Epidermis peeled from dorsal surface and sides of gynostemium (1 flower)	Flower remained fresh for an extended period		
26 (dates not listed)	*Phalaenopsis esmeralda* Rch. *f.* (Fig. 2-8J)	Stigmatic surface wounded (2 flowers) Pollinated (1 flower)	All 3 flowers wilted after 2–3 days		
27A (dates not listed)	*Phalaenopsis cornu-cervi* Bl. et Rchb. *f.*	Stigmatic surfaces of 9 flowers were wounded	All flowers wilted after 2–4 days		

(continued)

Table 2-2. (continued).

27B (dates not listed)	(Fig. 2-12A, 2-12B)	Gynostemia of 3 flowers cut under the stigma	Flowers remained fresh for 6 days and wilted on the 7th day	wounding of other parts. In two experiments wounding the dorsal surface of the gynostemium at stigma height was also effective (experiments 25G and 29E) 6. Both river sand and wounding the perianth, did not cause swelling of the ovary and/or gynostemium
		Pollinated (1 flower)	Flower started to wilt after 2–3 days	
28 (5–7 Feb 1908)	*Rhynchostylis retusa* (Fig. 2-10B is representative)	Stigmas of 12 flowers wounded with needle	7 Feb: 6 flowers wilted and 6 are starting to wilt	
28 (31 Jan–6 Feb 1908)	(paper lists 31 Feb as date)	Stigmas of 12 flowers wounded with needle	4 Feb: 8 flowers wilted 5 Feb:10 flowers wilted 6 Feb: all 12 flowers wilted	
28 (5–8 Feb 1908)		Stigmas of 9 flowers wounded with needle	8 Feb: all 9 flowers wilted	
28 (8–10 Feb 1908)	(Fig. 2-10B)	Upper parts of stigmas of 12 flowers wounded with needle	10 Feb: all 12 flowers wilted	
29A (dates not listed)	*Dendrobium superbum*	Stigmas of 7 flowers wounded with needle	All 7 flowers closed after 1–2 days	
29B (dates not listed)	(Fig. 2-12G)	Uppermost part of stigma of 1 flower wounded	Flower closed after 3 days	
29C (dates not listed)		Lowermost parts of stigmas of 2 flower wounded	Flowers remained fresh for 6 days	
29D (dates not listed)		Gynostemia of 4 flowers removed	Flower closed after 4 days	
29E (dates not listed)		Epidermis peeled from dorsal surfaces and sides of gynostemia of 5 flowers	1 flower closed after 1 day. The other 4 were still open after 6 days	
30 (start 6 Feb 1908)	*Oncidium flexuosum* (Fig. 2-12F)	2 flowers pollinated	Both wilted after 3 days	
		Stigmas of 5 flowers wounded	All 5 flowers wilted after 3 days	
		Gynostemia of 5 flowers removed	Flowers still fresh after 5 days	
31 (8–10 Feb 1908)	*Dendrobium wardianum* (Fig. 2-12D)	Stigma of 1 flower wounded on 8 Feb	10 Feb: Flower closed	

(continued)

32 (8–10 Feb 1908)	*Vanda insignis* (Fig. 2-12C)	1 flower pollinated. The stigma of 1 flower wounded with needle	10 Feb: Both flowers start to discolor and wilt
33 (start on 6 Feb 1908)	*Vanda tricolor*	Stigmas of 6 flowers wounded	Flowers were still fresh after 6 days
34 (4–10 Feb 1908)	*Aerides odorata* (Fig. 2-9)	4 Feb: 3 flowers pollinated 4 Feb: Stigmas of 3 flowers wounded 6 Feb: 2 flowers pollinated Stigmas of 2 flowers wounded	8 Feb: Slow wilting 8 Feb: Flowers unchanged 10 Feb: Flowers start to wilt 10 Feb: Flowers unchanged
35 (6–12 Feb 1908)	*Calanthe veratrifolia* (Fig. 2-12E)	3 flowers pollinated Stigmas of 4 flowers wounded	8 Feb: Flowers start to wilt 12 Feb: All flowers wilted
36 (4–12 Feb 1908)	*Dendrobium antennatum* (Fig. 2-8A)	2 flowers pollinated Stigmas of 2 flowers wounded	12 Feb: All 4 flowers still unchanged (i.e., still fresh)
37 (9–19 Feb 1908)	*Trichoglottis geminata* (Fig. 2-11E)	9 Feb: 2 flowers pollinated and stigmas of 2 flowers wounded 14 Feb: Stigmas of 2 flowers wounded	14 Feb: All 4 flowers did not change substantially (i. e., still fresh) 19 Feb: Flowers did not change (i.e., still fresh).
38 (27 Feb 3 Mar 1908)	*Dendrobium macrophyllum* (Fig. 2-11A, 2-11D)	3 flowers pollinated and stigmas of 2 flowers wounded	3 Mar: All flowers unchanged (i.e.,still fresh)
39 (26 Feb–2 Mar 1908)	*Arachnanthe sulingi* (Fig. 2-11B)	3 flowers pollinated and stigmas of 4 flowers wounded	2 Mar: All flowers unchanged (i.e., still fresh)
40 (16–27 Apr 1908)	*Renanthera maingayi* (Fig. 2-11C)	2 flowers pollinated and stigmas of 2 flowers wounded	27 Mar: All flowers unchanged (i.e., still fresh)

[a] The orchid names in this table are those used by Hans Fitting. Please see appendix 2 for updated nomenclature.

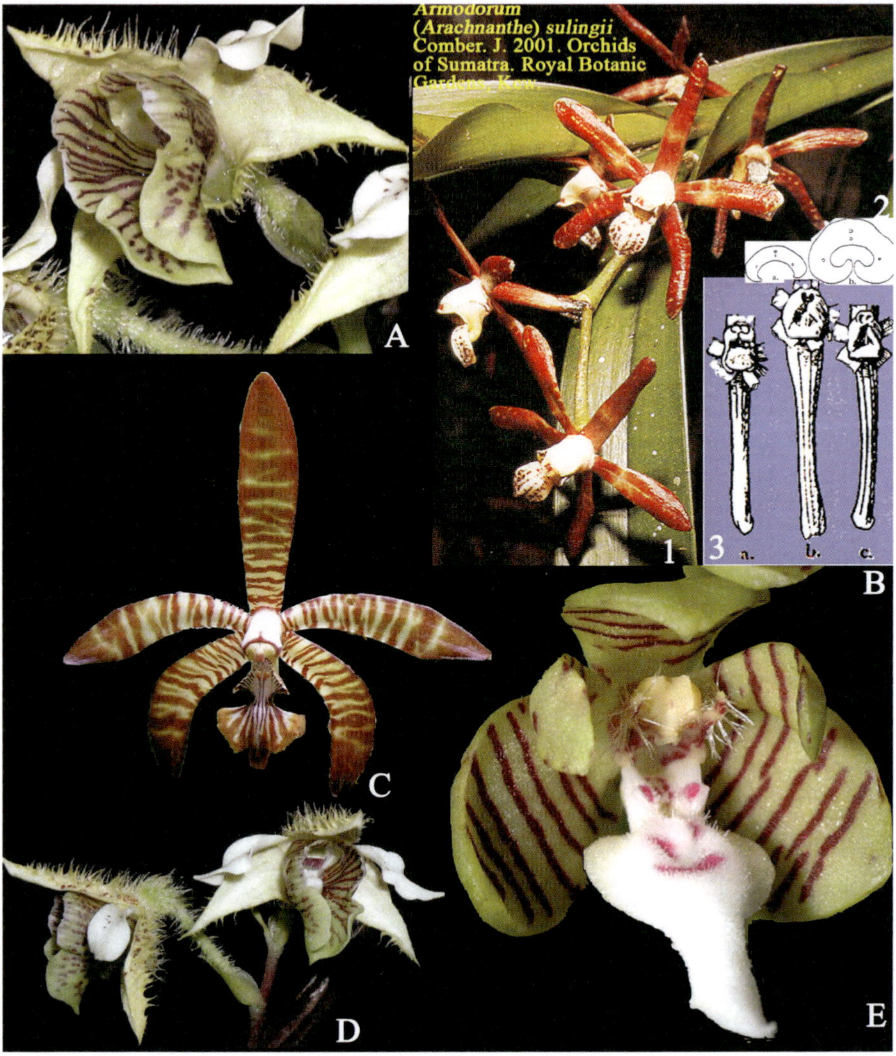

Fig. 2-11. Orchids used by Hans Fitting in wounding experiments. **A.** *Dendrobium macrophyllum*, **B.** *Armodorum* (*Arachnanthe*) *sulingii*. **1.** Plant in bloom. **2.** Cross-section through gynostemium and stigma: left, before pollination; right, a few days after pollination. **3.** Front view of gynostemium: a, before pollination; b, seven days after pollination with living pollen; c, seven days after pollination with steam-killed pollen. **C.** *Arachnanthe clarkei*. **D.** *Dendrobium macrophyllum*. **E.** *Trichoglottis geminata* (Fitting, 1909; C, T. W. Yam and J. Arditti; E, courtesy Eric Hunt).

Flowers pollinated with pollinia which were submerged in chloroform for 30 min exhibited phenomena which were the same as those which are brought about by live pollen. Onset of such phenomena was also equally rapid (Table 2-3). The phenomena were reduced in intensity following pollination with pollen which was soaked in chloroform for 1 h (Table 2-3).

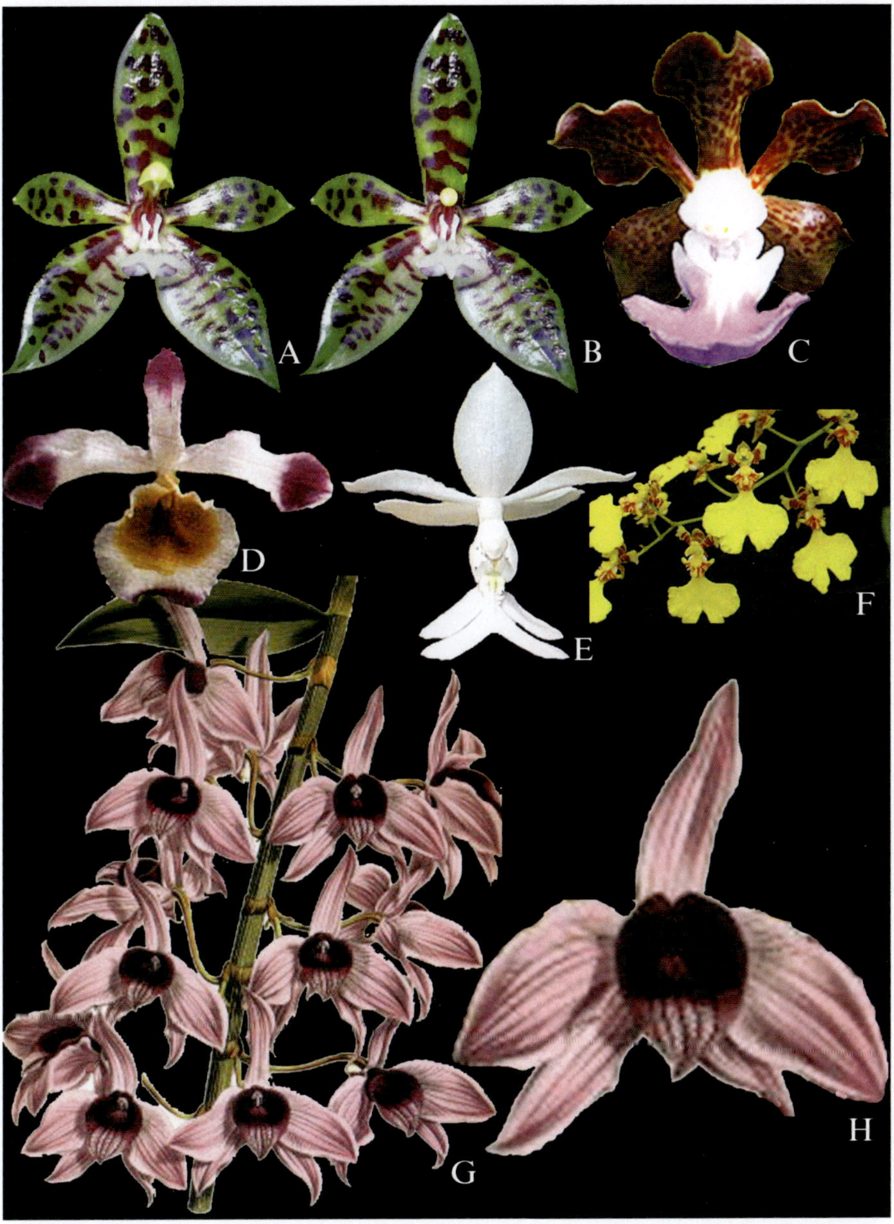

Fig. 2-12. Species Hans Fitting used in wounding experiments. **A.** Intact flower of *Phalaenopsis cornu-cervi*. **B.** Flower of *Phalaenopsis cornu-cervi* without gynostemium. **C.** *Vanda insignis*. **D.** *Dendrobium wardianum*. **E.** *Calanthe veratrifolia*. **F.** *Oncidium flexuosum*. **G.** *Dendrobium superbum* inflorescence. **H.** Flower of *Dendrobium superbum* without gynostemium (A, J. Arditti and T. W. Yam; B, the flower in A altered with Photoshop; C, T. W. Yam; D, J. Arditti; G, H, Orchid Album published in 1882).

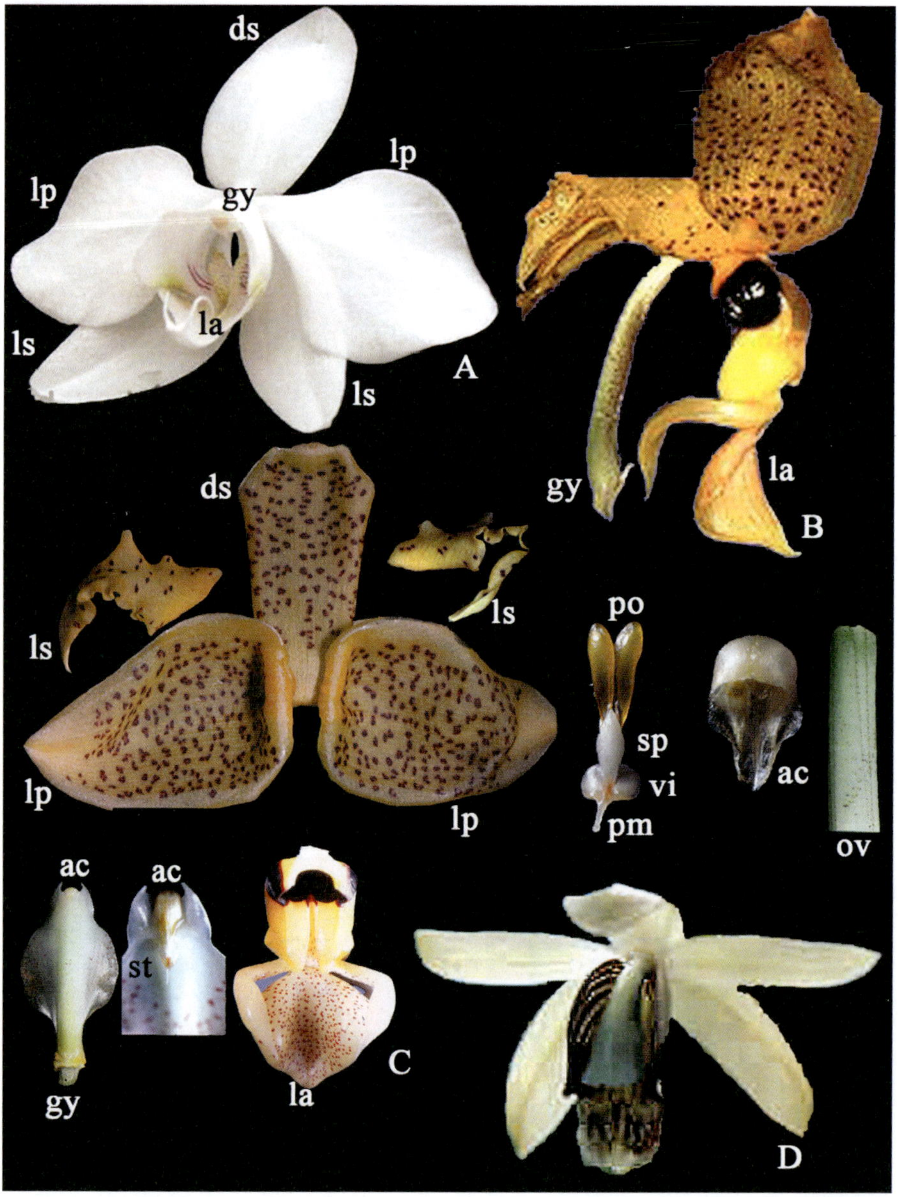

Fig. 2-13. Orchids used by Hans Fitting in experiments with live and dead pollinia and pollen extracts. **A.** *Phalaenpsis amabilis*. **B.** *Stanhopea*. **C.** Parts of *Stanhopea* flower. **D.** *Coelogyne swaniana*. Explanation of symbols: ac, anther cap; ds, dorsal sepal; gy, gynostemium; la, labellum; lp, lateral petal; ls, lateral sepal; ov, ovary; pm, pollinarium; po, pollinia; sp, stipe; st, stigma; vi, viscidium (A, the late Dr. Djunaidi Gandawijaja; B, C, D, J. Arditti and T. W. Yam).

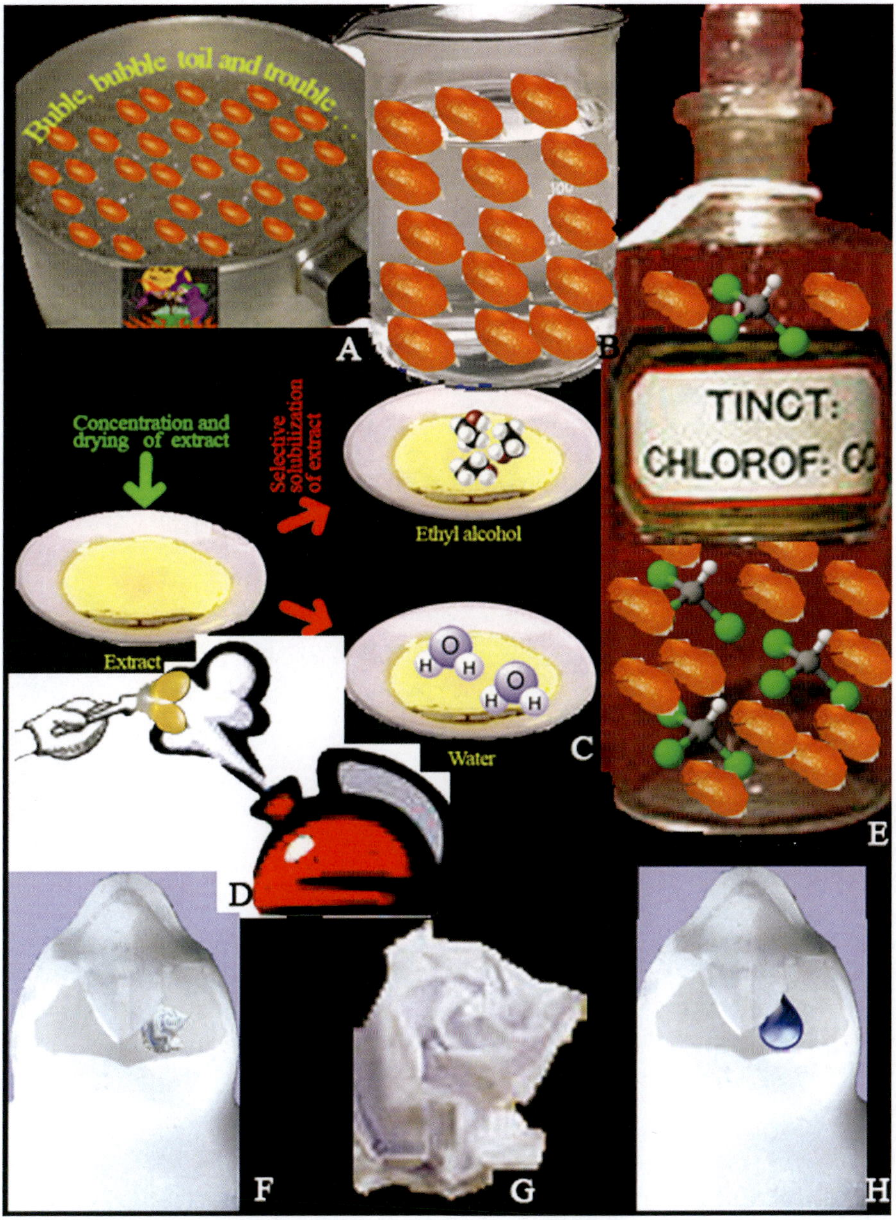

Fig. 2-14. Killing and extracting pollen. **A.** Pollen being killed and extracted with boiling water. **B.** Pollinia being extracted with cold water. **C.** Drying and extracting pollen extract. **D.** Steam-killing pollinia. **E.** Pollinia immersed in chloroform, **F.** Filter paper wad in stigma. **G.** Filter paper wad. **H.** Drop in stigma (A–H, prepared with a graphic program using separate images; C, The round plate-like objects are actually watch glasses which is what Fitting actually used; E, The chloroform bottle is old enough to be similar to the one Fitting may have had at his disposal; F, H, J. Arditti).

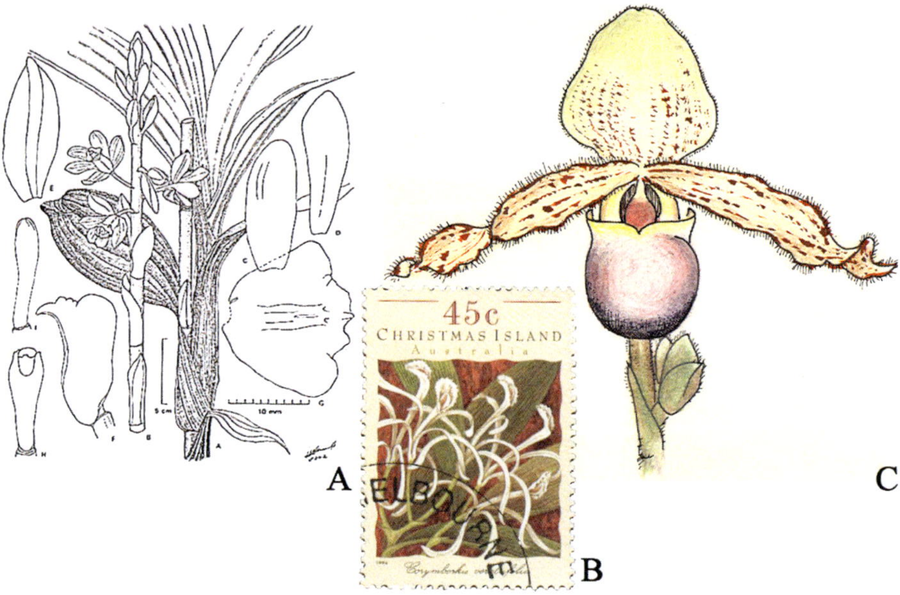

Fig. 2-15. Orchid used by Hans Fitting in experiments with inter- and intra specific pollination as well as with live and dead pollinia **A.** *Phaius amboinensis*. **B.** *Corymborkis veratrifolia*. **C.** *Paphiopedilum glaucophyllum* (compiled by J. Arditti).

Fitting's choice of killing agent was fortunate because IAA and the ethylene precursor, 1-aminocyclopropane-1-carboxylic acid (ACC) are not soluble or nearly insoluble in chloroform (Fig. 2-14E). This near insolubility may explain the different effects of pollinia which were soaked in chloroform for 30 or 60 min. Fitting gave no reasons for choosing chloroform, but since auxin was not known at the time it is clear that he did not make his choice on the basis of solvent/solubility characteristics. Had he chosen ethyl alcohol (which is usually more easily available and therefore would seem to be a more obvious choice) Fitting's result would have been different because auxin is ethanol soluble. Soaking the pollinia in ethanol would have extracted the auxin and ACC and rendered the pollen ineffective. This would have changed Fitting's findings and perhaps altered the course of his experiments.

Steam-killed pollinia (Fig. 2-14D) were as effective as live pollen (Table 2-3). However pollen killed by keeping it in boiling water (Fig. 2-14A; Table 2-3) did not bring about post-pollination phenomena. The reason for this is the solubility of auxin in hot water. This solubility is limited, but the exposure time was probably long enough to dissolve sufficient auxin from the pollen to render it ineffective. After these experiments Fitting concentrated and dried the extracts in watch glasses (Fig. 2-14C), extracted and/or redissolved them in several solvents and applied the resulting solutions to stigmas as drops (Fig. 2-14H) or in wads of filter paper (Fig. 2-14F, G; Table 2-3). A number of the extracts induced post pollination phenomena (Table 2-3). Cold water extracts (Fig. 2-14B) had a similar effect (Table 2-4).

Table 2-3. Hans Fitting's experiments with live and dead pollinia and pollen extracts at the Bogor Botanical Gardens in 1908 (Fitting, 1909a)[a].

Experiment number	Orchid	Description	Results	Fitting's conclusions	Current explanation
42A (13 Feb, 11:00 14 Feb, 11:00 1908)	*Phalaenopsis amabilis* (Fig. 2-13A)	2 pollinia submerged in chloroform for ½ h, air dried ¼ h and placed in stigma on 13 Feb 11:00	14 Feb, 06:00: Stigma closed, flower open; 11:00 flower closed. 18 Feb: Gynostemium strongly swollen down to its base, ovary starts to wilt without swelling. 19 Feb: Ovary wilted further, pollen grains collapsed, not germinated	1. The pollen does not have to germinate on the stigma to cause swelling of the gynostemium and wilting of the flower 2. Dead pollen does not function like live one 3. Dead pollen can cause the gynostemium to swell and the flower wilt, but it does not bring about swelling of the ovary 4. Swelling of the gynostemium and swelling of the ovary are not linked	Current knowledge is: (1) swelling of the gynostemium and the ovary are auxin effects, (2) wilting of the flower is caused by ethylene, (3) orchid pollinia contain IAA, and (4) there are good reasons to believe that orchid pollen contains ACC. IAA and ACC are insoluble in chloroform. Hence, it is not surprising that pollinia which were soaked in chloroform can bring about phenomena which are caused by auxin and ethylene. Swelling of the overy requires continuous supply of auxin which cannot be provided by dead pollinia
42B (13 Feb Feb, 12:00–14 Feb, 10:00 1908)		2 pollinia submerged in chloroform for ½ h, air dried ¼ h and placed in stigma of flower (cut and placed in water in a laboratory) at 12:00 on 13 Feb	13 Feb, 18:00: Start of inward movement of stigma edges. 14 Feb 06:00: Stigma completely swollen (i.e., closed; 08:00: Flower half closed; 10:00: Pollen grains collapsed, not germinated. Pollen grains collapsed and contents rough. Untreated pollen is well rounded and transparent		
42C (13 Feb 11:00)		1 pollinium placed in chloroform for ½ h, air dried for ¼ h and suspended in water			
43A (14 Feb 11:00–19 Feb 1908)	*Phalaenopsis amabilis* (Fig. 2-13A)	Flower pollinated with pollen subjected to 3 min of 98–99°C steam	15 Feb 06:00: Flower starts to close, stigma not closed; 11:00: Flower and stigma closed. 19 Feb: Ovary not swollen, but wilted. Flower abscised after a few days	1. Steam-killed pollen could cause wilting and stigmatic closure 2. The active principle in the pollen is heat stable	Fitting's conclusions are consistent with what is known about IAA and ACC at present

(continued)

Table 2-3. (continued).

Experiment number	Orchid	Description	Results	Fitting's conclusions	Current explanation
43B (14 Feb 11:00–19 Feb 1908)		Flower pollinated with pollen subjected to 3 min of 98–99°C steam	15 Feb 06:00: Flower half closed, stigma closed 19 Feb: Ovary not swollen, but wilted. Flower abscised after a few days		
43B (16 Feb 17:00–23 Feb 1908)		Flower pollinated with pollen subjected to 25 min of 98–99°C steam, allowed to stand for 3 h and subjected to an additional 3 min of steam	17 Feb 06:00: Flower open stigma half closed; 12:00: 12:00: Ovary swollen (i.e., closed), flower starts to close. 18 Feb 06:00: Flower wilted. 23 Feb: Ovary not swollen, but wilted Examination of the pollen in experiments showed that the pollen grains were collapsed and that pollen tubes did not develop		
44 (start on 17 Feb at 12:00)	*Phalaenopsis amabilis* (Fig. 2-13A)	2 flowers that were open for 2 days were pollinated with pollen subjected to 98-99° steam, 4 h to a moist environment and placed for 3 min in boiling water. This experiment was repeated twice	Flowers remained fresh	Steam does not kill the pollen which is unlikely, or the boiling water either extracts or destroys the active principle	ACC is water soluble. IAA is very sparingly soluble in water it at all, but both must have come out of the steam-killed pollen into the boiling water
45 (18 Feb 08:00–19 Feb 12:00 1908	*Phalaenopsis amabilis* (Fig. 2-13A)	Five pairs of pollinia were extracted by placing them in 2 ml of boiling distilled water for 3 min. After that the pollinia were removed from the water which was reduced in volume through evaporation to 2–3 drops. The experiments were carried out with 3 flowers who were open for several days			

(continued)

45A Start18 Feb 08:00 1908		Wad of filter paper wetted with distilled water was placed in stigma	19 Feb 12:00: Flower still open	1. The active principle can be extracted from pollen with hot water	ACC is water soluble. IAA is sparingly soluble or insoluble in water
45B Start 18 Feb 08:00 1908		Wad cf filter paper wetted with pollen extract was placed in stigma	19 Feb 06:00: Stigma swollen (i.e. closed). Flower starts to close; 12:00: Flower closed	2. The extracted pollen lost its ability to affect the gynostemium and perianth	Fitting probably extracted both with boiling water and did not have a single compound in his extract
45C Start 18 Feb 08:00 1908		Drop of pollen extract was placed on stigma with "Starnadel"	19 Feb 06:00: Stigma swollen (i.e., closed). Flower starts to close; 12:00: Flower closed	3. Since it is water soluble the active principle is chemical in nature	
46A1 (18 Feb 10:00 1908)	*Cymbidium finlaysonianum* (Fig. 2-8D)	2 flowers pollinated xenogamously with living pollen	19 Feb 07:00: Gynostemium very swollen near stigma. 20 Feb 07:00: More swollen; 21 Feb 07:00: The same. 23 Feb morning: Flowers abscised	Fitting concluded that killed and living pollen had the same effect on swelling of the gynostemium and wilting of the flower. This is not the case. Boiled pollen is dead, but it did not induce swelling	IAA is not soluble in chloroform. This means that pollen soaked in chloroform still contained auxin which caused swelling and should have initiated ethylene evolution and senescence of the perianth. It is hard to explain why it did not do that
46A2 (start 18 Feb 10:00 1908)	*Cymbidium finlaysonianum* (Fig. 2-8D)	2 flowers pollinated with pollen which was submerged in chloroform for 1 hour	19 Feb 07:00: Gynostemium swollen near stigma, but less than the one pollinated with live pollen. 20 Feb 07:00: more swollen; 11:00: 5 fresh looking flowers abscised. 21 Feb 06:00: Last flower abscised		

(continued)

Table 2-3. (continued).

Experiment number	Orchid	Description	Results	Fitting's conclusions	Current explanation
46B (start 20 Feb 11:00 1908)		2 flowers pollinated xenogamously with living pollen	21 Feb 08:00: Gynostemium very swollen near stigma. 21 and 22 Feb: Swelling intensified. 24 Feb: Both flowers abscised		
47A (Start 20 Feb 11:00–end 24 Feb 1908)	*Stanhopea* sp. (Fig. 2-13B)	1 flower open for 2 days pollinated with pollen which was kept in chloroform for 1 h	21 Feb 06:00: Gynostemium somewhat swollen near stigma. 22 Feb: The same. 23 Feb 07:00: Flower started to wilt		IAA was not extracted with chloroform. It was extracted by boiling water, ACC was not extracted with chloroform
47B (Start 20 Feb 11:00–end 24 Feb 1908)		1 flower open for 2 days pollinated with pollen extracted for ½ h with boiling water	21 Feb 06:00, 22 Feb 06:00: No change in the gynostemium. 23 Feb 07:00: Flower started to wilt 24 Feb: Stigma closed and gynostemium swollen in A; Stigma open, gynostemium not swollen in B		
48A (Start 30 Jun 19:00 1908)	*Stanhopea insignis*	1 flower pollinated geitonogamously	2 Jul: Stigma swollen (i.e., closed), flower closed. 4 Jul: Gynostemium closed		IAA was not extracted from pollinia with chloroform
48B (Start 30 Jun 19:00 1908)		1 flower pollinated with pollen which was submerged in chloroform for 1 h	2 Jul: Stigma swollen (i.e., closed), flower closed. 4 Jul: Gynostemium swollen		
48C (Start 30 Jun 1908)		1 flower unpollinated	2 Jul: Gynostemium not swollen, flower closed		

(continued)

49A (21 Feb–23 Feb 1908)	*Coelogyne swaniana* Rolfe (Fig. 2-13D)	21 Feb 11:00: 3 flowers pollinated with normal pollinia	23, 24 Feb 07:00: All 3 flowers unchanged gynostemium tips widened		IAA was not extracted from pollinia with chloroform
49B		21 Feb: 3 flowers pollinated with pollen which was soaked in chloroform for 1 h	23, 24 Feb 07:00: All 3 flow-unchanged, gynostemium tips widened		
50A (25 Feb–13:00–27 Feb 07:00 1908)	*Dendrobium superbum* (Fig. 2-12G)	25 Feb 13:00: 2 flowers which were open for a few days were pollinated with pollen that was killed by subjecting it to steam for ¼ h	26 Feb 07:00: Gynostemium swollen, both flowers closed 29 Feb 07:00: Gynostemium swollen even more, flowers wilted		Nothing was extracted from steam-killed pollen
50B		25 Feb 13:00: 1 flower pollinated normally	27 Feb 07:00: Gynostemium swollen, flowers closed and wilted		
51A (21 Feb 10:00–29 Feb 1908)	*Arachnanthe sulingi* Benth. (Fig. 2-11B)	A. 1 flower pollinated with living pollen B. 4 flowers pollinated with pollen soaked in chloroform for 1 h C. 3 flowers pollinated with pollen subjected to steam for ¼ h	22 Feb 07:00: Gynostemia of all three flowers strongly swollen 25 Feb 07:00: Swelling of gynostemia intensified 27 Feb 07:00: No change 28 Feb 07:00: All C flowers and 2 B flowers abscised 29 Feb: 2 B flowers abscised All A flowers remained on the plant with swollen ovaries and wilted perianth		

(continued)

Table 2-3. (continued).

Experiment number	Orchid	Description	Results	Fitting's conclusions	Current explanation
52A (experiments started on 18 Feb 12:15	*Rhynchostylis retusa* (Fig. 2-7A)	4 flowers pollinated with pollen which was soaked in chloroform for ½ h	19 Feb 07:00: Gynostemium swollen. 20 Feb 06:00: swelling intensified, flowers start to wilt		Chloroform and steam did not extract IAA and/or ACC from the pollen, boiling water did
52B		4 flowers pollinated with pollen subjected to steam from 09:00 to 09:15 and 12:00 to 12:15	19 Feb 07:00: Gynostemium swollen. 20 Feb 06:00: flowers start to wilt		Since IAA is only sparingly soluble in water it is possible that Fitting did not extract or at least did not always extract all of it from the pollen
52C		2 flowers pollinated with pollen placed in boiling water for 3 min	20 Feb 07:00: Flower completely unchanged (i.e., fresh)		The IAA-like or ethylene-like effects he observed in some of his experiments could be due to residual IAA or other substances which remained in the pollen after it was killed and extracted with boiling or hot water
53A	*Aerides odorata* (Fig. 2-9A, 2-9D, 2-9E)	16 Feb 12:45: 5 flowers pollinated with pollen which was soaked in chloroform for ½ h	17 Feb 09:00: Gynostemium starting to swell. 18 Feb 11:00: Gynostemium very very swollen		
53B		17 Feb 12:45: 4 flowers pollinated with pollen subjected to steam from 09:00 to 09:15 and 12:00 to 12:03 (this is likely a typographical error, the time was probably 12:30)	18 Feb 09:00: Gynostemium swollen. 19 Feb: swelling even more pronounced		

(continued)

53C	21 Feb 11:00: 4 flowers pollinated with pollen soaked in chloroform for 1 h	22 Feb 07:00: All gynostemia swollen 23 Feb 07:00: Swelling of gynostemia more pronounced, flowers starting to wilt. 24 Feb: Same as on 23rd Feb. 25 Feb 07:00: Swelling of gynostemia of normally pollinated flowers is more pronounced than in the others
53D	21 Feb 11:00: 3 flowers pollinated with pollen subjected to steam for ¼ h	
53E	21 Feb 11:00: 1 flower pollinated normally	
53F	16 Feb 13:00: 4 flowers pollinated with pollen placed in boiling water	17 Feb 09:00: Gynostemia not swollen. 18 Feb 07:00: Same. 19 Feb for 3 min 07:00: Gynostemia swollen to a very limited extent. 20 Feb 07:00: Same
53G	7 pollinia were boiled in water (volume not given) for 3 min the extract (which was probably concentrated) was filtered and applied to stigmas of 4 flowers with the tip of a “Starnadel” at 09:30 on 18 Feb	18 Feb 18:00: Gynostemia start to swell 19 Feb 08:00: Swelling is intensified 20 Feb 08:00: Swelling is intensified further
53H1	18 Feb 11:00: dry filter wads placed in stigmas of 3 flowers	21 Feb: Flowers fresh
53H2	18 Feb 11:00: filter paper wads wetted with distilled water placed in stigmas of 3 flowers	

(continued)

Table 2-3. (continued).

Experiment number	Orchid	Description	Results	Fitting's conclusions	Current explanation
53H		18 Feb 11:00: filter paper wads wetted with pollen extract placed in stigmas of 3 flowers	19 Feb 08:00: Gynostemium swollen. 20 Feb: Gynostemium swelling intensified		
53I		18 Feb 11:00: 4 flowers pollinated with pollen extracted twice with boiling water	20 Feb 06:00: Flowers fresh 21 Feb 07:00: Gynostemium swollen slightly. 22 Feb: Same		
54	*Phalaenopsis cornu-cervi* (Fig. 2-12A)	24 Feb 16:30: 2 flowers open for a day pollinated with pollen subjected to steam for ¼ h	25 Feb 07:00: Stigmas swollen, flowers fresh. 29 Feb 07:00: Flowers half closed, wilting. 2 Mar 07:00: Flowers abscised, perianth not green		
55A	*Phalaenopsis violacea* (Fig. 2-6F1)	18 Feb 10:00: 3 flowers pollinated with pollen which was soaked in chloroform for 1 h	18 Feb 18:00: Stigmas closed, flowers open. 19 Feb 06:00: Gynostemium swollen, flower half closed, start of senescence. 20 Feb 06:00: Closing of flowers and yellowing progressing. 22 Feb 06:00: Same. 24 Feb 18:00: Flowers abscised, perianth not green		

(continued)

55B	19 Feb 11:00: flower which was open for a day pollinated with pollen which was subjected to steam for ¼ h	19 Feb 18:00: Stigma half closed, flower unchanged. 20 Feb 06:00: Gynostemium swollen, flower half closed, start of yellowing. 21 Feb 06:00: Flower completely closed. 24 Feb 18:00: Flower abscised, perianth not green
55C	20 Feb 11:00: Flower open for 1 day pollinated with pollen which was extracted twice with boiling water	21 Feb 06:00: Flower unchanged. 22 Feb 06:00: Stigma not closed, flower starts to close and turn yellow. 23 Feb 06:00: Flower wilted 24 Feb 16:00: Flower abscised
55D	25 Feb 13:00: Flower open for 1 day pollinated with pollen which was extracted twice with boiling water	26 Feb 06:00: Stigma open, flower starts to close. 27 Feb 06:00: Flower half closed, yellowing, stigma open. 28 Feb 06:00: Same. 1 Mar 06:00. Flower abscised
55E	27 Feb 11:00: Flower open pollen which was extracted first for 3 min with boiling water, then placed in new water and heated for 1 h in a water bath	28 Feb–29 Feb 06:00: Flower unchanged. 1 Mar 06:00: for 1 day pollinated with Start of floral closing and yellowing; 18:00: Flower closed, stigma open. The flower abscised after a few days without greening

[a] The orchid names in this table are those used by Hans Fitting. Please see Appendix 2 for updated nomenclature.

Table 2-4. Hans Fitting's experiments at the Bogor Botanical Gardens in 1908 on the extraction with cold water of an active principle from pollinia (Fitting, 1909a)[a].

Experiment number	Pollen source	Orchids pollinated and description of experiment	Results	Fitting's conclusions	Current explanation
60 (no dates listed)	*Aerides odorata* (Fig. 2-9A)	Pollinia, 20 pairs were extracted by soaking them in cold water for 2 h The extract was reduced to a small volume by placing it in a water bath for 5 min, Cotton wads saturated with the extract were placed in stigmas of *Aerides odorata, Cymbidium finlaysonianum,* and *Phalaenopsis amabilis*	Wads saturated with extract caused swelling of the gynostemia. Distilled water did not	The active principle can be extracted from pollinia by cold water in a relatively short time This active principle may be located not deep inside the pollen grains but at, near or on their surface or between them	Auxin is only sparingly soluble in cold water. Therefore these findings are puzzling. See text
61 (no dates listed)	*Phalaenopsis amabilis* (Fig. 2-13A)	Pollinia, 20 pairs were extracted by soaking them in 2 ml of cold water for 2 h. The extract was Reduced to a small volume By placing it in a water bath, Extract placed in stigmas of *Phalaenopsis violacea, Aerides odorata, Cymbidium finlaysonianum,* and *Phalaenopsis amabilis*	The extract caused swelling of the gynostemia		

[a]The orchid names in this table are those used by Hans Fitting. Please see Appendix 2 for updated nomenclature.

These extraction and redissolving experiments (Fitting, 1909a) showed (Table 2-5) that the active principle could be extracted by water and ethanol.

The extraction experiments suggested to Fitting that the effects of pollen are due to a chemical and this lead him to test a number of substances (Table 2-6; Fitting, 1909a). Fitting did not give any reasons for selecting the substances he tested. The selection of substances (Table 2-6) makes no sense at present and may not have been logical even in 1907. Therefore it is possible that he simply used whatever was easily available to him in Bogor at the time. Of the substances he used only 5% sucrose had an effect (Table 2-6) on a single flower of *Phalaenopsis amabilis*. This effect is hard to explain and one is tempted to invoke a dictum which may not have been unfamiliar to Fitting: *Ein Versuch, kein Versuch* (one experiment, no experiment). One possible explanation is that the sucrose acted as an osmoticum which caused an influx of water into the stigma and gynostemium.

Inter-Specific and Inter-Generic Pollination

Fitting's carried out all of his initial work by pollinating flowers with pollen of the same species. After that he studied the effects of pollinia of one orchid species on the stigmas of another and found that intra and inter taxon pollination had the same effects (Table 2-7). This is not surprising at present, but must have been a new and interesting finding at the time. Fitting's finding that some extracts of vegetative organs and a floral segment brought about post pollination phenomena (Table 2-8; Fitting, 1909a) must have been puzzling. At present these effects can be explained by the presence in the extracts of: (1) auxin and/or its precursors, and/or (2) ethylene precursors and/or substances which can induce its production.

Effects of Pollen of Non-Orchidaceous Plants

Having established that orchid pollinia have the same effects regardless of whether flowers are pollinated with: (1) their own pollen (i.e., self pollinated), (2) pollen from another flower of the same species, (3) pollen from flowers of other orchid species, (4) pollen from other orchid genera. Fitting took the next logical step. He determined whether pollen from several non-Orchidaceous plants (Fig. 2-16; Table 2-8) had the same effects as orchid pollinia. What he found was that the non-Orchidaceous pollen caused orchid flowers to wilt and senescence (all effects of ethylene), but did not bring about swelling of the gynostemium and stigmatic closure (both are auxin effects). What this means is that non-Orchidaceous pollen can induce ethylene evolution but does not contain any or enough IAA to bring about auxin effects. This is not surprising in view of the fact that orchid pollinia contain very high auxin levels (for a review see Avadhani et al., 1994), perhaps higher than in any other pollen.

Table 2-5. Hans Fitting's experiments at the Bogor Botanical Gardens in 1908 on the chemical nature of the active principle in pollinia (Fitting, 1909a)[a].

Experiment number	Description of experiment and its results	Fitting's conclusions	Current explanation
62 (no dates listed)	*Aerides odorata* pollinia, 20 pairs, mixed with acid-washed sea sand and glycerin were ground with a glass rod. More glycerin wad added, the solids were separated by filtration and suspended in 99.5% ethanol for 24 h. A flocculent gelatinous precipitate formed and was separated by filtration, washed with ethanol and air dried. Part of the precipitate was dissolved in water, absorbed into cotton wads and assayed. The rest was dissolved in water, filtered, precipitated with absolute ethanol and applied to flowers of *Phalaenopsis violacea, Aerides odorata* and *Cymbidium finlaysonianum*. The precipitate had no effect	The active principle is not an enzyme	Fitting discarded the glycerin used to grind the pollinia and the ethanol employed to produce and wash the precipitate. IAA is soluble in ethanol. Thus he probably dissolved IAA in the ethanol and discarded it along with other alcohol soluble substances. Glycerin soluble substances were also discarded. He probably also dicarded ACC
63 (no dates listed)	*Aerides odorata* pollinia, 22 pairs, were extracted by placing them in water for 2 h. The solution was filtered and reduced in volume on a water bath. The yellowish, glasslike, transparent precipitate which formed was active like pollen. A Lassaigne assay failed to detect the presence of nitrogen	The active principle is most probably "stickstoffrei" (i.e., nitrogen-free)	Both IAA and ACC contain nitrogen. Lassaigne's assay was probably not sensitive enough
64 (no dates listed)	*Aerides odorata* pollinia, 51 pairs, and 22 pairs of *Phalaenopsis amabilis* pollinia were extracted by placing them in cold water for 4 h. The solution was filtered and reduced in volume on a water bath. A Lassaigne assay showed that only traces of nitrogen were present	The active principle is most probably "stickstoffrei" (i.e., nitrogen-free)	Both IAA and ACC contain nitrogen Lassaigne's assay not sensitive enough That is why it failed to detect nitrogen when the sample consisted of 22 pairs of pollinia and detected only traces when 73 pairs were assayed

(continued)

65 (dates as listed in next column)	*Phalaenopsis amabilis* pollinia, 38 pairs were extracted by placing them in cold water for 5 h. The extract, filtered twice and reduced to a small volume on a water bath, was very active on *Aerides* flowers. Addition of absolute ethanol (AE) produced a flocculent white precipitate which was washed with AE. The filtrate was reduced to a small volume, AE was added again to form a precipitate. This was repeated until no additional precipitate was formed. The filtrate was dried in a watch glass and formed a glassy, transparent golden yellow precipitate. This was washed with AE, dissolved in water and dried to form a glassy, transparent, colorless precipitate. The extracts were assayed with *Aerides odorata* and *Phalaenopsis amabilis*. "The results were most interesting!" A. Assays with *Aerides odorata* 27 Feb 07:00: Cotton wads saturated with the fraction precipitated by AE placed in stigmas of 4 flowers — 1 Mar: Flowers unchanged, still fresh. Gynostemium not swollen 27 Feb 07:00: Cotton wads saturated with the fraction not precipitated by AE placed in stigmas of 4 flowers — 27 Feb 18:00: Gynostemium starts to swell; 28 Feb 07:00: Gynostemium very swollen. 28 Feb: Same as 27th B. Assays with *Phalaenopsis amabilis* 26 Feb 06:00: Cotton wads saturated with the fraction precipitated by AE placed in stigmas of 2 flowers — 27 Feb 06:00: Flowers unchanged; 27 Feb 18:00: Gynostemium unchanged, flowers half closed; 29 Feb 18:00: Flowers wilted, gynostemium not swollen 26 Feb 18:00: Cotton wad saturated with the fraction not precipitated by AE placed in stigma of one flower — 27 Feb 11:00: Stigma half swollen; 27 Feb 18:00: Stigma completely swollen; 28 Feb: Flower completely closed, gynostemium swollen	1. The cold water extract contains two active principles 2. One of the active principles can be precipitated with AE, the other cannot 3. The principle which cannot be precipitated with AE is the one which causes swelling of the gynostemium and shortens the life span of flowers 4. The fraction which can be precipitated with AE does not cause swelling of the gynostemium, but it does shorten the life span of flowers	The cold water extract may have contained more than two principles IAA is soluble in ethanol. It causes swelling of the gynostemium and initiates ethylene evolution which shortens the life span of flowers The fraction which can be precipitated with AE probably brings about ethylene evolution Fitting was concerned with precipitates and some AE and water-soluble fractions, but he did not seem to consider the possibility that his precipitates and soluble fractions may have contained more than one substance He also discarded some fractions as for example the glycerin used to homogenize pollinia

(continued)

Table 2-5. (continued).

Experiment number	Description of experiment and its results		Fitting's conclusions	Current explanation
66 (dates as listed in next column)	*Phalaenopsis amabilis* pollinia, 27 pairs were extracted with cold water, the extract was fractionated as in experiment 65 and the dried concentrate was dissolved in two drops of water			
	A. 13 Apr 07:00: One flower treated with the fraction precipitated by AE	14 Apr 07:00: Flower half closed, gynostemium unchanged 15 Apr. 07:00: Flower completely closed, gynostemium unchanged		
	15 Apr 07:00: Two flowers treated with the fraction precipitated by AE	16 Apr 07:00: one flower closed and the other still open		
	B. 15 Apr 07:00: One flower was treated with the fraction not precipitated by AE	16 Apr 07:00: Gynostemium swollen, flower closed 16 Apr 18:00: Both flowers closed, gynostemium unchanged		
	14 Apr 07:00. Two flowers treated as above	15 Apr 07:00: Gynostemium swollen, flower half closed		
	(The experiments are listed here in the same order as in Fitting's paper.)			
67 (dates as listed in next column)	*Aerides odorata* pollinia, 18 pairs, were extracted with cold water for 4 h, the extract was dried over a water bath and the residue was extracted several times with AE. The alcohol solution was dried over a water bath and produced glassy transparent residue. The AE soluble and AE insolble fractions were assayed on *Aerides* flowers		The active principle is less soluble in AE than in water (but changed his mind later) The active principle is not an enzyme	IAA is very soluble in ethanol and sparingly soluble in water

(continued)

	28 Feb 13:00: Cotton wads saturated with the AE soluble fraction which was redissolved in water were placed in stigmas of 4 flowers	29 Feb 06:00: Gynostemium swollen 1 Mar 07:00: The same		
	28 Feb 13:00: Cotton wads saturated with the AE insoluble fraction which was redissolved in water were placed in stigmas of 4 flowers	29 Feb 06:00: Gynostemium swollen 1 Mar 07:00: The same		
Several unnumbered experiments and observations	• *Phalaenopsis amabilis* pollinia which were soaked in AE for 4 hours retained their activity (but see below)		The active principle is not very soluble in AE	IAA is very soluble in ethanol
	• Extracts obtained by extracting *Vanda tricolor* and *Zygopetalum makayi* with AE for 24 hours were active.		The active principle is soluble in ethanol	IAA is ethanol soluble
	• A 24 h chloroform:ethanol (99:1) extract had some activity			IAA is not chloroform soluble,
	• Cold water extracts of *Phalaenopsis* and *Aerides* pollinia are weakly acidic			As the name indole acetic acid indicates, IAA is an acid
	• When pollinia extract was heated in a crucible it turned brown-black and smelled like burned carbohydrates			Extract may have contained sugar(s)
	• After the extract was heated to glowing the remnant was a very small traces of white ash			Not surprising
	• The brown-black residue after heating and the white ash are inactive			IAA and ACC were destroyed by the heating
	• Fehling's solution was not reduced by the extract even after it was hydrolyzed for an extended period with 1% or stronger hydrochloric acid			The extract did not contain reducing sugars, or not enough to be detected with this test
	• Filtered extract is a clear solution			
	• The extract did not contain droplets and solid particles			An aqueous extract would not contain particles
	• The dried extract residue dissolves easily in water			Not surprising since the original extraction was with water
	• There are no lipid droplets in the extract			To be expected, the original extract was aqueous. Lipids are not water soluble.

(continued)

Table 2-5. (continued).

Experiment number	Description of experiment and its results	Fitting's conclusions	Current explanation
	• Both the AE soluble and insoluble fractions of the *Phalaenopsis* extract produced a copious flocculent precipitate when treated with lead acetate (PbAc). The AE soluble and very active fraction of dried extract of *Aerides* pollinia did not form such precipitate on treatment with PbAc. The AE insoluble and water soluble fraction of the extract did form a precipitate when PbAc was added. This precipitate became shiny and crystalline in areas where the watch-glass was rubbed with a glass rod. As a heavy metal lead probably precipitated proteins	The active principle cannot be precipitated with PbAc. Cold water extracts of pollinia contain more than one substance. Fitting gave no reasons for his PbAc assay	IAA and ACC cannot be precipitated with PbAc. Pollinia do contain more than one substance The active principle is not an (1) inorganic salt (2) organic compound, (3) a common hydrolysable carbohydrate, (4) a fat This conclusion excludes just about all compounds that could be extracted from pollinia

[a] The orchid names in this table are those used by Hans Fitting. Please see Appendix 2 for updated nomenclature.

Table 2-6. Hans Fitting's experiments at the Bogor Botanical Gardens in 1908 on the effects of various substances on orchid flowers (Fitting, 1909a)[a].

Substance applied to stigma	Orchid	Results
Citric acid, 0.5%	*Aerides odorata*, 2 flowers	No effects
Dextrin, 2%	*Aerides odorata*, 4 flowers	No effects
	Phalaenopsis amabilis, 2 flowers	No effects
Diastase (starch hydrolyzing enzyme)	*Cymbidium finlaysonianum*, 2 flowers	No effects
	Rhynchostylis retusa, 6 flowers	No effects
Ferric ($FeCl_3$, waters of hydration not indicated) or ferrous chloride ($FeCl_2$, waters of hydration not indicated), Fitting does not indicate which, 1%	*Rhynchostylis retusa*, 6 flowers	No effects
Malic acid, 0.5%	*Aerides odorata*, 2 flowers	No effects
Manitol, 5%	*Aerides odorata*, 4 flowers	No effects
	Cymbidium finlaysonianum, 2 flowers	No effects
Oxalic acid, 0.5%	*Aerides odorata*, 2 flowers	No effects
Potassium nitrate (KNO_3), 10%	*Rhynchostylis retusa*, 6 flowers	No effects
Sodium carbonate, 1%	*Aerides* (probably) *odorata*, 3 flowers	No effects
Succinic acid, 0.01%	*Aerides odorata*, 2 flowers	No effects
0.5%	*Aerides odorata*, 2 flowers	No effects
Sucrose, 0.1%	*Aerides* (probably) *odorata*, 3 flowers	No effects
	Cymbidium finlaysonianum, 3 flowers	No effects
0.5%	*Aerides* (probably) *odorata*, 3 flowers	No effects
	Cymbidium finlaysonianum, 3 flowers	No effects
5%	*Aerides odorata*, 3 flowers	No effects
	Arachnanthe sulingi, 4 flowers	No effects
	Phalaenopsis amabilis, 1 flower	Flower wilted after two days
	Rhynchostylis retusa, 4 flowers	No effects
Tartaric acid, 0.5%	*Aerides odorata*, 2 flowers	No effects

[a] The orchid names in this table are those used by Hans Fitting. Please see Appendix 2 for updated nomenclature.

Table 2-7. Hans Fitting's experiments on the effects of live and dead pollinia from one genus or species on other taxa at the Bogor Botanical Gardens in 1908 (Fitting, 1909a)[a].

Experiment number	Pollen source	Orchids pollinated and description of experiment	Results	Fitting's conclusions	Current explanation
56 (no dates listed)	*Phalaenopsis amabilis* (Fig. 2-13A)	Pollen soaked in chloroform or subjected to steam placed in stigmas of *Aerides odorata*, *Dendrobium superbum*, *Cymbidium finlaysonianum*	Pollen had same effects on the three species as on *Phalaenopsis amabilis* in previous experiments	1. The effects of killed pollen are not species specific 2. Pollen which brings about swelling of its own gynostemium can have the same effect only on gynostemia whose own pollen can cause them to swell	Pollen brings about swelling of the gynostemium because it contains auxin and perhaps ethylene precursor(s) as well as possibly other hormones and/or induces production of ethylene and/or auxin and/or other substances by the stigma, rostellum and/or other parts of the column
		Filter paper soaked in pollen extract placed in stigma of *Aerides odorata*	Extract had the same effects as dead pollen of *Phalaenopsis amabilis* or *Aerides odorata*		
57 (no dates listed)	*Aerides odorata* (Fig. 2-9)	Pollen soaked in chloroform or subjected to steam placed in stigmas of *Phalaenopsis amabilis, Cymbidium finlaysonianum, Dendrobium superbum, Rhynchostylis retusa, Phalaenopsis violacea* and *Arachnanthe sulingii*	Same as in experiment 56	3. Pollen which does not bring about the swelling of its own gynostemium can cause the gynostemia of other species to swell if their own pollinia can do so 4. The active principle in pollen is widespread	
		Filter paper soaked in pollen extract placed in stigma of *Phalaenopsis amabilis* and *Cymbidium finlaysonianum*	Same as in experiment 56		
		Dead *Aerides* pollen placed in stigmas of *Spathoglottis filuata*, *Corymbis disticha* and *Eulophia macrostachya*	Same effect as live pollen of treated species, but no notable swelling of gynostemia		

(continued)

58 (no dates listed)	*Calanthe veratrifolia* (Fig. 2-12E) *Spathoglottis filuata*, *Phajus amboinensis* (Fig. 2-15A) and *Corymbis disticha* (Fig. 2-15B)	Live and steam-killed pollen placed in the stigma of *Aerides odorata*	Same effect as live pollen of treated species
59 (no date listed)	*Paphiopedilum glaucophyllum* (Fig. 2-15C)	Steam-killed pollen placed in stigma of *Phalaenopsis amabilis*	Same effect as live pollen of treated species

[a] The orchid names in this table are those used by Hans Fitting. Please see Appendix 2 for updated nomenclature.

Table 2-8. Hans Fitting's experiments at the Bogor Botanical Gardens in 1908 on the effects on orchid flowers of pollen from plants other than orchids (Fitting, 1909a)[a].

Experiment number	Description of experiment and its results		Fitting's conclusions	Current explanation
68	Cotton wads containing extracts of: (a) leaves, (b) perianth segments, (c) fruits and (d) pollinia were applied to stigmas of two flowers of *Aerides odorata*. This experiment was repeated twice.		Only extract d caused swelling of the gynostemium	All parts of the plant contain IAA. Fitting's extracts a-c may not have contained enough auxin
69	Cotton wads containing an extract of gynostemia without pollinia were were placed in stigmas of *Phalaenopsis amabilis*.			
	Two flowers were treated at 08:00 on 8 Apr	10 Apr 07:00: Flowers closed 11 Apr: Flowers wilted, gynostemium not swollen		
	One flower was treated at 07:00 on 9 Apr	10 Apr 18:00: Flowers closed 11 Apr 18:00: Flowers wilted, gynostemium not swollen		
70	Pollen of *Eucharis grandiflora* (Amaryllidaceae), *Alpinia hookeriana*, *Canna* sp. (Cannaceae), and *Hedychium* sp., (last two, (Zyngiberaceae) had no effects when applied to stigmas of *Aerides*			The pollen did not contain enough IAA and could not induce ethylene production
71	A. *Hedychium* sp. pollen was placed in stigmas of *Phalaenopsis amabilis*			
	1 Mar 07:00: One flower treated	2 Mar 07:00: Flower closed, stigma open 3 Mar 18:00 Flower wilted, stigma open	Since the pollen acted without germinating it is possible to assume that dead pollen will also be active	The pollen did not contain enough IAA, but could induce ethylene production
	2 Mar 06:00: One flower treated	3 Mar 06:00: Flower unchanged. 18:00: Flower closed, stigma open 4 and 5 Mar: Same		
	7 Apr 07:00: One flower treated	9 Apr 07:00: Flower unchanged 10 Apr 06:00: Flower closed 11 Apr: Flower closed, stigma open		

(continued)

	8 Apr 07:00: One flower treated	10 Apr 06:00: Flower closed 11 Apr: Flower closed, stigma open	
	B. *Hedychium* sp. pollen was placed in stigmas of *Phalaenopsis violacea*.		
	29 Feb 07:00: One flower treated	1 Mar 06:00: Flower starts to close 3 Mar 06:00: Flower closed, stigma open	
	7 Apr 07:00: One flower treated	10 Apr 06:00: Flower half closed and turning yellow	
	Pollen failed to germinate in both (A and B) sets of experiments, but shortened the life span of flowers of both species		
72	A. *Hedychium* sp. pollen which was subjected to steam for 5 min was placed in stigma of newly opened *Phalaenopsis amabilis* flower		
	7 Apr 07:00: Pollen applied	9 Apr 07:00: Flower unchanged 10 Apr 07:00: Flower starts to close 11 Apr 18:00: Flower wilted, stigma open	Pollen did not contain any or enough IAA to induce closing of the stigma, but it did bring about ethylene evolution
	B. *Hedychium* sp. pollen which was subjected to steam for 5 min was placed in stigma of newly opened *Phalaenopsis violacea* flower		
	7 Apr 07:00: Pollen applied	8 Apr 18:00: Flower started to turn yellow 10 Apr 06:00: Flower half closed and turned yellow 11 Apr 07:00: Flower completely closed, stigma open	
	The pollen shortened the life span of flowers of both species. Both the application of the pollen and the pollen itself did not wound the stigma		

(continued)

Table 2-8. (continued).

Experiment number	Description of experiment and its results	Fitting's conclusions	Current explanation
No number	Pollen of *Begonia geogensis* did not cause wilting of flowers of *Phalaenopsis* species		Pollen did not contain any or enough IAA to have an effect and did not induce ethylene evolution
73	*Impatiens rodrigesi* pollen was applied to stigmas of *Phalaenopsis violacea* flowers 7 Apr 07:00: Pollen applied — 8 Apr 06:00: Flower starts to turn yellow 9 Apr 07:00: Flower is yellow, starts to close 10 Apr 07:00: Flower half closed 11 Apr: Flower fully closed, stigma open Pollen germinated partially. Experiment could not be repeated due to lack of flowers		Pollen did not contain any or enough IAA to induce closing of the stigma, but it did bring about ethylene evolution
74	*Hibiscus rosa-sinensis* var. *genuinus* Hochr. pollen was applied to stigmas of *Phalaenopsis* flowers A. 7 Apr 08:00: Pollen applied to stigmas of three flowers of *Phalaenopsis amabilis* — 8 Apr 07:00: Tip of gynostemium and stigma swollen in one flower, all 3 flowers open 8 Apr 18:00: All 3 flowers closed 9 Apr 06:00: All 3 flowers closed Stigmas of 2 flowers open Pollen did not germinate	1. The pollen of *Hibiscus rosa-sinensis* shortened life span of flowers and caused stigmatic closure 2. Swelling of the gynostemium is less pronounced than after pollination with orchid pollen 3. The pollen did not germinate	This pollen contains a limited concentration of IAA and seems to induce low levels of ethylene evolution

(continued)

9 Apr 07:00: Pollen applied to stigma of one flower of *Phalaenopsis amabilis*	10 Apr 07:00: Flower open, stigma half closed 11 Apr 07:00: Flower half closed 12 Apr 07:00: Flower completely closed, stigma half closed
B. 7 Apr 08:00: Pollen applied to stigma of one flower of *Phalaenopsis violacea*	8 Apr 18:00: Stigma half closed, flower starts to close 9 Apr: Flower completely close. Pollen did not germinate
8 Apr 07:00: Pollen applied to stigma of one flower of *Phalaenopsis violacea*	8 Apr 18:00: Stigma almost completely closed 9 Apr 06:00: Stigma completely closed. Flower starting to turn yellow 10 Apr 07:00: Flower half closed 11 Apr 07:00: Flower fully closed
9 Apr 07:00: Pollen applied to stigmas of two flowers of *Phalaenopsis violacea*	10 Apr 07:00: One stigma half closed, the other slightly closed 11 Apr 07:00: Flowers yellowing, half closed. One stigma three quarter closed, the other half closed
9 Apr 07:00: Pollen applied to stigma of one flower of *Phalaenopsis violacea*	10 Apr 07:00: Flower unchanged 11 Apr 07:00: Flower yellowing half closed 12 Apr 07:00: Flower fully closed stigma open 13 Apr Same as above
15 Apr 07:00: Pollen applied to stigma of one flower of *Phalaenopsis violacea*	16 Apr 07:00: Flower unchanged 17 and 18 Apr 07:00: Stigma somewhat swollen. Stigma closed

(continued)

Table 2-8. (continued).

Experiment number	Description of experiment and its results		Fitting's conclusions	Current explanation
	C. 15 Apr 08:00: Pollen applied to stigma of one flower of *Phalaenopsis esmeralda*	16 and 17 Apr 07:00: Stigma half closed. Flower unchanged		
	13 Apr 08:00: Pollen applied to stigma of one flower of *Phalaenopsis esmeralda*	14 Apr 07:00: Stigma closed. Flower unchanged		
	The order in which these experiments are listed here is the same as in Fitting's original paper			
75	*Hibiscus rosa-sinensis* var. *genuinus* Hochr. pollen which was steamed for five minutes was applied to stigmas of *Phalaenopsis* flowers			
	A. 8 Apr 07:00: Pollen applied to stigmas of two flowers of *Phalaenopsis amabilis*	9 Apr 07:00: Flowers unchanged 10 and 11 Apr 07:00: Flowers closed. Stigmas closed 10 Apr 07:00: Flower open. Stigma half closed	Living and dead *Hibiscus* pollen have the same effect. *Hibiscus* pollen contains the same substance as orchid pollinia	
	B. 9 Apr 07:00: Pollen applied to stigmas of one flower of *Phalaenopsis violacea*	10 Apr 18:00: Flower yellowed. Stigma half closed 11 Apr 06:00: Both flower and stigma half closed		
No number	Sap of floral segments of *Hibiscus* applied to stigmas of flowers of *Phalaenopsis violacea* and *Phalaenopis esmeralda* had no effects			

[a] The orchid names in this table are those used by Hans Fitting. Please see Appendix 2 for updated nomenclature.

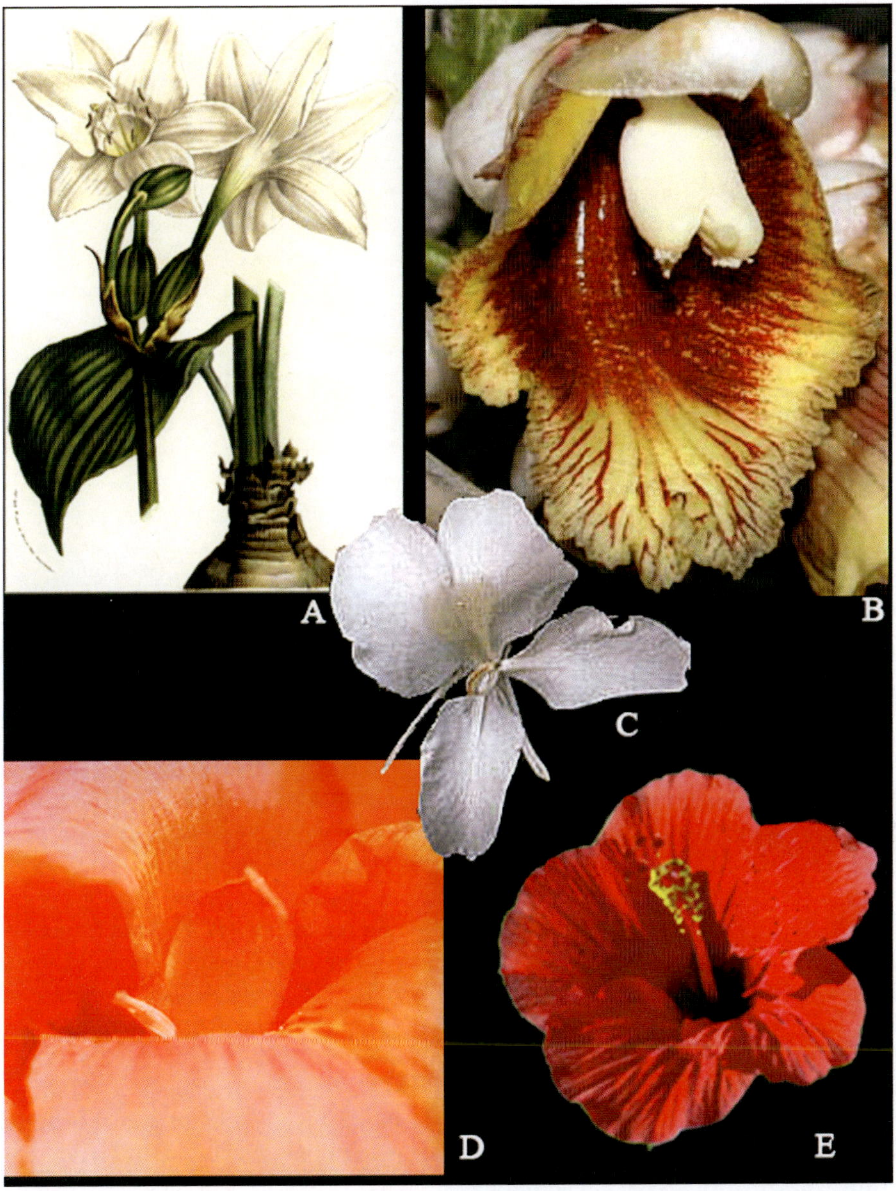

Fig. 2-16. Plants whose pollen was applied to orchids by Hans Fitting. **A.** *Eucharis grandiflora* Planch., Amarillidaceae. **B.** *Alpinia hookeriana*, Zingiberaceae. **C.** *Hedychium* sp., Zingiberaceae. **D.** *Canna* sp., Cannaceae showing anther. **E.** *Hibiscus rosa-sinensis*, Malvaceae (A, Lemaire, 1854; B, courtesy Aoki Shigenobu, Maebashi, Japan; C, E, J. Arditti; D, courtesy The Digital Flora of Texas).

Effects of the Location of Pollen Insertion

Fitting placed the pollen in stigmas in his initial experiments (Tables 2-1–2-8). In another series of experiments (Fig. 2-17; Table 2-9) he placed living and dead quarter pollinia sections inside gynostemia. Both the living and dead sections induced swelling of gynostemia and ovaries, stigmatic closure and wilting of the flowers regardless of where they were placed. Induction of these post-pollination phenomena is an indication that auxin and ACC diffused from the pollen section and that there was ethylene evolution. The latter could have been caused by auxin from the pollinia, the wounding of the gynostemium which must have been caused by the insertion of the sections, or ACC which diffused from the pollen.

The rostellum is a major site of ethylene evolution in pollinated orchid flowers (see Avadhani et al., 1994 for a review). Therefore it is not surprising that a pollinium section placed below it caused post-pollination phenomena.

A pollinium section placed in a stigma following removal of the stigmatic fluid also brought about post-pollination phenomena which are induced by auxin or ethylene. The latter is not surprising since removal of the fluid may have injured the stigma and caused ethylene evolution. Auxin-induced phenomena are more difficult to explain because movement of the hormone from the pollen section to the stigma requires a fluid. One possible explanation is that the stigma produced new fluid following the removal. Another possibility is that moisture from stigmatic cells facilitated diffusion of auxin from the pollen. It is also possible that Fitting did not remove all the fluid from the stigma.

These experiments were followed by studies of the nature of the active principle and lead Fitting to suggest that the active substance in orchid pollinia was a hormone (Table 2-5) which came to be known as *Pollenhormon* (Fitting, 1909a, b, c, 1910, 1911, 1912, 1921, 1936).

Fitting's experiments required a very large number of flowers. Even in Bogor he could not find enough flowers of any one orchid and used many species (Tables 2-1–2-4). He liked to work with *Phalaenopsis* (Figs. 2-5, 2-6, 2-10) and *Cymbidium* (Fig. 2-8) because the flowers are large and the phenomena are easy to observe. On his return to Germany he repeated some of the experiments with another large-flowered orchid with easily observable post-pollination phenomena, *Cattleya* (Fitting, 1910). After his experiments with *Cattleya* (Figs. 2-15, 2-18, 2-19) Fitting seems to have lost interest in orchids and pursued other avenues of research. He never worked with orchids again although he did mention them in several papers (Fitting, 1911, 1912, 1921, 1936).

One reason why Fitting abandoned orchids suggested by Professor Frits Warmolt Went (1903–1990) is that "he was too much steeped in the 'stimulus' concept" (letter dated June 19, 1974 to J. A.). Because of that Went concluded that Fitting's "work on orchid pollination, which could easily have led to an explosion on plant hormone work, did not have any lasting effect." In a way Went

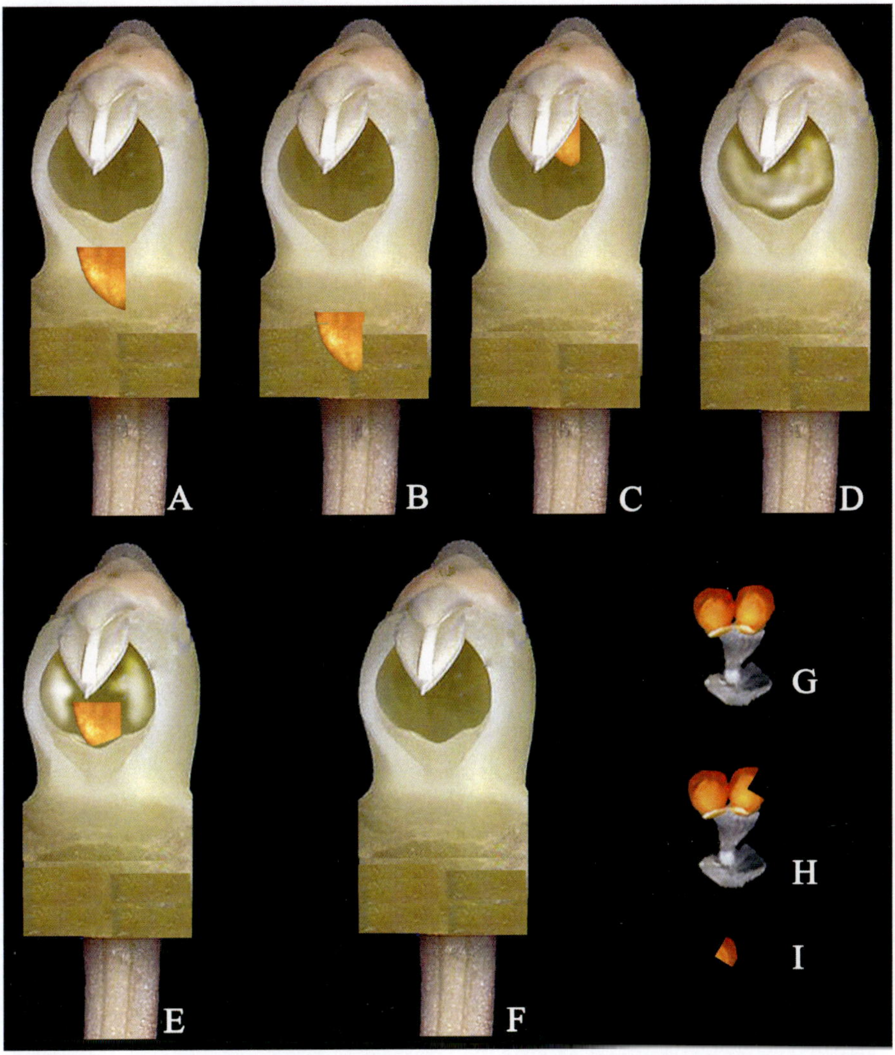

Fig. 2-17. Approximations of where Fitting placed pollinia quarters in gynostemia. The pollinia quarters were placed on the outside of the gynostemia in this illustration so that they and their approximate locations can be seen. Fitting placed them inside where they would not be visible except in sections. **A.** One quarter of living pollinium inserted deep into stylar canal without disturbing the stigma. **B.** One quarter of steam-killed pollinium inserted deep into stylar canal down to the region of the attachment of perianth segments. **C.** One quarter of steam-killed pollinium inserted just below the rostellum (only the lower portion of the quarter pollinium is showing). **D.** Stigmatic fluid removed from stigma. **E.** One quarter of steam-killed pollinium placed in stigma following removal of the stigmatic fluid. **F.** Intact unpollinated gynostemium. **G.** Intact pollinarium. **H.** Pollinarium with one quarter of a pollinium removed. **I.** One quarter of a pollinium (Computer generated from existing images by J. Arditti).

Table 2-9. Hans Fitting's experiments at the Bogor Botanical Gardens in 1908 on the effects on orchid flowers of inserting pollen in the gynostemium in locations other than the stigma (Fitting, 1909a)[a].

Experiment number	Description of experiment and its results		Fitting's conclusions	Current explanation
76	One quarter of a living pollinium was inserted deep into the stylar canal without disturbing the stigma			
	13 Feb 11:00: Pollinia inserted in gynostemia of four flowers	14 Feb 06:00: Stigmas start to close 14 Feb 11:00: Stigmas closed Perianth starts to close 14 Feb 14:00 (original paper lists this erroneously as 06:00, or it could be 15 Feb 06:00): Gynostemium swollen, flowers closed 16 Feb 11:00: Ovaries start to swell and turn green	1. Pollen, dead or alive need not be placed in the stigma to be active. It can be active even if placed in the stylar canal 2. Pollen can be active even if placed in the tip of the stigma 3. The influence which emanates from the stigma and affects the perianth is the same as the one which results from wounding	IAA in the inserted pollinia was effective in/from the insertion site and was also transported to a site where it initiated ethylene evolution Both pollination and wounding induce ethylene evolution
77	One quarter of a pollinium killed by steaming inserted deep into the stylar canal of *Phalaenopsis amabilis* down to the region of the attachment of the perianth segments without disturbing the stigma		4. The influence is produced even by pollen placed in stigmas after the removal of the stigmatic fluid	IAA diffused from pollen into the stigma even in the absence of stigmatic fluid
	A. 18 Feb 10:00: Pollinium inserted in gynostemium of one flower	19 Feb 07:00: Gynostemium tip and stigmatic area swollen. Flower half closed 19 Feb 12:00: Flower fully closed 21 and 22 Feb 06:00: Flower wilted. Ovary not swollen	5. Dead pollen can cause swelling of the ovary when placed deep in stylar canal close to the ovary	IAA from the pollen diffuses down into the ovary and causes the swelling

(continued)

	B. 28 Feb 08:00: Pollinium inserted in gynostemium of one flower	28 Feb 18:00: Flower starts to close 23 Feb: Flower abscised 29 Feb 06:00: Flower fully closed. Stigma starts to close 1 Mar 07:00: Flower wilted. Stigma almost fully closed 3 Mar 07:00: Ovary not swollen and wilted
	C. 28 Feb 08:00: Pollinium inserted in gynostemium of one flower	29 Feb 06:00-1 Mar 07:00: Same as in B above
78	A. 27 Feb 11:00: One quarter of a steam-killed pollinium inserted just below the rostella in stigmas of two *Phalaenopsis* flowers	28 Feb 07:00: Stigma half closed. Flower starts to close 28 Feb 18:00. Stigma and flower fully closed 3 Mar 06:00: Flower abscised
	B. 24 Feb: The stigmatic fluid was removed from the lowermost parts of the stigmas of two (probably *Phalaenopsis*) flowers	27 Feb 11:00: Flowers still fresh
	27 Feb 11:00: One quarter of a steam-killed pollinium was placed in the stigma	28 Feb 07:00: Flower and stigma start to close 28 Feb 18:00 (original paper states erroneously 06:00): Flower and stigma fully closed 3 Mar: Flower abscised

[a] The orchid names in this table are those used by Hans Fitting. Please see Appendix 2 for updated nomenclature.

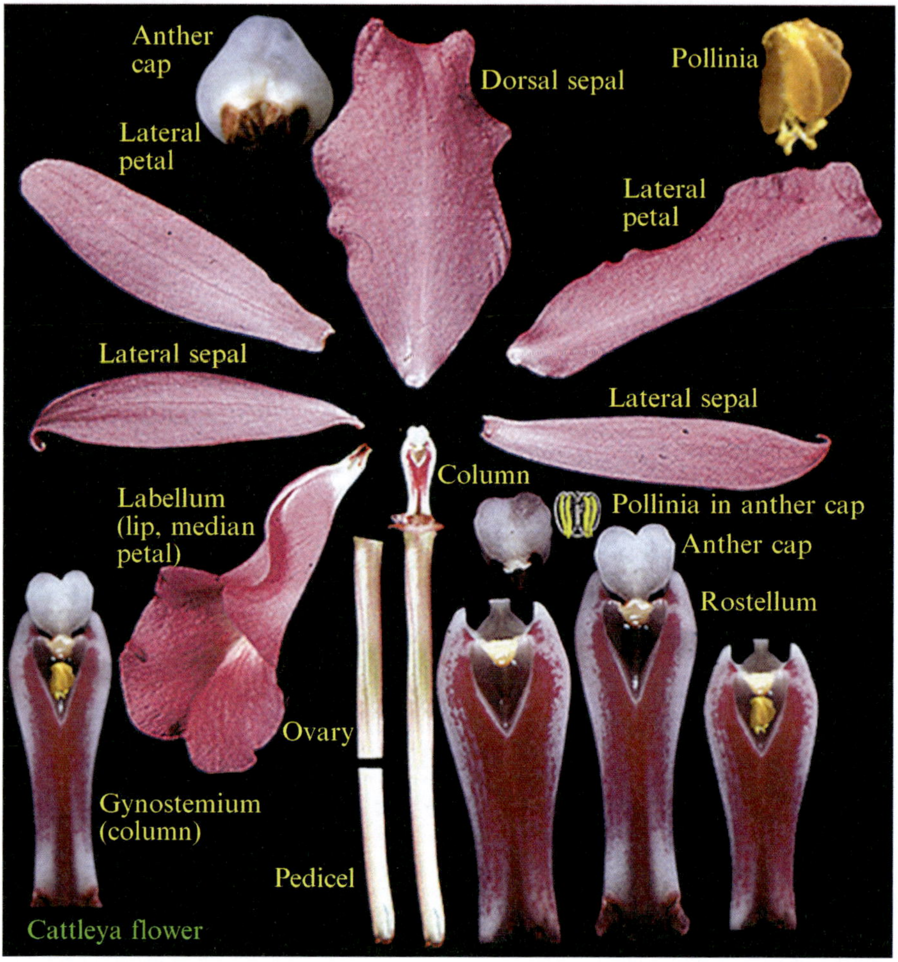

Fig. 2-18. *Cattleya* flower and its parts (Photographs and montage by J. Arditti).

was right because: (1) save for one paper from Japan (Morita, 1918), (2) a short period of work (eight papers) in Germany by Friedrich Laibach (1885–1967; Fig. 2-25) and his associates (1930–1934, see below), (3) research on ovulation (Heslop-Harrison, 1957; Magli, 1958; Dolcher, 1961a, b, 1967) which was only tangentially related to his research, and (4) several references in the first book on plant hormones (Went and Thimann, 1937), Fitting's work on orchids lay almost forgotten until one of us (J. A.) became interested in post-pollination phenomena of orchids (for a review see Avadhani et al., 1994).

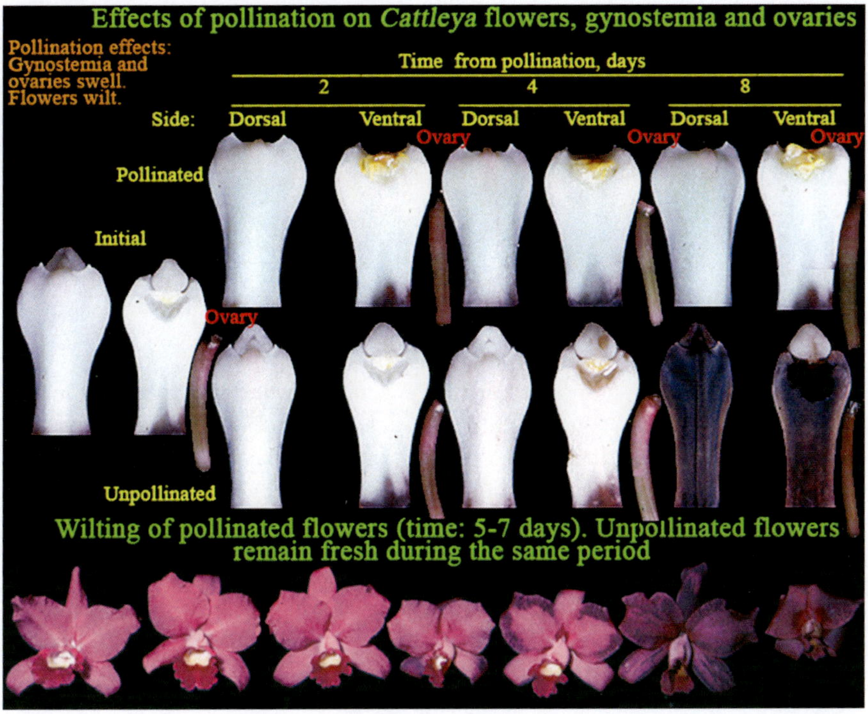

Fig. 2-19. Post-pollination phenomena in *Cattleya* (*Guarianthe*) flowers as Hans Fitting probably observed them. (source: the photographs were taken as part of research on post-pollination phenomena in orchid flowers in J. Arditti's laboratory; for a review see Avadhani et al., 1994).

In a conversation with one of us (J. A.) Went indicated that he did not make a connection between Fitting's research and his own work with *Avena* coleoptiles which led to the discovery of auxin. This is easy to understand because there is no obvious connection between the bending of coleoptiles and swelling of gynostemia and wilting of flowers. The connection was established only after auxin was discovered and its effects were studied extensively.

Ovary Swelling and Greening of Perianth

Post-pollination phenomena in orchid flowers are expressed in the: (1) perianth (sepals and petals) which move, senesce, and change color and structure and/or function, (2) gynostemium which swells, (3) stigma which closes, and (4) ovary which swells while the ovules develop internally. Fitting observed and describes all of these and attempted to determine if one can be induced without the other, especially 1–3 (Tables 2-1–2-9). However, he was also interested in the swelling of the ovary (4 above) and the effects of pollen on it. To study this effect he pollinated flowers with living and dead pollen (Table 2-10) and concluded that pollinia can cause swelling of ovaries in some flowers, but not in others and that pollen tubes may be required for an ovary to swell. These conclusions were not entirely accurate (for a review see Avadhani et al., 1994).

Confirming Experiments

Fitting's experiments with *Cattleya, Odontoglossum* and *Zygopetalum* and European orchid flowers (Figs. 2-20, 2-21) on his return from Bogor involved pollination with their own and each other's living and dead pollen. The results of these experiments confirmed his findings in Bogor. *Zygopetalum mackayi* flowers exhibited phenomena similar to those of *Phalaenopsis violacea* in that perianth segments turned green and persisted on the fruit (Fig. 2-12, 2-13). It is reasonable to assume that the green segments carry out photosynthesis. Other studies carried out by Fitting in Strasbourg were designed to characterize the active principle in orchid pollen. They were physiological and chemical in nature, ahead of their time and modern (Table 2-11).

Classification of Phenomena

After concluding his experiments in Strasburg, Fitting separated the pollination induced phenomena into four categories and listed the factors which bring about or affect them (Table 2-12, 2-13). This classification is still valid at present, but it has been augmented by research since then (for a review see Avadhani et al., 1994).

Hormone

Fitting apparently read widely and seems have been familiar with literature on subjects other than plants because he knew of a suggestion by the British physiologist Ernest Starling (1866–1927) that substances he called hormones affect development in animals (Table 2-14). He concluded correctly that "the active substances in pollinia are hormones" (Fitting, 1910; Table 2-14). *Pollenhormon*, became a name associated with the hormone in orchid pollen. More recent research showed that Fitting was only partially right because orchid pollen contains (Fig. 2-24D) more than one hormone. It does contain a high concentration

Table 2-10. Hans Fitting's experiments at the Bogor Botanical Gardens in 1908 on swelling of orchid ovaries and greening of the perianth (Fitting, 1909a)[a].

Experiment number or observation	Description of observation of experiment and its results		Fitting's conclusions	Current explanation
Swelling of the ovary	Pollinia dead or alive do not generally cause noticeable swelling of ovaries in: *Coelogyne swaniana, Cymbidium finlaysonianum, Phalaenopsis amabilis, Phalaenopsis cornu-cervi, Phalaenopsis violacea* and *Stanhopea* sp. The only exception occurred in experiment 76: Ovaries of *Phalaenopsis amabilis* started to swell		Pollinia can cause swelling of ovaries in some flowers, but not in others. A pollen tube may be required for ovary swelling in the latter	Live pollinia which produce or cause the production of IAA may be required for ovary swelling in some orchids
Swelling of the ovary	*Arachnanthe sulingi* was pollinated with dead pollinia (a, ovary of unpollinated flower; b, ovary of flower pollinated with dead pollinia)	Ovary (b) was swollen 7 days after pollen was placed in stigma		
Swelling of the ovary	*Rhynchostylis retusa* was pollinated with dead pollinia (a, ovary of unpollinated flower; b, ovary of flower pollinated with dead pollinia)	Ovary (b) was swollen 8 days after pollen was placed in stigma		
Swelling of the ovary	*Aerides odorata* was pollinated with dead pollinia (a, ovary of unpollinated flower; b, ovary of flower pollinated with dead pollinia)	Ovary (b) was swollen 6 days after pollen was placed in stigma		
Swelling of the ovary	One flower of *Arachnanthe sulingi* was pollinated with it own dead pollinia (a). Another flower was pollinated with dead pollinia of *Aerides odorata* (b)	Ovary of *Arachnanthe sulingi* pollinated with dead pollinia of *Aerides odorata* (b) became more swollen than the ovary of a flower pollinated with its own pollen (a)		

(continued)

Table 2-10. (continued).

Experiment number or observation	Description of observation of experiment and its results		Fitting's conclusions	Current explanation
Swelling of the ovary	One flower of *Rhynchostylis retusa* was pollinated with it own dead pollinia (a). Another flower was pollinated with dead pollinia of *Aerides odorata* (b)	Ovary of *Rhynchostylis retusa* pollinated with dead pollinia of *Aerides odorata* (b) became more swollen than the ovary of a flower pollinated with its own pollen (a)		
76	A plug made of cotton was inserted into the stylar canals of two *Phalaenopsis amabilis* flowers and pushed in deeply to form an obstacle for the growth of pollen tubes. The flowers were pollinated after that		Pollen tubes must reach ovaries to cause swelling	That or at least auxin must do so
	13 Feb 11:00: Flowers pollinated with living pollinia	14 Feb 11:00: Flowers half closed. Stigmas swollen 16 Feb 11:00: Flowers wilted. Ovaries not swollen 19 Feb 06:00: Ovaries wilted and neither green nor swollen. Pollen tubes reached but did not grow past the cotton plugs		
Greening of perianth	Dead pollinia or pollen extract did not bring about the greening of the perianth in *Phalaenopsis violacea* and *Phalaenopsis cornu cervi*. The flowers closed, turned yellow, wilted and abscised. In *Phalaenopsis violacea* the flower closes and perianth segments turn and seem to be wilting before they start to turn green 5–6 days after pollination			
Pollination effects	Pollen of *Cattleya labiata* had the same effects on *Phalaenopsis violacea* and *Phalaenopsis cornu-cervi* as their own dead pollen. Ovary did not become swollen and perianth did not turn green			

[a] The orchid names in this table are those used by Hans Fitting. Please see Appendix 2 for updated nomenclature.

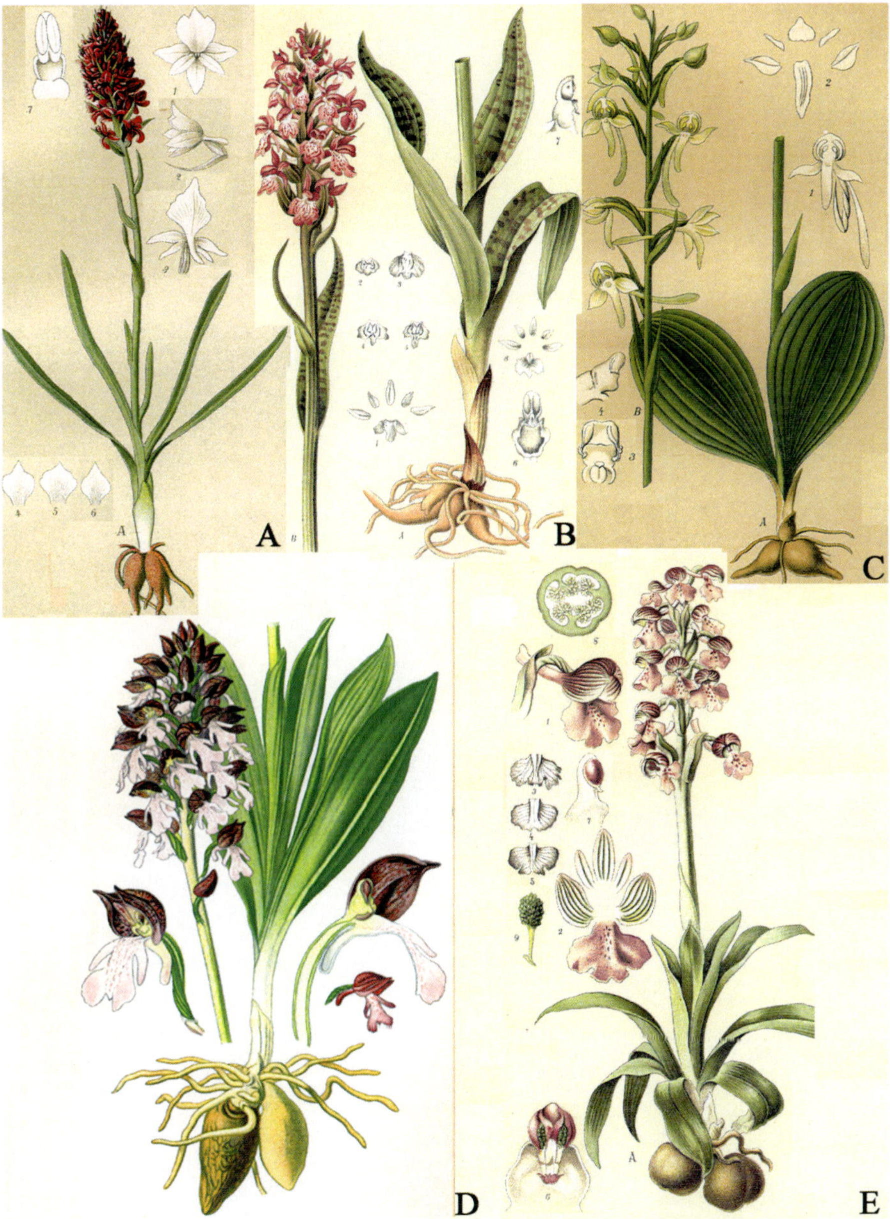

Fig. 2-20. European orchids used by Hans Fitting in his experiments in Strasbourg. **A.** *Gymnadenia conopea*. **B.** *Orchis latifolia*. **C.** *Platanthera bifolia* **D.** *Orchis fusca*. **E.** *Orchis morio* (sources: A–C, Schulze, 1894; D, Correvon, 1899).

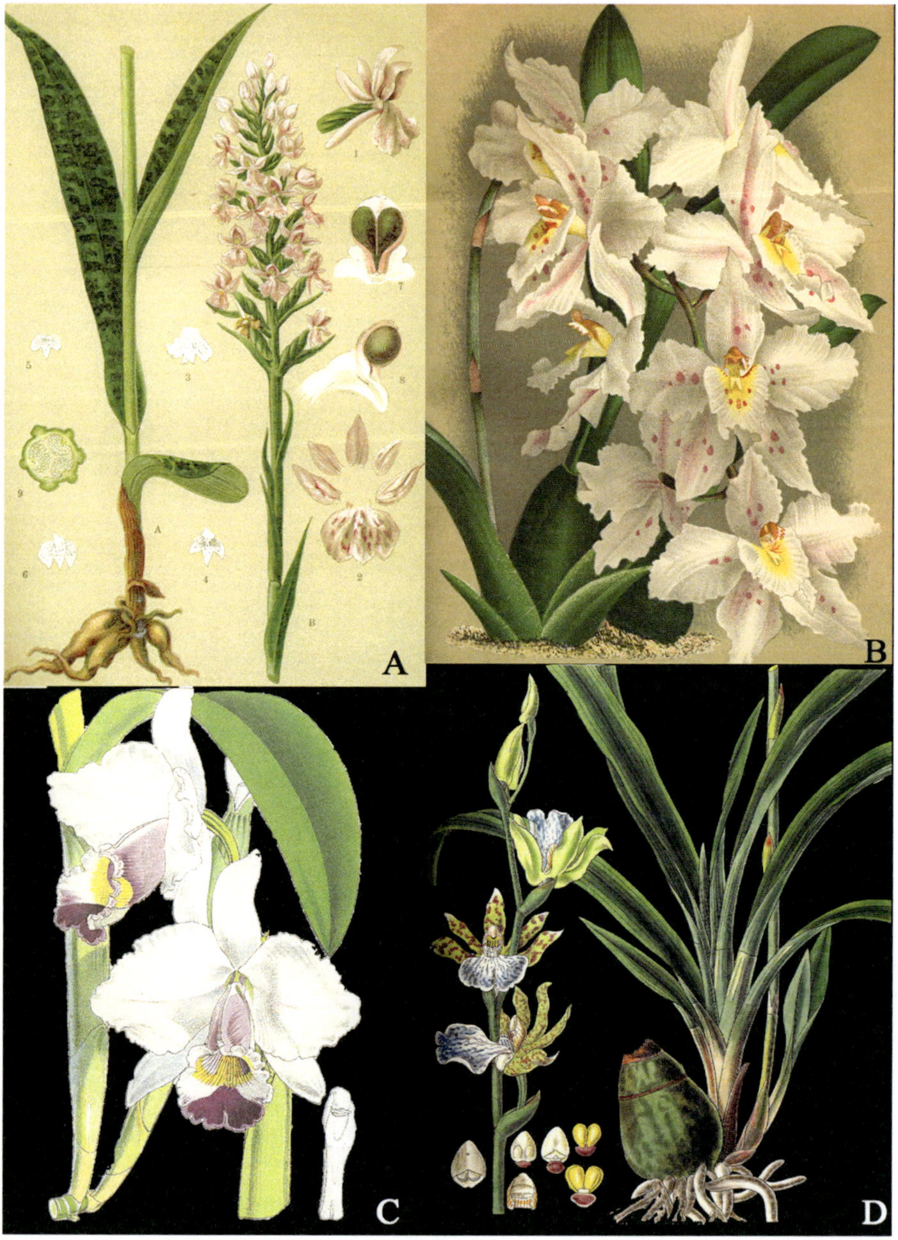

Fig. 2-21. Tropical and European orchids used by Hans Fitting for his experiments in Strasburg. **A.** *Orchis maculata*. **B.** *Odontoglossum crispum*. **C.** *Cattleya trianei*. **D.** *Zygopetalum mackayi* [sources: A, Schulze, 1894; B, a collection of lithographs from the 1800s owned by J. Arditti; C, D, plates 5504 (1865) and 2748 (1827) respectively from the Curtis Botanical Magazine].

Table 2-11. Treatments and/or factors which bring about and/or affect post pollination phenomena in orchids (Fitting, 1909a)[a].

Phenomena	Factors which affect it and/or induce phenomena
I. Premature aging of flower (i.e., reduced life span)	Sand in the stigma; saliva in the stigma of *Rhynchostylis*; dead pollen from the species being treated or another species; alcohol soluble and alcohol insoluble fractions of aqueous pollen extract; dead pollen extracted with water several times in *Phalaenopsis violacea*; dead or living pollen of *Hedychium* (Zingiberaceae), *Hibiscus* (Malvaceae) and *Impatiens* (Balsaminaceae) in Phalaenopsis
	Gynostemium extract, 5% sucrose and wounding of the gynostemium and/or stigma in *Phalaenopsis amabilis*
II. Stigmatic closure and swelling of the gynostemium	Living or dead orchid pollinia from the species being tested or other orchids selected at random; alcohol soluble fraction of water extract of pollinia; living or dead pollinia of *Hibiscus* (Malvaceae)
III. Swelling of the ovary	Dead pollinia or pollen extract cannot cause swelling of the ovary in *Cymbidium finlaysonianum. Coelogyne swaniana, Phalaenopsis amabilis, Phalaenopsis cornu-cervi, Phalaenopsis violacea* and brought about only limited swelling in *Aerides odorata, Arachnanthe sulingi* and *Rhynchostylis retusa*. Only when inserted very deep in the stylar canal and close to the ovary did dead pollen cause some swelling
IV. Greening of the perianth	This occurs only in some species and is caused by living pollen

[a] The orchid names in this table are those used by Hans Fitting. Please see Appendix 2 for updated nomenclature.

of auxin (Fig. 2-24B), but it also contains ACC (Fig. 2-24C). There is also evidence that orchid pollen contains gibberellins and cytokinins (for a review see Avadhani et al., 1994).

Kôichi Morita and Japanese Orchids

If Fitting's research on orchid pollination generated any interest it was not enough to spur research by others for a number of years. However, eight years after his last research paper on orchid pollinia (Fitting, 1910) a Japanese investigator, Kôichi Morita (we could find no information about him) published a paper describing research which repeated and extended Fitting's experiments with *Cymbidium virens* Lindl and other Japanese orchids (Morita, 1918; Fig. 2-22, 2-23, Tables 2-15, 2-16). He reported that pollination extended the life span of *Cymbidium virens* (this name was introduced by Reichenbach and referred erroneously to *C. virescens* Lindl.,

Table 2-12. Observations on flowering, pollination and post-pollination phenomena in orchids (Fitting, 1909b)[a].

Orchid or process	Observation or terminology
Active principle in pollen	Soluble in hot water [IAA is sparingly soluble in water, more so in hot than in cold]. It is found on the surface of the pollen [actually IAA is inside the pollen grains]. There are two substances in the cold water extract. One, alcohol insoluble, brings about swelling of the gynostemium and shortens the life span of flowers. The other, alcohol soluble, causes wilting of *Phalaenopsis amabilis* flower, but not swelling of the gynostemium [IAA and ACC are alcohol soluble]. Both retain their activity after prolonged boiling. Characteristics of the alcohol insoluble fraction are: (1) occurs inside the anther, (2) organic compound, (3) easily soluble in hot or cold water [ACC is, IAA is not], (4) not easily soluble in alcohol [both IAA and ACC are alcohol soluble], (5) cannot be precipitated from aqueous solution with alcohol, (6) is heat resistant, (7) does not reduce Fehling's solution even after being heated with hydrochloric acid, (8) does not precipitate with lead acetate and according to other assays does not contain nitrogen, (9) not an enzyme, (10) could not be replaced by citric acid, dextrin, diastase, malic acid, oxalic acid, succinic acid, 5% sucrose or tartaric acid. The same active principle is also present in pollen of non orchidaceous plants
Aerides odorata	Pollination shortens life span. Volcanic windblown sand placed in the stigma causes rapid closing and wilting of the flower. For effects of dead pollinia see "Pollen" entry in this table
	Gynostemium became swollen when larvae ate a hole in the stigma
Arachnanthe sulingi	For effects of dead pollinia see "Pollen" entry in this table
Calanthe	Ovary swells when parasitized by larvae of a gall causing insect (Fitting quotes Forbes, 1885 on this)
Coelogyne swaniana	For effects of dead pollinia see "Pollen" entry in this table
Stanhopea sp.	For effects of dead pollinia see "Pollen" entry in this table
Cymbidium finlaysonianum	For effects of dead pollinia see "Pollen" entry in this table
Dendrobium crumenatum	Flowers last for a day
Dendrobium superbum	Flowers last 14 days. Pollination shortens life span. Volcanic windblown sand placed in the stigma causes rapid closing and wilting of the flower (wounding effect?)
Floral segments and stigma	"There is a correlative link of unknown nature between the 'Wundreiz' (wound stimulus) and the other floral segments which brings about some of the changes." Swelling and closing of the stigma is the result of direct contact. [These are Fitting's conclusions. Actually IAA is transported from the stigma into the gynostenium perhaps even into the ovary. In addition IAA probably diffuses from pollen tubes.]
Floral senescence following pollination	Pollination induced floral senescence and other processes is "Autonome Postflorationvorgange" (induced post anthesis processes). Note: Later Fitting called these "Aitionom" processes

(continued)

Floral senescence without pollination	Natural flower senescence and other processes is "Induzierte Postflorationvorgange" (autonomic post anthesis processes)
Gynostemium	Removal of gynostemium above the base of the stigma has no effect of the life span of flowers Removal of gynostemium tissue on ventral side above the stigma has no effect of the life span of flowers Deep wounding which reached vascular or stigmatic tissue did reduce the life span of flowers
Larvae, insect	See *Aerides odorata* in this table
Liparis latifolia	Ovary swells when parasitized by larvae of a gall causing insect (Fitting quotes Treub, 1883 on this)
Oncidium incurvum	Pollination shortens life span. Volcanic windblown sand placed in the stigma causes rapid closing and wilting of the flower
Orchis mascula	Ovary swells after the flower is pollinated with *Cypripedium parviflorum* pollen (Fitting quotes Hildebrand, 1865 and Strassburger, 1866 on this)
Orchis morio	Ovary swells after the flower is pollinated with *Frtillaria* pollen (Fitting quotes Hildebrand, 1865 and Strassburger, 1866 on this)
Ovary	Swells even when flower is pollinated with pollen which cannot fertilize ovules
Ovary	Wilts and abscises with the blossom if the flower is not pollinated
Ovules	Not present in the ovary at the time of pollination (Fitting quotes Hildebrand, 1863a, 1863b, 1868 on this
Perianth and stigma	See "Floral segments and stigma" in this table
Perianth segments	Greening occurs only after ovaries start to swell. Cannot be brought about by dead pollen
Phalaenopsis amabilis	Flowers last for a month. Pollination shortens life span. Volcanic windblown sand placed in the stigma causes rapid closing and wilting of the flower. Removal of gynostemium above the base of the stigma has no effect of the life span of flower. For effects of dead pollinia see "Pollen" entry in this table
Phalaenopsis cornu-cervi	Flowers last for a month. Pollination shortens life span. For effects of dead pollinia see "Pollen" entry in this table
Phalaenopsis esmeralda	Follination shortens life span. Volcanic windblown sand placed in the stigma causes rapid closing and wilting of the flower
Phalaenopsis violacea	Flowers closes 1–2 days after pollination, turns yellow and starts to wilt. Then the ovary starts to swell and turns green. At this point wilting of yellowed perianth stops and starts to turn green. Volcanic windblown sand placed in the stigma causes rapid closing and wilting of the flower. For effects of dead pollinia see "Pollen" entry in this table

(continued)

Table 2-12. (continued).

Orchid or process	Observation or terminology
Pollen	Pollen causes its effects through a chemical agent. Dead pollen (killed by steaming or soaking in chloroform) can induce stigmatic closure and swelling of the gynostemium and even some, but not marked swelling and elongation of the ovary in *Aerides odorata. Arachnanthe sulingi* and, *Rhynchostylis retusa*. In *Cymbidium finlaysonianum*, *Coelogyne swaniana*, *Stanhopea* sp., *Phalaenopsis amabilis*, *Phalaenopsis cornu-cervi*, *Phalaenopsis violacea* and *Zygopetalum makay*i stigmatic closure and swelling of the gynostemium are not accompanied by swelling and elongation of the ovary. To have an effect on ovaries pollen must germinate and produce tubes which enter the ovary. For details about the active principle see "Active principle in pollen" in this table
Pollen tubes	For pollen tubes to have an effect on the ovary they must enter it. For details about the active principle see "Active principle in pollen" in this table
Pollinia	Some pollinia are more active that others
Rhynchostylis retusa	Flowers last for a month. Pollination shortens life span. Volcanic windblown sand placed in the stigma causes rapid closing and wilting of the flower. Removal of gynostemium above the base of the stigma has no effect of the life span of flower. For effects of dead pollinia see "Pollen" entry in this table
Sand	When volcanic blown sand was placed in stigmas the life span of flowers was reduced
Several	Perianth turns green after pollination and remains on the fruit until it ripens
Several	Pollination shortens life span of flowers
Several	Stigma closes 1–2 days after pollination
Stigma and floral segments	See "Floral segments and stigma" in this table
Stigmatic fluid or secretion	Removal by wiping it away reduces the life span of flowers
Vanda tricolor	Pollination shortens life span. Volcanic windblown sand placed in the stigma does not cause closing and wilting of the flower
Water (hot) extract of pollen	Active
Wounding	Incisions, scratches or punctures of the stigma shorten the life span of flowers. Wounding of the gynostemium has a minor effect unless it is severe. Deep wounding which reached vascular or stigmatic tissue did reduce the life span of flowers. Wounding causes its effects through the formation of a "Wundreiz" (wound stimulus)
"Wundreiz"	Wound stimulus caused by wounding of the stigma which reduces the life span of flowers and causes other effects
Zygopetalum makayi	For effects of dead pollinia see "Pollen" entry in this table

[a]The orchid names in this table are those used by Hans Fitting. Please see Appendix 2 for updated nomenclature.

Table 2-13. Flowering, pollination and post-pollination phenomena in orchids (Fitting, 1910)[a].

Orchid or process	Observation, terminology and comments by Fitting or [the authors of this review]	
Active principle	Fitting carried out very elaborate (for the time) experiments on the characteristics of the active principle and reported that it is: (1) insoluble in chloroform [neither is IAA], (2) more easily soluble in water than in ethanol [the reverse is true for IAA, but Fitting may have isolated a mixture of IAA and other substances including ACC], (3) insoluble in ethyl and petroleum ether, (4) neither a reducing sugar not a polymer which produces reducing sugars on hydrolysis, (5) not a lipid, resin, wax, cholesterol, carbohydrate, glycoside, tannin, mucilage, enzyme, protein, an acid which is soluble in water, but not in alcohol, (6) not a salt, and (7) a nitrogen containing substance, (8) leached from pollinia suspended in water, but some activity remains in the pollen, (9) present, produced and releases by pollen tubes, (10) located on the surface of pollen grains inside the pollinia, (11) produced by pollen tubes	
Cattleya trianae	The active fraction of pollinia extract has the same characteristics as the active fractions of extracts Fitting obtained in Bogor in 1908	
Experiments with German native orchids to determine if pollen of temperate climate species has the same effects as that of those from the tropics		
Gymnadenia conopea	Four flowers wilted in 16 days if unpollinated and in 13 days if pollinated	These are experiments Fitting carried out in Strassburg after he returned from Indonesia
Orchis fusca	Four flowers wilted in 16 days if unpollinated and in 7 days if pollinated	
Orchis fusca	Four flowers wilted in 12 days if unpollinated and in 6 days if pollinated	
Orchis latifolia	Four flowers wilted in 14 days if unpollinated and in 7 days if pollinated	
Orchis latifolia	Four flowers wilted in 14 days if unpollinated and in 8 days if pollinated	
Orchis maculata	Four flowers wilted in 13 days if unpollinated and in 6 days if pollinated	
Orchis maculata	Four flowers wilted in 14–16 days if unpollinated and in 10 days if pollinated	
Orchis morio	Four flowers wilted in 13 days if unpollinated and in 6 days if pollinated	
Orchis morio	Four flowers wilted in 10 days if unpollinated and in 6 days if pollinated	
Platanthera bifolia	Four flowers wilted in 17 days if unpollinated and in 15 days if pollinated	
Injections of water or pollen extract into *Odontoglossum*	Both water- and extract-injected flowers wilted in 10–12 days and before untreated flowers. The extract did not cause swelling of the gynostemium. [The wilting is not surprising because it could have been due to wound ethylene. However, the lack of swelling is if the extract contained IAA.]	
Odontoglossum crispum	Gynostemium swells when pollinated with dead pollinia	

(continued)

Table 2-13. (continued).

Orchid or process	Observation, terminology and comments by Fitting or [the authors of this review]
Pollen	Both live and dead pollen shorten the life span of flowers, but living pollinia have a more rapid and more pronounced effect in some cases. Fitting concluded that the pollen contains a chemical principle
Pollinia	Pollinia of *Epipactis palustris, Orchis latifolia* and *Orchis mascula* brought about swelling of the gynostemium and shortening of the life span in flowers of *Oncidium sphacelatum* and *Oncidium sphegiferum*, but not *Orchis* species. Also pollinia of *Paphiopedilum callosum* brought about post-pollination phenomena in *Phalaenopsis amabilis* and pollinia *Paphiopedilum barbatum* had the same affect on *Oncidium sphacelatum*
Tropical and German native orchids	There are four main phenomena which are induced by pollination: (I) Closing and wilting of perianth, (II) Closing of stigma and swelling of gynostemium, (III) Swelling of ovary, and (IV) Greening of perianth segments [this is limited to a few species]. Pollination also reduces the life span on flowers. Gynostemium turns green in many species. The effects of pollination on German native orchids are not pronounced
Zygopetalum mackaii and other species	Pollination with live pollen increases the life span of perianth segments by causing them to change into leaf like photo-synthetic structures which remain on the ovary until it ripens. Dead pollen and wounding of the stigma do not have this effect

[a] The orchid names in this table are those used by Hans Fitting. Please see Appendix 2 for updated nomenclature.

Table 2-14. Fitting's proposal that plants produce hormones (Fitting, 1910) with comments by the authors of this review between brackets.

Original German	English (free) translation
Starling[a] ... [mentioned but not cited, but see below] hat für das Tier vorgeschlagen alle derartigen Stoffe, die im eigenen Stoffwechsel des Organismus erzeeugt, ohne Nahrugsstoffe zu sein als Reizstoffe die Enwicklung oder sonstige Lebenstätigkeit des Organismus beeinfluseen mit dem Namen Hormone (von ό□μάω ich reize, rege an) zu bezeichnen. Ich möchte vorschlagen, diesen Terminus auch für die Pfanzen zu verwenden ... In diesem Sinne also wären wohl die Reizstoffe der ungekeimten Pollinien echte Hormone	Starling[a] [see below] suggested for animals substances which are not nutrients and function as stimuli for development or other life functions. He named them hormones (from ό□μάω to stimulate, to activate). I would like to suggest the same term for plants ... In this sense the stimulating [i.e., active] substances in the pollinia are true hormones. [This is the first use of "hormone" relative to plants]

[a] Ernest Starling (1866, London-1927, on a ship near Kingston, Jamaica), An English physiologist is credited with coining the term hormone (from the Greek horman which means to set in motion) in The Croonian Lectures on the chemical correlation of the functions of the body which were delivered at the Royal College of Physicians and subsequently published in the Lancet (2: 339–341, 1905). Actually the term was suggested to him by the Cambridge Physiologist, William B. Hardy (1864–1934) during a visit to his laboratory.

the correct name now is *C. goeringii* var *goeringii*) flowers (Table 2-17). This is interesting because if pollination extends the life span of flowers it is usually by causing the perianth to turn green and persist on the fruit. In this case Morita reported that the flowers wilted eventually. Morita also obtained additional information regarding the solubility of the active principle and the effects of a number of substances. (Table 2-15). His approach was innovative because it was quantitative since he measured the elongation of the gynostemium (Fig. 2-24A) whereas Fitting's reports were only descriptive in this respect.

Relating Pollenhormon to Auxin

Despite Fitting's and Morita's efforts the identity of the active principle (or principles) in pollen was not established. The reasons could be the: (1) lack of appropriate technology, (2) fact that Fitting and Morita were ahead of their time, (3) state of plant physiology which had yet to include the hormone concept, and (4) lack of continued interest (both Fitting Morita did not publish any additional papers on the subject). The discovery of auxin by Frits Warmolt Went (18 May 1903–15 May 1990) in 1926 (Went, 1926, 1990) changed all that by: (1) firmly establishing the hormone concept in plant physiology (Went and Thimann, 1937), (2) providing the means to assay and quantify auxin (Fig. 2-26A, 2-26B, 2-26D),

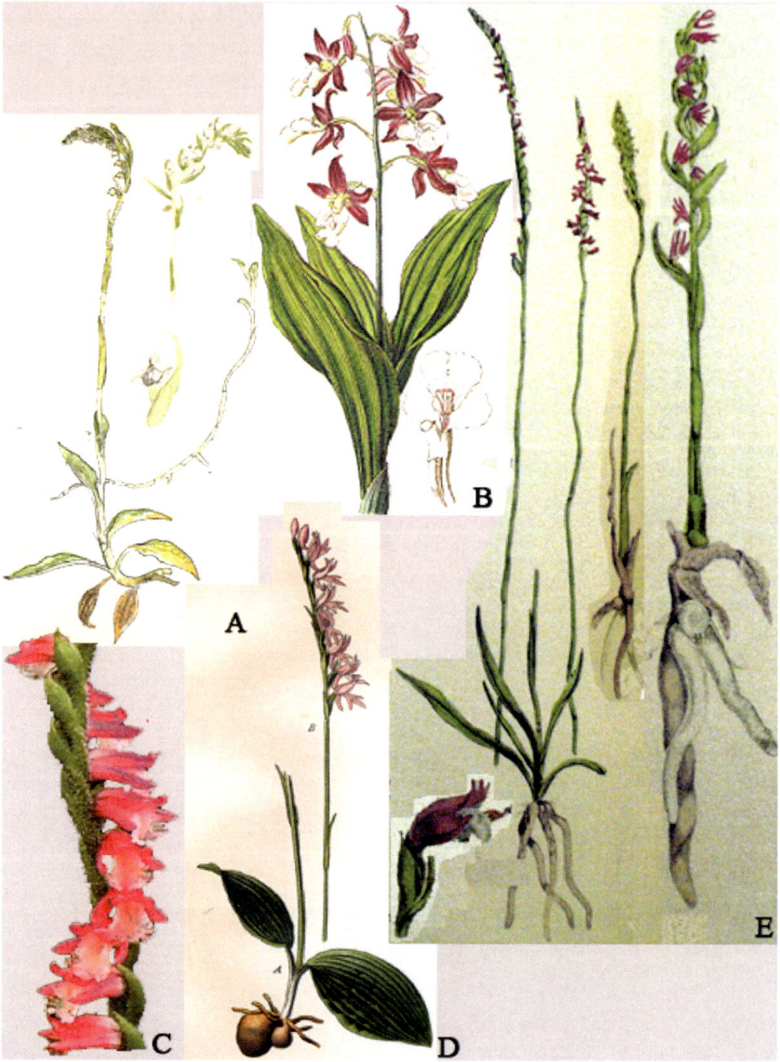

Fig. 2-22. Some of the orchids used by Kôichi Morita in his experiments. **A.** *Goodyera repens*. **B.** *Calanthe discolor*. **C, E.** *Spiranthes australis*. **D.** *Gymnadenia cucullata*. (sources: A, C, J. Arditti's collection; B, Botanical Register 26: plate 55, 1840; D, E, Correvon, 1899).

and (3) spurring further research (for a review of early research on auxin see Went and Thimann, 1937).

The identity of the active principle was established by Friedrich Laibach (1885–1967; Fig. 2-5D) and his associates (Laibach, 1930, 1932, 1933a, b; Laibach and Kornmann, 1933; Laibach and Maschmann, 1933; Mai, 1934; Maschmann and Laibach, 1993). They used Went's *Avena* coleoptile bioassay (Fig. 2-26, Tables

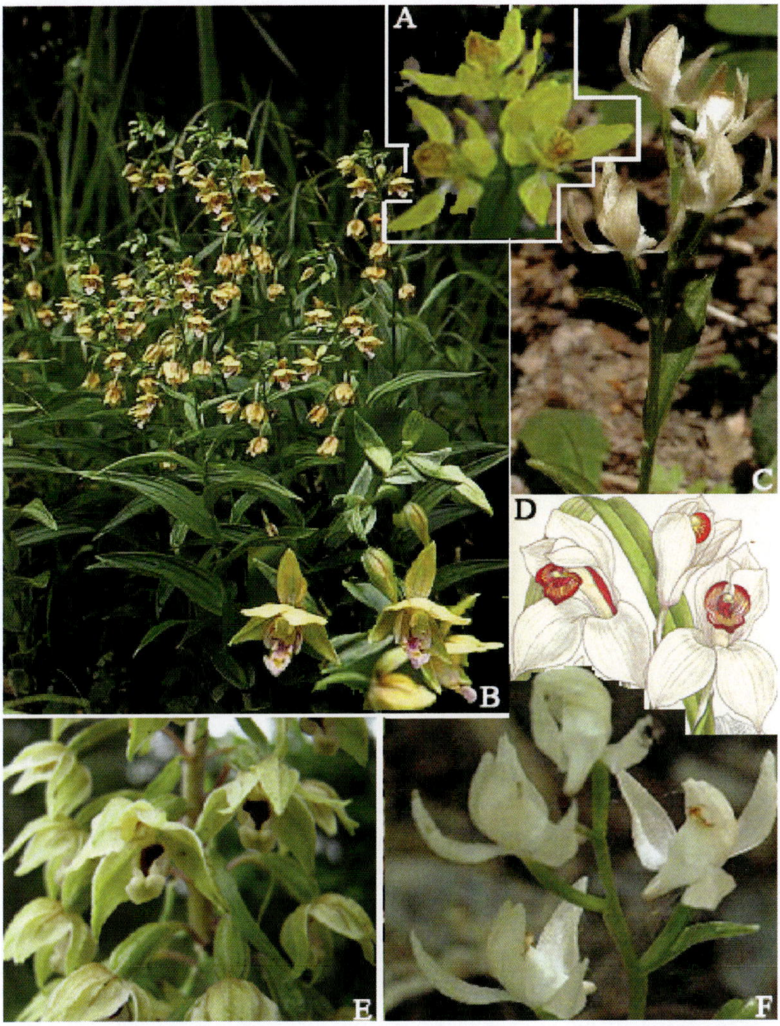

Fig. 2-23. Several of the orchids used by K. Morita for his experiments. **A.** *Epipactis falcata.* **B.** *Epipactis thunbergii.* **C, F.** *Epipactis erecta.* **D.** *Cymbidium virens.* **E.** *Epipactis papillosa* (A, C, E, F, J. Arditti's collection; D, plate 8151 from Curtis Botanical Magazine, vol. 135, 1907).

2-18, 2-19) and showed that Fitting's *Pollenhormon* had the same effects as auxin and its content in pollinia of different orchids is not the same (Fig. 2-26). They also demonstrated that its solubility in hot and cold water is different (Fig. 2-6). Actual determinations of auxin in orchid pollinia have shown that some orchids contain as much as 100 µg IAA g pollen^{-1} (R. Müller, 1953). This may be the highest or at least one of the highest concentrations of auxin in any plant tissue.

Table 2-15. Koichi Morita's experiments on the pollination of *Cymbidium virens* (Morita 1918)[a].

Experiment number	Description of experiment and its results		Morita's conclusions	Current explanation
Morita cites Fitting's papers (1909, 1910) and seems to have based his work on them. He determined the effects of pollination on the life span of *Cymbidium virens* flowers (Table 2-17) and other Japanese orchids (Table 2-19) thereby extending Fitting's work to temperate climate species.				
	Pollination extended the life span of *Cymbidium virens*, but what he observed may have been perianth greening as in some *Phalaenopsis* species. The gynostemium of this species did become swollen and elongated (Fig. 2-24A), the stigma closed and the ovary increased in size. It was 18 mm long and 2.5 mm in diameter before pollination and 38 mm long and 4 mm in diameter 7 days after pollination.			
1	24 Feb, 10:00: One flower of *Cymbidum virens* was pollinated with pollinia killed by soaking them in chloroform for ½ h	27 Feb, 10:00: Stigma started to close 28 Feb, 10:00; Stigma closed even more 1 Mar, 10:00: No change 16 Mar, 13:00: Flower starts to wilt	Dead pollen is as active as living pollinia The active substance is not deactivated by exposure to heat or chloroform	
2	25 Feb, 11:00: One flower of *Cymbidum virens* was pollinated with pollinia killed by steaming them for 10 min	27 Feb, 10:00: Stigma started to close 1 Mar: Stigma closed even more 16 Mar: Flower starts to wilt		
3	Six pairs of *Cymbidium virens* pollinia were extracted with warm water for 20 min and the extract was reduced to two drops on a water bath. Cotton wads saturated with extract were placed in stigmas			
	26 Feb, 10:00: Two flowers treated	1 Mar, 10:00: Stigma started to close 3 Mar, 10:00: Stigma closed even more Next days: No change		
4	Eight pairs of *Cymbidium virens* pollinia were extracted with cold water for 22 h and the extract was reduced to a small volume by placing it in a water bath for 7 min. Cotton wads saturated with extract were placed in stigmas			
	27 Feb, 10:00: Two flowers treated	3 Mar, 10:00: Stigma closed Next days: No change	The active principle can be extracted with both hot and cold water	

(continued)

5	Pollinia, 23 pairs were ground in a mortar with 5 ml water and allowed to stand for 22 h. After filtering the filtrate was reduced to 2 ml		
	6 Mar, 11:00: Two flowers treated	8 Mar, 11:00: Stigma starts to close 9 Mar, 13:00: Stigmatic closure is more pronounced Next days: No change	
6	Pollinia, 11 pairs were submerged in 3 mm of glycerine for 2 days. Both glycerin and the pollinia were used to treat flowers.		
	9 Mar: Glycerin was used to treat two flowers	No change	
	Pollinia were washed with distilled water and placed in the stigma of one flower	Pollinia had the usual effect	Glycerin could not extract the active substance
7	Pollinia, 10 pairs were placed in 5 ml ether for 24 h. The ether was reduced to two drops in volume. These drops were mixed with water and the mixture was used to treat flowers		The active substance can be extracted with ether and ethanol
	15 Mar, 13:00: Two flowers treated	16 Mar, 11:00: Stigmas start to close, but not as intensely as those treated with aqueous extract	
8	Pollinia, 10 pairs were extracted with 4 ml absolute ethanol for 14 days. The ethanol was reduced to a small volume which was used to treat flowers		
	7 Apr, 11:00: Two flowers treated	12 Apr, 10:00: One stigma closed, the other still open	
No number	Extract from experiment 5 was injected into the tip of the gynostemium		To be effective the active substance must be in the stigma

(continued)

Table 2-15. (continued).

Experiment number	Description of experiment and its results		Morita's conclusions	Current explanation
9	River sand boiled in hydrochloric acid for 5 min and washed several with distilled water was used to treat flowers			
	28 Feb, 11:00: Sand placed in stigma of one flower	3 Mar, 10:00: No change 7 Mar, 11:00: Flower starts to wilt 15 Mar: Flower wilted	1. Pollination has the following effects on flowers of *Cymbidium virens*: (a) shorten the life span of flowers, (b) cause stigmatic closure, (c) bring about swelling of the gynostemium, (d) induce swelling and elongation of the ovary	
10	Glass powder treated like the sand in experiment 9 was applied to stigmas 3 Mar, 11:00: Flowers treated	Next days: No change		
No number	Spores of saprophytic fungi placed in stigmas had no effects			
12, 13, 15, 16	Pollinia of *Dendrobium fimbriatum* var. *oculatum, Calanthe veitchii, Epipactis falcata* and *Calanthe discolor* were placed in stigmas of *Cymbidium virens* and caused stigmatic closure		2. Pollen killed with vapor or chloroform causes stigmatic closure, but does not reduce the life span of flowers	
14	Pollinia of *Paphiopedilum argus, Paphiopedilum boxallii* and *Paphiopedilum lathamianum* placed in stigmas of *Cymbidium virens* had no effects		3. Aqueous, ethanol and ether extracts of pollen cause stigmatic closure 4. Sand grains placed in the stigma and wounds in the gynostemium had no effects	
No number	Pollen of *Thea japonica, Prunus mume, Narcissus jonquilla, Brassica campestris, Hyacinthus orientalis* and *Salix thunbergiana* placed in stigmas of *Cymbidium virens* had no effects		5. Pollen of other orchids is effective 6. Some organic substances are effective	Morita confirmed Fitting's findings
No number	The following did not cause stigmatic closure: 5% NaCl; 1% K_2CO_3; 1% KNO_3; 0.025N, 0.05N and 0.1N acetic acid; 0.025N, 0.05N and 0.1N butyric acid; 2.5% and 5% glucose; 0.025 N formic acid; 2,5%, 5% and 10% fructose, 10% maltose and 1%, 2.5%, 3% and 5% sucrose. The following had an effect: 0.05N and 0.1N formic acid; 10% glucose; pure oleic, palmitic and stearic acids and 10% sucrose			

[a] The orchid names in this table are those used by Koichi Morita. Please see Appendix 2 for updated nomenclature.

Table 2-16. Koichi Morita's experiments on the effects of pollination on the life span of Japanese orchids (Morita 1918)[a].

Species	Life span (days)		Remarks
	Unpollinated	Pollinated	
Calanthe discolor	14	10	There was no swelling of gynostemia or closing of stigmas in any of these orchids. However the ovary enlarged
Cymbidium virens	14–25	30–40	
Epipactis erecta	8–10	7	
Epipactis falcata	8–12	8–12	
Epipactis papillosa	7	7	
Epipactis thunbergii	7–10	7–10	
Goodyera repens	8–10	7	
Gymnadenia cucullata	8–10	7	
Platanthera yatabei	7–10	7–10	
Spiranthes australis	10	5	

[a]The orchid names in this table are those used by Koichi Morita. Please see Appendix 2 for updated nomenclature.

Table 2-17. Effects of pollination on the life span of *Cymbidium virens* flowers (Morita, 1918)[a].

Unpollinated					Pollinated					
Full bloom (a)	Start to wilt (b)	Elapsed time, d (c: a to b)	Fully wilted (d)	Elapsed time, hr (e: b to d)	Full bloom (f)	Pollinated (g)	Start to wilt (h)	Elapsed time, d (i: g to h)	Fully wilted, d (j)	Elapsed time, d (k: h to j)
22 Mar	7 Apr, 10:00	16	10 Apr, 10:00	36	22 Mar	22 Mar, 10:00	21 Apr, 08:00	32	3 May, 08:00	12
22 Mar	15 Apr, 10:00	24	21 Apr, 10:00	144	22 Mar	22 Mar, 10:00	21 Apr, 08:00	32	1 May, 08:00	10
23 Mar	7 Apr, 10:00	6	13 Apr, 10:00	144	23 Mar	26 Mar, 11:00	26 Apr, 09:00	30	29 Apr, 11:00	3
23 Mar	10 Apr, 10:00	9	17 Apr, 16:00	176	29 Mar	31 Mar, 10:00	29 Apr, 11:00	29	1 May, 08:00	3
	Total elapsed time, a to d					Total elapsed time f to j				
	22 Mar-10 April: 17 days and 12 hours					22 Mar- 3 May: 42 days				
	22 Mar-21 April: 30 days					22 Mar- 1 May: 42 days				
	23 Mar-13 April: 12 days					23 Mar-26 Apr: 33 days				
	23 Mar-17 April: 16 days and 8 hours					29 Mar-29 Apr: 32 days				

[a] The orchid name in this table is the one used by Koichi Morita. Please see Appendix 2 for updated nomenclature.

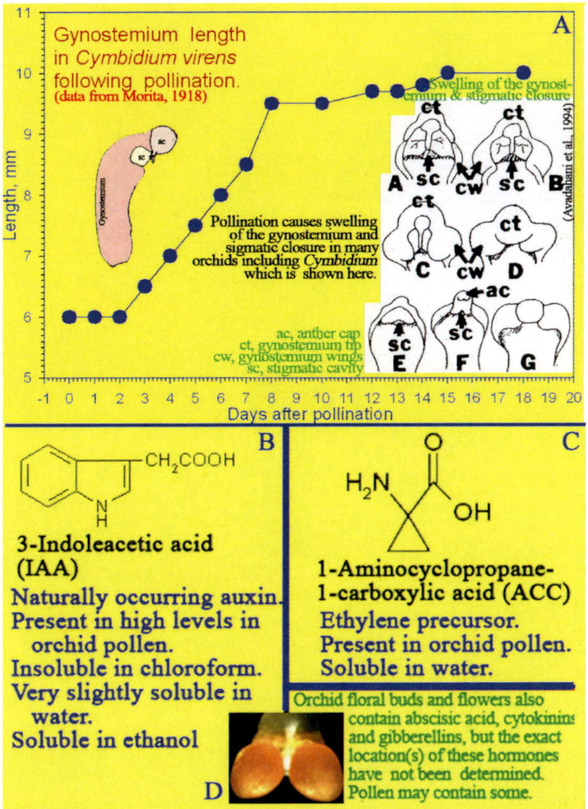

Fig. 2-24. Pollination effects on the gynostemium in *Cymbidium virens* following pollination and hormones which are present in orchid pollen. **A.** Elongation (graph) and swelling (line drawings) of the gynostemium of *Cymbidium virens*. **B.** Auxin, 3-Indoleacetic acid. **C.** 1-Aminocyclopropane-1-carboxylic acid (ACC), precursor of ethylene. **D.** *Phalaenopsis* pollinia, highly magnified (A, plot based on data from Morita, 1918 and drawings by Emma Web who was a technician in J. Arditti's laboratory at the time; B,C, computer generated; D, J. Arditti).

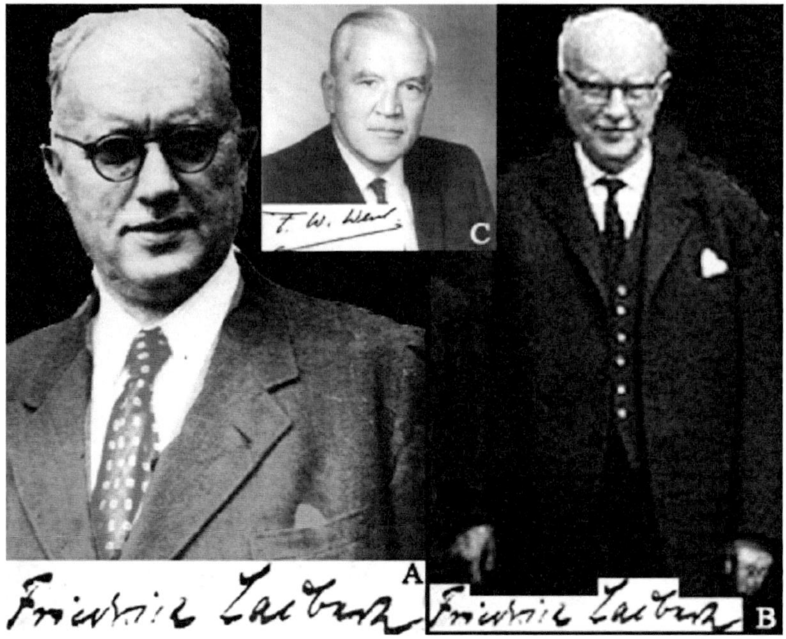

Fig. 2-25. Students of auxin. Friedrich Laibach (1885–1967) before or during (**A**) and after (**B**) World War II. **C.** Frits W. Went (A, courtesy the late Mrs. Sigrid Fitting; B, part of a group photograph taken at the First International Symposium on Arabidopsis Research, 21–24 April, Göttingen, Germany, courtesy Elliot M. Meyerowitz, California Institute of Technology; see also Meyerowitz, 2001; C, courtesy the late F. W. Went).

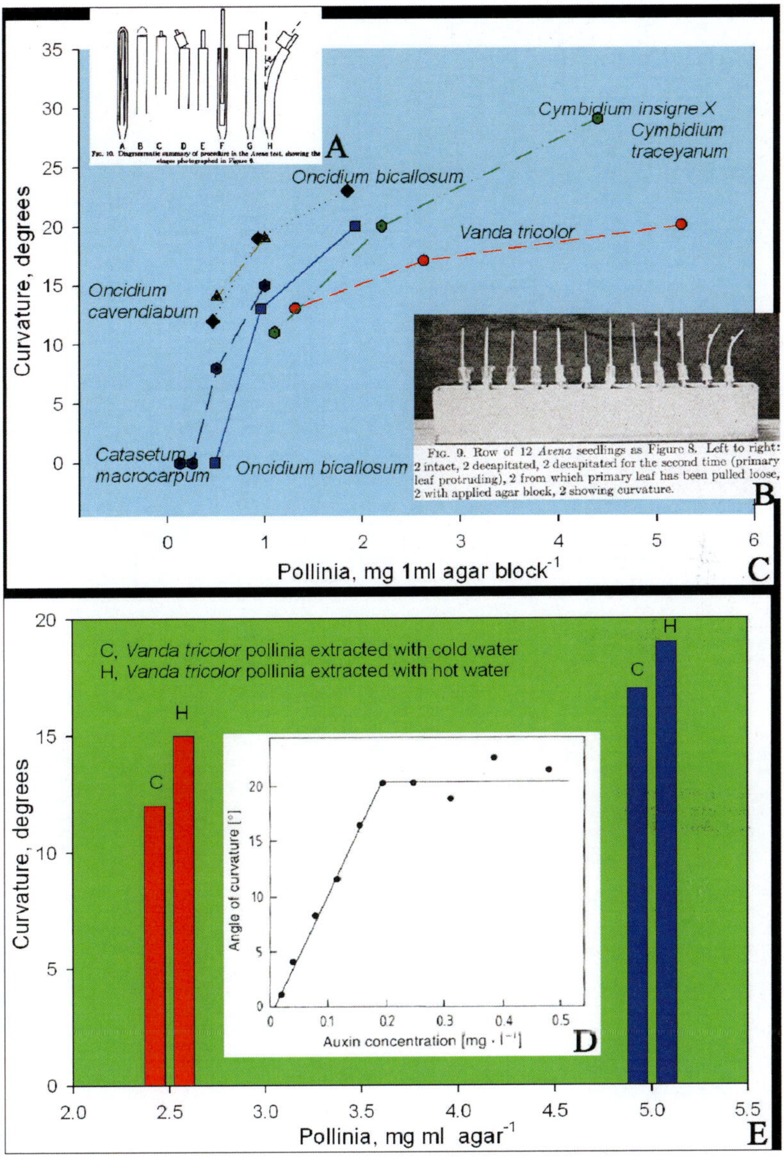

Fig. 2-26. *Avena* coleoptile assays. **A**. Diagrammatic representation of the assay. **B.** An early assay at the California institute of Technology in Pasadena. Laibach probably used similar assays. **C.** Assay of auxin content in pollinia of several orchids. **D.** Standard curve for auxin levels. **E.** Comparison assays of hot and cold water extracts of *Vanda tricolor* pollinia (A, B, Went and Thimann, 1937 (this is a classic volume and the first book on the subject by the discoverer of auxin, the late Frits W. Went and one of the early workers in the area who became a very prominent and influential American plant physiologist, the late Kenneth V. Thimann; the illustrations were taken from a copy owned by J. Arditti which is autographed by both authors; C, E, graphs based on data by Laibach; E, modern version of a graph in Went and Thimann, 1937).

Table 2-18. Friedrich Laibach's (1885–1867) experiments on the nature of Fitting's *Pollenhormon* (Laibach, 1930, 1932, 1933a, 1933b; Laibach and Kornmann, 1933; Laibach and Masschmann, 1933; Mai, 1934; Maschmann and Laibach, 1933)[a].

Experiment number	Description of experiment and its results	Laibach's conclusions	Current explanation
Laibach used the term *Wuchsstoff* (growth substance) and defined as "Non nutritional substances, widely distributed in plants and animals which can promote the elongation of *Avena* coleoptiles (see also Table 2-14). He used F. W. Went's (1903–1990) *Avena* coleoptile method to assay for the presence and activity of growth substances (Laibach, 1932)			
1	Five pairs each of *Cattleya* sp. and *Coelogyne massangeana* were suspended in 0.5 ml of water for 15 min. After that both the water and the pollinia were mixed with 0.5 ml of 3% agar. When the agar solidified it was cut into small cubes, each containing a pollinium. These cubes were assayed with six coleoptiles, five of which which bent 25–40°. The cube was not attached well to the one which did bend	The substance which causes bending of *Avena* coleoptiles can be extracted easily from pollinia with water	Laibach seems to have ignored three possibilities: (1) not all of the active substance was extracted, (2) there may have been two substances, one which was extracted with the water and another which was not, and (3) there may have also been an inhibitor which was not extracted from the pollinia
2	Cubes remaining from experiment 1 were liquefied in a water bath, then the agar was allowed to solidify again, cut into cubes and assayed as before. All six coleoptiles bent 15–25°		
3	Five pairs of pollinia from *Cattleya* sp. were used to prepare agar cubes as in experiment 1 except that some contained pollinia and some did not. Six cubes with pollinia and an equal number without were assayed with *Avena* coleoptiles. Bending of pollinia was as follows: Cubes with pollinia: 15°, 10°, 18°, 28°, 26°, 5° [Ave: 17°]. Cubes without pollinia: 20°, 13°, 30°, 32°, 13°, 28° [Ave: 22.7°]		
4	Pollen of *Hibiscus syriacus* and *Hibiscus schizopetalus* was used to prepare agar cubes as in experiment 1. The extracts were assayed with five coleoptiles each with bending being 15–20° and 20–30° respectively	*Hibiscus* pollen contains growth substance	
5	Pollen of *Anoda cristata*, Malvacae was assayed and found to cause very limited bending	This species contains very little growth substance	

(continued)

6	Pollinia (75 mg) of several orchids (*Vanda tricolor*, *Coelogyne massangeana*, *Coelogyne speciosa*, and *Phalaenopsis amabilis*) were extracted as follows: (1) Suspended in 25 ml 70% ethanol under nitrogen at 70° C. (2) After filtration the pollinia were extracted in the same manner for a second time. (3) The alcohol was removed from the combined filtrates under nitrogen, 40° C and pressure of 15 mm mercury. (4) What remained was an aqueous extract which was acidified with 1.5 ml 2N acetic acid and extracted four times with 50 ml aliquots of peroxide free ether. (5) Following this the ether was shaken three times with 10 ml aliquots of 2.5% sodium carbonate. (6) The carbonate solution was acidified with acetic acid and shaken with four 50 ml aliquots of ether. (7) After 2 h drying with sodium sulfate the ether was evaporated under nitrogen and the residue was stored in a desiccator. (8) To assay the residue it was dissolved in 0.6 ml water and 0.5 ml of this solution was mixed with 0.5 ml of 3% agar. Half of this was cut into cubes and assayed with seven *Avena* coleoptiles. Bending was 15°, 20°, 10°, 8°, 0°, 25°, 10° [Ave: 12.6]		The growth factor has the same characteristics as the substance isolated by Kögl and Hagen Smit	Kögl and Hagen Smit isolated auxin (but there is more to this story)
7	*Anoda cristata* pollen (500 mg) was extracted as in experiment 6. Results of the assays were negative		Experiments 7–10 show pollen of orchids and Malvaceae contain a substance which can be extracted with hot water and which can induce swelling of the gynostemium and stigmatic closure. Substance(s) extracted from animal organs which cause bending of *Avena* coleoptiles can also bring about swelling of the gynostemium in tropical orchids	This experiment is not impressive for several reasons not the least of which is the lack of a cotton wad only control. In fact it is possible to ask if the experiment was necessary
8	*Vanda tricolor* 8 Sept 1932 10:00 One flower self pollinated	10 Sept 1932 10:00 Gynosteium swollen and yellow. Lip yellow, Stigma closed		
	Three flowers untreated	Unchanged		
	Cotton wads impregnated with extract of guinea pig liver placed in stigmas of three flowers	Slight color changes. Gynostemium slightly swollen, Stigma slightly closed		

(continued)

Table 2-18. (continued).

Experiment number	Description of experiment and its results		Laibach's conclusions	Current explanation
	Cotton wads wetted with tap water placed in stigma of one flower	Unchanged		
9	Agar blocks from experiment 6 were placed in stigmas of *Phalaenopsis amabilis* flowers			
	(a) 12 Sept 1932: Blocks placed in stigmas of two flowers	13 Sept 1932: Stigmas half closed 14 Sept 1932: Stigma fully closed. Gynostemium somewhat swollen and turning yellow. Perianth wilting		IAA was present in these blocks. No agar only controls
	(b) 12 Sept 1932: Cotton wad saturated with water soluble, ether insoluble fraction from experiment 6 placed in stigma of one flower	13 Sept 1932: Flower unchanged 14 Sept 1932: Flower started to wilt, gynostemium became yellowish, but did not swell, stigma open		This extract probably induced some ethylene evolution No controls
	(c) 12 Sept 1932: One flower was self-pollinated	Flower was wilting, gynostemium was turning yellow and was very swollen. Stigma was closing		
	(d) 14 Sept 1932: Cotton wads saturated with chicken liver were placed in stigmas of two flowers	16 Sept 1932: Stigma half closed		Why was this experiment carried out?

(continued)

10	(a) 12 Sept 1932 12:00: Cotton wads saturated with extracts as shown below were placed in stigmas of tree flowers	14 and 15 Sept 1932: Slight swelling of the gynostemia of all nine flowers		No dry cotton and solvent only controls
	Ether soluble pollen extract.			
	Water soluble fraction of ether soluble extract			
	Water soluble fraction of ether in soluble extract			
	(b) 13 Sept 1932: One flower self-pollinated	14 Sept 1932: Gynostemium very swollen		
	(c) 14 Sept 1932: Flowers were treated as listed below			
	Cotton wads impregnated with liver extract were placed in stigmas of two flowers	14 Sept 1932: Gynostemium slightly swollen		Were the liver extract experiments necessary?
	Agar block containing liver extract was placed in the stigma of one flower	20 Sept 1932: Gynostemium very very swollen	It is possible that *Pollenhormon* and *Wuchsstoff* (a word used at the time for auxin) are one and the same substance	Reasonable conclusion
	Cotton wad wetted with tap water was placed in the stigma of one flower	No change		No dry cotton wad control

[a] The orchid names in this table are those used by Laibach and his associates. Please see Appendix 2 for updated nomenclature.

Table 2-19. Friedrich Laibach's (1885–1867) and Ernst Maschmann's experiments on the nature of *Wuchsstoff* (growth substance) in orchid pollinia (Laibach and Maschmann, 1933a).

Experiment number	Descriptions of experiments and their results		Laibach's conclusions	Current explanation
This paper reports on large number of experiments some of which repeat previous work yet again. Therefore only some of the experiments will be summarized in this table				
Pollinia were extracted by macerating them in water and keeping the mixture at 70°C for 15–30 min. The solid material was separated from the liquid through centrifugation or filtration. Assays were carried out as described in Table 2-15 (Laibach, 1932)				
Several *Avena* assays	Extract was incorporated in agar block and assayed with Went's *Avena* coleoptile	Extracts caused bending (Fig. 2-24, 2-26)	*Wuchsstoff* content in orchid pollinia is higher than in any other plant assayed up to that time	This may be still true
Assays of pollinia and extracts	Dead pollinia and extracts were applied to stigmas of their own or different orchids	The results confirmed the presence of *Wuchsstoff* in pollinia		These repetitions of of experiments which were repeated several times before may not have been needed
Several extractions of *Wuchsstoff* from pollinia showed that the substance was soluble in water, ethanol, acetone and ether and was acidic in nature. Based on this and other findings the conclusion was that the substance has the same characteristic as auxin				Orchid pollinia do contain auxin, but Laibach may have extracted more than one substance
Irradiation with UV light (313–365 nm) had no effect on the extract				
Drying pollinia under illumination and the dark did not reduce the activity or their extract				
Avena coleoptile assays and experiments with *Phalaenopsis* flowers showed that the *Wuchsstoff* was also present in pollen tubes				
Laibach and Maschmann reached the following conclusions on the basis of their experiments:				This is to be expected
	1. All extracts which caused bending of *Avena* coleoptiles also induced swelling of gynostemia, and conversely extracts which did not cause bending had no effects on gynostemnia			
	2. The growth induced by the extracts was to increase in cell size, not cell division			An auxin effect
	3. Transport of the growth substance was mainly basipetal with very little of it moving upwards or sideways			Polar transport is a characteristic of Auxin
	4. The active principle is soluble in water, ethanol, acetone and ether. It is insoluble in aliphatic and aromatic organic solvents. The substance is acidic. It can be removed with ether from sodium bicarbonate solution and after acidification from the basic aqueous phase			IAA has similar characteristics
	5. Oxidation with hydrogen peroxide reduced or eliminated the activity of the substance			
	6. Prolonged heating at 100°C [and also autoclaving for 10 min at 130°] does not reduce or eliminate the activity of the substance			

Conclusions

There can be no question that the *Pollenhormon* is the auxin indoleacetic acid (IAA; Fig. 2-24B) or a mixture of hormones which included IAA. It is also possible that Fitting's extracts contained ACC (Fig. 2-24C). Had Fitting continued to work with orchids he may have discovered auxin. Some of the pollen effects he reported are known at present to be induced by ethylene. In 1901 the Russian plant physiologist Dimitry Neljubov (1879–1926) showed that ethylene affected the growth of pea seedlings (Neljubov, 1901), but it was too much of a stretch in 1909 to connect abnormal growth of pea seedlings and wilting of orchid flowers. Even had he made the connection there was no technology at the time that would have allowed Fitting to measure ethylene evolution by pollinated orchid flowers.

In summary, there can be no question that Fitting introduced the term "hormone" into plant science (Table 2-14). Fitting even had auxin in his extracts, but did not discover it because he stopped his orchid research too soon. Frits W. Went discovered auxin.

Dedication

I dedicate my efforts to the memory of Professor Grover C. Stephen, my first and best department chair at the University of California, Irvine, one of my few deans at UCI worthy of respect and most of all my life long friend – Joseph Arditti, Professor Emeritus.

Acknowledgments We thank Jean Miller and Kathryn Kjaer for assistance with literature searches and for obtaining papers for us through interlibrary loan and Dr. Gunther Gerlach for many valuable suggestions.

Literature Cited

Avadhani, P. N., H. Nair, J. Arditti, and C. S. Hew. 1994. Physiology of orchid flowers. Pages 189–362 *in* J. Arditti (ed.), *Orchid biology, reviews and perspectives*, Vol VI. Wiley, New York.

Brooke, J. and G. Bone. 1950. *The wild orchids of Britain*. Bodley Head, London.

Correvon, H. 1899. *Albun des orchidées del'Europe centrale et septentrionale*. Libraire O. Doin, Paris.

Correvon, H. 1923. *Albun des orchidées del'Europe*. H. Correvon "Floraire," Chéne-bourg, Geneva.

Darwin, C. 1904. *The fertilization of orchids by insects*, 2nd edition, 7th printing. John Murray, London.

Dolcher, T. 1961a. Relazioni ormonali nello sviluppo dell'ovario delle orchidee. I. Effeto in vivo del trattamento con auxine. *Nuovo Giornale Botanico Italiano* 68: 213–215.

Dolcher, T. 1961b. Relazioni ormonali nello sviluppo dell'ovario delle orchidee. II. Azione delle auxine in segmenti isolati dell'asse florale. *Nuovo Giornale Botanico Italiano* 68: 216–219.

Dolcher, T. 1967. Relazioni ormonali nello sviluppo dell'ovario delle orchidee. 3. Induzione delle frutificazione mediante trattamenti auxinici. *Nuovo Giornale Botanico Italiano* 101: 41–57.
Fitting, H. 1909a. Die Beeinflussung der Orchideenblüten durch die Bestäubung und durch andere Umstände. *Zeitschrift für Botanik* 1: 1–86.
Fitting, H. 1909b. Die Beeinflussung der Orchideenblüten durch die Bestäubung und durch andere Umstände (abstract). *Botanische Jahrbuch für Systematik* 43: 24.
Fitting, H. 1909c. Entwicklungsphysiologische Probleme der Fruchtbildung. *Biologisches Centralblatt* 19: 193–205, 225–239.
Fitting, H. 1910. Weitere entwicklungsphysiologische Untersuchungen an Orchideenblüten. *Zeitschrift für Botanik* 2: 225–267.
Fitting, H. 1911. Untersuchungen über die vorzeitige Entblätterung von Blüten. *Jahrbuch für Wissenschaftliche Botanik* 49: 187–266.
Fitting, H. 1912. Über eigenartige Farbänderungen von Blüten and Blütenfarbstoffen. *Zeitschrift für Botanik* 4: 81–105.
Fitting, H. 1921. Das Verblühen der Blüten. *Die Naturwissenschaften* 9: 1–9.
Fitting, H. 1936. Die Hormone als physiologische Reizstoffe. *Biologisches Centralblatt* 56: 69–86.
Forbes, H. O. 1885. A naturalist's wandering in the Eastern Archipelago. A narative of travel and explorations from 1878–1883.Harper & Brothers, Franklin Square, New York.
Halbsguth, W. 1962. Hans Fitting zum 85 Geburtstag. *Forschungen und Fortschritte* 36: 122–124.
Halbsguth, W. 1974. Hans Fitting 23 April 1877 bis 6 Juli 1970. *Berichte der Deutsche Botanische Gesellschaft* 86: 577–586.
Heslop-Harrison, J. 1957. The physiology of reproduction in *Dactylorchis*. I. Auxin and the control of meiosis, ovule formation and ovary growth. *Botanisk Notiser* 110: 28–49.
Hildebrand, F. 1863a. On the impregnation in orchids as a proof of the two different effects of the pollen. *Annals and Magazine of Natural History* Ser. 3, 12: 167–194.
Hildebrand, F. 1863b. Die Fruchtbildung der Orchideen, ein Beweis für doppelte Wirkung des Pollen. *Botanische Zeitung* 21: 325–332.
Hildebrand, F. 1865. Bastardirungsversuche an Orchideen. *Botanische Zeitung* 23: 245–249.
Hildebrand, F. 1868. Notizen über die Geschlechtsverhältnisse brazilianischer Pflanzen. Aus einem Briefe von Fritz Müller. Sta Catarina, 12 Sept 1867. *Botanische Zeitung* 26: 113–116.
Laibach F. 1930. Untersuchungen über die Postfloration tropischer Orchideen. *Planta* 9: 341–387.
Laibach, F. 1932. Pollenhormon und Wuchstoff. *Berichte der Deutsche Botanische Gesellschaft* 50: 383–390.
Laibach, F. 1933a. Wuchstoffversuche mit lebenden Orchideenpollinien. *Berichte der Deutsche Botanische Gesellschaft* 51: 336–340.
Laibach, F. 1933b. Versuche mit Wuchstoffpaste. *Berichte der Deutsche Botanische Gesellschaft* 51: 386–392.
Laibach, F., and P. Kornmann. 1933. Zur Frage des Wuchstofftransportes in der Haferkoleoptile. *Planta* 21: 396–418.
Laibach, F. and E. Maschmann. 1933. Über den Wuchstoff der Orchideenpollinien. *Jahrbücher für Wissenschaftliche Botanik* 78: 399–430.
Lawler, L. J. 1984. Ethnobotany of the Orchidaceae. Pages 27–149 *in* J. Arditti (ed.). *Orchid biology, reviews and perspectives* Vol. III. Cornel University Press, Ithaca, New York.
Lemiare, C. 1854. *Flore des serres et des jardins de lEurope*.Volume 9, Plate 957. Louis van Houtte, Gent, Belgium
Mai, G. 1934. Korrelationsuntersuchungen an entspreiteten Blattstielen mittels lebender Orchideenpollinien als Wuchstoffquelle. *Jahrbücher für Wissenschaftliche Botanik* 79: 681–713.
Magli, G. 1958. Possibilitá di sostituire con auxine l'azione del polline per lo sviluppo degli ovuli delle orchidee. *Nuovo Giornale Botanico Italiano* N. S. 65: 401–417.
Maschmann, E., and F. Laibach. 1933. Über Wuchstoffe. *Biochemische Zeitung* 255: 446–452.
Medvedev, Z. A. 1971. *The rise and fall of T. D. Lysenko*. Anchor Books, Doubleday and Co, Garden City, NY.
Meyerowitz, E. M. 2001. Prehistory and history of *Arabidopsis* research. *Plant Physiology* 125: 15–19.

Morita, K. 1918. Influences de la pollinisation et d'autres actions extéreurs sur la fleur du *Cymbidium virens* Lindl. *Botanical Magazine* (Tokyo) 32: 39–52.
Müller, F. 1868. Ueber Befruchtungserscheinungen bei Orchideen. *Botanische Zeitschrift* 26: 630–632.
Müller, F. 1886. Biologische Beobachtung an brasilianischen Orchideen. *Verhandlungen der Botanischer Verein der Provinz Brandenburg* 28: IV (page number is in Roman numeral).
Müller, R. 1953. Zur quantitativen Bestimmung von Indolyl-essigsäure mittels Papierchromatographie und Papierelektrophorese. *Beiträge zur Biologie der Pflanzen* 30: 1–32.
Neljubov, D. 1901. Über die horizontale Nutation der Stengel von *Pisum sativum* und einiger anderer Pflanzen. *Beihefte zum Botanischen Zentralblatt* 10: 128–139.
Parkinson, S. 1640. *Theatrum botanicum*. N. P. London.
Schulze, M. 1894. Die Orchidaceen Deutschlands, Deutsch-Oesterreichs und der Schweiz. Fr. Eugen Köhler's Verlag, Gers.-Untermhaus.
Tran, H., H. Vu, A. Mahunu, D. Chien, J. Arditti and R. Ernst. 1995. Chlorophyll formation in flowers and fruits of *Phalaenopsis* (Orchidaceae) species and hybrids following pollination. *American Journal of Botany* 82: 1089–1094.
Treub, M. 1883. L'actions des tubes polliniques surf le développement des ovules ches les Orchidées. *Annales du Jardin Botanique de Buitenzorg* 3: 120–127.
W. F. H. B. 1897. Fritz Müller. *Nature* 56: 546–548.
Went, F. W. 1926. On growth accelerating substances in the coleoptile of *Avena sativa*. *Proceedings of the Koenigliche Akademie der Wetenschappen* 30: 10–19.
Went, F. W. 1974. Letter to Joseph Arditti from the Desert Research Institute at the University of Nevada, Reno dated 19 June 1974 (this letter, filed in the post pollination file, part of a large collection of reprints on many aspects of orchids, was given by J. A. to the Singapore Botanic Gardens).
Went, F. W. 1990. Orchids in my life. *In* J. Arditti (ed) *Orchid biology: reviews and perspectives*, Vol IV. Timber Press, Portland, OR, pp. 21–36.
Went, F. W., and K. V. Thimann. 1937. *Phytohormones*. Macmillan, New York.

Appendix 1

Letters from Prof. Hans Fitting to Prof. Joseph Arditti
Translated and annotated by Dr. Hubert Kurzweil
Comments are between brackets.
The letters are part of a reprint collection donated by Prof. Arditti to the Singapore Botanic Gardens.

Bonn, 2 Nov 1969

Dear Dr. Arditti,

Many thanks for the reprint of the extremely interesting paper '*Effects of Auxin*' by you and Mr. Robert L. Knauff, and your equally valuable friendly letter of 23rd Oct 1969. In both of these you acknowledged in an unusually friendly and appreciative way my own orchid work, which I have done many years ago. This was indeed very good for me, now that I am 93 years of age!

Unfortunately I cannot send you the reprints that you requested as already for a few years I have not got any left, and also because I had to give up my scientific work.

I wish you great success with the continuation of your research,

With my best wishes,

H. Fitting

I hope that you can read my letter; already for some time I am no longer allowed to use the typewriter!!

Bonn, 10th November 1969

Dear Dr. Arditti,

This is my reply to your very friendly letter of 12th Nov. 1969 [two days after he wrote this one?; one of these two dates must be wrong] in which you ask "what made you observe the orchid flowers"? While I was preparing my lectures on pollination biology (1903–1907) as a 'Privatdozent' at the University of Tübingen, I was very much interested in a theory that is very common in the older German literature on pollination biology, namely that the pollinia of many exotic orchids act like a poison during cross-pollination! At the time I was not able to find out who first proposed this theory! A long time after my trip to the tropics (namely in about 1920), I got to know volume II (1921, letters) of the new edition of Fritz Müller's "Work, letters and life" (newly published by Alfred Müller, 1915–1921, in three volumes). Müller lived in Brazil for a long time, wrote several important publications on pollination biology and is the founder of the "biogenetic rule" ["biogenetic law"]. I am sure you know about him (biogeography by E. Loens, *Berichte der Deutschen Botanischen Gesellschaft* 15, 1897). In volume II (which contains his letters) there is a very important letter, written on 1st January 1867 and addressed to Charles Darwin!! It provides the solution to my above-mentioned mystery!! On pages 104–128 he elaborates on the toxicity of the orchid pollinia! In view of your own orchid work you should definitely read this important letter! According to my notes (unfortunately these are not very clear), Fr. Müller reported in 1886 something similar in male flowers of *Catasetum* [I don't know what he means with "(gleich mir, 1920)"; this could mean that Fitting observed the same in 1920]. This was also published it in *Verhandlungen des Botanischen Vereins der Provinz Brandenburg*, 1886, volume 28, page IV (I don't know if this is also written in his letters in 1921). But what is this apparent "poison" all about? This was the starting point for my orchid research in 1907 and 1908 in Buitenzorg! Unfortunately I did not mention this in my first orchid publication in 1909a, as I thought that the apparent toxicity of orchid pollinia was already well-known in Germany at the time! The result of my Buitenzorg work: Mueller's poison is obviously my pollen hormone!! [Pollenhormon]

In 1930 and the following years Laibach confirmed that β – Indolylessigsäure (IES [English IAA]) – that was discovered long time after 1907/08 – triggers the postfloral development of the orchids with all of its consequences, just like my pollen hormone also does! But to date there has not been exact chemical proof that my pollen hormone actually is the IAA!! As we have seen repeatedly in the meantime,β – Indolylessigsäure (IAA) causes so many different reactions that one should not be surprised that it also acts similar to my pollen hormone. The exact proof, that my pollen hormone is actually IAA would obviously be very difficult. Therefore I do not really support .... 'convincion' [there is one word illegible; could be "Ihre" = your], that the pollen hormone and "Auxin" are the same. And besides: what actually is the so-called "Auxin" chemically? A few years ago a Dutch colleague, in a publication in *Acta Botanica Neerlandica* (unfortunately I forgot the volume and the page number!), inserted a big question mark after "Auxin"!

On a personal level I can assure you that I very much enjoyed my orchid work in 1907/08 in the magnificent Buitenzorg Botanical Garden (as you say, it has "inspired" me); and also, that the substantial German Buitenzorg-scholarship which was awarded to me in 1907 (– 6,000 German Mark!!) was put to good use for our beautiful *Scientia Amabilis*. So I returned to Europe in 1908 with the great feeling, that I had done something really worthwhile with this substantial scholarship, both important and valuable, namely that I had discovered the first plant hor….. [cut off, I suppose he means hormone].

I will be especially grateful to receive, as already promised, your article on "Fitting's pollen hormones", which I will certainly be very happy about and which I regard as a great honour.

I hope that you can read my bad but age-related hand-writing .... [partly cut off; maybe he means "Writing"] is very difficult and gives me lots of pain! My heart is weak and requires constant medical supervision!!

With friendly regards,

....

Postscriptum: Please do not take the following comment for immodest! It is rather important [in this sentence there is one word which I cannot read]!. My Danish colleague P. Boysen-Jensen said to me in Copenhagen in 1950, that my orchid publication (1909a) has inspired him to study the chemical conductivity of the phototropic impulse in Pfeffer's Institute in Leipzig and inspired by the same (see also the historical notes on p. 21–23 in P.B. Bell's book "Darwin's biological work etc."; Cambridge; 1959). And Boysen-Jensen's short, but deservedly so widely accepted work inspired Went to his great work on the Auxins, which you appreciate so much with reason.

10th December 1969

Dear Dr. Arditti,

Many thanks for your friendly letter of 1st December. I was motivated by this letter to read carefully through all of my orchid publications once again. Therefore I can now give you additions to my earlier letters! Making these additions was rather difficult as I have been suffering from heart problems for several years (but the diagnosis of my doctors is only 'age-related heart'). Therefore I am allowed only to go on short walks of 1/2–3/3 [maybe he means 3/4] hours. Unfortunately I can no longer visit the library at the university to look at old and current literature, as my home is situated in the southern part of Bonn, far away from the Institute of Botany and the library; and I had to sell my own very large library in 1952. Therefore, when replying to queries, I have to rely on my mostly superficial notes and on my aging memory.

But I would like to draw your attention to a few important questions!

Question I. How I got my results in Bogor is actually already contained on pages 1–2 of my paper in 1909a, if you read it carefully!!: namely through detailed description of the postfloral processes and their causes. At first I did not deal with the so-called "poison" of the great Fritz Müller which you can see from the fact that in all of my papers in 1909a, 1909b, 1910 I never mentioned Müller's "poison" and pointed out that this poison is apparently my pollen hormone. {I only mentioned Hans Winkler's paper of 1905 briefly (see Fitting, 1909a, p. 69) where he refers to the pollen "poison"}. Therefore the toxicity of the pollinia was only a very minor problem of my work, and it was quite a coincidence that I found the pollen hormone in the course of my developmental-physiological studies; but I have discovered it!!, simply through systematically and well thought-through experimenting.

Question II. From where and since when did botany in Germany know about the "toxicity" of pollinia? To answer this question you should read Fritz Müller's paper carefully (cited in my paper in 1909a; 1868, p. 629ff.), maybe he mentions the "poison". The thick book of Kerner von Marilaun (1895), that I have recently looked through carefully, does not make any mention of Fritz Müller's orchid work and his "poison"; did he not know about them??

Comment: Fritz Müller's paper in *Verhandlungen des Botanischen Vereins der Provinz Brandenburg*, 1886, volume 28, also seems to be mentioned in his work "Work, letters and life" on page 471 (my own rather superficial note says 8th May 1886, probably it refers to a letter of Fritz Müller??).

Question III. Who suggested that the pollen hormone and the much later discovered Indolylessigsäure (IAA) are probably the same? (I prefer the term IAA over the ambiguous term "Auxin" = growth hormone). This was of course Laibach and his students! At least they were able to show that IAA has got the same effect on orchid flowers as the pollen extract.

Question IV. But is that the proof, that my pollen hormone is actually the same as IAA? If I understood your paper of 1969 correctly, you showed that the postfloral processes consist of partial processes which are in their principle (partly chemically) actually quite different from IAA (I was not able to read any of the papers that you cited in 1969, and therefore do not know their contents). Therefore I repeat what I said already earlier, namely that we need exact chemical proof that the pollen hormone indeed contains IAA! Laibach has not done this! Of course I don't know if you would agree with me. Perhaps one should look through the papers of Laibach and his students again carefully; but I don't think that one would find something important.

[there is something missing,]

.............; I have not got them here – my reprint collection is at the Institute of Botany!!

I was most intrigued by your mention that you will go to Bogor soon! What will be the garden (and the primitive institute), now after the Dutch have left Java?? – I wish you all the best for this wonderful trip!!

At last some literature from my notes:

I. Dolcher, Tullio. 1961. Relazioni ormonali nello sviluppe dell'ovario delle Orchidee. *Nuovo giornal botan. ital.* 68(1–2): 213–215.
II. Dolcher, Tullio. 1961. Azione delle auxine in segmenti isolati dell' asse floral. *Nuovo giornal botan. ital.* 68(1–2): 216–219.

I have not read any of these two papers, and therefore do not know what their content is!

With best wishes,
Hans Fitting

Comments in brackets are by the translator.

Appendix 2

Plant Names Used in this Chapter

Plant names used in the chapter are listed in this appendix. Currently accepted names are listed in bold face. Names follow largely the World Checklist of Monocotyledons (The Board of Trustees of the Royal Botanic Gardens, Kew. Published on the Internet; http://www.kew.org/wcsp/monocots/ accessed 16 October 2006).

Orchids

***Aerides falcata* Lindl. & Paxton**
***Aerides odorata* Lour.**
Arachnanthe clarkei (Rchb. f.) Rolfe = ***Esmeralda clarkei* Rchb. f.**
Arachnanthe sulingi (Blume) Benth. = ***Armodorum sulingi* (Blume) Schltr.**
***Armodorum sulingi* (Blume) Schltr.**
Bletia verecunda (Salisb.) R. Br. = ***B. purpurea* (Lam.) DC.**
***Calanthe discolor* Lindl.**
***Calanthe × veitchii* Hort.**
Calanthe veratrifolia (Willd.) R. Br. ex Ker Gawl. = ***C. triplicata* (Willemet) Ames**
***Catasetum* sp.**
***Cattleya labiata* Lindl.**
***Cattleya trianae* Linden & Rchb. f.**
***Coelogyne asperata* Lindl.**
***Coelogyne massangeana* Rchb. f.**
***Coelogyne pandurata* Lindl.**
***Coelogyne speciosa* (Blume) Lindl.**
***Coelogyne swaniana* Rolfe**
Corymbis disticha (Breda) Lindl. = ***Corymborkis veratrifolia* (Reinw.) Blume**
***Corymborkis veratrifolia* (Reinw.) Blume**
***Cymbidium finlaysonianum* Lindl.**
Cymbidium sanguinolentum Teijsm. & Binn. = ***C. chloranthum* Lindl.**
Cymbidium virens Rchb. f., sphalm. for *C. virescens* Lindl. = ***C. goeringii* (Rchb. f.) Rchb. f. var. *goeringii***
***Cypripedium acaule* Aiton**
***Dendrobium antennatum* Lindl.**
***Dendrobium crumenatum* Sw.**
Dendrobium fimbriatum Hook. var. *oculatum* Hook. = ***D. fimbriatum* Hook.**
***Dendrobium macrophyllum* A. Rich.**
Dendrobium superbum Rchb. f. = ***D. anosmum* Lindl.**

***Dendrobium wardianum* R. Warner**
Epipactis erecta (Thunb.) Sw. = ***Cephalanthera erecta* (Thunb.) Blume**
Epipactis falcata (Thunb.) Sw. = ***Cephalanthera falcata* (Thunb.) Blume**
***Epipactis palustris* (L.) Crantz**
***Epipactis papillosa* Franch. & Sav.**
***Epipactis thunbergii* A. Gray**
Eulophia macrostachya Lindl. = ***E. pulchra* (Thouars) Lindl.**
***Goodyera repens* (L.) R. Br.**
***Gymnadenia conopsea* (L.) R. Br.**
Gymnadenia cucullata (L.) Rich. = ***Neottianthe cucullata* (L.) Schltr.**
Liparis latifolia Lindl. = ***Stichorkis latifolia* (Lindl.) Pfitzer**
***Masdevallia glandulosa* Königer**
***Odontoglossum crispum* Lindl.**
***Oncidium flexuosum* Lodd.**
***Oncidium incurvum* Barker ex Lindl.**
***Oncidium sphacelatum* Lindl.**
Oncidium sphegiferum Lindl. = ***O. divaricatum* Lindl.**
Orchis fusca Jacq. = ***O. purpurea* Huds.**
Orchis latifolia L. = ***Dactylorhiza incarnata* (L.) Soó subsp. *incarnata***
Orchis maculata L. = ***Dactylorhiza maculata* (L.) Soó**
***Orchis mascula* (L.) L.**
***Orchis morio* L.**
***Paphiopedilum argus* (Rchb. f.) Stein**
***Paphiopedilum barbatum* (Lindl.) Pfitzer**
Paphiopedilum boxallii (Rchb. f.) Pfitzer = ***P. villosum* (Lindl.) Stein var. *boxallii* (Rchb. f.) Pfitzer**
***Paphiopedilum callosum* (Rchb. f.) Stein**
***Paphiopedilum glaucophyllum* J.J. Sm.**
Paphiopedilum lathamianum, name not traced
Phaius amboinensis Blume = ***P. terrestris* (L.) Ormerod**
***Phalaenopsis amabilis* (L.) Blume**
***Phalaenopsis cornu-cervi* (Breda) Blume & Rchb. f.**
Phalaenopsis esmeralda Rchb. f. = ***P. pulcherrima* (Lindl.) J.J. Sm.**
***Phalaenopsis regnieriana* Rchb. f.**
***Phalaenopsis violacea* H. Witte**
***Platanthera bifolia* (L.) Rich.**
Platanthera yatabei Maxim. [nom. nud.]
Renanthera × maingayi (Hook. f.) Ridl. = ***Arachnis × maingayi* (Hook. f.) Schltr.**
***Rhynchostylis retusa* (L.) Blume**
Spathoglottis filuata, name not traced
Spiranthes australis (R. Br.) Lindl. = ***S. sinensis* (Pers.) Ames**
***Stanhopea insignis* Frost**
***Trichoglottis geminata* (Teijsm. & Binn.) J.J. Sm.**
***Vanda insignis* Blume**

***Vanda tricolor* Lindl.**
***Vanilla* sp.**
Zygopetalum mackayi Hook. = ***Z. maculatum* (Kunth) Garay**

Other Plants

Alpinia hookeriana Valeton (Zingiberaceae) = ***A. latilabris* Ridl.**
***Anoda cristata* Schltdl. (Malvaceae)**
Begonia geogensis (Begoniaceae), name not traced
Brassica campestris L. (Brassicaceae) = ***B. rapa* L.**
***Canna* sp. (Cannaceae)**
***Eucharis* x *grandiflora Planch. & Linden* (Alliaceae)**
***Hedychium* sp. (Zingiberaceae)**
***Hibiscus rosa-sinensis* L.(Malvaceae)**
***Hibiscus schizopetalus* (Mast.) Hook. f. (Malvaceae)**
***Hibiscus syriacus* L. (Malvaceae)**
***Hyacinthus orientalis* L. (Asparagaceae)**
Impatiens rodgrigesi (Balsaminaceae), name not traced
***Narcissus jonquilla* L. (Alliaceae)**
***Prunus mume* Siebold. & Zucc. (Rosaceae)**
Salix thunbergiana (Salicaceae) = ***S. gracilistyla* Miq.**
Thea japonica (Theaceae) = ***Camellia japonica* L.**

3

Are Orchids Mentioned in the Bible?

ARNOLD S. DUNN AND JOSEPH ARDITTI

Contents

Introduction 142
Aaron's and Moses's Staffs 142
Mustard Seeds versus Orchid Seeds 146
Dudaim 149
Post Biblical Jewish Writings 154
Conclusion 156
Dedications 156
Literature Cited 156

Kull, Arditti and Wong (eds.), *Orchid Biology: Reviews and Perspectives, X*,

Introduction

Only 100–130 plants are mentioned in the Bible, some in general terms such as "thorns," "thistles," "briars", "grass," and similar vegetation (Blackman, 1983; Felix, 1974a; Musselman, 2005; Souvay, no date). Others are mentioned by specific names such as "shoshana" (which has been interpreted to mean "lily," but in present day Hebrew is used for "rose" as is the term "vered"), "shoshanat ha'amakim" (identified as *Lilium candidum*), "havazelet ha-Sahron" (*Pancratum maritimum*) and grape vines. There are over 525 references to trees in the Bible of which 22 can be recognized at present (Musselman, 2003). They include cedars of Lebanon, carobs, apples, almonds, date palms, figs and others (Musselman, 2003). Of these the date palm, fig, olive, pomegranate and tamarisk are also mentioned in the Koran (Musselman, 2003). Some biblical plant names have been applied to several species at present. An example of this is "lily" which may be a red tulip (*Tulipa*), anemone, white madonna lily (*Lilium candidum*), hyacinth (*Hyacinthus orientalis*) and crocus (the genus *Crocus*; Hepper, no date). Orchids are most probably not mentioned in the Bible. However, various biblical scholars of both the new and old testaments have claimed that some plants which are mentioned in the Bible are orchids.

One book about the plants of the Bible does not even mention orchids (Henslow, 1906). Another has stated that "although there are 70 kinds of orchids native to the Holy Land, it is not probable that Jesus' listeners were at all acquainted with their dust-like seeds. These apparent dust motes probably would not have been recognized as seeds" (Moldenke and Moldenke, 1952; for physical characteristics of orchid seeds see Arditti and Abdul Ghani, 2000). The view of an American orchid expert (Ames, 1942) is that "there is not any indication of the use of orchids by the Hebrews either for medicine or decorations. This is not strange because the country of the Bible people was not an orchid country." Still, a number of writers have suggested that other plants, even if they are referred to by their own names, are Orchidaceous in nature (Chatin, 1868; Sugaya, 1999; for reviews see Lawler, 1984). The logic behind these suggestions is neither obvious nor clear. It is actually strained and even contrived. Two of the attributions seem to be due to misunderstandings.

Aaron's and Moses's Staffs

Orchis purpurea Huds (Figs. 3-1C and 3-2C) has ovoid tubers and numerous short and thick roots. It has three to six leaves which are oblong or oblong-lanceolate, bright, grey-green, glabrous and shiny. They are 10–22 cm long and 2–5 cm wide. The inflorescence is robust, round, solid, 20–40 cm and 60–90 cm tall. Flowers are numerous, unscented, 15–20 mm in size, with a dark red, helmet like dorsal sepal and a pale reddish labellum with darker spots. This widely distributed species is found in the British Isles, Bulgaria, Caucasus, Czech Republic, Crimea, Denmark, Egypt, France, Greece, Hungary, Israel, Italy, probably Jordan, Netherlands,

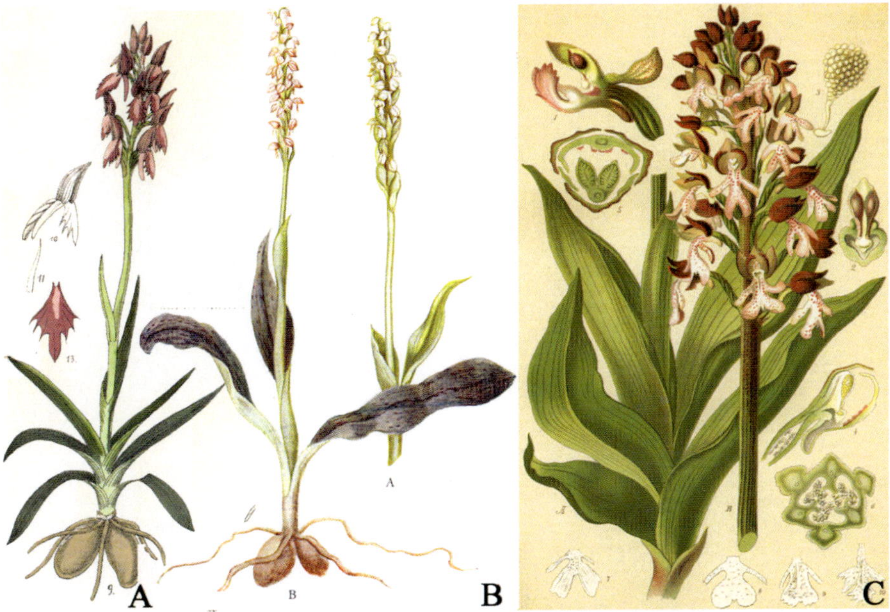

Fig. 3-1. Three *Orchis* species. **A.** *Orchis sancta.* **B.** *Neotinea maculata* (*Satyrium maculatum*). **C.** *Orchidspurpurea* (A, Camus and Camus, 1929; B, Godfery and Godfery, 1933; C, Schulze, 1894).

Northwest Africa, Poland, Portugal, Romania, parts of Russia, Slovakia, and Spain (Schulze, 1894; Camus, Bergon, and Camus, 1908; Keller and Schlechter, 1928; Camus and Camus, 1929; Godfery, 1933; Hayek, 1933; Summerhayes, 1968; Danesch and Danesch, 1969; Täckholm and Drar, 1969; Füller, 1972; Landwehr, 1977; Sezik, 1984; Hossain and El-Gadi, 1985).

A suggestion that "the rods of Moses and Aaron (Exodus 4:2, 17, 20; Numbers 17:2–10) have been identified as *Orchis purpurea* Huds" (Lawler, 1984) seems to be based on a misunderstanding. *Flora Magica* (Teirlinck, 1930)[1] presents several passages which refer to Aaron's rod (pp. 162–163) and on subsequent pages (163–164) lists several plants which have vernacular names that include the words "rod of Aaron." This list includes *Orchis purpurea* (which *Flora Magica* presents incorrectly as *Orchis purpurea* L. rather than the correct *Orchis purpurea* Huds.). *Flora Magica* does not suggest that Aaron's staff or rod (Fig. 3-3F) was made of *Orchis purpurea*. It merely lists this species as having a common name which involves the words "Aaron's rod."

Even if *Flora Magica* did suggest that Aaron's rod was made of *Orchis purpurea*, the suggestion would make no sense for two reasons. First, even if the inflorescence of *Orchis purpurea* is robust, it is still herbaceous and not rigid and hard enough to

[1] We thank Dr. Barbara Gravendeel of the Netherlands for obtaining for us copies of the relevant pages from *Flora Magica* and translating passages from it.

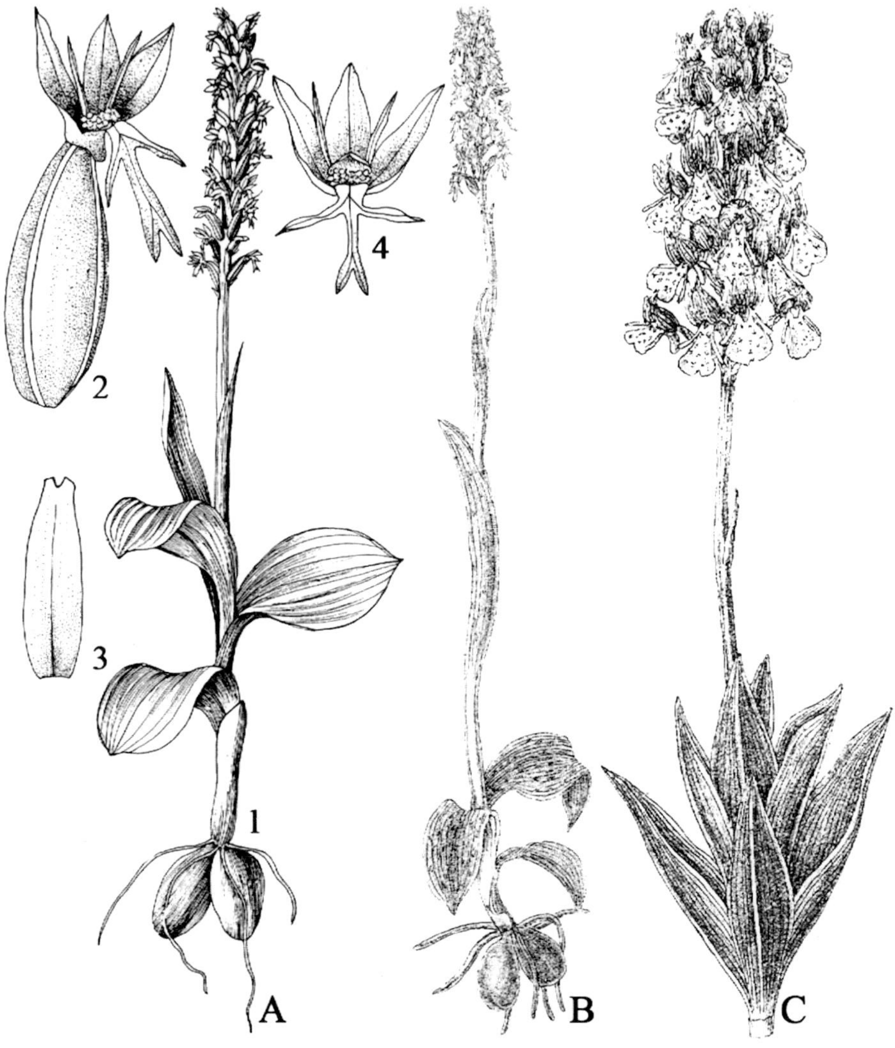

Fig. 3-2. Drawings of two orchids. **A, B.** *Netinea maculata* (*Satyrium maculatum*). **C.** *Orchis purpurea* (Hossain and El-Gadi, 1985; Camus et al., 1908).

be used as a rod or staff. Second, Egyptian paintings show that their staffs and rods were as long as the height of a human being or longer (Fig. 3-3A, 3-3E, 3-3G, and 3-3H) whereas inflorescences of *Orchis militaris* are at most 90 cm tall (Schulze, 1894; Camus et al., 1908; Keller and Schlechter, 1928; Camus and Camus, 1929; Godfery, 1933; Hayek, 1933; Summerhayes, 1968; Danesch and Danesch, 1969; Täckholm and Drar, 1969; Füller, 1972; Landwehr, 1977; Sezik, 1984; Hossain and El-Gadi, 1985).

An argument in favor of the idea that Aaron's rod may have been or was made of *Orchis purpurea* could be that the inflorescence axis is "round, solid, glabrous,

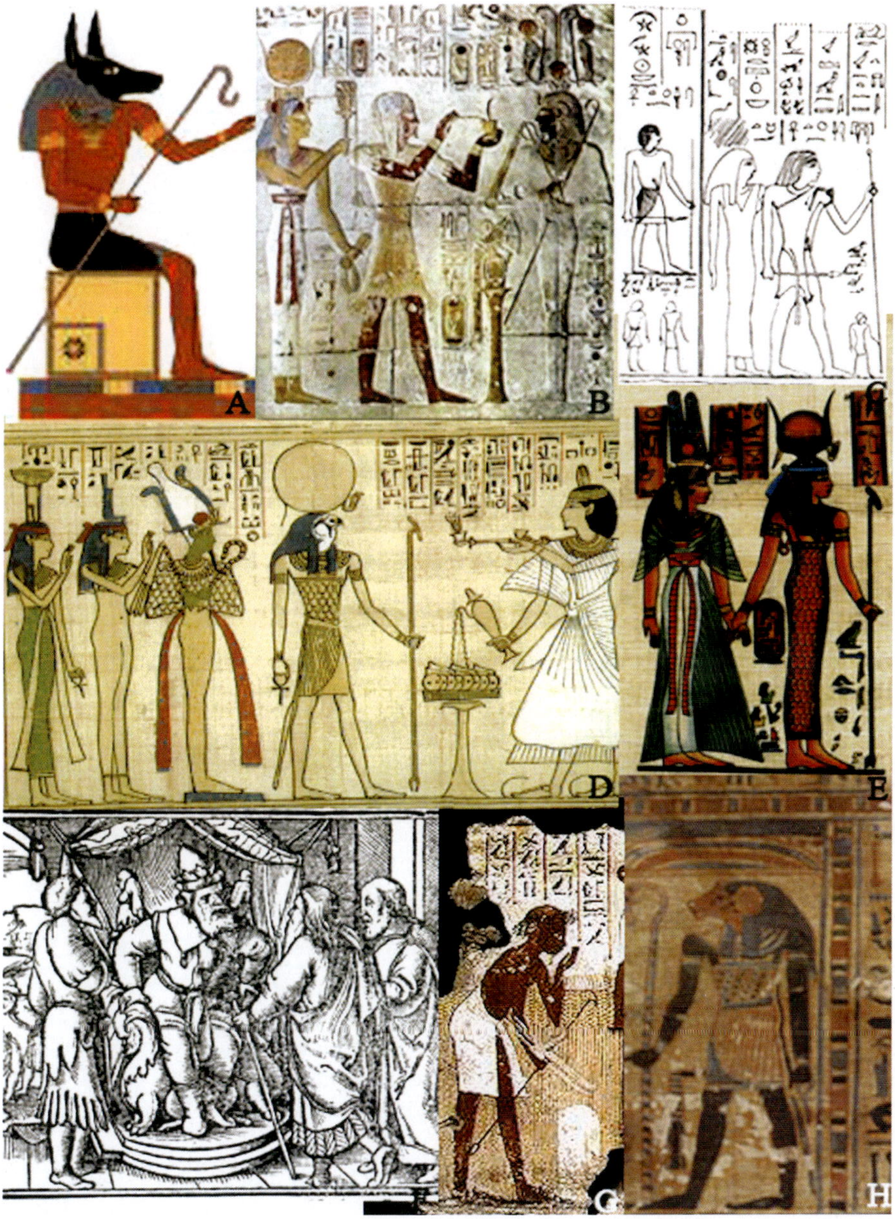

Fig. 3-3. Egyptian (**A-E, G, H**) and Aaron's (**F**) staffs
(A, http://www.besttreasurehunts.com/popup_image.php/pID/84;
B, http://www.touregypt.net/featurestories/priest2.jpg;
C, http://showcase.netins.net/web/ankh/niankh.jpg
D, http://www.symbolworld.org/learning/history/egypt/egypt-syms_files/sebek.gif;
E, http://www.directfromegypt.com/2004/misc_items/14_elgana_s.jpg;
F, http://www.metmuseum.org/collections/images/ep/images/ep50.70.L.jpg;
G, http://www.ancientegypt.co.uk/life/explore/images/fiegrain.jpg;
H, http://www.albanyinstitute.org/exhibits/Images/permanent/coffindetail.cat.web.JPG).

pale green below, angular, channeled and often dark dull purple above, sometimes with lines of green cells like a *pattern on a snake's skin*" (Godfery, 1933 with emphasis added). If this is what Aaron's rod looked like, it may have appeared as a snake to Pharaoh and his attendants. But then if this is the case, one could argue that the rods of Pharaoh's attendants were also made or could have been made of *Orchis purpurea*. And if all rods were made of *Orchis purpurea*, how did Aaron's orchid eat the Pharaoh's attendants' orchids? There are no known cannibalistic orchids. Given these facts and considerations the only possible conclusion is that *Orchis purpurea* is not referred to in the Bible.

Mustard Seeds versus Orchid Seeds

Orchid seeds (Figs. 3-4–3-7) are very light and small and produced in very large numbers (up to 4,000,000 per fruit have been counted). They are the smallest known seeds. The longest orchid seed, that of the neotropic species *Epidendrum secundum* is 6.0 mm long. It is 120 times longer than the shortest seed which is only 0.05 mm long and produced by the New Caledonian *Anoectochilus imitas*. The "widest" (the "width" is actually a diameter) orchids seeds, those of the New Guinean *Dendrobium insigne* (0.9 mm) an the South East Asian (Cambodia, Laos, Malaysia, Thailand, Vietnam). *Galeola nudifolia* (0.93 mm) seeds are 90 times "wider" than the "thinnest" ones which are the *Gastrodia* (a genus found in Australia, China, East Asia, Japan, Korea, New Zealand) type and measure 0.01–0.1 mm. Weight of only a few orchid seeds is known: 0.4–90 μg. Orchid embryos are even smaller and lighter. As a result the air space inside orchid seeds is 0.4–97% of the volume inside the seed coat. Because of that orchid seeds can float in the air or water for extended periods and travel distances as long as 2,000 km (for a review see Arditti and Abdul Ghani, 2000).

Mustard seeds are spherical, yellow, brown or white in color, 1.6–3 mm long, 1.2–2.1 mm wide, and weigh approximately 4 mg (Fig. 3-7). They are larger than the average orchid seed, solid without large air space and do not float like the seeds of orchids. In fact there is absolutely no physical resemblance between mustard and orchid seeds. Still, there is a statement that "it has been suggested that mustard seed of Jesus' parables may have been orchid seed (Moldenke and Moldenke, 1952)." The relevant passages in the Bible are:

> Then he said, "To what shall we liken the kingdom of God? Or, with what parable shall we picture it? (Mark 4:30) "It is like a mustard seed, which when it is sown on the ground, is *smaller than all of the seeds on earth* (Mark 4:31 with emphasis added).
>
> In another parable he said to them: "The kingdom of heaven is like a mustard seed, which a man took, and sowed in his field (Matthew 13: 31) "Which indeed is *the least of all the seeds*; but when it is grown it is greater than the herbs and becomes a tree, so that the birds of the air come and nest in its branches. (Matthew 13:32 with emphasis added).

The logic behind the suggestion "that the mustard seed of Jesus's parables may have been orchid seed" (for a review see Lawler, 1984) seems to be that if the New

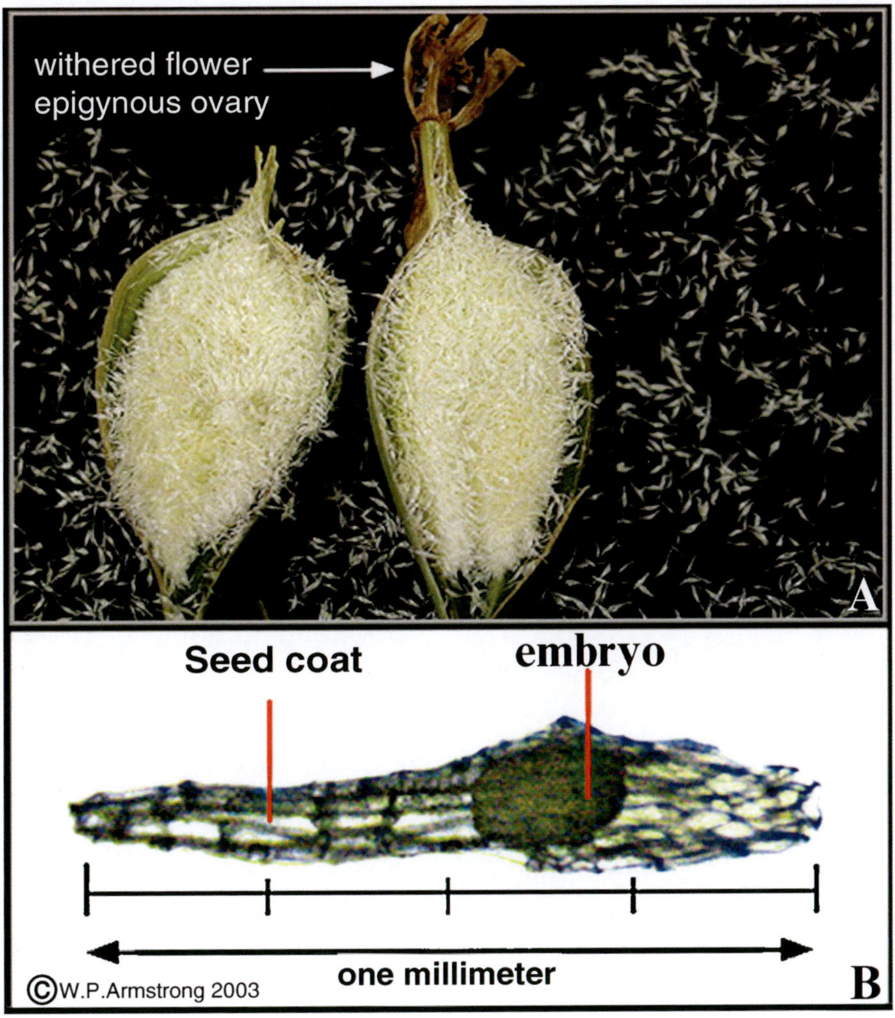

Fig. 3-4. Ripe orchid fruit with mature seeds (**A**) and (**B**) a single seed (Courtesy W. P. Armstrong).

Testament states that mustard has "smaller than all seeds on earth" then the actual reference must be to orchids seeds because they are the smallest in the world. To put it differently, the tortured logic seems to be that since the Bible (new and/or old testaments) cannot be wrong, mustard must mean orchid.

Not all biblical botanists accept this argument. Some agree that "surely the mustard seed was not really the smallest because the seed of the black orchid is smaller as already noted in medieval times by e.g., Albertus Magnus. But the mustard seed was in Jewish tradition proverbial for smallness" (Medema and Musselman, no date). There is no reference to Magnus's observation that the seeds of the "black orchid" are smaller than those of orchids. The references to the "proverbial

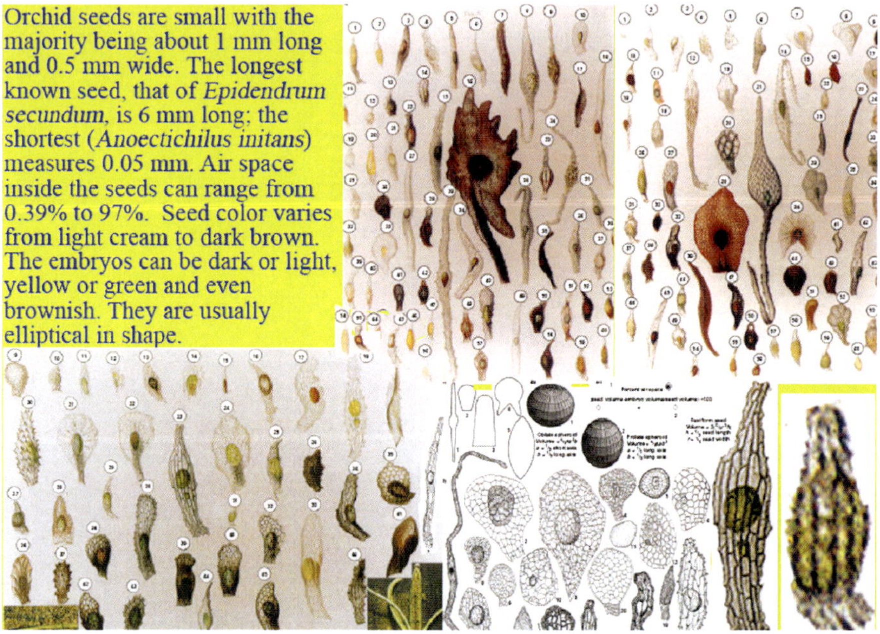

Fig. 3-5. Paintings and drawings of orchid seeds (Arditti and Abdul Ghani, 2000).

smallness" are Matthew 17:20 ("... if you have faith as a mustard seed...") and Sura 21, 48 in the Koran. In other words, the Bible (in this case Matthew 17:20) is used to provide proof for itself. The argument seems circular.

More logical suggestions are that: 1) "the seed of both black and white mustard is similar in size, about 1.0–3.0 mm (⅛ in.) so it is not the smallest seed, but it is the smallest seed of those which 'you plant in the ground'" (Medema and Musselman, no date). This argument which makes sense, does not attempt to claim that "mustard" means "orchid." It merely points to the fact that mustard has the smallest seeds among the crops cultivated by farmers in biblical times and lands, and 2) "... the seeds of *Brassica* [*nigra*] are small, and were probably the smallest seeds known to the common country folk comprising Jesus' audience in Galilee, yet they are far from being 'the least of all seeds'.... Mustard was probably a commonly cultivated garden herb in Jesus' day, as it is still in the Orient, for its oil. Its seeds would, therefore, be very familiar to country folk" (Moldenke and Moldenke, 1952).

Altogether it is reasonable to assume that both Mark and Matthew meant "mustard" rather than "orchid" when they said or wrote "mustard." Suggestions that "mustard" meant "orchid" have no basis in science, fact and biblical scholarship. In fact it seems that just as with the rod of Aaron, the suggestion that the reference to mustard seed was actually to orchid seeds is due to a misunderstanding by botanists who are not Bible experts and students of the Bible who are not plant scientists.

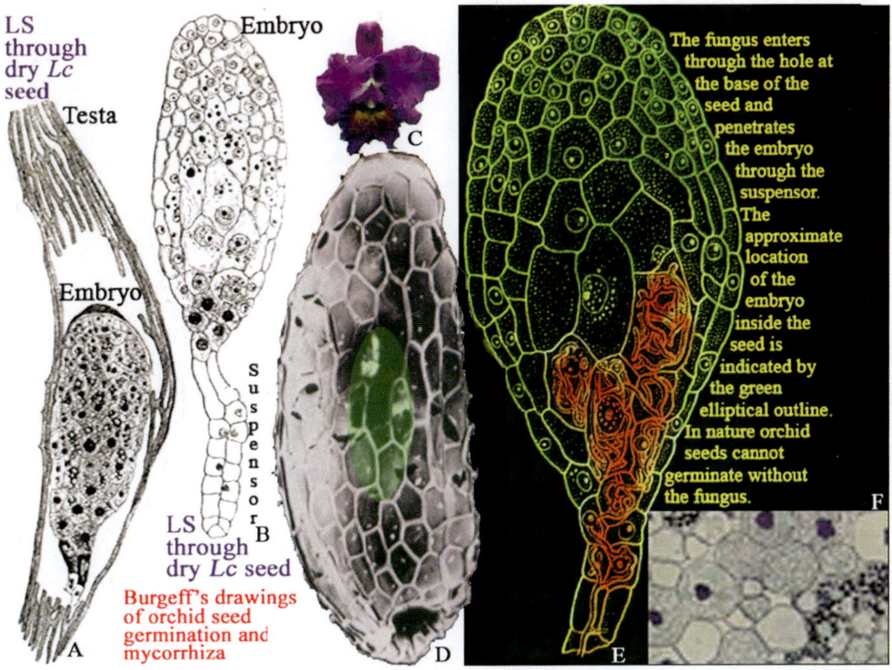

Fig. 3-6. Orchid seeds and embryos. **A.** Longitudinal section through a seed showing embryo and inclusions in its cells: starch (dark spots) and oil droplets (translucent grayish-white spots). **B.** Longitudinal section through *Laeliocattleya* (*Lc*) embryo. **C.** *Laeliocattleya* (*Lc*) flower. **D.** Scanning electron microscope photograph of seed showing location of embryo. **E.** Embryo being colonized by a fungus. **F.** Fungus pelotons in orchid seed cells (A, B, E, drawings from Burgeff, 1936; C, D, Joseph Arditti (scanning electron microscope photograph by technician Allison Oliva), "embryo" and montage, computer generated).

Dudaim

At one time or another the Hebrew word Dudaim was considered to be:

- Banana (Fig. 3-8C, 3-8D, and 3-8F) for no apparent reason. This view is no longer held or espoused by biblical botanists.
- A melon, either *Cucumis melo* or *Citrullus vulgaris* (at one time) or (currently) *Citrullus lanatus* (Figs. 3-8E, 3-9A, B, C, E, and F; Henslow, 1906; Moldenke and Moldenke, 1952; Musselman, 2005; Souvay, no date), because of a melon named *Cucumis dudaim* by Linnaeus. This view has also been discarded.
- Love-apple (Fig. 3-8) or mandrake, *Mandragora officinarum* or *Mandragora autumnalis* (Figs. 3-9A, D, and 3-10G–3-10J) because that is how the Hebrew word was translated (Hertz, 1975). The fruit is described or illustrated as being red or "the size of a large plum, quite round, yellow and full of soft pulp. The fruit is still considered in the East as a love- charm. This explains Rachel's intent to obtain it" (Hertz, 1975; Figs. 3-8–3-10). The Greek word is μανδραγθρας.

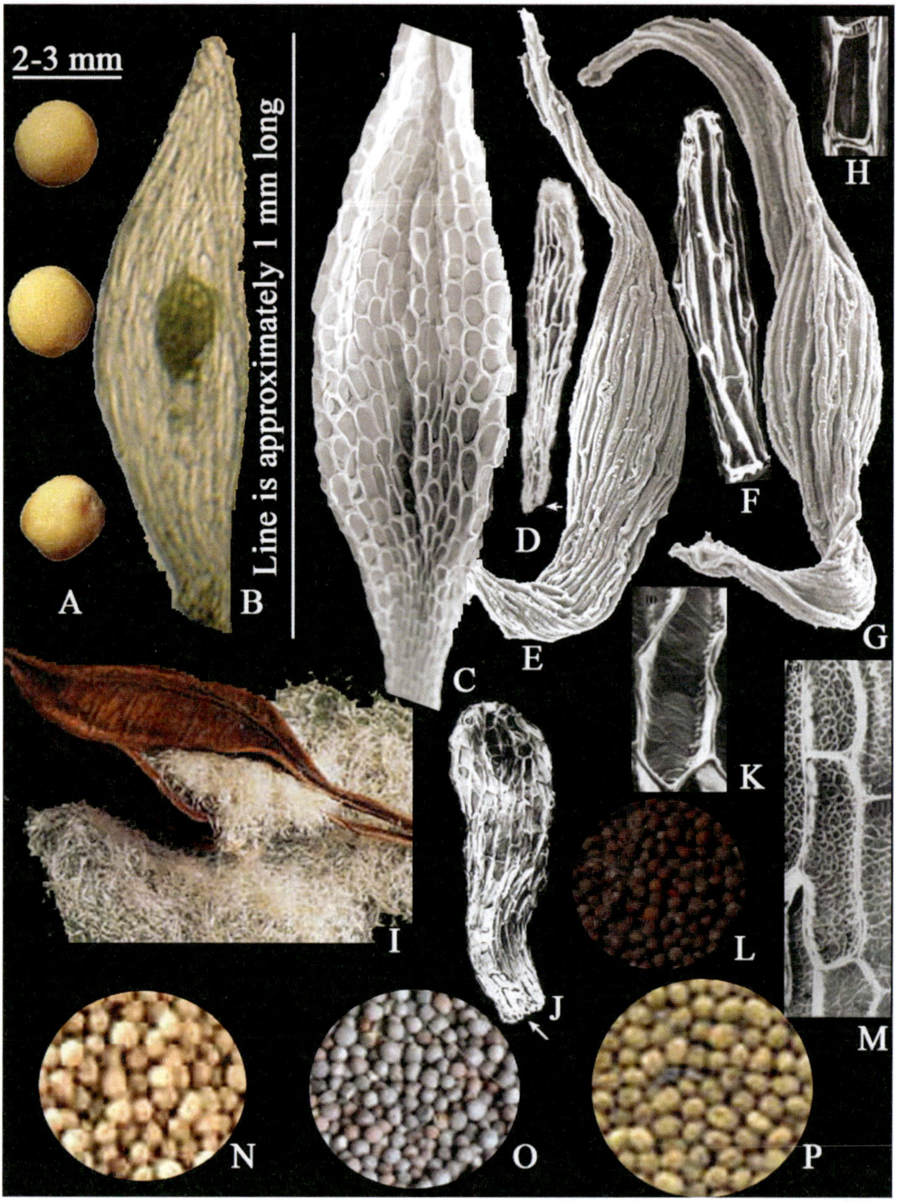

Fig. 3-7. Mustard and orchid seeds. **A, L, N–P.** Mustard seeds. These seeds vary in color from dark brown (K) to yellow (O), light yellow (M) and gray (N). They measure about 1.6–3 mm in diameter with 2 mm being most common. Mustard seeds are solid and not balloon-like. Orchid seeds are ballon-like. Air space inside the seed coat (testa) can be as high as 90%. **B.** Light microscope photograph of an orchid seed. **C–G, J.** Scanning electron microscope photographs of orchid seeds. **H, K, M.** Scanning electron microscope photographs of individual cells of seed coats of orchid seeds. The walls of some seed coats are smooth, others are reticulated. **I.** Ripe fruit and mature seeds (A, L, N–P. World Wide Web; B, I, courtesy Markus and Martin Axelsson, Sweden; C, E, G, courtesy Tea Meula, OARDC, Ohio State University; D, F, H, J, K, M, Allison Oliva, a technician in J. Arditti's laboratory at the time).

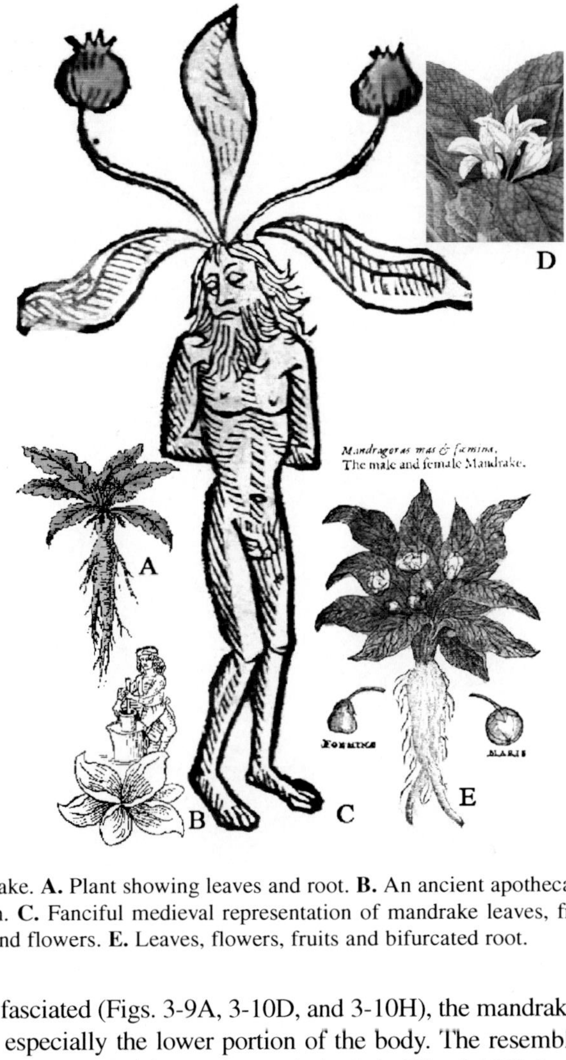

Fig. 3-8. Mandrake. **A.** Plant showing leaves and root. **B.** An ancient apothecary making a mandrake preparation. **C.** Fanciful medieval representation of mandrake leaves, fruits and man-like root. **D.** Leaves and flowers. **E.** Leaves, flowers, fruits and bifurcated root.

Being often fasciated (Figs. 3-9A, 3-10D, and 3-10H), the mandrake root resembles a human form, especially the lower portion of the body. The resemblance to humans was often exaggerated and elaborated upon in herbals (Fig. 3-10G, I, and J). In Roman times it was believed that humans who uprooted these plants died. Therefore dogs were tied to the plants and made to pull them out (Fig. 3-10I). These dogs were believed to die after that. It was also believed that the plants shrieked as they were being pulled out. Shakespeare immortalized this belief in *Romeo and Juliet* as:

> Shrieks like the mandrake torn out of the earth,
> That living mortals hearing them run mad.

For reasons which are neither given nor buttressed by citations or justified in any way, two sources assert that mandragores (i.e., dudaim) are salep (Fig. 3-10), a drink or pudding, that is made from orchid tubers:

> Dudaim de la Bible, demandé par Rachel à Lia, représenté pour les uns par les bulbes de l'*Orchis sancta* [Fig. 3-1A] ou du *Satyrium maculatum* [Fig. 3-1B] de la Palestine, pour

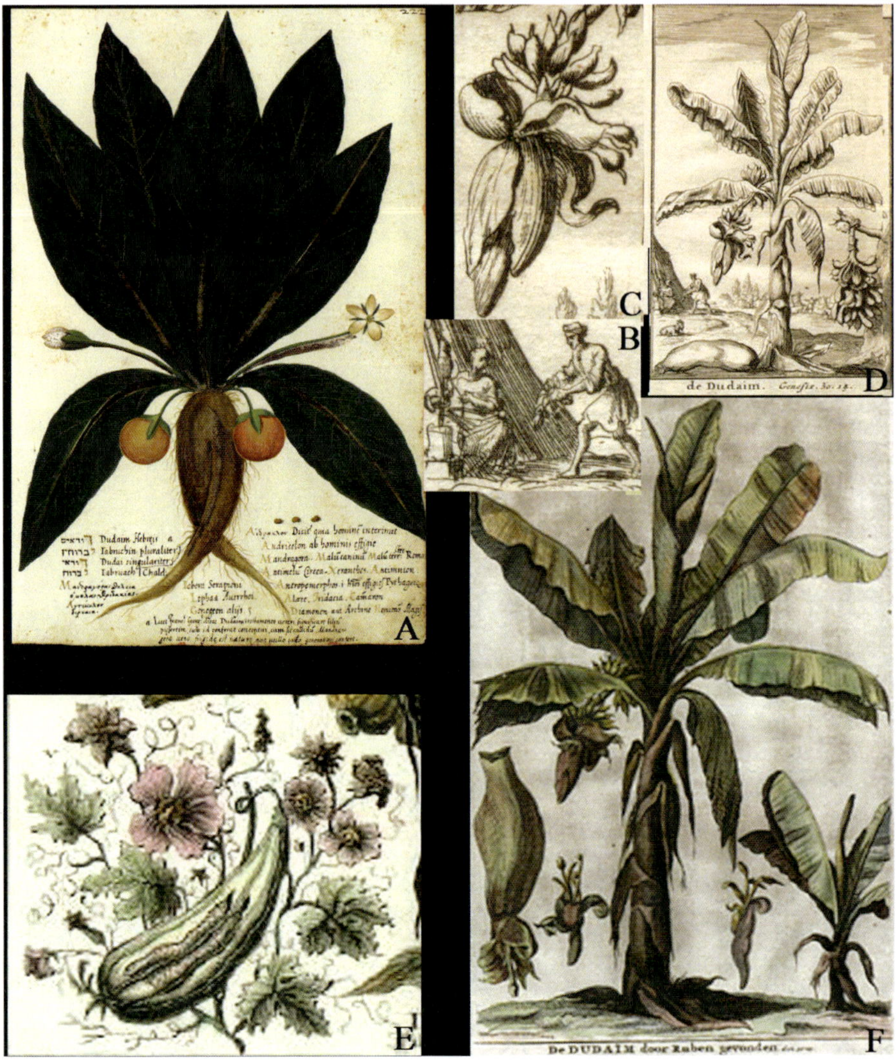

Fig. 3-9. Several representations of Dudaim. **A.** Dudaim as Mandrake. **B.** Dudaim being brought to Leah. **C, D, F.** Dudaim as banana. **E.** Dudaim as melom (Dutch books from the 1700s and 1800s).

d'auitres par le Salep, par les fruits du *Cucumis dudaim* ou *odoratissimus* de l'Inde et de la Perse…" In free translation "Dudaim of the Bible that is requested by Rachel of Leah is for some the bulbs of *Orchis sancta* [Fig. 3-1A] or of *Satyrium maculatum* [Fig. 3-1B], but according to others for the fruits of *Cucumis dudaim* or *Cucumis odoratissimus* from India or Persia (Chatin, 1868).

… les madragores, (les dudaim) de la Bible sont du Salep…" which translates into "the mandragores (the dudaim) of the bible are salep" (Sugaya, 1999; an e-mail request for clarification or citation, elicited no reply from Dr. Sugaya).

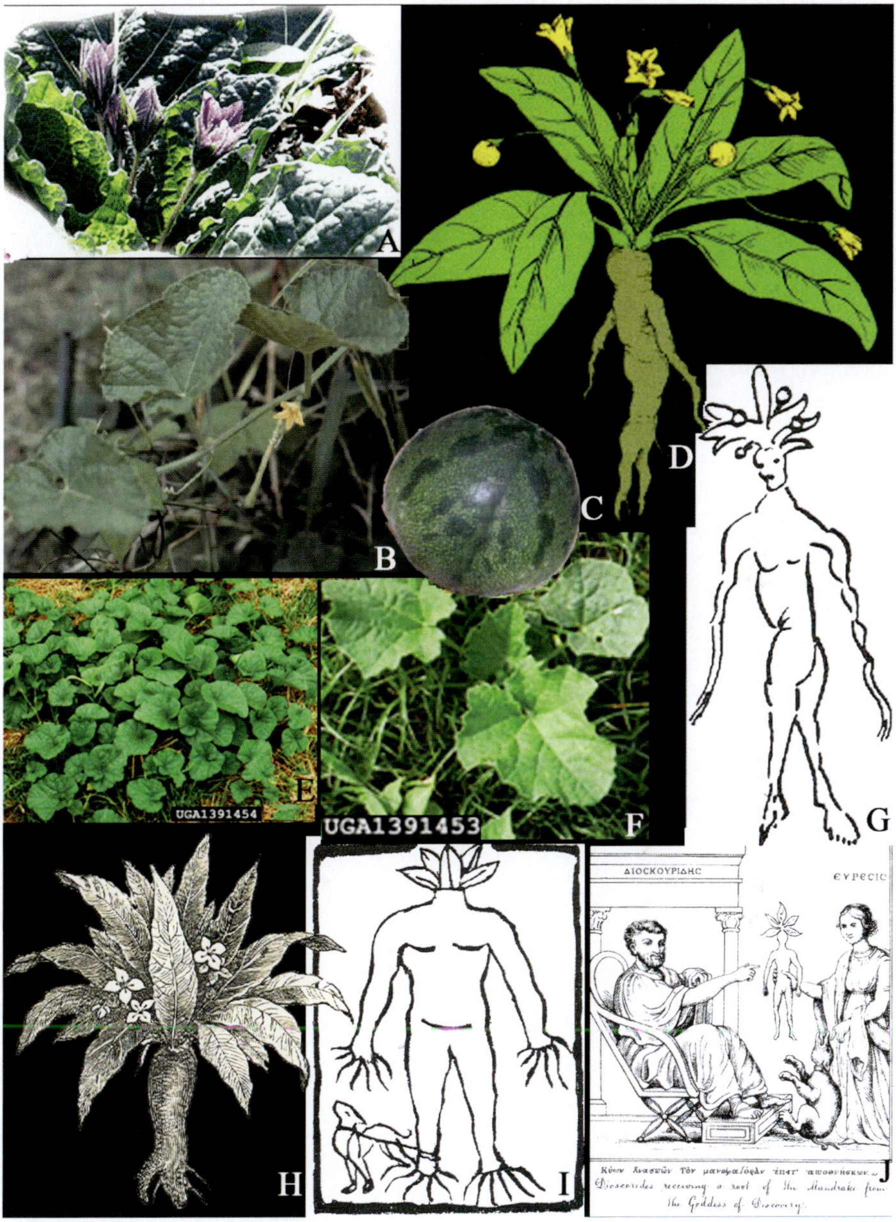

Fig. 3-10. Different interpretations of Dudaim. **A-C, E, F.** *Cucumis melo*, a melon. **D, G-I.** Mandrake. **J.** Dioscorides receiving a Mandrake root from the Goddess of Discovery (A, C, John D. Byrd, Mississippi State University, www.forestryimages.org; D, World wide web; B, Digital Flora of Texas; E, F, Larry Allain @ USDA-NRCS PLANTS Database and USDA; G, J, Gunther, 1968; H, Henslow, 1906; I, Herbal of Apuleius Barbarus).

One possible, even if far fetched, explanation for suggesting that dudaim are salep (Fig. 3-10), may be found in Genesis 30: 15–16 (Hertz, 1975): "And Reuben went in the days of wheat harvest, and found mandrakes in the field, and brought them unto his mother Leah. Then Rachel said to Leah: 'Give me, I pray thee, of thy son's mandrakes' (14). And she said unto her, 'It is a small matter that thou hast taken away my husband? and wouldest thou take away my son's mandrakes also?' and Rachel said, 'Therefore he shall lie with thee to-night for thy son's mandrakes" (15). "And Jacob came from the field in the evening and Leah went out to meet him, and said: 'Thou must come in unto me; for I have surely hired thee with my son's mandrakes.' And he lay with her that night" (16).

These verses confirm that in Biblical times as well as later, mandrake was considered to have sexual and/or aphrodisiac attributes. And, since some orchid roots or tubers are used to make salep (Fig. 3-11), it is possible that the writers cited above (Chatin, 1868; Sugaya, 1999) or at least one of them assumed that dudaim refers to orchids. The offering of dudaim to "my beloved" in the Song of Solomon 7:14 also suggests an association with sex. If only one writer made this assumption, it was probably the earlier one (Chatin, 1868) and this lead to the later statement. Interestingly, Genesis 30: 14–16 also shows that marital and family relations and intrigues during biblical times were not all that different from those at present.

As with Aaron's rod and mustard seeds, it seems that the assumption that dudaim are orchids was made not because of any biblical writings, but due to incorrect assumption and logic by relatively modern writers.

Post Biblical Jewish Writings

Volume XIII of the *Encyclopedia Judaica* has a table of plants mentioned in the Bible and the Babylonian Talmud (a code of law developed by rabbis working in the academies of Babylonia between *c*. 220 CE and *c*. 500 CE; Fig. 3-1), and the Jerusalem Talmud (developed by rabbis living in Israel between *c*. 200 CE and *c*. 375 CE). Both Talmuds discuss and elaborate the Mishnah, an earlier compilation of legal oral traditions dating back to *c*. 200 BCE and first redacted in 200 CE. The Mishnah of the Jerusalem Talmud is, however, often not the same as that found in the Babylonian Talmud. The above *Encyclopedia Judaica* table contains no mention of *Orchis* sp. in the Hebrew Bible, but does refer to a plant called *khalbetzin* (חלבצין) in two Mishna references (*Sheviith* 7:1 and 7:2); *khalbetzin* is described in the *Encyclopedia Judaica* reference as "flower with edible bulb" (Felix, 1974b). This description can apply to any biblical root crop (but in this case not to potatoes since they are native the New World).

Contrary to the *Encyclopedia Judaica* list the term *khalbetzin* does not appear in *Sheviit* 7:1 of the Babylonian and the Jerusalem Talmuds. *Khalbetzin* is found, however, in Mishna 7:2 of both Talmuds, but is given different translations. In a translation of the Jerusalem Talmud (*The Talmud of the Land of Israel*), *khalbetzin* is listed as an orchid (Avery-Peck, 1991). In the Babylonian Talmud, however,

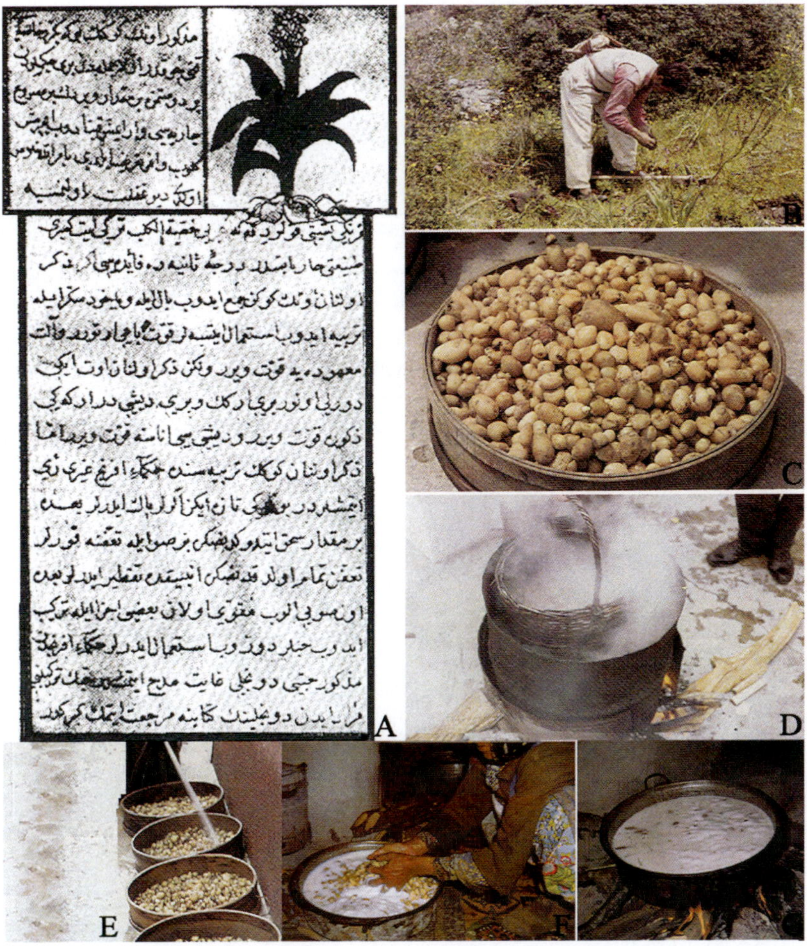

Fig. 3-11. Salep. **A.** Name (upper left) of orchid (upper right) and description (bottom) of salep in a handwritten Turkish book by Mahmet Ali published in 1691–1692 and entitled *Tercüme-i Cedide Fil-Havasil Müfrede*. "The plant is called Testiculu di canis (dog testicles) in Latin, Hüsyet-ül kelb (dog testicles) and It Kesesi (dog testicles) in Turkish [Husye=kese=testicolo=test icle; kelb=it=canis=dog]. First it warms the body. Second, it increases sexual prowess and strengthens sexual organs when blended with honey and sugar. Thc plant occurs in two forms, male and female. The male plants is used by men and the female is used by women. European doctors use different methods for blending. Fresh roots are usually fermented in aqueous medium after they are cleaned. They distill the mixture after fermentation is completed. The distillate is blended with other drugs and used in the form of pills. Dunceli, a European doctor praises these pills. Those who want to learn the composition of these pills should refer to Dr. Dunceli's book. Alternately those interested can refer to the pills chapter in the book *Jewels of Medicine* which was translated into Turkish by Dr. E.E. Sezik. The pills described here are made of salep which is a holy substance" (translation by Dr. Eczaci Ekrem Sezik from Sezik, 1984). **B.** Salep collectors in the Turkish province of Muğla digging for tubers. **C.** Orchis tubers for sale in the Milas Bazaar, Turkey. **D.** Tubers being boiled. **E.** Washing tubers, **F, G.** Boiling tubers in the Turkish province of Kastamonu. The cleaned tubers are submerged in milk or buttermilk and boiled like potatoes (Courtesy Dr. E. E. Sezik).

khalbetzin is translated as an asphodel (*Asphodelus*, Liliaceae). Jastrow's dictionary of the Babylonian and the Jerusalem Talmud and the Misha translate *khalbetzin* as *Ornithogalum* (Jastrow, 1975) which also belongs to the Liliaceae, not the Orchidaceae.

No reason is given by Avery-Peck for the identification of *Khalbetzin* as an orchid (Avery-Peck, 1991). "Flower with an edible bulbs" is too broad a statement to justify identification of any plant which had bulb, tubers, corm or swollen roots. Also, among the few orchids whose tubers ("bulbs") are reported to have been eaten are *Gastrodia elata* and several *Cymbidium* species (Lawler, 1984) neither of which is found in the Middle East. *Orchis* tubers are used to prepare salep (Fig. 3-10) in the Middle East, but are not eaten as such (Lawler, 1984; Sezik, 1984). Even if *khalbetzin* is an orchid it should be noted that the tractate Sheviith in the Mishna (Blackman, 1983) is an expansion of biblical laws rather than a part of the Bible. In the extensive index to the Soncino Hebrew-English Babylonian Talmud (Slotski, 1990) there is no reference to orchids.

Conclusion

There is no evidence that orchids are mentioned on the Bible despite the fact that there are several native species in the Holy Land (Dafni, 1981).

Dedications

For Doris Dunn, Ph.D., and Jonathan and David.–A.S.D.
For Liza and Stuart Krassner because of Jonathan.–J.A.
Both of us have sons named Jonathan.

Literature Cited

Ames, O. 1942. The origin of the term orchis. *American Orchid Society Bulletin* 11: 146–147.

Arditti, J. and A. K. Abdul Ghani. 2000. Numerical and physical properties of orchid seeds and their biological implications. Tansley Review No. 110. *New Phytologist* 145: 367–421.

Avery-Peck, A. J. 1991. *The Talmud of the Land of Israel, volume 5 Shebiit*. The University of Chicago Press, Chicago.

Barbarus A. *ca* 1000. Herbal (1925 printing). Roxburghe Club, Oxford.

Blackman, P. 1983. *Mishnayot, volume VII*. Judaica Press, Gateshead, UK.

Burgeff, H. 1936. *Samenkeimung der Orchideen und Entwicklung ihrer Keimpflanzen*. Verlag von Gustav Fischer, Jena.

Camus, E. G. and A. Camus. 1929. *Iconographie des orchdées d'Europe et du bassin Méditerranéen*. Paul Lechevalier, Paris.

Camus, E. G., P. Bergon and A. Camus. 1908. *Monographie des Orchidées de l'Europe, de l'Afrique septentrionale, de l'Asie Mineure et des Provinces Russes transcaspiennes*. Jacques Lechevalier, Paris (this remarkable book was published in 175 mimeographed copies in some of which the illustrations were hand colored; when available copies sell for thousands of dollars).

Chatin, A. 1868. L'histoire naturelle médicale a l'exposition universelle. Pages 295–368 *in* M. M. Chevalier (ed.), *Rapports de jury international, Exposition Universelle de 1867 a Paris*. Imprimerie Administrative de Paul Dupont, Paris.

Dafni, A. 1981. *Orchids of Israel*. Massada Publishing House, Tel Aviv.

Danesch, E., and O. Danesch. 1969. *Orchideen Europas. Südeuropa*. Verlag Hallwag, Bern.

Felix, J. 1974a. Flowers of the Bible. Pages 1366–1367 *in* C. Roth (ed.), *Encyclopaedia Judaica, volume VI*. Keter Publishing, Jerusalem.

Felix, J. 1974b. Plants. Pages 614–627 *in* C. Roth (ed.), *Encyclopaedia Judaica, volume XIII*. Keter Publishing, Jerusalem.

Füller, F. 1972. *Die Orchideen Deutschlands, 3. Die gattungen Orchis und Dactylorhiza*, 2nd ed. Die Neue Brehm-Bücherei, A. Ziemsen Verlag, Wittenberg Lutherstadt.

Godfery, M. J., and H. M. Godfery. 1933. *Monograph & Iconograph of Native British Orchidaceae*. Cambridge University Press, Cambridge.

Gunther, R. T. (ed.). 1968. *The Greek Herbal of Dioscorides Illustrated by a Byzantine A. D. 512 Englished by John Goodyear A. D. 1655*. Reprint of the first priting in 1933.

Hayek, A., and F. Markgraf. 1933. *Prodromus florae peninsulae Balcanicae*. Verlag der Repertorium Specierum novarum Regni Vegetabilis, Dahlem bei Berlin.

Henslow, G. 1906. *The Plants of the Bible: Their Ancient and Medieval History, Popularity Described*. Masters & Co., Ltd., London.

Hepper, F. N. No date. Plants in the Bible *in Holmans bible dictionary*. http://www.biblicalgardens.org/spsMy Web/Hepper_PlantsInTheBible.htm.

Hertz, J. H. 1975. *The Pentateuch and Haftorahs*. Soncino Press, London.

Hossain, A. B. M. E., and A. El-Gadi. 1985. Orchidaceae. *In* S. M. H. Jafri and A. El-Gadi (eds.), *Flora of Libya*. The National Academy for Scientific Research, Tripoli, Lybia.

Jastrow, M. 1975. *A dictionary of the Targumim, the Talmud Babli and Yerushalmi, and the Midrashic literature*. Judaica Press, New York.

Keller, G., and R. Schlechter. 1928. *Monographie und iconographie der orchideen Europas and des Mittelmeergebites*. Verlag der Repertorium Specierum novarum Regni Vegetabilis, Dahlem bei Berlin.

Landwehr, J. 1977. *Wilde orchideeën van Europa*. Vereiniging tot Behoud van Natuurmonumenten in Nederland, 's-Graveland, The Netherlands.

Lawler, L. J. 1984. Ethnobotany of the Orchidaceae. Pages 27–149 *in* J. Arditti (ed.), *Orchid Biology: Reviews and Perspectives*. Cornell University Press, Ithaca, NY.

Medema, H. P. and L. J. Musselman. No date. The parable of the mustard seed. http://www.odu.edu/webroot/instr/sci/plant.nsf/pages/mustard.

Moldenke, H. and A. Moldenke. 1952. *Plants of the Bible*. Ronald Press, New York.

Musselman, L. J. 2003. Trees in the Koran and the Bible. *Unasylva* 213: 47–52 and also on WWW, http://www.the-tree.org.uk/Sadred%20Grove/Articles/K&B/koran&bible.htm.

Musselman, L. J. 2005. All the plants in the Bible. http://www.odu.edu/webroot/instr/sci/plant nsf/pages/all bibleplantslist.

Neusner, J. (ed.). 1991. *The Talmud of the Land of Israel, volume 5*. Shebiit. The University of Chicago Press, Chicago. This Talmud is also known as the Jerusalem Talmud.

Schulze, M. 1894. *Die Orchidaceen Deutschlands, Deutsch-Oesterreichs unde der Schweiz*. Fr. Eugen Köhler's Verla, Gers.-Untermhaus.

Sezik, E. E. 1984. *Orkidelerimiz, Türkye'nin Orkideleri*. No publisher listed.

Souvay, C. L. No date. Plants in the Bible *in Catholic Encyclopedia*. http://www.newadvent.org/cathen/12149.htm.

Sugaya, N. 1999. Les sciences médicales dans Bouvard et Pecuchet de Gustave Flaubert. Doctoral Thesis in French Literature, vol. II. University of Paris VIII, Vincennes à Saint-Denis, France.

Summerhayes, V. S. 1968. *Wild Orchids of Britain*. Collins, London.

Tackholm, V., and M. Drar. 1969. Orchidceae in *Flora of Egypt* Vol. IV. The Public Organization for Books and Scientific Appliances, Cairo University Press, Cairo.

Teirlinck, I. 1930. *Flora magica*. De Sikkel, Anwerp, The Netherlands.

MORPHOLOGY

4

Food-Hair Form and Diversification in Orchids

KEVIN. L. DAVIES

Contents

Introduction 160
Food-Hairs and Pseudopollen 161
Terminology 163
Occurrence of Pseudopollen 165
Trichome Morphology 169
Development of Pseudopollen 170
Food Content and Ecology 170
Trichomal Secretion of Resin-Like and Waxy Materials 172
Trichomal Elaiophores 173
Evolution of Food-Hairs and Food-Hairs as Taxonomic Characters 176
Conclusions 177
Glossary 178
Literature Cited 180

Kull, Arditti and Wong (eds.), *Orchid Biology: Reviews and Perspectives, X*,

Introduction

Orchids are renowned for their diverse and often elaborate pollination strategies (van der Pijl and Dodson, 1969; van der Cingel, 2001). Some reward pollinators with food (e.g., nectar, food-hairs and oils), floral fragrances and other compounds such as resin-like substances and wax (van der Pijl and Dodson, 1969; Dressler, 1990, 1993; van der Cingel, 2001) and these rewards, in turn, reinforce pollinator foraging behaviour (van der Pijl and Dodson, 1969; Proctor and Yeo, 1975; Dressler, 1990; Proctor, Yeo, and Lack, 1996; van der Cingel, 2001). Many, however, produce no rewards whatsoever, and here attraction by mimicry and deceit tend to predominate (Porsch, 1908; van der Pijl and Dodson, 1969; Ackerman, 1984; Neiland and Wilcock, 1998, 2000; van der Cingel; 2001). In fact, some one-third of orchid species attract potential pollinators solely by deceit (Ackerman, 1984) and it is thought that deceptive pollination evolved from reward-mediated pollination systems (Ackerman, 1986). The former may involve complex mimicry strategies such as food-fraud, pseudocopulation and pseudoantagonism (van der Pijl and Dodson, 1969; Dressler, 1990; van der Cingel, 2001) and once attracted to the flower by olfactory and visual cues, the precise configuration of the floral parts, the presence of honey guides and tactile stimuli provided by floral hairs and papillae ensure that orientation of the insect upon the flower is optimal for pollination.

Although the rewardless condition is common amongst orchids, a significant number of species, nonetheless, produce food rewards (van der Pijl and Dodson, 1969; Dressler, 1990, 1993; van der Cingel, 2001). Many angiosperm families reward pollinators with pollen (Proctor and Yeo, 1975; Proctor et al., 1996). However, that of epidendroid orchids is bound within pollinia and is thus inaccessible to foraging insects (van der Pijl and Dodson, 1969; Dressler, 1990, 1993; van der Cingel, 2001). Even so, floral, food rewards such as nectar, food-hairs and floral oils play an important role in the successful pollination of many orchids (van der Pijl and Dodson, 1969; Proctor and Yeo, 1975; Dressler, 1990; Proctor et al., 1996; van der Cingel, 2001) and their effectiveness in the attraction of pollinators has been convincingly demonstrated for a number of species (Dafni and Ivri, 1979; Inoue, 1986; Johnson and Bond, 1997; Johnson and Nilsson, 1999; Neiland and Wilcock, 1994, 1998, 2000; Smithson, 2002). Moreover, they have been shown to be potent even in small quantities (Ackerman, Rodriguez-Robles, and Meléndez, 1994) and Neiland and Wilcock (1998) have reported that species that offer rewards often double their chances of developing fruit and seed. However, reward production and the subsequent processes of fruit- and seed-maturation are costly both in terms of materials and energy expenditure and this may outweigh the benefits (Southwick, 1984; Pyke, 1991; Ackerman et al., 1994; Meléndez-Ackerman, Ackerman, and Rodriguez-Robles, 2000 and references therein). Despite the cost, floral rewards, nevertheless, generally confer evolutionary advantage.

Although primitive orchid species reward pollinators with pollen (Kocyan and Endress, 2001; Sugiura, Miyazaki, and Nagaishi, 2006), nectar is the most common food reward in Orchidaceae and, in some species this is secreted by trichomes.

Thus, floral food-hairs can conveniently be divided into four main categories: protein- or starch-containing trichomes including pseudopollen that are nibbled or gathered by insect pollinators (Janse, 1886; Porsch, 1905, 1906; Beck, 1914; van der Pijl and Dodson, 1969; Davies and Winters, 1998; Davies, Winters, and Turner, 2000; Davies, Roberts, and Turner, 2002; Davies, Turner, and Gregg, 2003a; Davies and Turner, 2004a, b, c; Matusiewicz, Stpiczyńska, and Davies, 2004; Davies and Stpiczyńska, 2006), resin-secreting trichomes (Porsch, 1905; von Kirchner, 1925; Macpherson and Rupp, 1935; van der Pijl and Dodson, 1969; Roberto Vásquez and Dodson, 1982; Davies et al., 2002, 2003a; Davies, Turner, and Gregg, 2003b; Davies and Turner, 2004a; Flach et al., 2004; Matusiewicz et al., 2004; Davies and Stpiczyńska, 2006; Davies, Stpiczyńska and Turner, 2006), oil-secreting, trichomal elaiophores (Vogel, 1974; Buchmann, 1987; Toscano de Brito, 2001; Mickeliunas, Pansarin, and Sazima, 2006) and nectar-secreting trichomes such as those lining the nectar spurs of *Aeranthes arachnites* (Thouars) Lindl., *A. grandiflora* Lindl. (Roberts, 2001), *Platanthera bifolia* L. (Stpiczyńska, 1997), *P. chlorantha* (Custer) Rchb. (Stpiczyńska, 2003; Stpiczyńska et al., 2005) and *Gymnadenia conopsea* (L.) R. Br. (Stpiczyńska and Matusiewicz, 2001). This fourth category, however, lies beyond the scope of this paper.

Food-Hairs and Pseudopollen

The development of food-hairs (including pseudopollen) in orchids represents a major stride in their evolution. Pseudopollen is a mealy material superficially resembling pollen and is produced as food-hairs become detached or fragment to form individual or small groups of cells. It is significant that flowers that produce pseudopollen tend to lack nectar (van der Pijl and Dodson, 1969) and, with the exception of some notable examples such as members of the *Maxillaria grandiflora* (Humb., Bonpl. & Kunth) Lindl. and *Maxillaria lepidota* Lindl. alliances, fragrance perceptible to humans (Flach et al., 2004; Singer et al., 2006). Moreover, a mentum, possibly a vestigial nectary spur, is usually present in pseudopollen-producing and rewardless species of *Maxillaria* Ruiz & Pav., indicating that pseudopollen may have replaced nectar as the main reward and that the production of nectar preceded the rewardless condition in that genus (Davies et al., 2003a). Indeed, Davies et al. (2005) estimate that some 56% of *Maxillaria* spp. are rewardless, 16–23% produce pseudopollen, 13% produce wax or a viscid, resin-like material rich in lipids and aromatic amino acids and only 8% produce nectar.

Food-hairs, including pseudopollen-forming hairs, usually occur upon the labellum (Figs. 4-1–4-13). They contain rich reserves of food, mainly protein, and are gathered or nibbled by insects. Of those orchid species that offer food rewards, most offer one type only. In some cases, however, several types of food reward may occur in a single species. For example, *Maxillaria violaceopunctata* Rchb.f. and *M. lepidota* produce both food-hairs and a resin-like substance (Davies et al., 2003a; Matusiewicz et al., 2004) and food-laden papillae and viscid material have

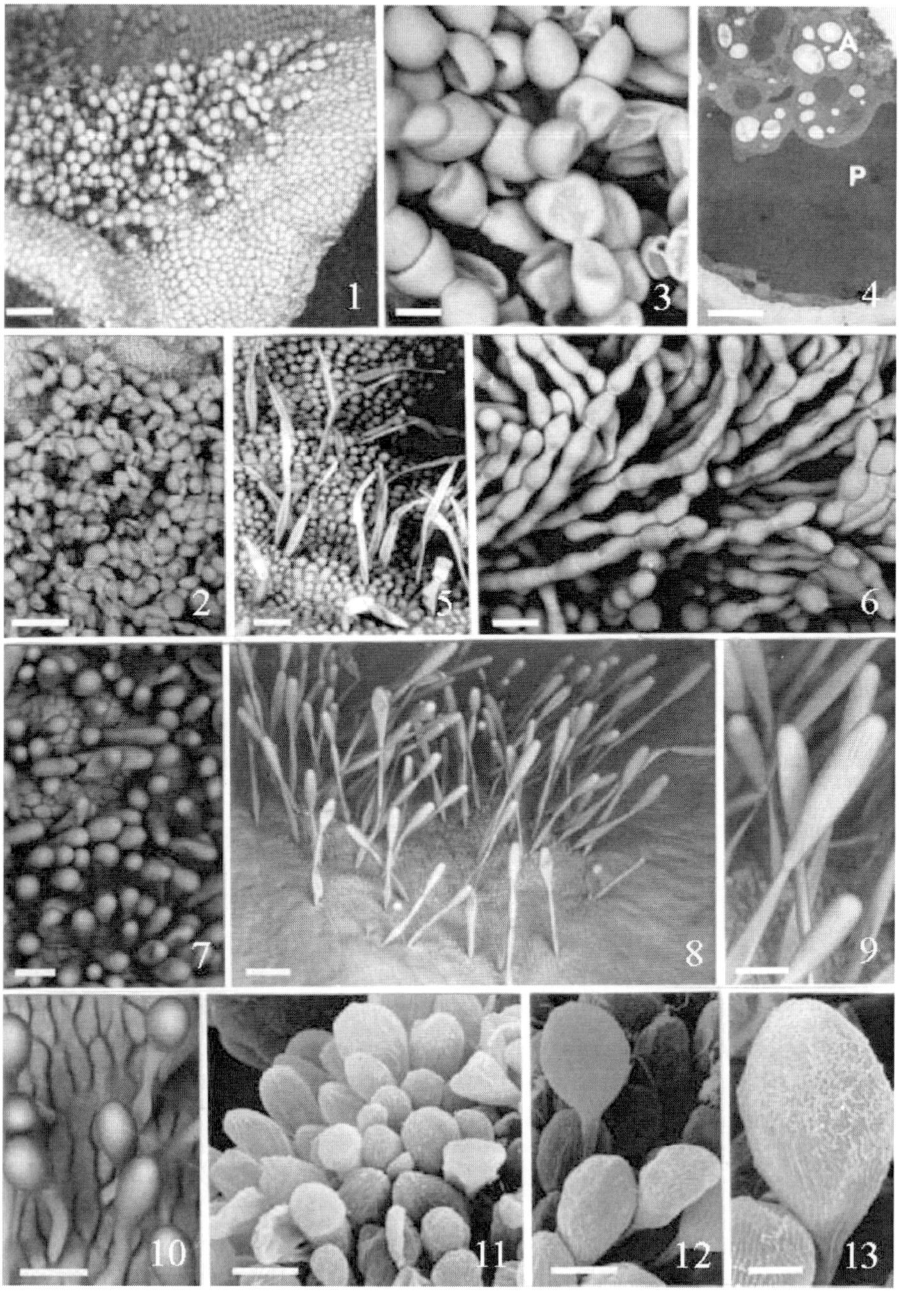
1
3
A
P
4
2
5
6
7
8
9
10
11
12
13

also been reported for *M. acutifolia* Lindl. and *M. tenuibulba* E.A. Christenson (Davies and Turner, 2004a).

Terminology

In recent years, there has been much debate over the terms '*food-hairs*' and '*pseudopollen*'. Lindley (1842), in his description of *Eria paniculata* Lindl., based upon the Latin diagnosis of Wallich (1830), referred to 'a line of fine white powder formed all along the middle of the lip' consisting of 'pyriform bodies…of the same nature as those which occur on *Polystachya*' (cited in Beck, 1914). He goes on to say 'They are in fact extremely short hairs of one joint tapering to the base; in appearance they resemble fine powdery dust'. Perhaps the first to compare the appearance of individual food-hair cells with pollen grains, however, was Janse (1886). Whilst Janse simply referred to such trichomes as '*Haare*', the title of his paper '*Imitirte Pollenkörner bei Maxillaria sp.*' clearly implies imitation or mimicry. Subsequent German literature (e.g., Porsch 1905) refers to food-laden hairs as '*Futterhaare*' but by 1914, Beck in his thorough work on the pseudopollen of *Eria* Lindl., was already interchangeably using the terms '*Futterkörper*' (food bodies), '*Pollennachahmung*' (counterfeit pollen), '*Scheinpollen*' (pollen look-alike) and

Figs. 4.1–4.13. Examples of food-hairs and pseudopollen in orchids. Fig. 4-1: Labellum of *Polystachya foliosa* (Hook.f.) Rchb.f. (sect. *Polystachya* Rchb.f.) showing uniseriate, moniliform, pseudopollen-forming hairs. Scale bar = 100 µm. Fig. 4-2: Labellar surface of *Maxillaria johniana* Kraenzl. (*M. grandiflora* (Humb., Bonpl. & Kunth) Lindl. alliance) with similar pseudopollen-forming hairs. Scale bar = 100 µm. Fig. 4-3: Detail of above showing uniseriate, moniliform trichomes consisting of ellipsoid cells. Scale bar = 25 µm. Fig. 4-4: Transmission electronmicrograph of section through trichome cell of *Maxillaria sanderiana* Rchb.f. (*M. grandiflora* (Humb., Bonpl. & Kunth) Lindl. alliance) showing cluster of amyloplasts containing starch grains (A) and amorphous protein body (P). Scale bar = 2.5 µm (by kind permission of the editor of *The Annals of Botany*). Fig. 4-5: Labellar surface of *Maxillaria ochroleuca* Lodd. ex Lindl. (*M. splendens* Poepp. & Endl. alliance) showing few-celled trichomes that are said to be gathered by a species of *Trigona*. Scale bar = 100 µm. Fig. 4-6: Uniseriate, moniliform trichomes of *Maxillaria discolor* (Lodd. ex Lindl.) Rchb.f. (*M. discolor* (Lodd. ex Lindl.) Rchb.f. alliance). These are said to be collected by the stingless bee *Trigona fulviventris*. Scale bar = 50 µm. Fig. 4-7: Labellar food-hairs of *Polystachya cultriformis* (Thouars) Spreng. (sect. *Cultriformes* Kraenzl.). Scale bar = 50 µm. Fig. 4-8: Labellar surface of *Polystachya maculata* P.J. Cribb (sect. *Cultriformes* Kraenzl.) showing 2–4-celled, labellar hairs with clavate terminal cells. Scale bar = 250 µm. Fig. 4-9: Detail of above. Scale bar = 100 µm. Fig. 4-10: Clavate, labellar food-hairs of *Polystachya campyloglossa* Rolfe (sect. *Affines* Kraenzl.). Scale bar = 25 µm. Fig. 4-11: Unicellular, pseudopollen-forming, labellar hairs of *Eria ridleyi* Rolfe (sect. *Mycaranthes* Rchb.f.). These, eventually, become detached from the labellar surface. Scale bar = 25 µm. Fig. 4-12: Similar labellar hairs in *Eria paniculata* Lindl. (sect. *Mycaranthes* Rchb.f.). Scale bar = 25 µm (by kind permission of the editor of *The Annals of Botany*). Fig. 4-13: Detail of labellar hair of *Eria oblitterata* (Blume) Rchb.f. (sect. *Mycaranthes* Rchb.f.) showing presumed wax deposits towards apex. Scale bar = 10 µm (by kind permission of the editor of *The Annals of Botany*).

'falscher Pollen' (false pollen). Thus, the various equivalents of the term *'pseudo-pollen'* were probably at first simply used to describe the pollen-like appearance of the labellar hairs to the human eye, much in the same way that the term *'farina'* was used to describe their flour-like consistency. When *'pseudopollen'* is used in this way, the prefix *'pseudo'* can be loosely translated as *'like'*, *'resembling'* or *'similar to'*. However, recently, the term *'pseudopollen'* has increasingly taken on its alternative meaning of *'false pollen'* thus implying mimicry and deceit. As a result, some authors such as Singer and Koehler (2004) have expressed their dissatisfaction with the term since, rather than merely describing the appearance of the trichomes, it infers that visiting insects are actually deceived into collecting them as though they were pollen. Singer and Koehler (2004) argue that since bees systematically visit, pollinate and gather the labellar trichomes of *Maxillaria brasiliensis* Brieger & Bicalho throughout the whole flowering period, the trichomes are somehow used during the life cycle of the pollinator. They claim that pollination here contrasts markedly with that found in truly deceptive orchids in that the latter are usually pollinated over short periods and generally display low fruit set. Singer (2002, personal correspondence), in referring to *M. brasiliensis* and *Polystachya concreta* (Jacq.) Garay & H.R. Sweet, states that *Trigona* bees 'visited all available flowers and filled their corbiculae with trichomes. This behaviour was observed for several days and in two different locations (whereas) bees which pollinate deceptive orchids clearly learn (within a few days) that the flowers are rewardless and stop their visits'. As a result, Singer and Koehler propose that the term *'pseudopollen'* be abandoned. Lately, so-called *'pseudopollen'* has been described from the anther connective of several members of the Theaceae (Tsou, 1997) and the use of this same term to describe structures morphologically unrelated to food-hairs has the potential to cause further confusion. Nevertheless, on balance, the continued use of *'pseudopollen'* is favoured, not only for historical reasons but also because, regardless of whether it actually deceives pollinators, it so aptly describes the farinaceous, pollen-like appearance of the trichomes.

The terms *'food-hairs'* and *'pseudopollen'* have often been used interchangeably. The former is a generic term, the latter more specific. Based on our current understanding, it is proposed that we clearly distinguish between these structures as follows: Food-hairs are epidermal trichomes. They *always contain food materials* and are *gathered or nibbled* by visiting insects. Many structures formerly described as food-hairs have been shown, on further investigation, to be food-laden papillae (Davies and Turner, 2004a). However, the distinction between papillae and hairs is not always clear and is often simply a matter of degree (Esau, 1965; Davies and Stpiczyńska, 2006). In fact, it is now known that pseudopollen-forming trichomes of certain species (e.g., *Maxillaria sanderiana* Rchb.f. and representatives of *Eria* sect. *Mycaranthes* Rchb.f.) actually develop from labellar papillae (Beck, 1914; Davies et al., 2000; Davies and Turner, 2004c). Unless gnawed by insects, food-hairs that do not produce pseudopollen tend to remain *attached* to the labellum. By contrast, pseudopollen is a *farinaceous, pollen-like material* that is usually formed as particular types of food-hair, during the course of their development, either become *detached* from the labellar surface or *fragment*. The complete hair, individual cells or chains or clusters of cells may be gathered by insects. It is not, however, always easy to establish whether trichomes naturally become detached from the labellum as they mature and so, in the absence of unequivocal evidence that this is the case, trichomes otherwise resembling pseudopollen-forming hairs and containing food are perhaps best referred to as '*food-hairs*'.

Equally important is that distinction be made between pseudopollen and labellar trichomes or other labellar structures often loosely referred to as '*pseudopollen*' but best termed '*pseudostamens*'. Whereas true pseudopollen resembles pollen and usually *rewards* pollinators, pseudostamens resemble a tuft of stamens and serve to *attract* insect visitors. Pseudostamens are said to occur in *Arethusa bulbosa* L., *Calopogon tuberosus* (L.) Britton, Sterns & Poggenb., *Pogonia ophioglossoides* (L.) Ker Gawl. (Thien and Marcks, 1972), *Calypso bulbosa* (L.) Oakes (Gumprecht, 1977), *Maxillaria camaridii* Rchb.f., *M. pulchra* (Schltr.) L. O. Williams (Davies and Turner, 2004a), *Dendrobium delacourii* Guillaumin (K.L. Davies, 2005, unpublished data) and possibly *Cleistes divaricata* (L.) Ames (Gregg, 1982, 1984, 1991a, b). Singer and Koehler (2004) have suggested that the labellar hairs of *M. camaridii*, which, incidentally, are not easily detached (Davies and Turner, 2004a), may also be harvested by insect visitors.

Of course, not all labellar hairs are food-hairs and the latter are distinguished on morphological grounds and in that they contain *elevated* levels of food substances. Such hairs are often squat (Fig. 4-7) or moniliform (Figs. 4-1–4-3, 4-6) with relatively short and wide or rounded cells, whereas others may have swollen tips (Figs. 4-8–4-13). Furthermore, not all hairs gathered by insects are floral in origin, nor seemingly contain food materials. For example, Singer (2002, personal correspondence) has observed the stingless bees *Trigona* and *Partamona* spp. (Meliponini) collecting trichomes from the stems of nectariferous, terrestrial orchids such as *Aspidogyne* Garay and, based solely upon the ease with which they become detached, Davies and Turner (2004c) have speculated that the peculiar, branched hairs found on the pedicellate ovary and the abaxial surface of the perianth of species of *Eria* sect. *Mycaranthes* may also be collected by insects and possibly used for nest-building.

Food-hairs, then, by definition, are floral trichomes and/or papillae that contain elevated concentrations of foods such as protein, starch and lipids (Porsch, 1905; van der Pijl and Dodson, 1969; Davies et al., 2000, 2002 2003a; Davies and Turner, 2004a, b, c). Pseudopollen, on the other hand, is formed when a particular type of food-hair fragments or becomes detached from the labellum.

Occurrence of Pseudopollen

Pseudopollen conforming to the above definition has so far been recorded for surprisingly few genera (Table 4-1). These include *Maxillaria* Ruiz & Pav. (Janse, 1886; Porsch, 1905; van der Pijl and Dodson, 1969; Davies and Winters, 1998;

Table 4-1. Summary of labellar features of principal taxa studied to date and for which histochemical data are available.

Taxon	Labellar feature	Foods present			Resinous or waxy secretion	Remarks	References
		Protein	Starch	Lipid			
Dendrobium unicum Seidenf. (N. Thailand)	Pseudopollen. Trichomes comprise stalk and multicellular 'head'	+	+	(–)	–	Starch is main food reserve. Lipid droplets not visible at light microscopy level but visible using TEM	Davies and Turner, 2004b
Eria pilifera Ridl. (S.E. Asia)	Unicellular, clavate trichomes	+	+	–	–		Davies and Turner, 2004c
Eria Lindl. section *Mycaranthes* Rchb.f. (S.E. Asia)	Pseudopollen. Unicellular, clavate trichomes	+	+	–	–	Cell wall with presumed wax	Davies and Turner, 2004c
Cymbidium lowianum (Rchb.f.) Rchb.f. (Burma, China, Thailand)	Papillae	+	+	–	+	Only protein present at elevated concentrations in papillae. Labellar secretion contains lipid	Davies et al., 2006
Cymbidium dayanum Rchb.f. (N. India, China, S.E. Asia)	Unicellular, clavate trichomes				+	Labellar secretion contains lipid	Davies et al., 2006
Maxillaria acuminata Lindl. alliance (Neotropics)	Papillae – conical or obpyriform	+	(+)	+	+	Labellar secretion contains lipids and aromatic amino acids	Davies et al., 2003b
Maxillaria discolor (Lodd. ex Lindl.) Rchb.f. alliance (Neotropics)	Simple, 3–6-celled food-hairs or moniliform, pseudopollen-forming trichomes	+	(+)	(+)	(+)	Protein is main food reserve. Starch, lipid and resinous secretion present in some species. Last contains lipids and aromatic amino acids	Davies et al., 2003a
Maxillaria grandiflora (Humb., Bonpl. & Kunth) Lindl. alliance (Neotropics)	Moniliform, pseudopollen-forming trichomes	+	(+)	(–)	–	Protein is main food reserve. Most contain starch and lack lipid	Davies et al., 2000; Davies and Turner, 2004a

(continued)

Maxillaria lepidota Lindl. alliance – *M. longissima* Lindl. and *M. lepidota* Lindl. (Neotropics)	Moniliform, pseudopollen-forming trichomes	+	(+)	+	(–)	Labellar secretion present in *M. lepidota*	Davies and Turner, 2004a; Matusiewicz et al., 2004
M. reichenheimiana Endres & Rchb.f. and *M. pseudoreichenheimiana* Dodson (Neotropics)	Simple, 2-celled food-hairs	+	–	–	–	Cell wall stains for lipids	Davies and Turner, 2004a
Maxillaria parkeri (Spreng.) Hook. alliance – *M.* cf. *setigera* Lindl. (Neotropics)	Pseudopollen. 2–3-celled, moniliform trichomes	+	+	–	–		Davies and Turner, 2004a
Maxillaria rufescens Lindl. alliance (Neotropics)	Papillae	+	+	+	+	Papillae along median axis of labellum are larger, clavate and contain more starch. Secretion contains lipids	Davies and Turner, 2004a
Maxillaria splendens Poepp. & Endl. alliance (Neotropics)	Simple, 3–4-celled food-hairs	+	+	–	–	No foods detected in hairs of *M. buchtienii* Schltr. Histochemical results based on *M. tenuis* C. Schweinf.	Davies and Turner, 2004a
Xylobium leontoglossum (Rchb.f.) Benth. ex Rolfe (Neotropics)	Papillae				+	Lipids present in labellar secretion	Davies and Stpiczyńska, 2006
Xylobium cf. *corrugatum* (Lindl.) Rolfe (Neotropics)	Papillae				+	Lipids present in labellar secretion	Davies and Stpiczyńska, 2006
Xylobium squalens (Lindl.) Lindl. syn. *X. variegatum* (Ruiz & Pav.) Garay & Dunst. (Neotropics)	Papillae				+	Spirit-preserved material. Not tested for foods	Davies and Stpiczyńska, 2006

(continued)

Table 4-1. (continued)

Taxon	Labellar feature	Foods present: Protein	Starch	Lipid	Resinous or waxy secretion	Remarks	References
Xylobium latilabium C. Schweinf. (Neotropics)	Moniliform, 4–8-celled, pseudopollen-forming trichomes				–	Spirit-preserved material. Not tested for foods	Davies and Stpiczyńska, 2006
Teuscheria wageneri (Rchb.f.) Garay (Neotropics)	Moniliform, 2–10-celled, pseudopollen-forming trichomes				+	Spirit-preserved material. Not tested for foods	Davies and Stpiczyńska, 2006
Polystachya Hook. section *Polystachya* Rchb.f. (Africa)	Moniliform, pseudopollen-forming trichomes	+	(+)	–	–	Most contain starch	Davies et al., 2002
Polystachya Hook. section *Affines* Kraenzl. (Africa)	Simple, unicellular or 2–4-celled food-hairs with sub-clavate to clavate apices	+	(+)	–	(–)	Most contain starch. Labellar secretion of *P. campyloglossa* Rolfe contains aromatic amino acids	Davies et al., 2002
Polystachya Hook. section *Caulescentes* Kraenzl. (Africa)	Simple, 2–4-celled food-hairs with sub-clavate to clavate apices	+	(–)	–	–	Most lack starch	Davies et al., 2002
Polystachya Hook. section *Cultriformes* Kraenzl. (Africa)	Simple, 2–4-celled food-hairs with sub-clavate to clavate apices	+	(+)	–	–	Most contain starch	Davies et al., 2002
Polystachya Hook. section *Humiles* Summerh. (Africa)	Simple, 2–4-celled food-hairs with sub-clavate to clavate apices	+	–	–	–		Davies et al., 2002
Polystachya Hook. section *Polychaete* Cribb (Africa)	Simple, 2–4-celled food-hairs with sub-clavate to clavate apices	+	(+)	–	–	Presence of starch variable	Davies et al., 2002
Polystachya Hook. section *Superpositae* Kraenzl. (Africa)	Simple, 2–4-celled food-hairs with sub-clavate to clavate apices	+	(–)	–	–	Most lack starch	Davies et al., 2002

+ and – indicate presence of and absence of food substances, respectively. Those symbols in parentheses indicate typical results but they may vary between species or individuals of the same species.

Davies et al., 2000, 2003a; Davies and Turner, 2004a), *Polystachya* Hook. (Porsch, 1906; Beck, 1914; Davies et al., 2002) and *Eria* (Beck, 1914; Davies and Turner, 2004c), as well as a small number of individual species such as *Dendrobium unicum* Seidenf. (Kjellsson and Rasmussen, 1987; Davies and Turner, 2004b). Other taxa reported to produce pseudopollen include *Gastrodia sesamoides* R. Br. (Jones, 1981, 1985), *Cephalanthera austiniae* (A. Gray) A. Heller (van der Cingel, 2001) and *Sobralia liliastrum* Lindl. (Romero, 1998). Nevertheless, it is speculated that many more orchid species produce food-hairs or pseudopollen since the diverse nature of these structures may well mean that, in the past, many types have gone unnoticed (Davies and Turner, 2004b).

Trichome Morphology

Food-hairs, including pseudopollen, show great diversity of form. In certain species of *Polystachya* and in *Eria* sect. *Mycaranthes* (Beck, 1914; Davies et al., 2002; Davies and Turner, 2004c), they may be unicellular and clavate (Figs. 4-11–4-13). Alternatively, they may be uniseriate and 2- to few-celled as in other species of *Polystachya* (Figs. 4-7–4-10; Davies et al., 2002) or multicellular as in some species of *Maxillaria* (Figs. 4-2–4-3, 4-6; Davies et al., 2000; 2003a; Davies and Turner, 2004a), *Polystachya* sect. *Polystachya* Rchb.f. (Fig. 4-1; Porsch, 1906; Beck, 1914; Davies et al., 2002) and *Dendrobium unicum.* (Kjellsson and Rasmussen, 1987; Davies and Turner, 2004b). Moniliform hairs predominate amongst members of the *Maxillaria grandiflora* alliance (Figs. 4-2–4-3), *Maxillaria discolor* (Lodd. ex Lindl.) Rchb.f. alliance (Fig. 4-6), *Maxillaria lepidota* alliance and members of *Polystachya* sect. *Polystachya* (Fig. 4-1; Janse, 1886; Porsch, 1905, 1906; Beck, 1914; van der Pijl and Dodson, 1969; Davies and Winters, 1998; Davies et al., 2000, 2002, 2003a; Davies and Turner, 2004a) and these become detached or fragment forming individual or short chains of cells. The labellar hairs of the *M. splendens* Poepp. & Endl. alliance are much smaller and comprise a few elongated cells only (Fig. 4-5; Davies and Winters, 1998; Davies et al., 2000, 2003a; Davies and Turner, 2004a). Peculiar trichomes, comprising a stalk and a multicellular 'head' occur upon the labellum of *Dendrobium unicum* and these 'heads' break up into multicellular clusters called 'granulae'. Certain species of *Xylobium* Lindl. and *Teuscheria* Garay also bear moniliform hairs but it is not known whether they contain foods (Davies and Stpiczyńska, 2006). Vogel (1979) has argued that such hairs, even if devoid of food, can still attract pollinators by mimicry. Pseudopollen is also said to occur in the closely related *Rudolfiella aurantiaca* (Lindl.) Hoehne (Braga, 1977) but Davies and Stpiczyńska (2006) failed to find it. Singer (2006, personal correspondence), instead, identified putative elaiophores in this species and oil has been recorded from the callus and lateral lobes of the labellum of *Rudolfiella picta* (Schltr.) Hoehne (K.L. Davies, 2006, unpublished data).

Development of Pseudopollen

Hitherto, the development of pseudopollen has been studied in detail only for *Maxillaria sanderiana* Rchb.f. (Davies et al., 2000). In this species, uniseriate, moniliform hairs develop from obpyriform labellar papillae. Each cell of the moniliform trichome contains a large, centrally placed, homogeneous, protein body which, on the basis of its profile, is thought to have intravacuolar origins (Fig. 4-4). This is enclosed by parietal cytoplasm containing a nucleus, mitochondria, plastids, dictyosomes, endoplasmic reticulum (ER) and small oil bodies. The basal cell undergoes repeated cell division and this results in the formation of a moniliform trichome. At first, numerous plasmodesmata maintain the cytoplasmic continuity of adjoining cells but these, together with the primary pit fields, become occluded as the individual trichome cells mature. The protein body enlarges and gradually displaces the cytoplasm. Eventually, dictyosomes and ER disappear and the plastids differentiate and form amyloplasts containing starch grains and plastoglobuli, but few lamellae. By now, the cell wall is distinctly lamellate. Finally, dissolution of the middle lamella results in the separation of the trichome cells.

Food Content and Ecology

According to earlier accounts, food-hairs contain aleurone grains, oil droplets (Porsch, 1905) or starch (van der Pijl and Dodson, 1969). Aleurone grains, by definition, have two components; a proteinaceous substrate with a crystalloid body (protein crystal) and a globoid body (double phosphate of calcium and magnesium with an organic radical; Esau, 1965). Porsch (1905), for example, reported aleurone grains, in *Maxillaria rufescens* Lindl. However, Davies and Turner (2004a) were unable to verify the claim although they did in fact detect protein, starch and oil droplets in that species. Indeed, recent histochemical studies have, in general, failed to confirm the presence of aleurone grains, in the stricter sense of the term, in *Eria*, *Maxillaria* and *Polystachya* spp. (Davies et al., 2000, 2002, 2003a; Davies and Turner, 2004a, c). Thus, it is probable that the term '*aleurone*' was formerly used loosely instead of '*protein body*'. Nevertheless, it has not been possible to eliminate totally the possibility that the food-hairs of some species do indeed contain true aleurone grains. For example, on one occasion, a single protein body with associated spheroidal, apparently crystalloid body and spherical, lipid globules was observed for food-hairs of *Maxillaria villosa* (Barb. Rodr.) Cogn. (Davies et al., 2003a).

The main food material found in orchid food-hairs, such as those of *Maxillaria* spp., is protein (Davies et al., 2000, 2003a; Davies and Turner, 2004a). Typically, this occurs as a discrete, homogeneous, protein body containing elevated concentrations of aromatic amino acids (Davies et al., 2000, 2003a; Davies and Turner, 2004a). The protein body is thought to have intravacuolar origins and is probably formed as water is withdrawn from the vacuole (Buttrose, 1963; Davies et al., 2000). Food-hairs of some orchid species lack obvious protein bodies and here, aromatic amino acids are usually distributed throughout the cytoplasm. This occurs in *Polystachya* (Davies et al., 2002) and *Eria* (Davies and Turner, 2004c). The food-hairs of a significant number of orchids also contain amyloplasts (Davies et al., 2000, 2002, 2003a; Davies and Turner, 2004a, b, c). However, it would appear that the amount of starch they contain can vary even between individuals of a single species and this is probably related to development. Moreover, in *Maxillaria acutifolia* and *M. tenuibulba*, starch grains are more concentrated within the larger, clavate papillae found along the median, longitudinal axis of the labellum than in peripheral papillae (Davies and Turner, 2004a). Although the papillae do not become detached from the labellar surface, it is nonetheless possible that they are nibbled by visiting insects. *Dendrobium unicum* is atypical in that the main food reward is starch not protein and each cell of the granula contains a single starch grain that occupies almost its entire volume (Davies and Turner, 2004b). Food-hairs and pseudopollen of some *Maxillaria* (Davies et al., 2000, 2003a; Davies and Turner, 2004a) and *Polystachya* spp. (Davies et al., 2002), as well as *D. unicum* (Davies and Turner, 2004b), also contain lipid droplets but their paucity and small size would indicate that these are unlikely to be important as food rewards.

Small, stingless bees (Meliponini) such as species of *Melipona*, *Partamona*, *Plebeia*, *Tetragonisca* and *Trigona* are considered to be the main pollinators of *Maxillaria* spp. (Singer and Cocucci, 1999; Roubik, 2000; Singer and Koehler, 2004). Although several early observations indicated that food-hairs and pseudopollen of *Maxillaria* spp. are gathered by insect pollinators (Dodson and Frymire, 1961; Dodson, 1962; van der Pijl and Dodson, 1969), there has been, in the absence of unequivocal evidence, a tendency to dismiss these reports as unreliable (e.g., Roubik, 2000). Recently, however, Singer (2004, personal correspondence; Singer and Koehler, 2004) has reported species of *Trigona* systematically visiting, pollinating and gathering the labellar hairs of *Maxillaria brasiliensis* and *M. ochroleuca* Lodd ex Lindl. (Fig. 4-5). Singer has also reported *Trigona fulviventris* both pollinating and gathering moniliform, pseudopollen-forming trichomes (Fig. 4-6) from the flowers of *M. discolor* (Lodd. ex Lindl.) Rchb.f. (Singer, 2004, personal correspondence) and a species of *Trigona* gathering 'trichomes' (presumably the large, clavate, starch-laden papillae) 'from the median region of the lip' of *M. rufescens* (Singer et al., 2004). The halictid bee *Dialictus* aff. *creberrimus* has also been observed both pollinating and gathering starchy pseudopollen from the nectarless flowers of *Polystachya flavescens* (Lindl.) J.J. Sm. (Goss, 1977), whereas a small xylocopid bee (*Exoneura* sp.) is known to pollinate and gather pseudopollen from the labellar callus of *Gastrodia sesamoides* (Jones, 1981, 1985).

Morphology of food-hairs and the chemical composition of the food that they contain probably play an important part in pollinator selection. Wasps, for example, do not consume pollen (Duncan, 1929 – cited in Proctor and Yeo, 1975) and are thus unlikely to gather pseudopollen. Similarly, the nutritional value of starch alone, as in *D. unicum* (Davies and Turner, 2004b), is probably insufficient nourishment for solitary bees but may supplement the diet of eusocial bees that have access to other food sources (Roubik, 2004, personal correspondence). Most food-hairs and papillae are abundant in protein, but this is probably available only to those insect visitors with chewing mouthparts. Many trichomes contain more than one type of food and may, thus, have the potential to reward more than one species of pollinator (Davies et al., 2000).

Trichomal Secretion of Resin-Like and Waxy Materials

The labellar papillae of some orchid genera secrete a viscid, resin-like material. These genera include *Maxillaria* (Porsch, 1905; van der Pijl and Dodson, 1969; Davies et al., 2003a, b; Davies and Turner, 2004a; Flach et al., 2004; Matusiewicz et al., 2004; Singer et al., 2006), *Polystachya* (Davies et al., 2002), *Xylobium*, (Davies and Stpiczyńska, 2006), *Cymbidium* Sw. (Macpherson and Rupp, 1935; Davies et al., 2006) and *Eria* (von Kirchner, 1925). A similar secretion occurs in *Teuscheria wageneri* (Rchb.f.) Garay but it is not certain whether it has nutritive value (Davies and Stpiczyńska, 2006). Generally, however, these secretions are rich in lipids and aromatic amino acids (Davies et al., 2003b; Davies and Turner, 2004a) and further analysis (Flach et al., 2004) has shown that triterpenoids (cycloartenol derivatives) form the main component. Unlike floral oil, which is typically secreted by well-defined elaiophores, these secretions are usually produced by poorly-defined areas of the labellum. Orchids that produce labellar, resin-like secretions usually lack pseudo-pollen and fragrance but all three characters are present in *M. lepidota* (Matusiewicz et al., 2004) and both secretion and moniliform hairs occur in *Teuscheria wageneri* (Davies and Stpiczyńska, 2006) and some members of the *Maxillaria discolor* alliance (Davies et al., 2003a).

To date, the secretion of resin-like material has been studied at the cellular level only for *Maxillaria* cf. *notylioglossa* Rchb.f., since re-determined as *M. divaricata* (Barb. Rodr.) Cogn. (Davies et al., 2003b). Here, secretory, obpyriform papillae occur along the median, longitudinal axis of the labellum. These papillae are nucleate, their dense cytoplasm containing mitochondria with well-developed cristae, an extensive system of smooth endoplasmic reticulum (SER) and spherical lipid bodies. These lipid bodies may be associated with the SER or occur as plastoglobuli within plastids. As they increase in size, they develop an electron-transparent core yet their margins remain strongly osmiophilic. By now, the plastid envelope has ruptured and lipid bodies are released into the cytoplasm where they become associated with the plasmalemma and outer, tangential cell wall. This wall is permeable and lacks ectodesmata but at regions of greatest secretory activity, the epidermal cuticle is usually thinner than elsewhere. Resin-like material is secreted onto the epidermal surface and the lipid bodies it contains coalesce, the viscid film finally obscuring the micromorphological topography of the labellum.

Similar secretions can be found in the closely related *Maxillaria acuminata* Lindl. (Davies et al., 2003b). They also occur in *M. violaceopunctata*, *M. villosa*, *M. nasuta* Rchb.f. (Davies et al., 2003a) and *M. brasiliensis* (Flach et al., 2004), – all members of the *M. discolor* alliance, as well as amongst members of the *M. rufescens* alliance such as *M. acutifolia*, *M. tenuibulba* (Davies and Turner, 2004a) and *M. hedwigae* Hamer & Dodson (Singer et al., 2004). Singer and Koehler (2004) have noted labellar, resin-like material in *M. equitans* (Schltr.) Garay, and Flach et al. (2004) in *M. friedrichsthalii* Rchb.f. The cell walls of bicellular food-hairs of *M. reichenheimiana* Endres & Rchb.f. and *M. pseudoreichenheimiana* Dodson also selectively stain for lipid with alcoholic Sudan III solution (Davies and Turner, 2004a). *Maxillaria cerifera* Barb. Rodr. and *M. notylioglossa* Rchb.f., species which are related to *M. divaricata* and *M. acuminata*, on the other hand, secrete wax (Porsch, 1905; van der Pijl and Dodson, 1969; Senghas, 1993; Singer and Koehler, 2004; Davies et al., 2003b; Flach et al., 2004) and dangling strings of wax are also produced by bicellular glands along the sepal margins of *Pleurothallis schiedei* Rchb.f. (Pridgeon, 1992). Flaky deposits of presumed wax occur on the apices of food-laden, pseudopollen trichomes of species of *Eria* sect. *Mycaranthes* and this is curious since waxy pseudopollen is not recorded elsewhere (Davies and Turner, 2004c). This substance, too, may perhaps function as a supplementary reward thereby contributing towards pollinator selection. Alternatively, it may simply protect the trichomes from desiccation or, by reducing wettability, aid their dispersal.

Although resin-like material and wax are clearly secreted by morphologically similar papillae in *Maxillaria*, it is unlikely that the latter compound has nutritive properties and, as a result, a number of authors (e.g., van der Pijl and Dodson, 1969; Endress, 1994; van der Cingel, 2001) have proposed that both these secretions are used as 'bee glue' for building and repairing nests or sealing cells. Others, however, claim that the resin-like material has nutritive value (Davies et al., 2003a, b, 2006) and is perhaps gathered and ingested by insects. Again, the evidence for this is scant. However, Flach et al. (2004) have reported Meliponini gathering material secreted by the labella of *Maxillaria* spp. and Macpherson and Rupp (1935) have observed the stingless bee *Trigona hockingsii* gnawing at viscid material produced along the median axis of the lip of *Cymbidium iridifolium* A. Cunn. ex Lindl. (syn. *C. madidum* Lindl.), whereas Braga (1977) reported the vespid wasp *Stelopolybia* cf. *pallipes* gathering similar material from the callus of *Maxillaria pendens* Pabst.

Despite differences in the ways insects are thought to utilize these compounds, gas chromatography-mass spectrometry and nuclear magnetic resonance (NMR) analysis have revealed remarkable similarities between resin-like material and wax in *Maxillaria*. Here, the waxy material consists mainly of 3β-hydroxy-cycloart-24-en-26-al (Flach et al., 2004). Comparison of the resin-like and waxy, labellar secretions of *M. friedrichsthalii*, and *M. cerifera*, respectively, have also revealed that they both contain a similar cycloartenol derivative that is absent from *M. brasiliensis*. Conversely, the labellar secretion of this last species contains a cycloartenol derivative that is not present in that of *M. friedrichsthalii* and *M. cerifera* (Flach et al, 2004). Clearly, our understanding of the role of these epidermal structures and the secretions they produce is still far from complete.

Trichomal Elaiophores

Many orchids, most notably Oncidiinae (*sensu* Chase et al., 2003; Chase, 2005), reward potential pollinators with oil. This is produced in floral glands termed elaiophores and these may be of two types; trichomal elaiophores (oil-secreting hairs) and epithelial elaiophores (oil-secreting epidermal cells; Vogel, 1974). Production of floral oils in orchids is considered polyphyletic and is thought to have evolved at least five times in Oncidiinae alone (Silvera, 2002). Oils gathered by bees (Apidae) are either mixed with pollen and fed to developing larvae (Michener, 2000) or used in the nest to seal and waterproof cells (Endress, 1994; Silvera, 2002 and references therein). However, field observations are often difficult to interpret. For example, Porsch (1906) reported that hairs upon the inner surface of the labellar pouch of European *Cypripedium calceolus* L. were licked or nibbled by *Andrena* bees. However, the subsequent studies of Ziegenspeck (1936) showed that the hairs did not contain food. More recently, van der Cingel (2001) suggested that they may secrete oils, but as yet, there is no reliable evidence that insects feed upon them, whereas Nilsson (1978) speculated that they may retain pheromones from visiting bees or, owing to their highly light-refractive properties, somehow stimulate photo-active behaviour in insects.

Unlike *Cypripedium calceolus*, flowers of certain representatives of Cranichidinae Lindl. (Dressler, 1993), Satyriinae Schltr. (Garside, 1922), Coryciinae Benth. (Buchmann, 1987; Steiner, 1989, 1993; Pauw, 2006), Bulbophyllinae Schltr. (Pohl, 1935; van der Cingel, 2001) and Oncidiinae (*sensu* Chase *et al.*, 2003; Chase, 2005) are known to secrete oils and these are gathered by insects (van der Pijl and Dodson, 1969; Buchmann, 1987; Dressler, 1990; Singer and Cocucci, 1999; van der Cingel, 2001; Silvera, 2002; Stpiczyńska, Davies, and Gregg, 2007). For example, Dressler (1993) observed oil secretion in *Ponthieva racemosa* (Walter) C. Mohr (Cranichidinae) and Garside (1922) described the pollination of *Satyrium bicallosum* Thunb. (Satyriinae) by fungus-gnats and proposed that these insects lick the oily secretion produced by short, unicellular hairs at the base and sides of the spurs. Trichomal elaiophores also occur in *Disperis* spp. (Coryciinae; Buchmann, 1987) and here, the flowers are pollinated by specialized, oil-collecting, *Rediviva* bees (Steiner, 1989, 1993). It is the labellar callus, however, that secretes the non-volatile oil in the closely related genus *Ceratandra* Eckl. & F. A. Bauer and this secretion contains a complex set of lipids. *Ceratandra atrata* (L.) T. Durand & Schinz and *C. bicolor* Sond. ex Bolus are pollinated by the oil-gathering bee *Rediviva gigas*, whereas *C. grandiflora* Lindl. is pollinated by beetles (Steiner, 1998). *Pterygodium alatum* (Thunb.) Sw., *P. caffrum* (L.) Sw., *P. catholicum* (L.) Sw., *P. volucris* (L.f.) Sw., *Corycium orobanchoides* (L.f.) Sw. and *Disperis bolusiana* Schltr. ssp. *bolusiana* are all pollinated by female *Rediviva peringueyi* (Pauw, 2006). In each case, oil is secreted by the lip appendage, a structure referred to by Kurzweil et al. (1991) as 'the most bizarre floral structure known in the orchids'. In *P. catholicum*, floral oil is contained within narrow, vertical grooves on the abaxial surface of the lip appendage and these lead into channels which run vertically inside the lip appendage to half its length. Both grooves and channels are lined with secretory cells that correspond to epithelial elaiophores (Vogel, 1974). Tarsal segments 2–5 of the front legs of *Rediviva peringueyi* bear plumose hairs which absorb the floral oil (Whitehead and Steiner, 2001) and, although this is not consumed by the adult insect, it is fed to the larvae (Vogel, 1974; Cane et al., 1983; Vinson et al., 1997). Representatives of Bulbophyllinae such as *Bulbophyllum macranthum* Lindl.,

B. lobbii Lindl. and *B. campanulatum* Rolfe also produce floral oil (Pohl, 1935; van der Cingel, 2001) and, given the enormity of *Bulbophyllum* Thouars, some species are very likely to have trichomal elaiophores.

Buchmann (1987), who studied the *Ornithocephalus* 'group' (Oncidiinae *sensu* Chase et al., 2003; Chase, 2005), found that 50 species of *Ornithocephalus* Hook. and four species of *Zygostates* Lindl. (formerly Ornithocephalinae) have elaiophores. In *Ornithocephalus*, these are mainly of the trichomal type and occur upon the labellum although, in some, they are found upon the column or other parts of the flower (Toscano de Brito, 2001). The pollinators are mainly species of *Paratetrapedia* with *P. testacea* pollinating *O. avicula* Rchb.f. (now *O. ciliatus* Lindl.) and *O.* cf. *patentilobus* C. Schweinf. in Peru, whereas *P. calcarata* pollinates *O. bicornis* Lindl. and *O. powellii* Schltr. in Panama (van der Cingel, 2001; Toscano de Brito, 2001). Trichomal elaiophores also commonly occur upon the labellar callus and petals of *Zygostates* (Toscano de Brito, 2001). Often, as in *Phymatidium falcifolium* Lindl., the labellar, trichomal elaiophores are capitate. They may occur at the base of the lip as in *Hintonella mexicana* Ames, upon the callus as in *Chytroglossa marileoniae* Rchb.f. or, as in *Platyrhiza quadricolor* Barb. Rodr., at the base of the lip and upon the tabula infrastigmatica. In the monotypic *Dunstervillea mirabilis* Garay, hairs, thought to be trichomal elaiophores and lining the spur, closely resemble labellar hairs found in the monotypic *Papillilabium beckleri* (F. Muell. ex Benth.) Dockr. (Sarcanthinae Benth; Toscano de Brito, 2001). Recently, Chase and co-workers (2003, 2005) have shown that the former Ornithocephalinae Schltr. is nested within a more broadly defined Oncidiinae and the occurrence of elaiophores within this latter subtribe may be prone to parallelisms (Singer et al., 2006). Trichomal elaiophores also occur upon the apex of the labellum and base of the column of *Grobya amherstiae* Lindl. (Catasetinae *sensu* Chase et al., 2003) and, in this species, floral oils are gathered by the pollinator *Paratetrapedia fervida* (Mickeliunas et al., 2006).

Since epithelial elaiophores are more common than trichomal elaiophores (Singer et al., 2006), research to date has inevitably concentrated on those species of Oncidiinae that possess the former and many of these, such as *Oncidium* Sw., *Ornithophora* Barb. Rodr. and *Sigmatostalix* Rchb.f., are pollinated by oil-gathering bees (Dressler, 1990; Singer and Cocucci, 1999; van der Cingel, 2001) such as *Tetrapedia* (Buchmann, 1987; Singer and Cocucci. 1999) and *Centris* spp. The floral oil of members of the '*Gomesa* clade' *such as Baptistonia echinata* Barb. Rodr., *Oncidium longicornu* Mutel, *O. truncatum* Pabst, *O. kautskyi* Pabst, *O. hookeri* Rolfe, *O. welteri* Pabst, *O. amictum* Lindl., *O. trulliferum* Lindl., *O. cornigerum* Lindl., *O. pubes* Lindl. and *Ornithophora radicans* (Rchb.f.) Garay & Pabst is characterized by the presence of diacylglycerols in which the acetyl group is invariably in position 1 of the glycerol moiety and the fatty acid located in position 2. The long-chain fatty acid has either hydroxyl or acetoxy groups at position 3 and 7. The same is true of *Phymatidium tillandsioides* Barb. Rodr. (syn. *Phymatidium falcifolium* Lindl.) and *Zygostates lunata* Lindl. (Reis et al., 2000, 2003, 2006) although, remarkably, that of *P. delicatulum* is composed largely of linear hydrocarbons (Reis et al., 2006). The major component of the floral oil of *Ornithophora radicans* is (2S, 3'R, 7'R)-1-acetyl-2-(3', 7'-diacetoxy-eicosanoyl)-glycerol

(Reis et al., 2003) and acylglycerols are thought also to occur in the floral oils of *Oncidium loefgrenii* Cogn. and *Gomesa* R. Br. (Singer, 2006, personal correspondence). The oils of certain other members of the '*Gomesa* clade' such as *Oncidium enderianum* auct. and related taxa also contain diacylglycerol derivatives but here, the compounds are not exuded (Reis et al., 2000).

Species such as *Trichocentrum stipitatum* (Lindl.) M. W. Chase & N. H. Williams, (Silvera, 2002), *Oncidium compressicaule* Withner, *O. guianense* (Aubl.) Garay, *O. haitiense* Leonard & Ames, *O. osmentii* Withner and *O. quadrilobum* C. Schweinf. (Dod, 1976; now assigned to the genus *Tolumnia* Raf.) mimic the flowers of New World, vinaceous Malpighiaceae. These vines have floral, epithelial elaiophores whose secretion consists of 8-acetoxy-substituted free fatty acids with a carbon-chain length of $C_{14} – C_{20}$. Female *Centris* spp. scrape the flower surface with their tarsi and transfer the oil to the rear legs. The intoxicated bees then carry it to brood cells where it is fed to larvae (Vogel, 1974, 1990; van der Cingel, 2001). *Tetrapedia diversipes* has also been seen gathering oil and pollinating the flowers of *O. paranaense* Kraenzl. (Singer and Cocucci, 1999), whereas species of *Tetrapedia* have been observed pollinating *O. pumilum* Lindl. and visiting flowers of *Gomesa recurva* R. Br. (Singer, 2006, personal correspondence). Epithelial elaiophores have also been observed for *Lockhartia* Hook. (Oncidiinae), *Rudolfiella aurantiaca* (Maxillariinae *sensu lato* – formerly Bifrenariinae Dressler; Singer, 2006, personal correspondence; Singer et al., 2006), and *R. picta* (K. L. Davies, 2006, unpublished data), whereas putative elaiophores have also been noted upon the labellum of *Grobya galeata* Lindl. (Catasetinae *sensu* Chase et al., 2003) and many species of *Cyrtopodium* R. Br. (Catasetinae; Singer et al., 2006).

Interestingly, chemical analysis of *Tetrapedia* nests have revealed the presence of two important compounds, namely, 3,7-dihydroxy-eicosanoic acid (Tetrapedic acid A) and 3,7-dihydroxy-docosanoic acid (Tetrapedic acid B) – chemicals that could equally have been derived from Oncidiinae or Malpighiaceae (Singer et al., 2006)

Evolution of Food-Hairs and Food-Hairs as Taxonomic Characters

Food-hairs, it would seem, gradually replaced nectar as the main pollinator reward in orchids and flowers that offer rewards were, in turn, replaced by rewardless types (van der Pijl and Dodson, 1969; Vogel; 1979; Davies et al., 2003a). The most common type of epidermal papilla found amongst angiosperms, including the labella of Orchidaceae, is the conical papilla, characterized by its wide base, concave, lateral walls and rounded tip (Kay, Daoud, and Stirton, 1981; Davies and Winters, 1998; Davies et al., 2003b, 2006; Davies and Turner, 2004a). Obpyriform papillae, such as those that secrete resin-like material and waxes (Davies et al., 2003b), those containing abundant starch (Davies and Turner, 2004a), as well as those destined to become food-hairs (Davies et al., 2000), are ultimately derived from these, whereas moniliform, pseudopollen-forming trichomes, as found in some *Maxillaria* spp., probably evolved from simple, uniseriate, multicellular hairs (Davies et al., 2003a).

Epidermal structures such as the stomatal apparatus, trichomes and glands have long been used as characters in orchid taxonomy (Pridgeon, 1993). Consequently, one would expect food-hairs, including pseudopollen, to be equally useful. However, the occurrence of almost identical food-hairs in unrelated genera on different continents is indicative of parallelism. For example, the pseudopollen-forming, moniliform hairs of the Neotropical genus *Maxillaria* and members of the largely African genus *Polystachya* sect. *Polystachya* are virtually indistinguishable even though these taxa are mainly pollinated by Meliponini and halictid bees, respectively (Davies and Winters, 1998; Davies et al., 2000, 2002, 2003a; Davies and Turner, 2004a). Moreover, this type of hair also occurs in *Xylobium latilabium* and *Teuscheria wageneri* (Maxillariinae *sensu lato*) and may have arisen in response to similar pollination pressures (Davies and Stpiczyńska, 2006). Owing to the prevalence of parallelism, especially in the case of Maxillarieae, a number of authors have warned against the use of pollination-related characters in taxonomy (Benzing, 1986; Stern, Judd, and Carlsward, 2004; Dathe and Dietrich, 2006; Davies and Stpiczyńska, 2006). Even so, some types of labellar hairs may yet prove useful in this respect. Toscano de Brito (2001), for example, considers the morphology and position of trichomal elaiophores in the *Ornithocephalus* 'group' taxonomically valuable. Furthermore, laterally-compressed, paddle- or lollipop-shaped, non-food trichomes are common upon the labella of *Xylobium* spp. but have not yet been found in any other members of Maxillariinae *sensu lato* and, although food-hair diversity in *Polystachya* generally does not reflect currently accepted infra-generic relationships based upon gross vegetative and floral morphology, moniliform, pseudopollen-forming hairs in the genus are seemingly restricted to *Polystachya* sect. *Polystachya* (Davies et al., 2002). It would thus appear that the value of floral food-hairs as taxonomic characters depends largely upon the taxon under consideration.

Conclusions

Given the morphological diversity of orchid food-hairs, it is very likely that many types have been overlooked in the past, especially in genera where they were not expected to occur. What is certain is that the development of food-hairs, in particular pseudopollen, represents a major stride in the evolution of orchids. Inevitably, there remain large gaps in our understanding of food-hairs and those insects that gather them or their secretions. With few exceptions, we are currently unable to relate named food-hair foragers to particular orchid taxa and thus, unable to appreciate fully the significance of diverse food-hair morphology in pollinator selection. Although on occasion Meliponini have been observed gathering pseudopollen, we still do not know for certain whether this material is actually ingested or how it is used by the insect or colony, nor why starch is the main food reserve in the pseudopollen of some species whereas in the majority, it is protein. It is also important to consider the timing and duration of anthesis and how this is related to the life-cycle and behaviour of the pollinator and activity of the colony as well as the cost of food-hair production, both in terms of materials and energy expenditure. Meliponini are thought to gather wax from the labella of some *Maxillaria* spp. (Flach et al., 2004), but why they should need to do so remains a mystery since these stingless bees, like honey bees, can make their own wax, whereas euglossine bees and halictid bees neither make wax nor use it for nest building (Roubik, 2004, personal correspondence).

Clearly, much still remains to be discovered about food-hairs and this can best be achieved by an integrated, multidisciplinary approach involving anatomical and morphological studies of the flower, entomological field work and biochemistry. Without such collaboration, the full significance of orchid food-hairs and the evolutionary advantage that they confer will continue to elude us.

Acknowledgements The author is grateful to Rodrigo B. Singer (Depto Botânica – Instituto de Biociências UFRGS – Universidade, Bairro Agronomia – Porto Alegre – RS, Brasil), to Michael P. Turner, Christian Baars and Susannah Moore (Cardiff University, Cardiff, UK) and Alan Gregg (Swansea Botanical Complex, Swansea, UK) for help in preparing this manuscript.

Glossary

Abaxial surface	The morphological under surface of a leaf, petal etc.
Aleurone grain	A protein substrate comprising a protein crystal and globoid body, the latter component rich in calcium and magnesium.
Amyloplast	A plastid involved in the storage of starch.
Aromatic amino acid	Amino acids (components of protein) with a ring-like molecular structure, e.g., tryptophan.
Capitate	Having a swollen, knob-like or pin-head-like tip.
Clavate	Club-like.
Dictyosome	The botanical name for the Golgi apparatus; an organelle involved in secretion.
Ectodesmata	Plasmodesmata connecting the cell to the exterior.
Elaiophore	An oil-secreting gland. They may be oil-secreting hairs (trichomal elaiophores), or oil-secreting epidermal cells (epithelial elaiophores).
Endoplasmic reticulum (ER)	Stacks of interconnecting, parallel lamellae involved in synthesis. Rough endoplasmic reticulum (RER) bears ribosomes and is involved in the synthesis of protein, whereas smooth endoplasmic reticulum (SER) lacks ribosomes and synthesizes lipids. ER is characteristic of secretory cells.
Epidermal	Relating to the epidermis, the outermost layer of a plant which bears hairs, cuticle, stomata, glands etc.
Food-hairs	Epidermal hairs that contain relatively high concentrations of food materials, mainly protein but sometimes

	starch. They may be nibbled or gathered by potential insect pollinators.
Labellum	A modified petal often referred to as the 'lip'. It is often more colourfully marked than the other petals and forms a landing platform for insect pollinators.
Lamellae	Arrays of membranous sacs.
Lamellate	Layered.
Lipids	A class of organic compounds that includes fats, oils and waxes.
Mentum	A chin-like structure formed by the tepals and resembling a short, blunt nectar spur.
Mitochondria	Ovoid organelles involved in cell respiration and energy production.
Moniliform hairs	Multicellular hairs where the cells are arranged like a string of beads.
Nucleus	A relatively large organelle containing genetic material as DNA and involved in the co-ordination of cellular activities.
Obpyriform	Bluntly pear-shaped.
Osmiophilic	Staining intensely with osmium tetroxide.
Papillae	Small, nipple-like projections arising from epidermal cells.
Parallelism	Where two or more unrelated species have evolved independently but in a similar manner.
Parietal	Peripheral; around the edges.
Pheromones	In this context, sex attractants. These compounds, unlike hormones, are secreted into the environment by insects to attract mates.
Plasmodesmata	Narrow strands of cytoplasm that pass through pits in walls between adjacent cells. They allow cytoplasmic continuity and transfer of substances from cell to cell.
Plastid	A spherical or ovoid organelle. Some like chloroplasts contain chlorophyll and photosynthesize. Others like chromoplasts in petals and fruit contain yellow, orange or red carotenoids and attract insect pollinators or animals that aid seed dispersal, respectively. Amyloplasts contain starch; proteoplasts, protein and elaioplasts, oil.
Plastoglobuli	Lipid bodies within plastids.
Polyphyletic	Derived from more than one ancestral species.
Pseudoantagonism	Aggressive attacks by bees against flowers which they are thought to perceive as trespassers on their territory.
Pseudocopulation	A pollination strategy where a male insect attempts to copulate with a flower resembling the female of the species.
Pseudopollen	Food-hairs that resemble pollen and are gathered by insects. Some, seemingly lack food contents and may attract pollinators by mimicry.

Pyriform	Pear-shaped.
Sensu lato	In its broad sense.
Sensu	In the sense of.
Trichome	A hair.
Uniseriate	In reference to a multicellular hair in which the cells are arranged in single file.

Literature Cited

Ackerman, J. D. 1984. Pollination of tropical and temperate orchids. *In* K. W. Tan (ed.), *Proceedings of the Eleventh World Orchid Conference*. American Orchid Society, Miami, FL, pp. 98–101.

Ackerman, J. D. 1986. Mechanisms and evolution of food-deceptive pollination systems in orchids. *Lindleyana* 1: 108–113.

Ackerman, J. D., J. A. Rodriguez-Robles, and E. J. Meléndez. 1994. A meagre nectar offering by an epiphytic orchid is better than nothing. *Biotropica* 26: 44–49.

Beck, G. 1914. Die Pollennachahmung in den Blüten der Orchideen-Gattung *Eria*. Sitzungs *Berichte Akadamie der Wissenschaften in Wien* 123: 1033–1046.

Benzing, D. H. 1986. The genesis of orchid diversity: emphasis on floral biology leads to misconceptions. *Lindleyana* 1: 73–89.

Braga, P. I. S. 1977. Aspectos biológicos das Orchidaceae de uma campina da Amazônia Central. *Acta Amazonica* 7 (suppl 2): 1–89.

Buchmann, S. L. 1987. The ecology of oil flowers and their bees. *Annual Review of Ecology and Systematics* 18: 343–396.

Buttrose, M. S. 1963. Ultrastructure of developing aleurone cells of wheat grain. *Australian Journal of Biological Science* 16: 768–774.

Cane, J. H., G. C. Eickwort, F. R. Wesley, and J. Speilholz. 1983. Foraging, grooming and mate-seeking behaviours of *Macropis nuda* (Hymenoptera, Melittidae) and use of *Lysimachia* (Primulaceae) oils in larval provisions and cell linings. *American Midland Naturalist* 110: 257–264.

Chase, M. W. 2005. Classification of Orchidaceae in the age of DNA data. *Curtis's Botanical Magazine* 22: 2–7.

Chase, M. W., R. L. Barret, K. N. Cameron, and J. V. Freudenstein. 2003. DNA data and Orchidaceae systematics: a new phylogenetic classification. *In* K. M. Dixon (ed.), *Orchid Conservation, Natural History Publications*. Sabah, Malaysia: Kota Kinabalu, pp. 69–89.

Dafni, A. and Y. Ivri. 1979. Pollination ecology and hybridization between *Orchis coriophora* L. and *O. collina* Sot. ex Russ. (Orchidaceae) in Israel. *New Phytologist* 83: 181–186.

Dathe, S., and H. Dietrich. 2006. Comparative molecular and morphological studies of selected Maxillariinae orchids. *Wildenowia* 36: 89–102.

Davies, K. L., and M. Stpiczyńska. 2006. Labellar micromorphology of Bifrenariinae Dressler (Orchidaceae). *Annals of Botany* 98: 1215–1231.

Davies, K. L., D. L. Roberts, and M. P. Turner. 2002. Pseudopollen and food-hair diversity in *Polystachya* Hook. (Orchidaceae). *Annals of Botany* 90: 477–484.

Davies, K. L. and M. P. Turner. 2004a. Morphology of floral papillae in *Maxillaria* Ruiz & Pav. (Orchidaceae). *Annals of Botany* 93: 75–86.

Davies, K. L. and M. P. Turner. 2004b. Pseudopollen in *Dendrobium unicum* Seidenf. (Orchidaceae): Reward or Deception? *Annals of Botany* 94: 129–132.

Davies, K. L. and M. P. Turner. 2004c. Pseudopollen in *Eria* Lindl. section *Mycaranthes* Rchb.f. (Orchidaceae). *Annals of Botany* 94: 707–715.

Davies, K. L. and C. Winters. 1998. Ultrastructure of the labellar epidermis in selected *Maxillaria* species (Orchidaceae). *Botanical Journal of the Linnean Society* 126: 349–361.

Davies, K. L., C. Winters, and M. P. Turner. 2000. Pseudopollen: its structure and development in *Maxillaria* (Orchidaceae). *Annals of Botany* 85: 887–895.

Davies, K. L., M. P. Turner., and A. Gregg. 2003a. Atypical pseudopollen-forming hairs in *Maxillaria* (Orchidaceae). *Botanical Journal of the Linnean Society* 143: 151–158.

Davies, K. L., M. P. Turner., and A. Gregg. 2003b. Lipoidal labellar secretions in *Maxillaria* Ruiz & Pav. (Orchidaceae). *Annals of Botany* 91: 439–446.

Davies, K. L., M. Stpiczyńska, and A. Gregg. 2005. Nectar-secreting floral stomata in *Maxillaria anceps* Ames & C. Schweinf. (Orchidaceae). *Annals of Botany* 96: 217–227.

Davies, K. L., M. Stpiczyńska, and M. P. Turner. 2006. A rudimentary labellar speculum in *Cymbidium lowianum* (Rchb.f.) Rchb.f. and *Cymbidium devonianum* Paxton (Orchidaceae). *Annals of Botany* 97: 975–984.

Dod, D. D. 1976. *Oncidium henekenii* – bee orchid pollinated by bee. *American Orchid Society Bulletin* 45: 792–795.

Dodson, C. H. 1962. The importance of pollination in the evolution of the orchids of tropical America. *American Orchid Society Bulletin* 31: 525–534, 641–649, 731–735.

Dodson, C. H. and G. P. Frymire. 1961. Natural pollination of orchids. *Missouri Botanical Garden Bulletin* 49: 133–139.

Dressler, R. L. 1990. *The Orchids – Natural History and Classification*. Harvard University Press, London.

Dressler, R. L. 1993. *Phylogeny and Classification of the Orchid Family*. Dioscorides Press, Cambridge, MA.

Endress, P. K. 1994. *Diversity and Evolutionary Biology of Tropical Flowers*. Cambridge University Press, Cambridge.

Esau, K. 1965. *Plant Anatomy 2nd edition*. Wiley, New York.

Flach, A., R. C. Dondon, R. B. Singer, S. Koehler, E. Amaral Maria do Carmo, and A. J. Marsaioli, 2004. The chemistry of pollination in selected Brazilian Maxillariinae orchids: Floral rewards and fragrance. *Journal of Chemical Ecology* 30: 1045–1056.

Garside, S. 1922. The pollination of *Satyrium bicallosum* Thunb. *Annals of the Bolus Herbarium* 3: 147–154.

Goss, G. J. 1977. The reproductive biology of the epiphytic orchids of Florida 6. *Polystachya flavescens* (Lindley) J.J. Smith. *American Orchid Society Bulletin*. 46: 990–994.

Gregg, K. B. 1982. Reproductive biology of the orchid *Cleistes divaricata* (L.) Ames var. *bifaria* Fernald growing in a West Virginia meadow. *Castanea* 54: 57–58.

Gregg, K. B. 1984. Reproductive biology of the orchid *Cleistes divaricata* (L.) Ames growing in a West Virginia meadow. *American Journal of Botany* 71: 79.

Gregg, K. B. 1991a. Reproductive strategy of *Cleistes divaricata* (Orchidaceae). *American Journal of Botany* 78: 350–360.

Gregg, K. B. 1991b. Defrauding the deceitful orchid: Pollen collection by pollinators of *Cleistes divaricata* and *C. bifaria*. *Lindleyana* 6: 214–220.

Gumprecht, R. 1977. Seltsame Bestaübungsvorgänge bei Orchideen. Sonderdruck. *Die Orchidee* 28, 3.

Inoue, K. 1986. Experimental studies on male and female reproductive success: effects of variation in spur length and pollinator activity on *Platanthera mandarinorum* ssp. *hachijoensis* (Orchidaceae). *Plant Species Biology* 1: 207–215.

Janse, J. M. 1886. Imitirte pollenkörner bei *Maxillaria* sp. *Deutsche Botanische Gesellschaft Berichte* 4: 277–283.

Johnson, S. D. and W. J. Bond. 1997. Evidence for widespread pollen limitation of fruiting success in Cape wildflowers. *Oecologia* 109: 530–534.

Johnson, S. D. and L. A. Nilsson. 1999. Pollen carryover, geitonogamy and the evolution of deception in orchids. *Ecology* 80: 2607–2619.

Jones, D. L. 1981. The pollination of selected Australian orchids. *In* L. Lawler and R. D. Kerr, (eds.), *Proceedings of the Orchid Symposium, 13th International Botanical Congress*. Orchid Society of New South Wales, Sydney, Australia.

Jones, D. L. 1985. The pollination of *Gastrodia sesamoides* R. Br. in southern Victoria. *Victorian Naturalist* 102: 52–54.

Kay, Q. O. N., H. S. Daoud, and C. H. Stirton. 1981. Pigment distribution, light reflection and cell structure in petals. *Botanical Journal of the Linnean Society* 83: 57–84.

Kjellsson, G., and F. N. Rasmussen. 1987. Does the pollination of *Dendrobium unicum* Seidenf. involve "pseudopollen"? *Die Orchidee* 38: 183–187.

Kocyan, A., and P. K. Endress. 2001. Floral structure and development of *Apostasia* and *Neuwiedia* (Apostasioideae) and their relationships to other Orchidaceae. *International Journal of Plant Sciences* 162: 847–867.

Kurzweil, H., H. P. Linder, and P. Chesselet. 1991. The phylogeny and evolution of the *Pterygodium–Corycium* complex (Coryciinae, Orchidaceae). *Plant Systematics and Evolution* 175: 161–223.

Lindley, J. 1842. *Eria paniculata* (33). *Miscellaneous matter of the Botanical Magazine*, p. 38.

Macpherson, K., and H. M. R. Rupp. 1935. The pollination of *Cymbidium iridifolium* Cunn. *The North Queensland Naturalist* 3: 26.

Matusiewicz, J., M. Stpiczyńska, and K. L. Davies. 2004. Pseudopollen in the flowers of *Maxillaria lepidota* Lindl. (Orchidaceae). *Proceedings of the 53rd Meeting of the Polish Botanical Society*, Torun. p. 14.

Meléndez-Ackerman, E. J., J. D Ackerman, and J. A. Rodriguez-Robles. 2000. Reproduction in an orchid can be resource-limited over its lifetime. *Biotropica* 32: 282–290.

Michener, C. D. 2000. *The Bees of the World*. John Hopkins University Press, Baltimore.

Mickeliunas, L., E. R. Pansarin, and M. Sazima. 2006. Biologia floral, melitofilia, e influência de besouros Curculionidae no sucesso reprodutivo de *Grobya amherstiae* Lindl. (Orchidaceae: Cyrtopodiinae). *Revista Brasileira de Botânica* 29: 251–258.

Neiland, M. R. M., and C. C. Wilcock. 1994. Reproductive ecology of European orchids. *Proceedings of 14th World Orchid Conference, Edinburgh, UK HMSO* 1994.

Neiland, M. R. M., and C. C. Wilcock. 1998. Fruit set, nectar reward and rarity in the Orchidaceae. *American Journal of Botany* 85: 1657–1671.

Neiland, M. R. M., and C. C. Wilcock. 2000. Effects of pollinator behaviour on pollination of nectarless orchids: floral mimicry and interspecific hybridisation. *In* K. L. Wilson, and D. A. Morrison (eds.), *Monocots: Systematics and Evolution*, CSIRO, Melbourne 2000, pp. 318–326.

Nilsson, L. A. 1978. Anthecological studies of the Lady's Slipper, *Cypripedium calceolus* (Orchidaceae). *Botaniska Notiser* 132: 329–347.

Pauw, A. 2006. Floral syndromes accurately predict pollination by a specialized oil-collecting bee (*Rediviva peringueyi*; Melittidae) in a guild of South African orchids (Coryciinae). *American Journal of Botany* 93: 917–926.

Pohl, F. 1935. Zwei *Bulbophyllum*-Arten mit besonders bemerkenswert gebauten Gleit- und Klemfallenblumen. *Beihefte zum Botanischen Centralblatt* 53: 501–518.

Porsch, O. 1905. Beiträge zur 'histologischen' Blütenbiologie I. *Österreichische Botanische Zeitschrift* 55: 253–260.

Porsch, O. 1906. Beiträge zur 'histologischen' Blütenbiologie II. *Österreichische Botanische Zeitschrift* 56: 41–47, 83–95, 125–143, 176–185.

Porsch, O. 1908. Neuere Untersuchungen über die Insektenanlockungsmittel der Orchideenblüte. *Mittelungen Naturwissenschaftlichen Vereines für Steiermark* 45: 346–370.

Pridgeon, A. M. 1992. *The Illustrated Encyclopaedia of Orchids*. Weldon Publishing, Sydney, Australia.

Pridgeon, A. M. 1993. Systematic anatomy of Orchidaceae. Resource or anachronism? *Proceedings of the 14th World Orchid Conference*, Glasgow. 84–91. HMSO, Edinburgh.

Proctor, M., and P. Yeo. 1975. *The Pollination of Flowers*. Collins, London.

Proctor, M., P. Yeo, and A. Lack. 1996. *The Natural History of Pollination*. Harper Collins, London.

Pyke, G. H. 1991. What does it cost a plant to produce floral nectar? *Nature* 350: 58–59.

Reis, M. G., A. D. de Faria, V. Bittrich, M. C. E. Amaral, and A. J. Marsaioli. 2000. The chemistry of flower rewards – *Oncidium* (Orchidaceae). *Journal of the Brazilian Chemical Society* 11: 600–608.

Reis, M. G., A.D. de Faria, M. C. E. Amaral, and A. J. Marsaioli. 2003. Oncidinol – a novel diacylglycerol from *Ornithophora radicans* Barb. Rodr. (Orchidaceae) floral oil. *Tetrahedron Letters* 44: 8519–8523.

Reis, M. G., R. B Singer, R. Gonçalves, and A. J. Marsaioli. 2006. The chemical composition of *Phymatidium delicatulum* and *P. tillandsioides* (Orchidaceae) floral oils. *Natural Product Communications* 1: 757–761.

Roberto Vásquez, C. and C. H. Dodson. 1982. Icones Plantarum Tropicarum: Orchids of Bolivia. *In* C. H. Dodson (ed.) *Series 1, Fascicle 6, Plate 5.* Marie Selby Botanical Gardens, Sarasota.

Roberts, D. L. 2001. *Reproductive Biology and Conservation of the orchids of Mauritius.* Ph.D. thesis. University of Aberdeen, UK.

Romero, G. A. 1998. Venezuela, Orchid Paradise Caracas: Armitano Editores C.A.

Roubik, D. W. 2000. Deceptive orchids with Meliponini as pollinators. *Plant Systematics and Evolution* 222: 271–279.

Senghas, K. 1993. Subtribus Maxillariinae. *In* F. G. Breiger, R. Maatsch, and K. Senghas, (eds), Rudolph Schlechter: *Die Orchideen*: Blackwell Wissenschafts-Verlag, Berlin 28: 1727–1776.

Silvera, K. I. 2002. *Adaptive Radiation of Oil-rewarding Compounds Among Neotropical Orchid Species (Oncidiinae).* M.Sc. Thesis, University of Florida, Florida.

Singer, R. B. and A. A. Cocucci. 1999. Pollination mechanisms in four sympatric southern Brazilian Epidendroideae orchids. *Lindleyana* 14: 47–56.

Singer, R. B., and S. Koehler. 2004. Pollinarium morphology and floral rewards in Brazilian Maxillariinae (Orchidaceae). *Annals of Botany* 93: 39–51.

Singer, R. B., A. Flach, S. Koehler, A. J. Marsaioli, and E. Amaral Maria do Carmo. 2004. Sexual mimicry in *Mormolyca ringens* (Lindl.) Schltr. (Orchidaceae: Maxillariinae). *Annals of Botany* 93: 755–762.

Singer, R. B., A. J. Marsaioli, A. Flach, and M. G. Reis. 2006. The ecology and chemistry of pollination in Brazilian orchids: Recent Advances. *In:* J. da Silva (ed.) *Floriculture, Ornamental and Plant Biotechnology* Vol. IV, Global Science Books, Middlesex, pp. 569–582.

Smithson, A. 2002. The consequences of rewardlessness in orchids: reward-supplementation experiments with *Anacamptis morio* (Orchidaceae). *American Journal of Botany* 89: 1579–1587.

Southwick, E. E. 1984. Photosynthate allocation to floral nectar: a neglected energy investment. *Ecology* 65: 1775–1779.

Steiner, K. E. 1989. The pollination of *Disperis* (Orchidaceae) by oil-collecting bees in southern Africa. *Lindleyana* 4: 164–183.

Steiner, K. E. 1993. Oil orchids and oil bees in southern Africa – *Disperis* and *Rediviva. South African Orchid Journal* 24: 2–5.

Steiner, K. E. 1998. The evolution of beetle pollination in a South African orchid. *American Journal of Botany* 85: 1180–1193.

Stern, W. L., W. S. Judd, and B. S. Carlsward. 2004. Systematic and comparative anatomy of Maxillarieae (Orchidaceae), *sans* Oncidiinae. *Botanical Journal of the Linnean Society* 144: 251–274.

Stpiczyńska, M. 1997. The structure of the nectary of *Platanthera bifolia* L. (Orchidaceae). *Acta Societatis Botanicorum Poloniae* 66: 5–11.

Stpiczyńska, M. 2003. Nectar resorption in the spur of *Platanthera chlorantha* (Custer) Rchb. Orchidaceae – structural and microautoradiographic study. *Plant Systematics and Evolution* 238: 119–126.

Stpiczyńska, M. and J. Matusiewicz. 2001. Anatomy and ultrastructure of the spur nectary of *Gymnadenia conopsea* L. (Orchidaceae). *Acta Societatis Botanicorum Poloniae* 70: 267–272.

Stpiczyńska, M., C. Milanesi, C. Faleri, and M. Cresti. 2005. Ultrastructure of the nectary spur of *Platanthera chlorantha* (Custer) Rchb. (Orchidaceae) during successive stages of nectar secretion. *Acta Biologica Cracoviensia* 47: 111–119.

Stpiczyńska, M., K. L. Davies, and A. Gregg. 2007. Elaiophore diversity in three contrasting members of the Oncidiinae (Orchidaceae). *Botanical Journal of the Linnean Society* 155: 135–148.

Sugiura, N., Miyazaki, S. and Nagaishi, S. 2006. A supplementary contribution of ants in the pollination of an orchid, *Epipactis thunbergii*, usually pollinated by hover flies. *Plant Systematics and Evolution* 258: 17–26.

Thien, L. B., and B. G. Marcks. 1972. The floral biology of *Arethusa bulbosa*, *Calopogon tuberosus* and *Pogonia ophioglossoides* (Orchidaceae). *Canadian Journal of Botany* 50: 2319–2325.

Toscano de Brito, A. V. L. 2001. Systematic review of the *Ornithocephalus* group (Oncidiinae: Orchidaceae) with comments on *Hofmeisterella. Lindleyana* 16: 157–217.

Tsou, C. H. 1997. Embryology of the Theaceae – anther and ovule development of *Camellia*, *Franklinia* and *Schima. American Journal of Botany* 84: 369–381.

van der Cingel, N. A. 2001. *An atlas of orchid pollination: America, Africa, Asia and Australia.* A.A. Balkema, Rotterdam, Netherlands.

van der Pijl, L. and C. H. Dodson. 1969. *Orchid Flowers: Their Pollination and Evolution.* University of Miami Press, Coral Gables, FL.

Vinson, S. B., H. J. Williams, W. Frankie, and G. Schrum. 1997. Floral lipid chemistry of *Byrsonima crassifolia* (Malpighiaceae) and a use of floral lipids by *Centris* bees (Hymenoptera: Apidae). *Biotropica* 29: 76–83.

Vogel, S. 1974. Ölblumen und ölsammelnde Bienen. *Abhandlungen Akademie Wissenschaften Mathematisch-Naturwissenschaften Klasse Tropische und Subtropische Pflanzenwelt* 7: 1–267.

Vogel, S. 1979. Evolutionary shifts from reward to deception in pollen flowers. *In* A. J. Richards (ed.), *The Pollination of Flowers by Insects*. Academic Press, London, pp. 89–96.

Vogel, S. 1990. History of the Malpighiaceae in the light of pollen ecology. *Memoirs of the New York Botanical Garden* 55:130–142.

von Kirchner, O. 1925. Über die sogenannten Pollenblumen und die Ausbeutestoffe der Blüten. *Flora* 118/119: 312–330.

Wallich, N. 1830. *Plantae Asiaticae rariores*; or, Descriptions and figures of a select number of unpublished East Asian plants, by Nathaniel Wallich, Treuttel and Wèurtz London, Paris, Strasburgh.

Whitehead, V. B. and K. E. Steiner. 2001. Oil-collecting bees of the winter rainfall area of South Africa (Melittidae, *Rediviva*). *Annals of the South African Museum* 108: 143–277.

Ziegenspeck, H. 1936. Orchidaceae *In:* O. von Kirchner, E. Loew and C. Schröter (eds) *Lebensgeschichte der Blütenpflanzen Mitteleuropas.* Spezielle Ökologie der Blütenpflanzen Deutchlands, Österreichs und der Schweiz. Band I, Verlagsbuchhandlung Eugenulmer, Stuttgart.

POLLINATION BIOLOGY

5

Orchids Pollen Dispersal Units and Reproductive Consequences

ETTORE PACINI

Contents

Introduction 187
Types of Pollen Dispersal Units 187
Monad Pollen with Pollenkitt 190
Monad Pollen Grouped by Elastoviscin 190
Isolated Tetrads 194
Aggregated Pollen Tetrads 194
Group of Tetrads United Externally by a Thin Layer of Callose: Soft Pollinium Type A 194
Tetrads Grouped by Elastoviscin: Soft Pollinium Type B 194
Tetrads Grouped by Common Walls to Form a Massula: Soft Pollinium C 195
Tetrads Grouped in a Compact Pollinium 195
Consistency of the Different PDU Types 195
Anther Pollen Structure and PDU Types 197
Tapetum and Locular Fluid 197
Tetrad Types 200
Mechanisms of Tetrad Adhesion in Developing Pollinia and During Transport 200
Pollen Wall Modifications Aperture and Increase in Number of Pollen Grains per PDU 201
PDU and Fossil Records 202
Pollen Volume and Vacuolisation 202
Pollen Carbohydrate Reserves 202
Pollen Presentation 203
Generative Cell and Paternal Inheritance 203
Developmental Arrest in the Different PDU Types and Pollen Longevity 204
Female Part Adaptations According to PDU Types 206
PDU and Pollen Load 207
Pollen Load and Paternity 207
Neotenic and Changing Characters in Orchid PDU Evolution 208
Animals Commonly Involved in Orchid PDU Dispersal and Inherent Problems 208
Conclusions and Further Work 209

Kull, Arditti and Wong (eds.), *Orchid Biology: Reviews and Perspectives, X*,

Glossary ... 210
Literature Cited ... 212
Technical Appendix ... 216
Simple Cytological Methods for the Study of Orchid Pollen ... 216
Pollen Water Content ... 216
Pollen Volume and Hydration Status ... 216
Number of Grains per Anther or Pollinium ... 216
Light Microscopy ... 217
Histochemistry ... 217
Pollen viability ... 217
Scanning Electron Microscopy ... 218
Transmission Electron Microscopy ... 218

Introduction

Male and female reproductive organs vary widely in angiosperms, due to the number of ovules per ovary and the number of pollen grains in pollen dispersal units (PDUs), a term used to indicate the different ways in which ripe pollen is presented for dispersal (Pacini, 1997). Pollen may travel as a single grain, or *en masse,* as a compound pollen (Knox and McConchie, 1986).

Pollen forms aggregates by various means: (i) viscous fluids derived from tapetum activity and/or degeneration; (ii) filaments derived from tapetum activity, composed of sporopollenin and continuous with pollen exine; (iii) threads derived from other anther parts; (iv) common walls, i.e., contiguous pollen grains derived from the same meiocyte or from close meiocytes which share a wall (Pacini and Franchi, 1999; Hesse, Vogel, and Halbritter 2000).

The PDU is often a systematic character. Only one type of PDU is present in many angiosperm families. Orchidaceae have different PDU, some are unique, others are present in other groups (Pacini and Franchi, 2000). One type, the pollinium, was recognized by early 19th century by plant embryologists such as Reichenbach (1852) and Hofmeister (1861).

There are several descriptive papers on the ontogenesis and ripe pollen of the more common PDU type in Orchidaceae (Williams, 1970, 1972; Williams and Broome, 1976; Newton and Williams, 1978; Ackerman and Williams, 1980, 1981; Yeung, 1987a, b, c; Clifford and Owens, 1990; Zavada, 1990; Brown and Lemmon, 1991; Pandolfi, Pacini, and Calder 1993; Pandolfi and Pacini, 1995; Freudestein and Rasmussen, 1997). Few of these studies consider all the different PDU types and compare them with other reproductive features (Wolter and Schill, 1986; Schill and Wolter, 1986; Zee and Siu, 1990; Johnson and Edwards, 2000; Pacini and Hesse, 2002).

Orchids have several reproductive peculiarities, both on the male and female side, and some specific terms related to different aspects of reproductive parts are typical of this family (see the glossary). Some of these features probably originated as an adaptation of male and female parts to an increase in the number of grains contained in the PDU. Wolter and Schill (1986) presented a survey on orchid PDU from an ultrastructural point of view. In the present series of volumes (no. IV) there is another essay by Yeung (1987a) devoted to embryology and general aspects of orchid reproduction. The present author and different co-workers have written papers on PDU types, germination, and biological consequences (Pacini and Franchi, 1996, 1999, 2000; Pacini and Hesse, 2002; Pandolfi et al., 1993; Pandolfi and Pacini, 1995) which inspired the present review.

Types of Pollen Dispersal Units

On the basis of the number of aggregated pollen grains and sticking modalities, Pacini and Franchi (1998) recognized 13 PDU types in Angiosperms: ten in monocots (Pacini and Franchi, 2000) and eight in orchids (Pacini and Hesse, 2002), four of which are typical of this group. Orchidaceae is the Angiosperm family with the greatest number of PDU types.

Monad pollen is the starting point for all PDU types in orchids which form in two steps. The first is common to many other angiosperms families, that is, the passage from monad to tetrad pollen (Pacini and Franchi, 1998). The second step is more complex and articulated because it gives rise to different cytological mechanisms of tetrad combinations to form different PDU types with a high number of pollen grains. Schill, Dunnenbaum, and Neyer (1992) reported that the number of pollen grains in a pollinium range from 40,000 and 4 million.

Pollen usually loses water before dispersal to slow down the metabolism in response to contact with the environment (Pacini and Hesse, 2004). Full metabolism is restored after rehydration on the stigma. This complex process of water loss and gain is called harmomegathy (Pacini, 1990) and interferes with pollen aggregation; in fact monads have a furrow and accomodate changes in water content, whereas in PDUs consisting of many thousands of pollen grains, harmomegathy becomes impossible.

Orchid PDU types differ due to: (a) different types of gluing material derived from tapetum degeneration, i.e. a pollenkitt or elastoviscin; (b) monad or tetrad grains; (c) different ways of tetrads cohesion by common walls to form the massulae; (d) massulae or a more compact units; (e) the harmomegathic capacity of the grains. A short description and scheme of the different PDU types in orchids and some examples from the literature may be found in Figs. 5-1–5-6.

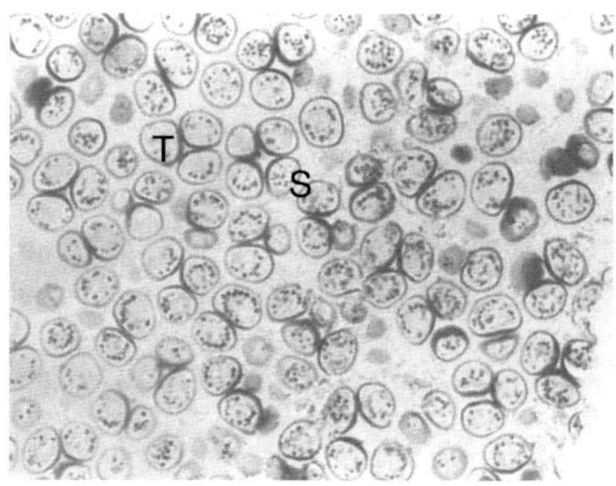

Fig. 5-1. Light micrograph of part of a *Pterostylis plumosa* anther at late microspore stage. Microspores are still in the position of the tetrad and different types of tetrads are recognizable. tetrahedric (T) and square (S). Intine has already formed and the furrow is oriented inwardly with respect to the tetrad. Microspore cytoplasm contains starch grains (amyloplasts); PAS.

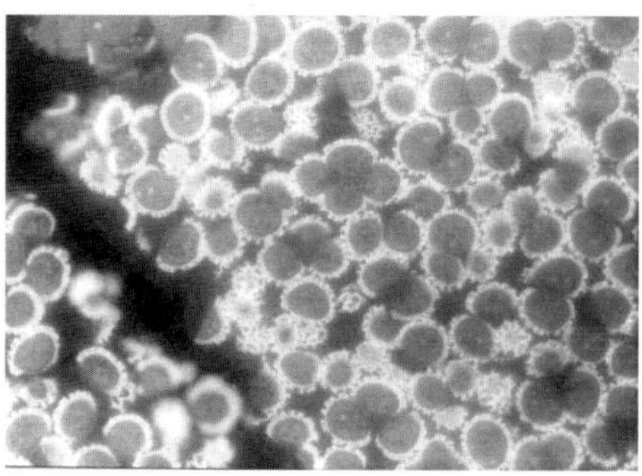

Fig. 5-2. Light micrograph of part of a *Pterostylis plumosa* anther at late microspore stage. Ornamented reticulate exine is present only in the outer part of the tetrad, Auramine O.

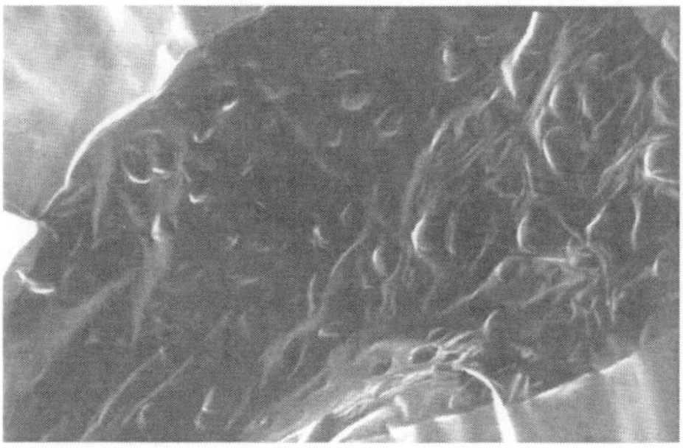

Fig. 5-3. Scanning electron micrograph of *Cypripedium calceolus* pollen grains engulfed in elastoviscin.

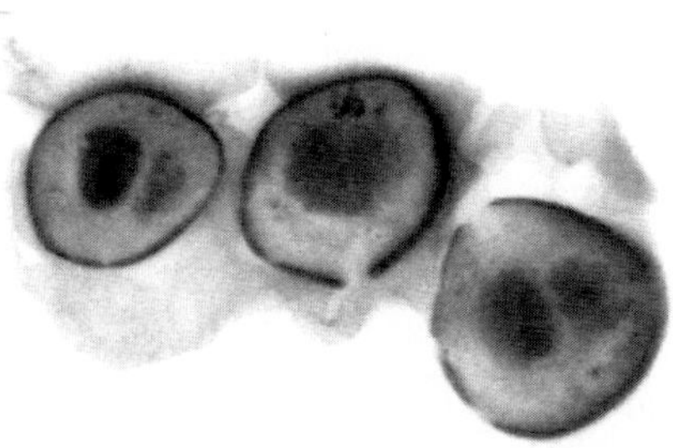

Fig. 5-4. Light micrograph of ripe pollen grains of *Paphiopedilum villosum* engulfed in elastoviscin. Pollen grains have a furrow and generative cell is spherical, Toluidine blue.

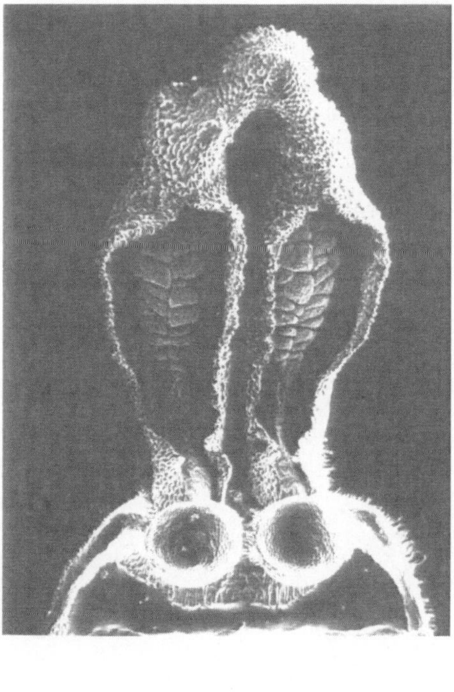

Fig. 5-5. Scanning electron micrograph of the androecium of *Ophrys insectifera* with two soft pollinia composed of a few dozen of massulae partially enclosed by the anther, two spherical viscidia are at the base of the pollinia enabling the pollinia to be taken by pollinators.

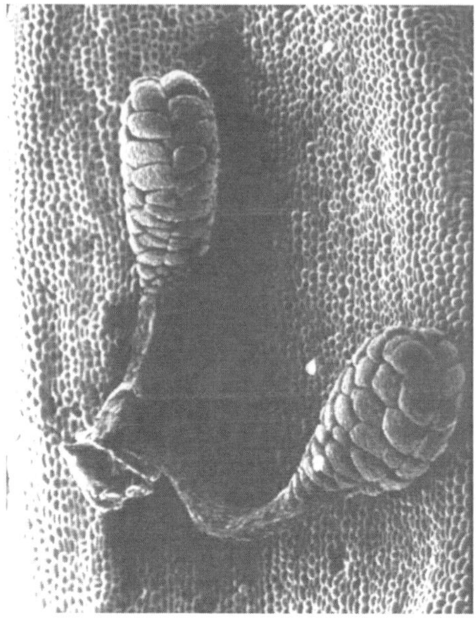

Fig. 5-6. Scanning electron micrograph of a pollinarium of *Orchis pyramidalis* with a viscidium connected to the pollinia by a stipe.

Monad Pollen with Pollenkitt

Monad pollen grains have an aperture (Fig. 5-1) of the furrow type (Punt et al., 1994), exine is reticulate (Fig. 5-2) and partially covered in pollenkitt (Fitzgeral et al., 1994, Fig. 13). In the case of *Pterostylis plumosa* (Pandolfi et al., 1993) the microspores of each tetrad persist in their relative position until anther opening (Figs. 5-1 and 5-2). PDUs of this type have been described in *Apostasia*, *Neuwiedia* (Kocyan and Endress, 2001), *Pterostylis plumosa* (Pandolfi et al., 1993), and *P. concinna* (Fitzgeral et al., 1994). This type of PdU is the commonest in the angiosperms and also the most primitive because pollenkitt is a primitive character and its absence is secondary (Pacini and Hesse, 2004).

Monad Pollen Grouped by Elastoviscin

The pollen grains have an aperture and are engulfed in elastoviscin (Fig. 5-3), an extremely viscous fluid, like chewing gum, found in different parts of orchid anthers in almost all the subfamilies of the Orchidaceae. It is homologous to pollenkitt (Wolter and Schill, 1985a, 1986, Table 5-1). This fluid is more abundant than pollenkitt because the pollen grains are totally engulfed in it (Fig. 5-3) and not in contact one with another as occurs when there is pollenkitt. Unlike pollenkitt, elastoviscin is not removed by non-osmicated fixative (Fig. 5-4). Elastoviscin is a product of degeneration of the tapetum cytoplasm and its main

Table 5-1. Table showing different types of histochemical techniques and their application to orchid pollen.

Stain or reagent	Specificity	Optics	References	Solvent and procedure	Comments
Polysaccharides					
PAS	Total insoluble polysaccha rides with two adjacent free hydroxyil groups	BF	O'Brien and McCully, (1981)	See O'Brien and McCully, (1981)	This is recommended to evidentiate the different intine layers
IKI (iodine potassium iodide or Lugol)	Starch	BF	Johansen, (1940)	Johansen, (1940)	This a simple and common technique to reveal starch in pollen and anther
Cacofluor white	ß-D-glucans (cellulose and callose)	UV	O'Brien and McCully, (1981)	0.1% in water. Stain from 20 s to 2 min. Rinse in water	Useful to evidentiate under the fluorescent microscope cellulose and callose
Aniline blue	1–3 – ß- glucans (callose)	UV	Currier and Strugger, (1956)	0.05% water soluble Aniline blue in phosphate buffer 0.07 M	Useful to evidentiate callose under the fluorescent microscope
Alcian blue 8GX	Acid polysaccharides	BF	Jensen, (1962)	1% in acetic acid 3%. Adjust to pH 2.5, mount directly with the stain	Recommended for poral and intine pectins
Rutenium red	Acid polysaccharides	BF	Johansen, (1940)	0.02–0.05% in water, mount directly with the stain	Recommended for poral and intine pectins
Lipids					
Auramine O	Lipids (including waxes and sporopollenin)	UV	Heslop-Harrison, (1977)	0.01% in tris HCL 0.05 M (pH 7.2)	Commonly used to detect sporopollenin and cuticle in fresh or fixed material under fluorescence microscope
Scarlet R	Surface lipids	BF	Heslop-Harrison and Heslop-Harrison, (1985)	Saturated in 70% ethanol	Recommended for pollenkitt

(continued)

Table 5-1. (continued).

Stain or reagent	Specificity	Optics	References	Solvent and procedure	Comments
Sudan dyes	Lipids	BF	Jensen (1962)	1% in water for 30 s. T = 37 °C. Acetic acid 1% in water for 30 s. T = 37°C. Rinse with water. Mount in jelly glycerol	To be used only on fresh material, especially for lipids reserves
Fluorol yellow 088	Lipids	UV	Reagan and Moffat, (1990)	0.01% in 50% PEG + 40% of glycerol	Very sensitive method for hand or cryostat sections
Proteins					
Aniline blue black	Total proteins	BF	Fisher, (1968)	1 g in 100 ml acetic acid 7%. Stain for 10 min at 50–60°C. Rinse in acetic acid 7%. Dry and mount with acetic acid 5% in glycerol	Can be used on whole pollen grains, cryostat sections or material fixed in aldehyde and embedded in resin
Coomassie Blue R250	Total proteins	BF	Heslop-Harrison et al. (1974)	0.25 mg in 100 ml of acetic acid 7%, stain for 3 min at 30°C. Rinse in acetic acid 7%, dry and mount in glycerol with acetic acid 5%	Can be used on whole grains, cryostat section or material fixed in aldehyde and resin embedded
Bromophenol Blue	Total proteins	BF	Mazia et al. (1953)	0–25 mg in 100 ml of acetic acid 7%. Stain for 3 min at 30°C. Rinse in acetic acid 7%. Dry and mount in glycerol with acetic acid 5%	Can be used on material fixed in aldehyde and embedded in glycerol or resin
1-anilino-8 – naphthalene sulfonic acid (1-ANS)	Total proteins	BF	Heslop-Harrison et al. (1984)	0.01% in 0.01 M phosphate buffer (pH 6.8) + 15% methanol	This can be used on fresh or aldehyde fixed material and embedded in resins
FITCH (Fluorescein iso-thiocyanate)	Total proteins	UV	Pearse, (1985)	1 mg/ml, in carbonate buffer 0.5 M (pH 9), stain for 30–60 s, rinse in water, dry and mount in glycerol or immersion oil	Can be used on fresh or aldehyde fixed material embedded in resin

(continued)

DNA					
Feulgen reaction	DNA	BF	O'Brien and McCully, (1981)	See O'Brien and McCully, (1981)	Material fixed with different methods embedded in wax and resins of high permeability
A ',6-diamidino – 2-phenylindole (DAPI)	DNA	UV	Goff and Coleman (1984)	0.01 mg/ml in water	Used for fresh and glutaraldehyde fixed material
Ethidium bromide	DNA	UV	Le Pecq and Paletti, (1967)	0.01% in water	Used for fresh and glutaraldehyde fixed material
General stains					
Azur B	Negatively charged metachromatic groups	BF	Flax and Himes, (1952)	0.25 mg/ml in 0.1 sodium citrate buffer (pH 4.0)	Best results with glutaraldehyde fixed material
TBO	Negatively charged metachromatic groups	BF	O'Brien et al. (1964)	0.05 in benzoate buffer (pH 4.4). Stain for 1–5 min. Rinse in water, dry and mount in glycerol or immersion oil	Recommended for different wall types
Lacto-phenol cotton blue	High density cytoplasm	BF	Darlington and La Cour (1960)	Used to determine the presence or absence of cytoplasm in pollen grains	Phenol 20 ml + lactic acid 29 ml + glycerol + distilled water + 0.05 cotton blue. Dissolve phenol in water heating moderately
Basic Fuchsin	Exine	BF	Faegri and Iversen (1989)	Often used to stain exine in fresh or fixed but not osmicated material	1% in water. Stain for 1–5 min. Rinse in water. Dry and mount in immersion oil or glycerol.
Acetolysis	Exine	BF	Faegri and Iversen (1989)	Faegri and Iversen (1989)	This method can be applied either directly to the whole PDU or sections. It destroys all cell components except exine. Use with caution

component, as in the case of pollenkitt, are spherosomes (Wolter, Seufertet, and Schill, 1988). Examples of this PDU may be found in *Cypripedium acaule* (Burns-Balogh and Hesse, 1988, Fig. 5-6), *Cypripedium calceolus*, *Paphiopedilum villosum* (Figs. 5-3 and 5-4) and *Apostasia wallachii* (Schill and Wolter, 1986, Figs. 5-1 and 5-2). Exine of *Apostasia wallachii* is reticulate whilst that of *Cypripedium acaule* and *C. calceoulus* is smooth; this means that for contrast elastoviscin viscosity, pollen-elastoviscin adhesion is higher in *Apostasia* than in *Cypripedium* because of the irregular exine surface.

Isolated Tetrads

From the examples in the literature, it is not possible to recognize pollen gluing material but it must be present because orchids are always entomophilous and tetrads must stick together and to insect bodies (Pacini, 2000). There are only two examples: *Epipactis microphylla* (Dressler, 1981, Fig. 3.22) and *Bletilla striata* (Wolter and Schill, 1986, Fig. 105). In *Epipactis* there is only one aperture per grain and it is of the furrow type.

Aggregated Pollen Tetrads

There is only one example: *Cleistes divaricata* (Gregg, 1991), and no data on the presence of furrows or other harmomegathic device. It is not clear how the tetrads stay together, but a separate category seems likely because the tetrads are not completely free but united by an unknown sticky fluid. This PDU is also characterized by the possibility of dispensing loosely aggregated tetrads on different pollinators, a process facilitated by certain floral structures (Gregg, 1991).

Group of Tetrads United Externally by a Thin Layer of Callose: Soft Pollinium Type A

The only example is *Polystachia pubescens* (Schlag and Hesse, 1993). No data are available on the presence of apertures or harmomegathic devices.

Tetrads Grouped by Elastoviscin: Soft Pollinium Type B

This PDU type is typical of *Neottia nidus-avis* (Buchner and Weber, 2000) and seems to be present in other Neottioideae (Yeung, 1987a). The pollinium is mealy and its tetrad content can be deposited on different stigmas. However, from these two papers it is not possible to understand if furrows or other harmomegathic devices are present.

Tetrads Grouped by Common Walls to Form a Massula: Soft Pollinium C

Tetrads forming the massulae of the pollinium are tightly packed with few, very reduced intercellular spaces (Fig. 5-6). Massulae are commonly pear shaped with the pointed part attached to the caudicle which is continuous with the stipe which bears the viscidium at one pole (Fig. 5-6). The external shape of the pollinium varies from spherical to pear-shaped (Schill and Pfeiffer, 1977; Johnson and Edwards, 2000).

Because they are loosely attached to the caudicle the massulae of a pollinarium can be dispersed on different stigmas depending on movements of the pollinator during reiterated visits to flowers. This PDU type is present in Spiranthoideae, Orchidioideae and Epidendroideae (Wolter and Schill, 1986) and also in Vandoideae according to Johnson and Edwards (2000). Harmomegathic devices are absent because the tetrads are compacted. Pollinia of this type remain protected in the anther until an insect touches the viscidium (Fig. 5-5). Only then is the pollinarium is removed from the flower.

The variations in this PDU type may concern wall structure: (a) exine is present around all the tetrads with some differences between inner and outer tetrads as in *Epidendrum scutella* (Cocucci and Jensen, 1969; Dressler, 1981, Fig. 3.22.f); (b) exine is present only on the outside of pollen grains of external tetrads (Figs. 5-7 and 5-8) as in *Epidendrum ibaguense* (Yeung, 1987b) and *Loroglossum hircinum* (Pandolfi et al., 1993). Possible variations are the presence of one or two viscidia. If there is only one, the two pollinia are kept together by the pollinator, as in *Loroglossum hircinum* (Fig. 5-6). If there are two it means that each pollinium may be dispersed separately by one or two visitors, as in *Pleurothallis eumecocaulon* (Stenzel, 2000) and *Ophrys insectivora* (Fig. 5-5).

Tetrads Grouped in a Compact Pollinium

There are four flat pollinia per pollinarium (Fig. 5-9). Exine is present only on the external surface of the outer tetrads. Harmomegathic devices are absent because of compact structure. This type of PDU is collected in one flower and completely deposited in another. It occurs in *Calypso bulbosa* (Proctor and Harder, 1995).

A similar structure is present in the dicot family Asclepiadaceae, but there are fewer grains per pollinium than in orchid pollinia, and external shape is totally different (Dannenbaum and Schill, 1991).

Consistency of the Different PDU Types

Wolter and Schill (1985b) drew up a table of four pollen forms for orchids, a kind of simplification of our PDU, and their presence in the subfamilies. PDU types

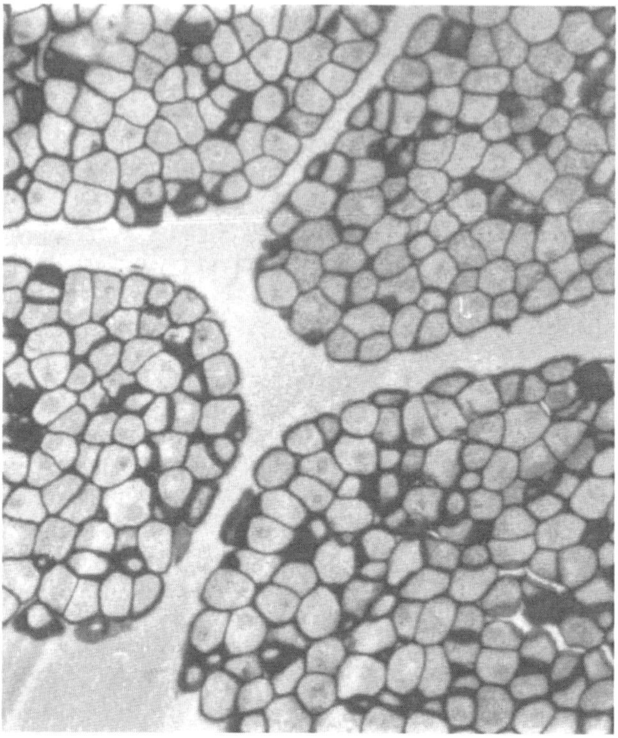

Fig. 5-7. Light micrograph of part of a ripe soft pollinium of *Ophrys sphegodes*. Few spaces are present between tetrads, PAS.

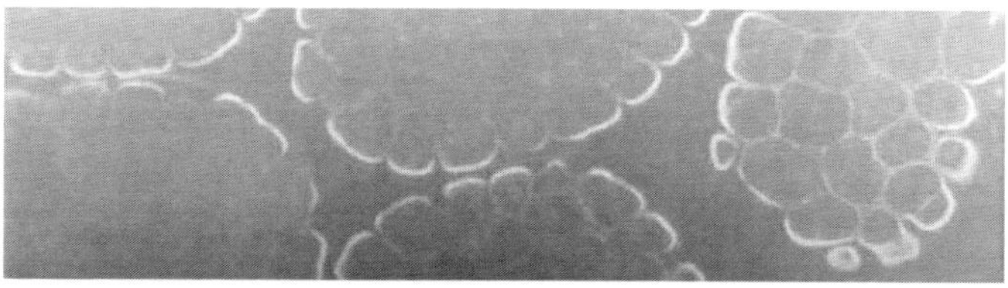

Fig. 5-8. Light micrograph of part of a ripe soft pollinium of *Ophrys sphegodes*. Exine is thick in the outer tangential walls of external microspores of outer tetrads and is reduced in radial and inner tangential walls. It is absent from inner microspores.

quite well respect the taxonomy of orchids, i.e., early orchids such as the subfamily Apostasioideae and early members of the Cypripedioideae, Spiranthoideae, and Orchidoideae have monad pollen and more evolved ones, such as the Epidendroideae and Vandoideae, only have different types of pollen presentation.

Soft pollinia types A, B and C and monad pollen grouped by elastoviscin are typical mechanisms of orchids (Pacini and Franchi, 2000).

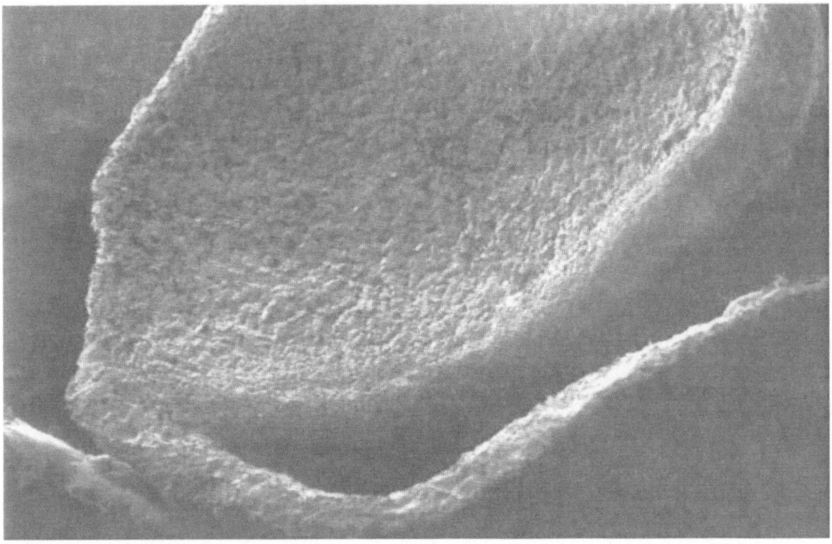

Fig. 5-9. Scanning electron micrograph of a compact pollinium of *Cattleya* sp. having flat shape and a long viscidium.

Anther Pollen Structure and PDU Types

Orchid PDU types influence anther structure. For example, the position of the mechanical layer and the wall thickening pattern are different in monad pollen and pollinia because of different types of pollen exposure. Freudestein (1991) reported different types of endothecial thickenings in orchids with different PDU types. Wall thickenings are different because in early orchids pollen is exposed inside the open anther, whilst in those having a pollinium, only part is exposed and visible from outside (Fig. 5-5), because the anther does open completely.

Some features related to the anther are considered in relation to PDU type in the following chapter.

Tapetum and Locular Fluid

Tapetum type is an embryological feature often with taxonomical significance (Davis, 1966) and the different types seem adaptations to certain pollination features (Pacini, 1997). Notwithstanding their different PDU types orchids always have a secretory tapetum, but may differ in walls persistence/disappearance and the products resulting from tapetal degeneration. In fact when the PDU is a monad, as in *Pterostylis plumosa*, tapetal cells loose their inner and radial walls (Pandolfi et al., 1993) as commonly occurs in angiosperms (Pacini, 1997), but these walls persist in *Cochlioda rosea* and *Stenoglottis longifolia* (Wolter and Schill, 1986,

Figs. 48 and 51) which have a pollinia as PDU type. Tapetal cells in angiosperms lose their walls to improve release of nutrients inside the loculus (Pacini, 1997). Persistence of tapetal cell walls in orchids with pollinia and the extreme reduction of locular fluid can be considered neotenic characters (Figs. 5-10–5-13) because both represent arrest at an early developmental stage and are present in early land plants, such as mosses.

In *Pterostylis plumosa* and orchids with monad pollen the locular space begins forming during meiotic prophase, as in many angiosperms, but grows less then in angiosperm (Pandolfi et al., 1993). This feature, too, can be considered neotenic.

The locular fluid is extremely reduced in the case of pollinia because of tetrad and massula packaging. This means that the transfer of nutrients from the tapetum to pollen developing in an anther must be efficient and very uniform. In such crowded loculi with so few spaces, nutrient translocation is facilitated by cytomictic channels between the microspores of the same tetrad, as in *Polystachia pubescens* (Schlag and Hesse, 1992, Fig. 5-10), and contiguous tetrads (Fig. 5-10). Because these channels, all the microspores of a loculus behave synchronously as a syncytium until the channels are sealed off after the first mitosis, pollen begins to ripen (Heslop-Harrison, 1968).

All tapetal cells have nutritive functions and degenerate producing pollenkitt and elastiviscin. This happens when PDU is monad with pollenkitt and monad pollen is grouped by elastoviscin. In the case of soft pollinium C and compact pollinium, instead, elastoviscin of the stipe, caudicle and viscidium are formed by non tapetal specialised cells while tapetal cells only have a nutritive function, degenerating without producing any residue.

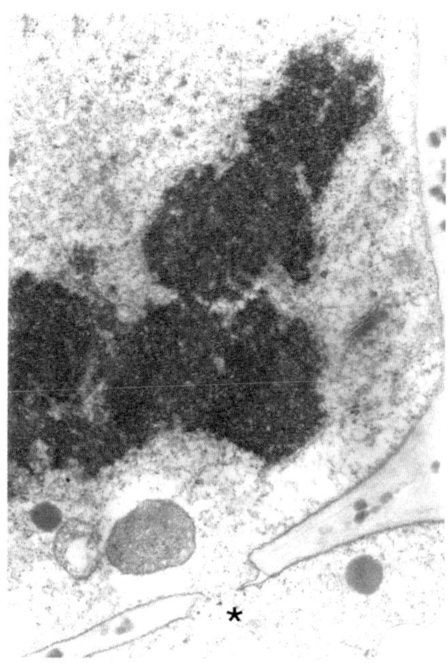

Fig. 5-10. Electron micrographs of a developing pollinium of *Orchis purpurea*. A cytomictic channel (asterisk) between two microspores of the same tetrad at late microspore stage.

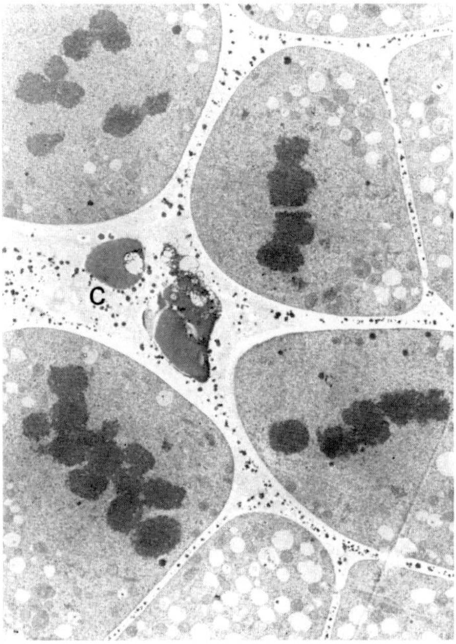

Fig. 5-11. Electron micrographs of a developing pollinium of *Orchis purpurea*. Synchronous pollen mitosis in grains next to the caudicle (C).

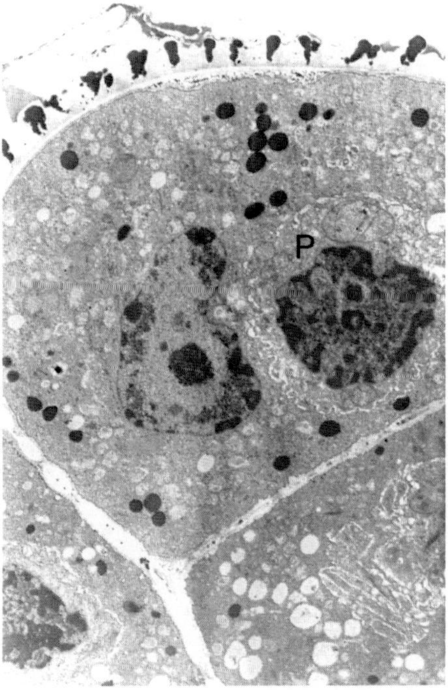

Fig. 5-12. Electron micrographs of a developing pollinium of *Orchis purpurea*. Part of an outer tetrad at maturity. Exine is present only in the external part of the tetrad. The generative cell is spherical in shape, contains a plastid (P) similar to those of the vegetative cell.

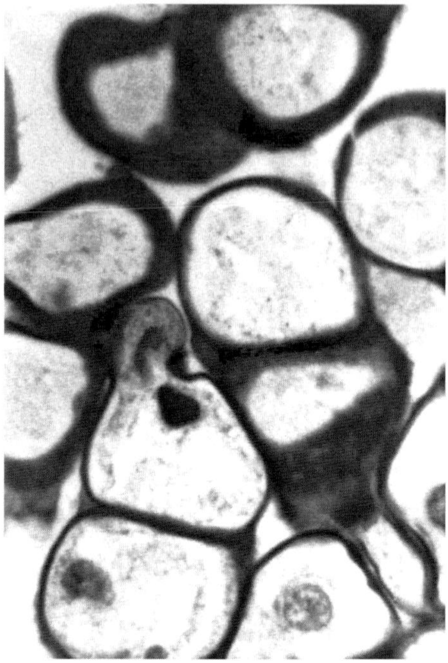

Fig. 5-13. Light micrograph of part of a *Loroglossum hircinum* pollinium 24 h after pollination. Tetrads are loosely arranged and a pollen tube has formed in the space. Generative cell now has the classical spindle shape.

Tetrad Types

Tetrad shape is a systematic character in angiosperm families (Davis, 1966) and is derived from the spatial orientation and relative positions of the two second meiotic division spindles (Brown and Lemmon, 1991). Orchidaceae have all six possible types: tetrahedral, decussate, square, rhomboidal, T-shaped, and linear, which have different tree dimensional shapes (Yeung, 1987a). The presence of all these tetrad types has a meaning in orchids with pollinia because their different shapes enable more compactness and smaller intertetrads spaces (Fig. 5-7). The percentages of the various tetrad types differ according to position: inside the pollinium, all the types are encountered; on the surface linear tetrads, disposed radially, are more common (Yeung, 1987a). The presence of different tetrad types in species with monad pollen, such as *Loroglossum hircinum* (Figs. 5-1 and 5-2), is apparently less understandable (Pandolfi et al., 1993), but from an evolutive point of view, features such as reduced locular fluid volume enable the formation of different types of pollinia (Pandolfi et al., 1993).

Mechanisms of Tetrad Adhesion in Developing Pollinia and During Transport

Pollinia are the most complex type of compound pollen, because the tetrads are kept together by various mechanisms. Yeung (1987a) discussed four types of mechanism responsible for tetrad aggregation. Here they are reported with some specific additions: (1) persistence of the callosic wall of the pollen mother cell; (2) persistence and or modification of certain intertetrad and intratetrad cytomictic channels (Fig. 5-10) enabling synchronous development (Fig. 5-11); (3) common and/or continuous exine and or intine (Figs. 5-7 and 5-8); (4) reduced increase in size of developing pollen grains irrespective of PDU type (Paldolfi et al., 1993); (5) the presence of tetrads of different types allowing more compact packaging; (6) the presence of intine bridges between contiguous tetrads formed as a consequence of cytomictic channel closure.

These adhesion mechanisms between tetrads are established during anther development. They mainly act during removal and transport. Few massulae are lost during collection by the pollinator and during the pollinator visits to different flowers. The tetrads of each massula stay together until deposited on a stigma. Strong adhesion between tetrads could subsequently be very negative because tetrads forming a massula must come apart to create space for pollen tube emission (Pandolfi and Pacini, 1995).

Pollen Wall Modifications Aperture and Increase in Number of Pollen Grains per PDU

Pollen wall structure and stratification change according to PDU types. Burns-Balogh (1983) proposed a theory on exine evolution in the Orchidaceae which reflects the different PDUs, though she did not consider this characteristic. Regular individual intine and exine are present around monad pollen. Different degrees of wall fusion appear with different PDU tetrad types. In the case of pollinia, wall structure varies according to the position of the tetrads in the pollinium. In orchids such as *Loroglopssum hircinum* (Pandolfi and Pacini, 1995), exine is only present on the outside of massule and not in a uniform layer but as a plaque for each microspore (Fig. 5-8). In *Epidendrum scutella* (Cocucci and Jensen, 1969) and *E. ibaguense* (Yeung, 1987b, c) ornamented exine is only present at the periphery of the pollinium and a foot layer is present in all tetrads. *Orchis purpurea* only has reduced exine on external tetrads (Fig. 5-12). Zavada (1990) described wall organization in some significant types of pollinia.

An aperture of the furrow type is present in monad pollen grains (Figs. 5-1 and 5-2) such as those of *Pterostylis plumosa* (Pandolfi et al., 1993) and cypripedioid orchids (Burns-Balogh and Hesse, 1988), but apertures are absent in more complex PDUs. The absence of a pore in ripe pollen means that the site for pollen tube emergence is determined only after pollen lands on the stigma. Studying *Loroglopssum hircinum* pollen germination, Pandolfi and Pacini (1995) observed that the site for pollen tube emission only became evident when the pollinium crumbled and a space was created between tetrads (Fig. 5-13).

PDU and Fossil Records

Although the exine of orchid pollen consists of the biopolymer sporopollenin, which has high biological and chemical resistance and fossilizes under special environmental conditions, orchid pollen is rarely encountered in the fossil record (Moore, Webb, and Collison, 1991). The reasons for this absence are essentially three: (1) orchids are almost exclusively entomophilous and produce little pollen; (2) their distribution is scattered, not crowded like weeds; (3) their pollen, especially if the PDU is a pollinium, produces loose and disorganized fossil debris (Wolter and Schill, 1985a). Only orchids with monads and tetrads can leave fossils resembling the original pollen, but they are similar to those of other monocots (Zavada, 1983).

Pollen Volume and Vacuolisation

Besides the reduced space of the loculus, there are also reduced volume increases of pollen during its development; these features are not specific to pollinia of different types but present throughout the family, are necessary for pollen packaging. Volume increase is only threefold in orchids whereas in the other angiosperms is from eleven fold to twenty fold (Pandolfi et al., 1993). This means that pollen growth is reduced as initial investment for reproduction. Pandolfi et al. (1993) were the first to notice that orchid pollen has a smaller volume than that of other angiosperms, irrespective of the PDU type. They also noticed reduced vacuolization during pollen development. Vacuolisation causes pollen to increase in volume of pollen (Pacini, 1994). Vacuoles are formed by different mechanisms but disappear due to formation of new cytoplasm and/or storage of carbohydrate or lipid reserves. Vacuoles with translucent content containing water and solutes are not found in ripe angiosperm pollen, except in orchids (Fig. 5-12). The presence of vacuoles in ripe pollen is another neotenic trait (Fig. 5-10–5-13).

Irrespective of orchid PDU type, extensive vacuolisation and increase in volume always occur before pollen tube emission (Pandolfi and Pacini, 1995).

Pollen Carbohydrate Reserves

Starch accumulates, once or twice during pollen development, and ripe pollen may be starchy or starchless (Franchi et al., 1996). Data on starch reserves are only available for a limited number of orchid species. Ripe pollen of *Pterostylis plumosa* (Pandolfi et al., 1993) contains few small starch grains. The amount of starch was greater just before pollen ripening, though completely absent in ripe *Loroglossum hircinum* pollen (Pandolfi and Pacini, 1995). The proportion of starch stored during orchid development is certainly much less than in other angiosperms because it is proportional to the low increase in volume. If ripe pollen is devoid of starch reserves it does not mean that carbohydrate reserves are missing: instead of in amyloplasts, they are stored in the cytoplasm (Franchi et al., 1996).

Pollen Presentation

Pollen presentation is a term used to indicate the way in which pollen is presented to dispersing agents (Pacini and Hesse, 2004). Pollen dispersal units of orchids are presented in different ways. Pollen is exposed in the fully opened anther when in monads and tetrads, irrespective of the adhesive material. When the PDU is a pollinium, presentation occurs inside the slightly opened anther (Fig. 5-5). As a consequence pollinators become charged with monad but must touch the viscidium in the case of different types of pollinium. A kind of vertical closed lip of the anther enables delivery of the pollinium when the pollinator accidentally touches the viscidium. The pollinium may decompose into massulae especially if the pollinium is collected several days after flower opening. The viscidium prevents contact between pollinator and pollen. It is so sticky that it adheres even to smooth surfaces.

Anthers such as those of *Cleistes divaricata* with tetrads covered by an unknown glue (Figs. 5-1–5-6) have mechanisms allowing pollen tetrads to be dispersed in different doses and at different times by the shaking caused by insects visiting the flower (Gregg, 1991). This mechanism is very similar to that of poricidal anthers of the Ericaceae and Violaceae (Faegri and van der Pijl, 1979).

The presence of a mechanical layer in different parts of the anther and different patterns of wall thickening allows total or partial opening of anthers of orchids with different PDU types (Freudestein, 1991).

Monad pollen is rarely lost during anther opening and exposure because it is firmly stuck with pollenkitt or elastoviscin (Fig. 5-14). When the PDU is a pollinium, slight loss of water occurs during the long male receptivity, causing detachment of some massulae from the caudicle and partial break up of the pollinium; this movement and the position of the stigma beneath the anther may sometimes lead to self-pollination. These phenomena were reviewed by Catling (1990).

Generative Cell and Paternal Inheritance

The generative cell of angiosperm pollen generally attains the classic spindle shape after the first haploid mitosis (Tanaka, 1993). Irrespective of the PDU type, orchid pollen always have a spherical generative cell (Pandolfi et al., 1993; Pandolfi and Pacini, 1995; Schlag and Hesse, 1992). Spherical generative cells may need less energy for their maintenance during the long pollen exposure (Pandolfi and Pacini, 1995). By the time the generative cell of orchids has entered the pollen tube, it is spindle shaped (Fig. 5-13).

	During development	Before exposure	During exposure
Monad pollen	Locular fluid conveys nutrients for pollen development 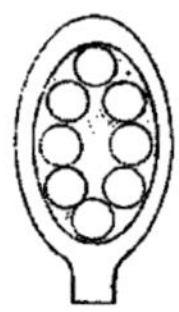	Nourishment from mother plant ceases; locular fluid is reabsorbed or evaporates Pollen grains lose water and decrease in volume; pollen shape changes from spherical to oval by infolding of furrow The pollenkitt (dotted) coats the grains before anther opens 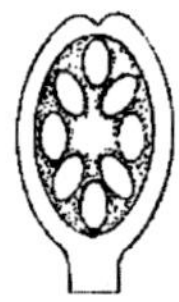	Anther open with wide stomium Pollen grains are totally exposed and loaded directly by pollinator Pollen grains may lose or gain water according to relative humidity and microenvironment of flower Pollenkitt (dotted) sticks grains to the anther 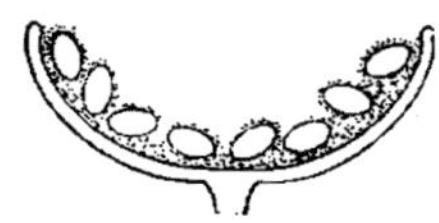
Pollinia	Little or no locular fluid, pollinium nourished directly by the mother plant via tapetal cells 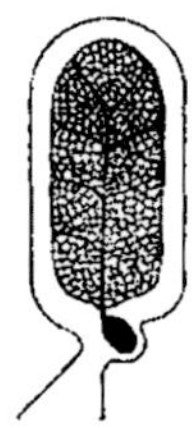	Nourishment from mother plant ceases 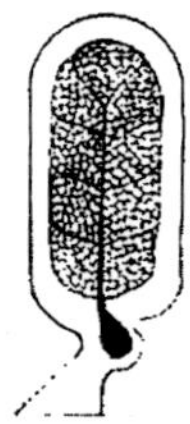	Anther opens with wide stomium Pollinium visible or not from outside, according to species and stomium size Viscidium (V) sticks pollinium to insect body on contact Pollinium sheltered by the anthe wall and may lose or gain water according to relative humidity of air 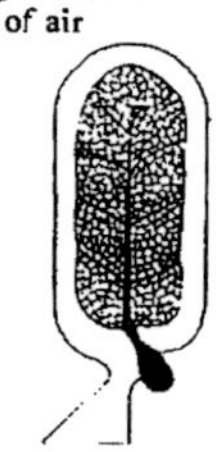

Fig. 5-14. Scheme of anthers of two extreme orchids from development to pollen exposure. Orchid anthers represented by horizontal lines; locular fluid represented by oblique lines is present only in orchids with monad pollen (Modified from Pacini and Hesse, 2002).

The presence of a plastid inside the generative cells of ripe *Orchis purpurea* pollinia, reported here for the first time (Fig. 5-12), cannot be considered an exception, because paternal plastid transmission has been reported in some orchids as *Catasetum discolor*, *Epidendrum scutella* and *Phaius tankervillae* at least (Sears, 1980).

Developmental Arrest in the Different PDU Types and Pollen Longevity

Footitt and Cohn (2001) coined the term developmental arrest to indicate a peculiar physiological state in which the development of animals and plants is arrested because the cycle of the plant is concluded and/or to survive to some environmental stress. The authors approached the problem from a biomolecular point of view. In both types of organisms arrest occurs by slowing of metabolism, and formation of some protective molecules.

Pollen is in an arrested developmental state to avoid damage during the passage in the atmosphere, but the degree of this arrest varies in orchids according to PDU type and the possibility of losing or gaining water. Pollen longevity is also linked to PDU type and it seems to depend on different factors: water content, carbohydrate reserves, microenvironment of the flower and pollen presentation mechanisms (Nepi, Franchi, and Pacini, 2001).

Unlike seed water content, pollen water content is a neglected physiological character. Partially dehydrated pollen (PDP), was recently defined as pollen with less than 30% at shedding and partially hydrated pollen (PHP) with a water content more than 30% (Nepi et al., 2001). More recently PDP was regarded as similar to orthodox seeds, and PHP as similar to recalcitrant seeds (Franchi, Nepi, and Pacini, 2002). This difference divides pollen physiology into two categories, PDP, namely orthodox pollen, survives desiccation better and PHP, namely recalcitrant pollen, cannot be stored or subjected to severe dehydration. PDP has furrows and other devices allowing volume decrease before dispersal and volume increase on the stigma. PHP is devoid of furrows.

Monad pollen of orchids, whether with pollenkitt or elastoviscin, is PDP and has a furrow allowing harmomegathic changes in volume and water content (Figs. 5-1–5-6). Pollinia of different types contain PHP because it does not undergo processes of dehydration or rehydration. These opposite processes would cause compression and/or distortion of pollen and cytoplasmic damage. Dehydration would cause compression of grains, especially those inside the massula. Rehydration would cause distortion, especially of external grains. We do not know if the two types of PDU with tetrads united by different substances and soft pollinia A and B have furrows or other mechanisms allowing harmomegathic changes (Figs. 5-1–5-6).

Pollen water content is known for few orchid species having type C pollinia and nothing is known for other PDU types. Franchi et al. (2002) listed water contents for six species and described the case of *Stanhopea tigrina,* in which pollen water content decreases during presentation.

The vegetative cell cytoplasm of pollen of a mature pollinium has another peculiar feature, namely small vacuoles even in the ripe stage (Fig. 5-12). This means that their developmental arrest is not as deep and intense as in other seed plants. Anther protection and vacuoles in ripe pollen suggest that a supply of water keeps the grains of a pollinium inside the anther wet during presentation. This is only a hypothesis and needs to be tested.

The presence of large amounts of sucrose and polysaccharides in the pollen cytoplasm is responsible for its longevity (Speranza, Calzoni, and Pacini, 1997). Pollen with a high water and sucrose content survive desiccation better than those with a low sucrose content (Dafni and Firmage, 2000). The pollinium of *Stanhopea tigrina* has 95.6 $\mu g\ mg^{-1}$ sucrose, which is high compared to other species with long-lived pollen (Nepi et al., 2001 and our unpublished data). The water content is 56.8% at onset of anthesis, decreasing to 34.3% after 1 week (our unpublished data). The fact that the pollinium is partially protected inside the anther, as well as the high sucrose content and other devices, probably protect it from invasion by bacteria and moulds. Despite the long presentation and high water content, pollinia are rarely infested by these organisms.

Pollen longevity seems to be due both to water and to carbohydrate content (Pacini, 1996; Speranza et al., 1997; Dafni and Firmage, 2000). The latter authors reported the longevity, in field conditions, of three orchid species having pollinia: *Gymnadenia conopsea*, *Dactylorhiza maculata* and *D. purpurella* to be 37, 40 and 51 days respectively. Stigma receptivity, that is proportionally shorter, is also reported.

One of the advantages of PHP is to emit a pollen tube a few minutes after landing on the stigma, by virtue of shorter rehydration time (Franchi et al., 2002). This is not the case with orchids because tube emission takes more than 24 h in *Loroglossum hircinum* (Pandolfi and Pacini, 1995). The reason for this delay springs from the fact there is no space for pollen tube growth in a ripe pollinium. They may only grow when inter tetrad spaces are formed (Pandolfi and Pacini, 1995). This means that the PHP of massulate orchids is not adapted for fast germination but rather for PDUs with large number of grains. The case of orchids also confirms that PDP is the primitive condition and PHP is more evolved. Gymnosperm pollen is almost exclusively PDP.

Borba and Semir (1999a, b) and Singer (2002) demonstrated that in some species of the genus *Bulbophyllum* and in *Trigonium obtusum* the slight loss of water that occurs in pollinia during dispersal allows them to fit into the stigma cavity. This mechanism is intended to avoid self and intraspecific pollination because pollinaria have different shape.

Female Part Adaptations According to PDU Types

Owing to the different orchid PDU types and their exposure mechanisms, the female counterpart has some adaptations. The stigma surface is papillated and there is a style in *Neuwedia*, *Apostasia* and those Apostasioideae having monad pollen (Kocyan and Endress, 2001). The stigma is a cavity having a shape matching that of the pollinium and the style is absent in the Oncidiinae (Dannenbaum, Walter, and Schill, 1989; Clifford and Owens, 1990). The stigma must differ according to PDU type because monad pollen can be deposited without any special disposition. Pollinia, especially of the compact type, need a defined and steric disposition to fill the stigma cavity.

The pollinium may contain hundreds of thousands of grains, but the female investment is reduced because ovules consist of few cells and the female gametophyte is not developed yet at pollination, at least in species having different types of pollinia. Most of female investment only occurs after receipt of the pollinium. Pollen tubes therefore await complete development of the female gametophyte. This may take as long as 3 months, as in *Epidendrum scutella* (Cocucci and Jensen, 1969).

PDU and Pollen Load

In many angiosperms there is a positive correlation between the number of pollen grains per anther and the number of ovules per ovary; this seems a rule with few exceptions and is named pollen/ovule ratio (Cruden, 1977; Pacini and Franchi, 1998). The higher the number of pollen dispersal units having different male sources falling on the stigma, the greater the possibility of a fruit having seeds with different paternity; in this case, male competition increases. Vice versa, if the ovary contains many ovules, female competition increases, especially when the number of pollen grains falling on the stigma decreases.

In angiosperms male competition occurs during pollen development, rehydration and germination (Ottaviano and Mulcahy, 1989). It has a genetic basis and also depends on chance, especially when pollen lands on the stigma, e.g., some grains rehydrate quickly because they are in contact with the stigma surface and others later because they are not in contact. In orchids having the pollinium as PDU, competition is extremely reduced during pollen development because of the presence of cytomictic channels. When they close, after the first haploid mitosis, the grains become separate and independent and competition becomes possible. In orchids having monads or tetrads, male competition is as in other angiosperms, i.e., continuous throughout development.

In monad pollen (PDP) the process of rehydration is long, but competition is strong, especially if the number of ovules per ovary is low and few pollen grains land on the stigma. Competition in orchids with pollinia is reduced because the number of pollen grains in a pollinium is not so different from the number of ovules in the ovary, i.e., the pollen to ovule ratio is close to one (Pacini and Franchi, 1998, 1999).

Pollen Load and Paternity

The consequences of pollination are different depending on whether insects transport monad pollen or a pollinia. If monad pollen is glued with pollenkitt, it can be dispersed on a higher surface of the stigma because of the low viscosity of this gluing material. In the case of monad pollen glued by elastoviscin, the high viscosity of this fluid keeps grains together if devices enabling pollen smearing are not present.

The evolution of the different PDU in orchids leads to the transfer of all the pollen of a flower to another flower. For this transfer only a single visit by a pollinator is needed. When a pollinator transports a type C pollinium (mealy) there are two possibilities: the whole pollinium is deposited on the stigma of another flower, or the massulae of the pollinium are dispersed on different flowers. In the case of compact pollinia the whole pollinarium, composed of four pollinia, is deposited on the stigma. The result of these different doses of pollen is differences in paternity and competition of male gametophytes. The greater the number of pollen grains having different sources, the greater the competition among grains. Nevertheless, male competition in orchids seems less than in species having only one ovule per ovary (Pacini and Franchi, 1999).

Proctor and Harder (1994) experimentally modified pollen load in three species of orchids with different types of PDU: *Cypripedium calceolus* (monad with elastoviscin), *Amerorchis rotundifolia* (mealy pollinia) and *Calypso bulbosa* (compact pollinia). Pollen load influenced the seed number/ovary in *Calypso* but not in *Cypripedium* and *Amerorchis*; the number of ovules/ovary was not considered.

Neotenic and Changing Characters in Orchid PDU Evolution

Neoteny is a term used in biology to indicate the persistence of juvenile characters in the adult stage. The involvement of this biological feature has been invoked in the success of angiosperms (Takhtajan, 1976). Orchids have different types of juvenile characters, like seed dispersed with embryo at globular stage consisting of few cells, i.e. arrested at an early stage of development. However neoteny also occurs in some anther characters (Table 3) in Pacini, and Hesse. Some of these features are typical of higher orchids, i.e., those with a pollinium, others, such as reduced pollen volume or a spherical generative cell in ripe pollen, are present in lower and higher orchids and were therefore necessary for the passage from individual pollen grains to the pollinium. Other features, like pollen dispersed with a high water content or vacuoles and tapetal cell wall persistence, are primitive characters because observed in early land plants (Figs. 5-10–5-13) as in spores of Bryophyta and certain Pteridophyta (Pacini, Franchi, and Hesse, 1985).

Animals Commonly Involved in Orchid PDU Dispersal and Inherent Problems

The different PDU types of orchids and devices enabling their dispersal make orchid pollen available for dispersal by different types of animals. Pollenkitt, elastoviscin and the viscidium do not have the same physicochemical features/properties and do not attach in the same amount to the pollinator body. Powdery pollen, covered in pollenkitt, typical of almost all zoophilous angiosperms, adheres to insect hairs, mammals fur and birds feathers. Viscidia and elastoviscin may adhere to smooth surfaces as bird bills, ants, and smooth parts of insects such as eyes, proboscis mouth parts and legs (Johnson and Edwards, 2000). Animals are attracted to flowers for nectar by scent (Gerlach and Schill, 1991) and also by deception (Dafni, 1987; Nilsson, 1992).

Orchids have exclusively zoophilous pollen dispersal, though Borba and Semir (1999a, b) reported that air currents at a particular speed are necessary for pollination to occur in *Bulbophyllum* and this is achieved trapping on insect in the flower column.

Viscidia and stipes enables transfer of pollinia without contact with the body of dispersing animals. This is important when pollinia are dispersed by Hymenoptera like bees and ants because pollen is known to be damaged by secretions of these animals (Beattie et al., 1985; Harris and Beattie, 1991). These secretions kill mould spores and bacteria, preventing their growth in the humid nests of these social insects. Because the number of orchid plants per unit area and flowers per plant is small, pollinia are transported for long periods of time before being deposited on a stigma. The pollinium is generally carried about 1–3 mm from the animal body minimising the effects of these glandular secretions.

Conclusions and Further Work

This review concerns the different types of PDU in orchids, their consequences for the female part and the reproductive steps after to pollination. It also considers neotheny in the light of reproductive features.

The concept of pollen dispersing units is quite recent and nobody had hitherto recognised all these types of PDUs in orchids or in a single family (Pacini and Hesse, 2002). The passage from tetrad pollen to the different types of pollinia is not yet clear and more studies are necessary to know the different steps leading to the different types of soft pollinia and hence to compact ones. Nor is clear the passage from PDP to PHP. On the female side it will be important to know whether the number of ovules increases with increasing number of grains per PDU. Another step that needs to be elucidated is the passage from pollenkitt to elastoviscin, as well as the chemical composition. The repercussions of diversification of PDU on the pistil is another interesting topic for future research.

A completely new line of research could be phytogeography in relation to PDUs, i.e., where species with primitive and more evolved forms of PDU are situated geographically. While there are many papers on the mechanisms of attraction of insects and dispersal of pollen, few experimental studies have been done on pollen longevity during presentation and dispersal. Pollinia are not exposed but are visible from through the stomium that may vary in width according to the species. The morpho–physiological aspects of this aperture are almost unknown. The same type of characters are unknown in apomictic and cleistogamous species. Another problem deserving attention is water content and carbohydrate composition in the different PDU, any variations during pollen presentation and any mechanisms to keep pollinia moist while awaiting dispersal in the anther.

Acknowledgements I am grateful to all my co-workers, especially those who worked on orchids. Special thanks to Serena Mugnaini who helped me prepare the manuscript and tables.

Glossary

Callose.	A polysaccharide, polymer of glucose, present as a molecular filter to separate cells having different fates, such as tetrads, in the male and female lines, and pollen tubes from the female tract.
Caudicle.	A mealy and often elastic protuberance where the massulae composing a pollinium adhere. It is a part of the pollen mass and is produced within the anther. It is continuous with the stipe.
Developmental arrested state.	A feature of the whole organism or of dispersing part of organisms, such as pollen and seeds. Development is arrested due to physico-chemicals modifications to metabolism.
Elastoviscin.	A highly viscous fluid only described in some Asclepiadaceae and Orchidaceae, i.e., two families which evolved the same pollen dispersal units independently (Wolter and Schill, 1986; Dannenbaum and Schill, 1991).
Female receptivity.	The period in which pollen is able to germinate on the stigma and to fertilize. It is determined by applying fresh pollen to stigmas at different times after flower opening.
Furrow.	External infolded part of the pollen grain where exine is reduced or even absent. Here intine is thicker. Poral intine is exposed during development but closes during partial dehydration of anther and pollen.
Harmomegathy.	A feature typical of many types of pollen, namely the ability to decrease in volume preliminary to the state of arrested development and to increase in volume on the stigma on rehydration with the water supplied by the female part. This change in volume is made possible by one or more furrows.
Locular fluid.	The fluid conveying the substances produced by the mother plant and transformed by the tapetum. It composition changes with developmental stage. It is reabsorbed and/or evaporated to allow pollen presentation and dispersal. Locular fluid is abundant in species having monad and tetrad pollen and extremely reduced in species with pollinia.
Loculus.	The cavity where pollen develops. In orchids it may be more or less crowded depending on the PDU type.
Male competition.	Competition between male gametes or gametophytes. Its strength depends on PDU type and number of ovules per ovary.
Massulae.	Aggregates of few tens of tetrads joined to the caudicle to form a pollinium.

Mechanical layer.	An external layer of the anther, normally consisting of cells dead at maturity, containing species-specific wall thickenings. Passive movements of these cells generated by inter-anther tensions and/or water loss cause anther opening and pollen presentation.
Meiocyte.	A cell undergoing the meiotic process.
Monad pollen.	Pollen dispersed in a single grains.
Neoteny.	Maintenance of a juvenile characteristic in the adult stage.
Pollen longevity.	The mean life of a pollen grain of a species. Water content, carbohydrate reserves and the possibility of carbohydrate inter-conversion determine the length of this physiological character. It is measured by observing whether pollen is viable at different times after flower opening.
Pollen viability.	The percentage of viable pollen grains in a population. Pollen viability is commonly detected by observing whether or not the plasma membrane is permeable to dye, assuming that only viable grains have a permeable plasma membrane.
Pollen presentation.	The way in which pollen is are exposed for dispersal.
Pollenkitt.	The product of tapetal degeneration deposited on pollen before anther opening. It contains lipids, carotene and other substances and imparts stickiness, color, smell, etc. to pollen.
Pollinarium.	The anther-derived container of pollen dispersal units.
Pollinium.	A compact and coherent mass of pollen grains, usually consisting of *massulae*, i.e. few tens of tetrads clumped together and attached to a caudicle.
Rostellum.	A flower part that separates anther from stigma.
Spherosomes.	Vesicles containing lipidic reserves originating from blebbing of smooth endoplasmic reticulum.
Sporopollenin.	A chemically and biologically resistant, elastic substance consisting of a mixture of carotene and carotenoid esters and constituting the exine, i.e. the external wall of pollen and spores of land plants.
Stipe.	A non-viscid linear part connecting pollinia to viscidium.
Tapetum.	Part of the inner anther, consisting of secretory cells, that nourishes and regulates pollen development. It degenerates producing pollenkitt or elastoviscin.
Viscidium.	(plural Viscidia). Sticky external part of orchid anther with pollinium-like PDU, responsible for anther adhesion to insect body.

Literature Cited

Ackerman, J. D., and N. H. Williams. 1980. Pollen morphology of the tribe Neottieae and its impact on the classification of the Orchidaceae. Grana 19. 7–18.

Ackerman, J. D., and N. H. Williams. 1981. Pollen morphology of the Chloraeinae (Orchidaceae. Diuridae) and related subtribes. Am. J. Bot. 68. 1392–1402.

Beattie, A. J., C. Turnbull, T. Hough, S. Jobson, and R. B. Knox. 1985. The vulnerability of pollen and fungal spores to ant secretion. Evidence and some evolutionary implications. Am. J. Bot. 72. 606–614.

Borba, E. L., and J. Semir. 1999a. Wind-assisted fly pollination in three *Bulbophyllum* (Orchidaceae) species occurring in the brazilian campos rupestes. Lindleyana 13. 203–21.

Borba, E. L., and J. Semir. 1999b. Temporal variation in pollinarium size after its removal in species of *Bulbophyllum:* a different mechanism preventing self-pollination in Orchidaceae. Plant Syst. Evol. 217. 197–204.

Brantjes, N. B. M. 1981. Ant, bee and fly pollination in *Epipactis palustris* (L.) Crantz (Orchidaceae). Acta Bot. Neerl. 30. 59–68.

Brown, R. C., and B. E. Lemmon. 1991. Pollen development in orchids. 1. Cytoskeleton and control of division plane in irregular patterns of cytokinesis. Protoplasma 163. 9–18.

Buchner, R., and M. Weber. 2000. PalDat*a* a palynological database. descriptions, illustrations, identification and information retrieval. http.//PALDAT.BOTANIK.UNIVIE.AC.AT/

Burns-Balogh, P. 1983. A theory on the evolution of the exine in Orchidaceae. Am. J. Bot. 70. 1304–1312.

Burns-Balogh, P., and M. Hesse. 1988. Pollen morphology of the cypripedioid orchids. Plant Syst. Evol. 158. 165–182.

Catling, P. M. 1990. Auto-pollination in the Orchidaceae. In: J. Arditti (ed.), Orchid biology. reviews and perspectives, Vol V. Comstock Publishing Associates, Ithaca, NY, pp. 121–157.

Clifford, S. C., and S. J. Owens. 1990. The stigma, style and ovarian transmitting tract in the Oncidiinae (Orchidaceae). morphology, developmental anatomy and histochemistry. Bot. Gaz. 151. 440–451.

Cocucci, A., and W. Jensen. 1969. Orchid Embryology. Pollen tetrads of *Epidendrum scutella* in the anther and on the stigma. Planta 84. 215–229.

Cruden, R. W. 1977. Pollen-ovule ratios: a conservative indicator of breeding systems in flowering plants. Evolution 31. 32–47.

Currier, H. B., and S. Strugger. 1956. Aniline blue and fluorescence microscopy of callose in bulb scales of *Allium cepa* L. Protoplasma 45. 552–559.

Dafni, A. 1987. Pollination of Orchids and related genera. evolution from reward to deception. In: J. Arditti (ed.), Orchid biology: reviews and perspectives, Vol IV. Comstock Publishing Associates, Ithaca, NY, pp. 79–104.

Dafni, A., and D. Firmage. 2000. Pollen viability and longevity. practical, ecological and evolutionary implications. Plant Syst. Evol. 222. 113–132.

Dannenbaum, C., and R. Schill. 1991. Die Entwicklung der Pollentetraden und Pollinien bei den Asclepiadaceae. Biblioteca Botanica 34. 1–138.

Dannenbaum, C., M. Walter, and R. Schill. 1989. Stigma morphology of the orchids. Bot. Jahrb. Syst. 110. 441–460.

Darlington, C. D., and L. F. La Cour. 1960. The handling of chromosomes (3rd revised ed.). George Allen & Unwin, London.

Davies, K. L., M. P. Turner, and A. Gregg. 2003. Atypical pseudopollen-forming hairs in *Maxillaria* (Orchidaceae). Bot. J. Linn. Soc. 43. 151–158.

Davis, G. L. 1966. Systematic embryology of the angiosperms. Wiley, London.

Dressler, R. L. 1981. The orchids. Natural history and classification. Harward University Press, Cambridge, MA.

Faegri, K., and J. Iversen. 1989. Textbook of pollen analysis (4th ed.). K. Faegri, P. E. Kaland, and K. Krzyrinski (eds.), Wiley, Chichester, UK.

Faegri, K., and L. van der Pijl. 1979. The principles of pollination ecology (3rd ed.). Pergamon, Oxford.

Fisher, D. 1968. Protein staining of ribboned epon sections for light microscopy. Histochemie 16. 92–96.

Fitzgeral, M. A., S. H. Barnes, S. Blackmore, D. M. Calder, and R. B. Knox. 1994. Pollen development and cohesion in a mealy and hard type of orchid pollinium. Int. J. Plant Sci. 155. 481–491.

Flax, M., and M. Himes. 1952. Microspectrophotometric analysis of metachromatic staining of nucleic acids. Physiol. Zool. 25. 297–311.

Footitt, S., and M. A. Cohn. 2001. Developmental arrest. from sea urchin to seeds. Seed Sci. Res. 11. 3–16.

Franchi, G. G., L. Bellani, M. Nepi, and E. Pacini. 1996. Types of carbohydrate reserves in pollen. localization, systematic distribution and ecophysiological significance. Flora 191. 143–159.

Franchi, G. G., M. Nepi, and E. Pacini. 2002. Partially hydrated pollen. taxonomic distribution and evolutionary significance. Plant Syst. Evol. 234. 211–227.

Freudestein, J. V. 1991. A systematic study of endothecial thickenings in the Orchidaceae. Am. J. Bot. 78. 766–781.

Freudestein, J. V., and F. N. Rasmussen. 1997. Sectile pollinia and relationships in the Orchidaceae. Plant Syst. Evol. 205. 125–146.

Gerlach, G., and R. Schill. 1991. Composition of orchid scents attracting euglossine bees. Bot. Acta 104. 379–391.

Goff, L. J. A., and A. W. Coleman. 1984. Elucidation of fertilization and development in a red alga by quantitative DNA microspectro-fluometry. Dev. Biol. 102. 1023–1024.

Gregg, K. B. 1991. Reproductive strategy of *Cleistes divaricata* (Orchidaceae). Am. J. Bot. 78. 350–360.

Harris, F. C. L., and A. J. Beattie. 1991 Viability of pollen carried by *Apis mellifera* L., *Trigena carbonaria* Smith and *Vespula germanica* (F.) Hymenoptera. Apidae, Vespidae. J. Aust. Ent. Soc. 30. 45–47.

Heslop-Harrison, J. 1968. Synchronous pollen mitosis and the formation of the generative cell in massulate orchids. J. Cell Sci. 3. 457–466.

Heslop-Harrison, J. and Y. Heslop-Harrison. 1985. Germination of stress-tolerant *Eucaliptus* pollen. J. Cell Sci. 73. 135–157.

Heslop-Harrison, J., R. B. Knox, and Y. Heslop-Harrison. 1974. Pollen wall proteins: exine held fractions associated with the incompatibility response in Cruciferae. Theor. Appl. Gen. 44. 133–137.

Heslop-Harrison, J., Y. Heslop-Harrison, and K. R. Shivanna. 1984. The evaluation of pollen quality and a further appraisal of the fluorocromatic (FCR) test procedure. Theor. Appl. Gen. 67. 367–375.

Heslop-Harrison, Y. 1977. The pollen-stigma interaction. pollen tube penetration in *Crocus*. Ann. Bot. 41. 913–922.

Hesse, M., S. Vogel, and H. M. Halbritter. 2000. Thread-forming structures in angiosperm anthers: their diverse role in pollination ecology. Plant Syst. Evol. 222. 281–292.

Hofmeister, W. 1861. Neue Beiträge zur Kenntnis der Embryobildung der Phanerogamen. II Monokotiledonen. Königlich sächsische Gesellschaft der Wissenschaften, Leipzig, Abhandlungen 7. 631–760.

Jensen, W. A. 1962. Botanical histochemistry: principles and practice. W. H. Freeman, San Francisco.

Johansen, D. A. 1940. Plant microtechnique. McGraw-Hill, New York.

Johnson, S. D., and T. J. Edwards. 2000. The structure and function of orchid pollinaria. Plant Syst. Evol. 222. 243–269.

Knox, R. B., and C. A. McConchie. 1986. Structure and function of compound pollen. In: Blackmore S., and I. K Ferguson (eds.), Pollen and spores: form and function. Linnaean Society Symposium Series 12, Academic, London, pp. 265–282.

Kocyan, A., and P. K. Endress. 2001. Floral structure and development of *Apostasia* and *Neuwiedia* (Apostasioideae) and their relationships to other Orchidaceae. Intern. J. Plant Sci. 162. 847–867.

Le Pecq, J. B., and C. Paletti. 1967. A fluorescent complex between ethidium bromide and nuclei acids. Physical-chemical characterization. J. Mol. Biol. 27. 87–106.

Mazia, D., P. Brewer, and M. Alfer. 1953. The cytochemical staining and measurement with mercuric bromophenol blue. Biol. Bull. 104. 527–540.

Moore, P. D., J. A. Webb, M. E. Collison. 1991. Pollen analysis (2nd ed.). Blackwell, Oxford.

Nepi, M., and G. G. Franchi. 2000. Cytochemistry of mature pollen. Plant Syst. Evol. 222. 45–62.

Nepi, M., G. G. Franchi, E. Pacini. 2001. Pollen hydration status at dispersal. cytophysiological features and strategies. Protoplasma 216. 171–180.

Newton, G. D., and N. H. Williams. 1978. Pollen morphology of the Cypripedioideae and the Apostasioideae (Orchidaceae). Selbiana 2. 169–182.

Nilsson, L. A. 1992. Orchid pollination biology. Trends Ecol. Evol. 7. 255–259.

O'Brien, T. P. N., N. Federn, and M. E. McCully. 1964. Polychromatic staining of plant cell walls by toluidine blue O. Protoplasma 59. 368–373.

O'Brien, T. P. N., and M. E. McCully. 1981. The study of plant structure - principles and selected methods. Termarcarphy Pty. Melbourne.

Ottaviano, E., and D. Mulcahy. 1989. Genetics of angiosperm pollen. Adv. Genet. 26. 1–64.

Pacini, E. 1990. Harmomegathic characters of Pteridophyta spores and Spermatophyta pollen. In: M. Hesse, and F. Ehrendorfer (eds.), Morphology, development and and systematic relevance of pollen and spores. Plant Syst. Evol. (Suppl. 5). 53–69.

Pacini, E. 1994. Cell biology of anther and pollen development. In: E. G. Williams, A. E. Clarke, and R. B. Knox (eds.), Genetic control of self-incompatibility and reproductive development in flowering plants. Kluwer, Dordrecht, The Netherlands, pp. 289–308.

Pacini, E. 1996. Types and meaning of pollen carbohydrate reserves. Sex. Plant Reprod. 9. 362–366.

Pacini, E. 1997. Tapetum character states: analytical keys for tapetum types and activity. Can. J. Bot. 75. 1488–1459.

Pacini, E. 2000. From anther and pollen ripening to pollen presentation. Plant Syst. Evol. 222. 19–43.

Pacini, E. and G. G. Franchi. 1999. Some cytological, ecological and evolutionary aspects of pollination. Acta Soc. Bot. Pol. 65: 11–16.

Pacini, E., and G. G. Franchi. 1999. Types of pollen dispersal units and pollen competition. In: C. Clément, E. Pacini, and J.-C. Audran (eds.), Anther and pollen: from biology to biotechnology. Springer Verlag, Berlin, pp. 13–20.

Pacini, E., and G. G., Franchi. 2000. Types of pollen dispersal units in Monocots. In: K. L. Wilson, and D. A. Morrison (eds.), Monocots, systematics and evolution. CSIRO, Melbourne, pp. 295–300.

Pacini, E., and M. Hesse. 2002. Types of pollen dispersal units in orchids. and their consequences for germination and fertilization. Ann. Bot. 89. 653–664.

Pacini, E., and M. Hesse. 2004. Cytophysiology of pollen presentation and dispersal. Flora 199. 273–285.

Pacini, E., G. G. Franchi, and M. Hesse. 1985. The tapetum: its form, function and possible phylogeny in embryophyta. Plant Syst. Evol. 149. 155–185.

Pandolfi, T., and E. Pacini. 1995. The pollinium of *Loroglossum ircinum* (L.) Rich. (Orchidaceae) between pollination and pollen tube emission. Plant Syst. Evol. 196. 141–151.

Pandolfi, T., E. Pacini, and D. M. Calder. 1993. Ontogenesis of monad pollen in *Pterostylis piumosa* (Orchidaceae Neottioideae) Plant Syst. Evol. 186. 175–185.

Pearse, A. G. 1985. Histochemistry, theoretical and applied, Vol 2. Analytical technology (4th ed.). Churchill Livingstone, Edinburgh.

Pecall, R. 1989. The unique pollination of *Leporella fimbriata* (Orchidaceae). pollination by pseudocopulating male ants (*Myrmecia urens*, Formicidae). Plant Syst. Evol. 167. 137–148.

Proctor, H. C., and L. D. Harder. 1994. Pollen load, capsule weight, and seed production in three orchid species. Can. J. Bot. 72. 294–255.

Proctor, H. C., and L. D. Harder. 1995. Effect of pollination success on floral longevity in the orchid *Calypso bulbosa* (Orchidaceae). Am. J. Bot. 82. 1131–1136.

Punt, W., S. Blackmore, S. Nilsson, and A. Le Thomas. 1994. Glossary of pollen and spore terminology. LPP Foundation, Utrect, Contribution series No 1.

Rcagan, S., and W. N. Moffat. 1990. Cytochemical analysis of pollen development in wild-type *Arabidopsis* and male-sterile mutant. Plant Cell 2. 877–889.

Reichenbach, H. G. 1852. De pollinis orchidacearum genesi ac structura. Leipzig.
Schill, R., and W. Pfeiffer. 1977. Untersuchungen an Orchideenollinien unter besonderer Berücksichtigung ihrer Feinskulpturen. Pollen et Spores 19. 5–18.
Schill, R., and M. Wolter. 1985. Ontogeny of elastoviscin in the Orchidaceae. Nord. J. Bot. 5. 575–580.
Schill, R., and M. Wolter. 1986. On the presence of the elastoviscin in all subfamilies of the Orchidaceae and the homology to pollenkitt. Nord. J. Bot. 6. 321–324.
Schill, R., C. Dunnenbaum, P. Neyer. 1992. Quantitative Untersuchungen an Orchideenpollinien. Bot. Jahr. Syst. Phlanzengeschichte und Planzengeographie 114. 153–171.
Schlag, M., and M. Hesse. 1993. Morphogenesis of the sporoderm in *Polystachia pubescens* (Orchidaceae). Grana 32. 22–28.
Schlag, M., and M. Hesse. 1992. The formation of the generative cell in *Polystachia pubescens* (Orchidaceae). Sex. Plant Reprod. 5. 131–137.
Singer, R. B. 2002. The pollination mechanism in *Trigonium obtusum* Lindl (Orchidaceae. Maxillariinae) Sexual mimicry and trap-flowers. Ann. Bot. 89. 157–163.
Sears, B. B. 1980. Elimination of plastids during spermatogenesis and fertilization in the plant kingdom. Plasmid 4. 233–255.
Speranza, A., G. L. Calzoni, E. Pacini. 1997. Occurrence of mono or disaccharides and polysaccharides reserves in mature pollen grains. Sex. Plant Reprod. 10. 110–115.
Steiner, K. E. 1998. The evolution of beetle pollination in a South African orchid. Am. J. Bot. 85. 1180–1193.
Stenzel, H. 2000. Pollen morphology of the subtribe Pleurothallinae Lindl. (Orchidaceae). Grana 39–108–125.
Takhtajan, A. I. 1976. Neoteny and the origin of flowering plants. In: C. B. Back (ed.), Origin and early evolution of angiosperms. Columbia University Press, New York, pp. 207–219.
Tanaka, I. 1993. Development of male gametes in flowering plants. J. Plant. Res. 106. 55–63.
Williams, N. H. 1970. Some observations on pollinaria in the Oncidinae. Am. Orchid Soc. Bull. 39. 32–43, 207–220.
Williams, N. H. 1972. Additional studies on pollinaria in the Oncidinae. Am. Orchid Soc. Bull. 41. 222–230.
Williams, N. H., and C. R. Broome. 1976. Scanning electron microscope studies of orchid pollen. Am. Orchid Soc. Bull. 45. 699–707.
Wolter, M., and R. Schill. 1985a. On acetolysis resistant structure in the orchidaceae (why fossil record of orchid pollen is so rare. Grana 24. 139–143.
Wolter, M., and R. Schill. 1985b. Ontogeny of the elastoviscin in the Orchidaceae. Nord. J. Bot. 5. 575–580.
Wolter, M., and R. Schill. 1986. Ontogenie von pollen, massulae und pollinien bei den orchideen. Tropische und subtropische Phlanzenwelt 56. 1–93.
Wolter, M., C. Seufertet, and R. Schill. 1988. The ontogeny of pollinia and elastoviscin in the anther of *Doritis pulcherrima* (Orchidaceae). Nord. J. Bot. 8. 77–88.
Yeung, E. C. 1987a. Development of pollen and accessory structures in orchids. In: J. Arditti (ed.), Orchid biology: reviews and perspectives, Vol IV. Comstock Publishing Associates, Ithaca, NY, pp. 197–225.
Yeung, E. C. 1987b. Mechanisms of pollen aggregation into pollinia in *Epidendrum ibaguense*. H. B. K. (Orchidaceae). Grana 26. 47–52.
Yeung, E. C. 1987c. The development and structure of the viscidium in *Epidendrum ibaguense* H.B.K. (Orchidaceae). Bot. Gaz. 148. 149–155.
Zavada, M. S. 1983. Comparative morphology of monocot pollen and evolutive trends of aperture and wall structure. Bot. Rev. 49. 331–379.
Zavada, M. S. 1990. A contribute to the study of pollen wall ultrastructure of orchid pollinia. Ann. Missouri Bot. Gard. 77. 785–801.
Zee, S. Y., and I. H. P. Siu. 1990. Studies on the ontogeny of the pollinium of a massulate orchid (*Peristylium spiranthes*). Rev. Palaebot. Palynol. 64. 159–164.

Technical Appendix

Simple Cytological Methods for the Study of Orchid Pollen

Methods for the study of orchid PDUs do not differ from those used to study pollen of other families (Nepi and Franchi, 2002), however here we present some methods to be used for peculiar orchids features. In my opinion the best results are achieved when various microscopical methods are used and coupled with biochemical methods to revealing some common plant molecules.

Pollen Water Content

Pollen water content can be measured if it is possible to collect enough pollen to weigh on an analytical balance. Thirty milligrams is the minimum quantity. Keeps the pollen at 110°C until constant weight. The percentage of water content is determined by the formula (FW-DW) × 100/FW where FW is fresh weight and DW is dry weight.

Pollen Volume and Hydration Status

Monad pollen of orchids is dispersed in a partially dehydrated state; the degree of dehydration and how they rehydrate are observed by comparing their morphology and volume in immersion oil, namely the dry state, and after full rehydration in water. Rehydration can be followed under the microscope and volume calculated with the formula $4/3\pi r^3$, if spherical, where r is the radius or $4/3\pi a^2 b$ if oval, where a and b are transverse and longitudinal semiaxes. This observation can be coupled with water content for the sake of completeness and to demonstrate changes in volume and shape.

Number of Grains per Anther or Pollinium

The number of pollen grains per anther, pollinium or massula can be used to obtain the pollen to ovule ratio, or to see intraspecies differences.

If pollen grains are in monads or tetrads, ripe anthers must be placed in a vial containing a fluid that removes pollenkitt. Ethanol is commonly used to speed the process and to facilitate pollen release. The vial is sonicated for a few minutes. Rinse with water and adjust the volume to a known quantity. Check whether the anther is empty under a stereomicroscope. Drop a known volume from the vial into a haemocytometer, count the different squares and calculate the total number of grains.

A special method was developed by Schill et al. (1992) who determined the number of pollen grains in a pollinium by reconstruction from serial sections.

Light Microscopy

This classical technique can be used routinely for morphological and developmental studies, it gives good results especially if coupled with histochemistry. I recommend staining sections with dyes which give good demonstrative images, as well as information about the chemical composition of walls, presence of vacuoles, reserve material etc.

Histochemistry

Ideally histochemistry should be carried out on fresh material to limit chemical alteration of the material. Nepi and Franchi (2000) described the best common methods to fix and embed anther and pollen, as well as the application of many histochemical tests. The most widely used and useful tests are reported in Table 5.1 with appropriate comments for orchid pollen.

Pollen viability

To measure pollen longevity it is necessary to know whether the pollen is able to fertilize. There are direct and indirect methods, with different accuracies, speeds and applications. The most widely method used is the FCR (fluoro-chromatic-reaction) because it is easy and the results are similar those obtained by *in vitro* germination (Heslop-Harrison, Helsop-Harrison, and Shivanna, 1984; Dafni and Firmage, 2000). Place 10 ml sucrose solution (from 5% to 30% according to the species) in a transparent vial, add fluorescein diacetate (20 mg in 10 ml acetone) slowly until persistent cloudiness of the solution. Disperse pollen to test in a drop of this solution, cover with a cover slide and observe by fluorescence microscope using a violet exciter filter transmitting a beam of violet-blue light. Grains with golden yellow fluorescence are considered viable because their membrane is permeable to the dye. Dead grains are not fluorescent or have faint fluorescence. Observation and count must be quick, no longer than 10 min, because the fluorescence fades.

This method does not give any problem when the PDU is a monad or a tetrad. Pollinia must be crumbled to observe the fluorescence of the single grains.

Scanning Electron Microscopy

Scanning electron microscopy is widely used for morphological, taxonomical and other studies. Problems arise from artifacts due to water loss and distortion during preparation or in the vacuum during observation. It is therefore necessary to check shape with a stereo microscope and/or in immersion oil.

Transmission Electron Microscopy

Transmission electron microscopy is useful to reveal submicroscopic details, especially if the aim is to follow the development of a structure. However, the research must be coupled with conventional histochemical tests and/or immunohistochemistry.

ART

6

A Rare Beauty: The Orchid in Western Art

KAREN E. QUINN

Contents

Orchids in Western Art 220
Literature cited 228
Appendix: List of Original Illustrations 229

Kull, Arditti and Wong (eds.), *Orchid Biology: Reviews and Perspectives, X*,

Orchids in Western Art

Unlike the rose or the lily, the orchid does not have a long history of representation in western art. While botanical illustrators depicted varieties of the flower as early as the ninth century (with earlier manuscripts probably lost; Jacquet, 1994), its first non-scientific appearance was late in the Middle Ages, and it was seen rarely thereafter until the nineteenth century. Today it still remains an unusual, but not unknown subject choice for artists.

In the South Netherlandish tapestry *The Unicorn in Captivity* (Fig. 6-1, Appendix), which dates to about 1495–1505, the mythological horned creature is surrounded by plants that include a wild orchid (Jacquet, 1994). The unicorn may represent a contented bridegroom, symbolically encircled by the bonds of matrimony (the fence). The vegetation, including the pomegranate, bistort, and orchid, were all associated with fertility, thus underscoring the theme of marriage (Barnet and Wu, 2005).

Flowers in European art at this time were typically symbolic and usually tied to Christian or earlier mythological iconography; they were not painted as independent subjects. A white lily, for example (also seen in the unicorn tapestry), was frequently used in compositions depicting the Virgin Mary, where it alluded to her purity, a general meaning the flower had acquired in antiquity (Hall, 1979). The orchid, however, did not have a religious connection, rather, its association with fertility dates back to the fourth century B.C. and came from its name, based on *orchis*, the Greek for *testis*, which referred to the plant's testiculate bulbs in certain species (Reinikka, 1995). The root, not the flower, was used medicinally as a curative or an aphrodisiac throughout the Middle Ages and well into the Renaissance. The orchid's form, name, and function thus became responsible for its sexual connotation and this meaning may have contributed to the paucity of its representation.

In the seventeenth century Dutch artists began to paint flowers as still life subjects for their own sake, celebrating the introduction of new species, most notably tulips, to Europe at this time. Luscious bouquets showcased the exoticism, rarity, and, by inference, the expense of the subject in combinations that could not be achieved naturally. The tulip, for example, a spring flower, would be displayed with early summer roses, or a hollyhock, which bloomed later still. Painters also adjusted the actual size of the flowers to suit their compositions. In a general sense flowers retained a symbolic meaning, representing the fleeting nature of life. Scholars have also explored the possible connotations of individual blossoms.

New species of orchids also made their way to Europe beginning in the seventeenth century, but slowly and sporadically. Artists do not seem to have responded to them in the same way as they responded to other flowers and they seldom appear in still lifes, although admittedly, it is sometimes difficult to identify every blossom in a given composition. One exception is *Lilies, Irises, Tulips, Roses, Orchids, Primroses, Peonies and other Flowers in a Sculpted Vase Decorated with the Figures of Amphitrite and Ceres, with a Branch of Flowers, a Stag Beetle and other Insects* painted in the 1620s by the Flemish artist Jan Brueghel the Younger (Fig. 6-2, Appendix). In this opulent floral still life in the Dutch tradition, Brueghel delighted in the number and variety of the flowers as well as the rich play of color. The individual blossoms are subordinate to the whole, but the orchids are visible to the left of center in the arrangement.

Although in the seventeenth century interest in the orchid in Europe began to shift from a vehicle for a broad range of loosely medicinal uses to scientific study and aesthetic appreciation, the great popularity of the orchid really began in the nineteenth century with the serious cultivation of the plant (Reinikka, 1995). In England, the Royal Botanical Gardens and the Horticultural Society of London, as well as several commercial entities helped to establish the groundwork for what would become known as "orchidomania." William Cattley's successful experiments in 1818 with the blooming of the plant that would become his namesake, the *Cattleya labiata*, were also a catalyst for the rise in enthusiasm for the orchid; the extraordinary beauty and large size of the blossoms of this species held special appeal for collectors and botanists, but also for the culture at large (Pridgeon, 1992; Reinikka, 1995).

Numerous botanical tomes not only contributed to advancing the serious study of orchids, but also to furthering the interest of the public. A wave of orchid books appeared throughout the nineteenth century and included John Lindley's *Genera and Species of Orchidaceous Plants* (London, 1830–1840) as well as his *Folia Orchidacae* (London, 1852–1859), James Bateman's *Orchidaceae of Mexico and Guatemala* (London, 1837–1843) and *Second Century of Orchidaceous Plants* (London, 1867), Charles Darwin's *Various Contrivances by which British and Foreign Orchids are Fertilized by Insects* (London, 1862), and Edward Rand's *Orchids. A Description of the Species and Varieties Grown at Glen Ridge, near Boston* (New York, 1876). Accurate and beautifully drawn illustrations by professional botanical artists accompanied many of these publications, continuing the tradition that dated back to antiquity. Women such as S. A. Drake (active 1818–1847) and Augusta Withers (active 1829–1865), who was "Flower Painter in Ordinary to Queen Adelaide," specialized in orchids (Robinson, 1999). Withers produced the images for Bateman's *Orchidacae of Mexico and Guatemala* and exhibited her studies at the Royal Horticultural Society.

In spite of its immense popularity, however, the orchid remained little known as a subject for non-scientific nineteenth-century artists. Quite possibly, its traditional association with sexuality in an age when floral symbolism became widely used prevented artists from featuring them. If mentioned in the widely-distributed dictionaries of flower meanings at all, the orchid was obliquely referred to as "a belle" (Stebbins, 2000). They thus appear only occasionally in the work of a handful of painters. George Henry Hall (1825–1913), for example, an American who trained in Europe and produced still lifes and figurative subjects, included orchids in a single work, *Still Life with Flowers* (1868, Fig. 6-3, Appendix). Influenced by the seventeenth-century Dutch tradition, Hall's composition similarly features a profusion of blooms of differing species rich in color and varying in size, shape, and texture that, according to one scholar, emphasize "God's natural handiwork" (Gerdts, 1981).

Known primarily as an animal painter and for his images of the American west, New York artist William Jacob Hays (1830–1875) executed a series of flowers studies toward the end of his career that included orchids (Fig. 6-4, Appendix). His oil sketches (preliminary works done in preparation for possible finished compositions) mostly date to the late 1860s and early 1870s and are related to the illustrations in botanical books, showing flowers at the various stages of blossoming usually identified by their Latin names, but cut off from their natural habitat (Keyes, 1984). The cultivation of orchids had reached the United States in the 1830s in Boston and had spread throughout the northeast in both private and commercial greenhouses and specimens would have been available to Hays. Additionally, Hays may even have grown orchids in his own conservatory (Keyes, 1984). Hays' studies have a richness of color and delicacy of modeling beyond their scientific accuracy. In 1871 he used these sketches as the basis for the unusual *Bouquet of Orchids*, a unique work in his oeuvre (Fig. 6-5, Appendix). Like Dutch still life painters, Hays used a variety of orchids that would not grow or bloom together in their natural habitats. It has been suggested that this lush display of blossoms may have been commissioned by an orchid collector (Stebbins, 2000).

The sole artist for whom orchid subjects were a specialty was Martin Johnson Heade (1819–1904), one of the most inventive American painters. In a career that spanned seven decades, he worked equally on landscapes, marines, and still lifes, producing unique compositions—evocative salt marsh scenes, dramatic thunderstorms at the shore, and exquisite, jewel-like combinations of tropical flowers and hummingbirds. The latter may be his most original contribution to American art.

Born in rural Pennsylvania, Heade trained in the late 1830s with folk artist Edward Hicks, best known for his many versions of the *Peaceable Kingdom*. Over the next two decades, Heade traveled around the United States, eking out a living making portraits. In 1858 he moved to New York City, where he met many of the painters of the Hudson River School, considered the first group of indigenous landscape artists in the United States. At this time, Heade began to turn to subjects other than portraiture, developing his own approach to both landscape and still life. In 1863 he traveled to Brazil, the first of three trips he made to Central and South America. He would later visit Nicaragua in 1866 and Colombia, Panama, and Jamaica in 1870. The motivation for the first trip was to research hummingbirds; Heade planned to produce a book based on the studies he made there which he would call *The Gems of Brazil*. The book was never published, but Heade used the material first for small paintings of the birds, and then for innovative compositions of the hummingbirds in exotic settings with native flowers, mostly orchids.

Heade's paintings of orchids and hummingbirds were a wholly new approach to the subject. Although ornithological illustrators had combined birds with flowers, and, in some instances had used landscape backgrounds, Heade integrated the hummingbirds and orchids into a believable (although not necessarily specific) tropical setting. Unlike earlier artists, he furthermore painted both the birds and the flowers accurately and to scale (Quinn, 1999). He did, however, take liberties with the specific hummingbirds and orchids he used in each composition; often the birds and flowers are indigenous to different parts of the world and would not be found together in their native habitats. The result was a different kind of art under the influence of science.

Many nineteenth-century artists showed a growing interest in the scientific world, but for Heade ornithology had been a lifelong pursuit. In *Forest and Stream*, the hunting and fishing magazine known today as *Field and Stream* to which Heade contributed regularly beginning in 1880, he wrote "From early boyhood I have been almost a monomaniac about hummingbirds" ("Didymus" [M. J. Heade] 1892). His decision to produce a book on them resulted from this passion; the choice to study them in Brazil can probably be attributed to his friendship with the Reverend James Cooley Fletcher, who had been first a missionary there, later secretary of the United States legation in Rio de Janeiro (Stebbins, 2000). Heade had met the well-connected Fletcher before he left for South America and Fletcher helped him once he arrived in Rio. In 1857 Fletcher had penned *Brazil and the Brazilians*, the best guidebook of its day. He also may have introduced the artist to Louis Agassiz, thus expanding Heade's acquaintances in the scientific world.

Although Heade's first orchid and hummingbird paintings date to 1871, he had likely seen orchids in the wild on this 1863 trip. In the unpublished manuscript for his book, *The Gems of Brazil*, probably begun around this time, he theorized that the shape of the beaks of certain hummingbirds was adapted to feed from "cup-shaped" orchids (Martin Johnson Heade papers, no date). In his preparation for his own publication and through his associates, Heade had become well-versed in the writings of Alexander von Humboldt, Louis Agassiz, Charles Darwin, and James Cooley Fletcher, all of whom described the flowers in their publications. Thus, his introduction to orchids as a subject came through his study of hummingbirds.

Still, one has to wonder what prompted Heade to showcase orchids most often with the hummingbirds. He had initially experimented with other flowers. In 1862 he painted a tiny composition of a single bird with a nasturtium and later that decade he featured North American ruby throats with branches of apple trees in full bloom. The series of small format hummingbird paintings related to the *Gems of Brazil* project sometimes included flowers, but in supporting roles and not orchids. The birds that Heade ultimately combined with orchids were tropical species, therefore, given his interest in science, he likely chose the orchids to enhance the realism of his works, even though the specific birds and flowers in a given composition might not be native to the same place. Heade also had the tendency to gravitate towards subjects that other artists avoided. In landscape, he produced some 150 modest and largely anonymous, but delicately luminous salt marsh scenes while other American painters focused on grand and readily recognizable natural landmarks such as Niagara Falls or Yosemite. That flower artists avoided orchids may have intrigued Heade. Most important for Heade, though, was the visual impact of the orchid and hummingbird combination: the flowers and birds play off each other in rich orchestrations of hues and Heade could create endless iterations of them.

Almost sixty orchid and hummingbird paintings by Heade are known today (Stebbins, 2000). As noted, the earliest are signed by the artist and dated 1871 in his hand. He continued to produce them for the rest of his career – the last in 1902. Many lack inscriptions by Heade, but style helps to determine their chronology. In general, in the earlier compositions Heade experimented with the number and types of birds and orchids, often choosing brilliantly colored examples of each. He established the general format of the compositions in his first efforts: flowers and birds are presented in the foreground, seeming large and detailed, but in fact painted to scale (Quinn, 1999). Placement varies. Initially, birds and flowers seem almost awkward, as if cut out and pasted in; they exist independently of each other. In *Cattleya Orchid and Three Brazilian Hummingbirds* (Fig. 6-6, Appendix) dated 1871, for example, the blossom is twisted in what seems like an unnatural manner at the left and the birds create a separate vignette at the right. The green and pink hues of the birds above the nest, a male and female amethyst woodstar (Calliphlox amethystina), contrast to the warmer tones of the red-tailed comet (Sappho sparganura) at the lower right. The vivid plumage of this larger bird is also juxtaposed to the subtler color of the pink orchid. In the earliest experiments, Heade delighted in the brilliance of each of his subjects, featuring them equally but distinctly in the painting.

Heade's compositions gradually became more integrated as the artist worked toward harmonizing the colors and unifying the relationships between his subjects. This is foreshadowed in *Cattleya Orchid and Three Brazilian Hummingbirds* with the deep pink neck feathers of the male amethyst woodstar echoing the lighter pink of the flower, thus drawing them together. As Heade continued to develop his theme over the next decade, he chose birds and flowers of similar hues. In *The Pink Orchid* (Fig. 6-7, Appendix; about 1875–1990), the two birds – a male amethyst woodstar at the left and a male Peruvian sheartail (Thaumastura cora) at the right – are close in color. Not only do the pinks and greens of the birds complement the large *Cattleya labiata*, but the centered arrangement of plant, buds, and blossoms leads the eye easily to the creatures on the right. Flowers and birds relate more gracefully in position and are chromatically connected. Even the detail of the cluster of tiny pink orchids at the far left, *Dendrobium stratiotes*, accentuates Heade's overall color scheme. The result is a more natural, if not more scientifically accurate view of the tropical microcosm.

The *Cattleya labiata* appeared in over forty-five of Heade's orchid and hummingbird paintings, making it by far his favorite species. In three known canvases he used the *Cattleya dowiana* and in five others the *Laelia purpurata*, the national flower of Brazil. *Dendrobium stratiotes* appear in a handful of compositions, but in a supporting role usually along with the *Cattleya labiata*, as in *The Pink Orchid*. In a single instance he included the *Aerides odorata* Lour., also with a *Cattleya labiata*.

Although Heade did not date any of the canvases in which either the *Cattleya dowiana* or the *Laelia purpurata* are featured, based on the evolution of his style they belong to the decade or so of his mature orchid and hummingbird pictures beginning about 1875 (Stebbins, 2000). In general, these paintings show the same concern for visual harmony that is evinced in *The Pink Orchid*. In *Orchids and Hummingbird* (Fig. 6-8, Appendix), for example, the deep magenta of the labellum of the two *Cattleya dowiana* blossoms is repeated in the shimmering neck of the amethyst woodstar, pulling the three subjects together. Likewise, the hues of the interior of the *Laelia purpurata* in *Two Hummingbirds above a White Orchid* (Fig. 6-9, Appendix) is picked up in the throat of the male amethyst woodstar.

No two orchid and hummingbird paintings are exactly alike, but Heade used the same orchid "models" over and over again in his compositions. In five dozen works, six different *Cattleya labiata* blossoms appear; the full bloom in *The Pink Orchid* was Heade's preference as it appeared in some twenty. The same two *Cattleya dowiana* and the identical *Laelia purpurata* occur in each of the paintings showcasing those species (three and five respectively). Apparently Heade worked from studies of flowers, leaves, and birds that he made in oil paint on unstretched canvas; these are strikingly similar to the sketches made by William Jacob Hays, who had a studio in the same building as Heade in New York City and may have influenced the older artist (Stebbins, 2000). A group of Heade's oil sketches survives, but not for every flower and unfortunately not for any of the six different *Cattleya labiata* in the finished paintings (see Stebbins, 2000). The *Laelia purpurata*, though, that Heade included in five compositions is the flower on the right of *Study of Laelia Purpurata and Another Flower* (Fig. 6-10, Appendix; the *Cattleya labiata* on the left was not used in any of the extant canvases). Not only is the shape and coloring of the orchid the same, but the size of each of the flowers corresponds exactly as curators and conservators at the Museum of Fine Arts, Boston have determined by carefully making tracings of a selection of Heade's blossoms. This would seem to indicate that Heade somehow repeatedly transferred the flower from the oil sketch, but examination of both the sketches (where they exist) and the finished works reveal no evidence of traditional methods such as pouncing (a process in which the design to be transferred is placed over the surface where the image is desired. Holes are pricked through the outline of the design and chalk or charcoal powder is rubbed through, creating a dotted drawing below. If Heade used this method, the extant sketches would show indication of pricking and they do not; Fulton, et al., 2002).

Recent research into the working methods of another nineteenth-century American artist, Fitz Henry Lane, has revealed a possible method of transfer for Heade's sketches, the optical device known as a camera lucida (Quinn, Kelberlau, and Woodword, 2006). Patented in England in 1807 by William Wollaston, the camera lucida consists of a prism attached to a movable arm that can be clamped onto a table. Looking through the prism the artist sees whatever is in front of him reflected onto a support such as paper or canvas below, which then can be easily traced. Compact and portable, the camera lucida was used for original work, such as accurately reproducing a landscape onto paper while sketching outdoors, but it could also be an aid in the enlargement, reduction, and one-to-one copying of designs. For the latter, the sketch to be transferred would be placed in front of the camera lucida at the same distance as the surface onto which the sketch was to be copied on the table below. Unlike pouncing, transferring in this manner does not affect the surface of the original sketch, which could explain why there is no marked evidence of a transfer method in Heade's work.

Heade's finished paintings are completely different from anything produced in western art – part still life and part landscape. Although he used his sketches for the flowers in his paintings, unlike William Jacob Hays, in the finished oils, he reintegrated the orchids into a seemingly natural habitat. Scholars have suggested that Heade may have been influenced by the recently published theories of Charles Darwin, including natural selection (Foshay, 1980; Foshay, 1981, 1984; Novak, 1995). Although Heade does not write about Darwin in any of his articles or personal letters, there is no doubt that he was aware of the controversial material, given his circle of associates and his own research into ornithology and botany (Stebbins,

2000). Whether he consciously intended the flower and hummingbird paintings to relate to Darwin's writings is unknown; up to a point the birds and flowers illustrate the theory of natural selection successfully, although Heade was not completely true to nature in that in some cases the choices of birds and flowers would not be found together in the wild. At the very least, science informed Heade's work, but the science was used as a tool towards an artistic end.

Heade's last dated orchid composition was painted in 1902. Although he enjoyed some success in his long career, Heade's orchids seemed to have had little impact on subsequent artists and the flower returned to its status as an unusual subject choice in the late nineteenth century in spite of its popular use in bouquets, as a lapel ornament, and even in jewelry (Davies, 1991; Loring, 1987). French painter Gustave Caillebotte (1848–1894), who raised orchids among other exotic plants, painted floral still lifes periodically, but much less often than his Impressionist colleagues and he seldom featured the orchid (Distel et al., 1994). In about 1893, however, he executed two sets of four canvases for the panels of doors in his own dining room; the subject was a view of his greenhouse including orchid plants, an unusual image of orchid cultivation (Fig. 6-11, Appendix). Occasionally, still life artists such as Paul de Longpré (1855–1911) or Charles Storer (1817–1907) worked with orchids. These two artists must have carefully studied the flowers since they also provided illustrations for botanical publications. Their compositions most often fluctuated between traditional floral arrangements and the scientific approach. Storer, a Boston artist, studied orchids in the late 1880s at the Langwater collection in North Easton, Massachusetts, at the Ames family estate and provided a single illustration for Henry Sander's *Reichenbachia* (1888–1892; Ferry, 2006). Some years later, he painted *Still Life of Orchids in an Urn* (1900; Fig. 6-12, Appendix), a more typical Victorian floral composition that relates back to seventeenth-century Dutch works.

An atypical example of a painter who worked with orchids in both traditional still life compositions and in landscape, German artist Ernst Papf (1833–1910) trained in Dresden, but immigrated to Brazil in 1867. Papf's arrival in Brazil does not overlap with Martin Johnson Heade's first South American trip, but it is possible that he knew Heade's work. In the 1890s Papf painted a group of tropical views that included orchids and other flowers in natural settings as in *Fleurs Exotiques* (1895; Fig. 6-13, Appendix). These are rare in that they recall Heade's orchid and hummingbird compositions. Unlike Heade's close ups, however, Papf's blooms are given some distance and featured less prominently, becoming part of a more pure landscape.

In the twentieth century, orchids continued to appear sporadically and for the most part they were subjected to an artist's interest in formal issues: color, line, or shape and, in many instances, degrees of abstraction. Emil Nolde (1867–1956), a German Expressionist who experimented with color, for example, painted the flower in a group of watercolors. In *Orchids and Anemones with Bronze Idol* (Fig. 6-14, Appendix), the orchids have taken on the heightened hues of the anemones and are recognizable only by their shape. Nolde exploited the fluidity of the water-based medium allowing the pigments to overlap and bleed into each other and avoided any attempt to recreate the modeling of the flowers. In a pen and ink drawing, Henri Matisse (1869–1954), who was also known for his experiments with color, but with a different approach than Nolde, pared the orchid down to an elegant outline (Fig. 6-15; Appendix). American photographer Edward Weston (1886–1958), was known for his series of still lifes of peppers and shells, but he seldom worked with flowers. He made negatives of an orchid only one time during his career in 1931 (Fig. 6-16, Appendix). The flower had been given to him (and for Weston, cost was very much an issue). His interest in its form is made explicit in the entry he wrote in his journal, "The orchid…was not so easy to do: too perfect—like trying to photograph a cathedral…I could spend a whole day or several days with it, could do something—in fact I have several negatives of much interest, but know there is so much more I could do, for it has possibilities far beyond a literal recording. I made several negatives in the tin funnel, direct sunlight, giving all the orchid's exotic quality, and a couple close up, more direct, powerful, abstract as it were: these latter are in the spirit I would go on with, given time. I don't know where I will attain another orchid!" (Weston, 1973). In *Orchid*, Weston closed in on the flower, centered it, and cropped it at the edges so that it fills the composition. While it is on the one hand recognizable as an orchid, it is also a series of abstract shapes – arcs and curves – tinged with diffused light.

Unlike Edward Weston, Georgia O'Keeffe (1887–1986) painted flowers throughout much of her lifetime, especially in the 1920s. Beginning in 1919 she did series of calla lilies, irises, petunias, poppies, and a wide range of other species. Like Weston, however, the orchid was a rarity in her oeuvre, and she produced only two known images of them—later, in 1941 (Lynes, 1999). Both *Narcissa's Last Orchid* (Fig. 6-17, Appendix) and *An Orchid* (Fig. 6-18, Appendix) were executed in pastel, a medium which O'Keeffe had first used in the 1910s. Typically, as she worked with a subject O'Keeffe would experiment with the juxtaposition of contrasting color, as seen in the white blossom against the tones of pink and brown in *Narcissa's Last Orchid*. In other compositions, though, including *An Orchid*, she limited her palette to explore the subtle variations of a single color, in this case white mixed with delicate touches of yellow and green. O'Keeffe also shifted the point of view of the flowers she studied, often closing in tightly and enlarging them. The resulting image can be almost abstract, as with *An Orchid*. She wrote of her subject, "A flower is relatively small. Everyone has many associations with a flower—the idea of flowers…Still in a way—nobody sees a flower—really it is so small—we haven't time…So I said to myself—I'll paint what I see—what the flower is to me but I'll paint it big and they will be surprised into taking time to look at it." (Lynes, 1999). The blossom was a vehicle for her experiments with color and composition, but it was also important to O'Keeffe that she present her still life in what she hoped was a new context.

Although she protested, beginning early in her career critics made Freudian interpretations of O'Keeffe's work, particularly noting sexual associations. Perhaps the longstanding sexual connotation of the orchid itself kept O'Keeffe from pursuing it further as a subject to avoid the obvious connection. Her friend, Charles Demuth, however, directly sought the visual parallels between orchids and sexual imagery. In discussing one of Demuth's works, the poet William Carlos Williams stated, "In my painting of orchids which Charlie did – the one called *Pink Lady*

Slippers (1918) he was interested in the similarity between the forms of the flowers and the phallic symbol, the male genitals" (Farnham, 1971). In addition, Demuth explored the color relationships of his subject, as in the watercolor *Wild Orchids* (Fig. 6-19, Appendix) where he sketched the flower in graphite and then carefully worked with pinks, reds, and violets varying the density of the medium on the paper, but also flattening the form of the flower. Similarly, later in the century Robert Mapplethorpe (1946–1989) would explore both the sensuality and the formal qualities of orchids in a series of photographs he made in the 1980s.

Contemporary artist Judy Cotton has embraced orchids as a subject on a number of different levels. She was first attracted to them in 1989 for their visual qualities. She found them "clear cut yet lush" and enjoyed the "sometimes complex yet sensual nature of their structure and how you have to unravel it when you draw them" (Cotton, 2007). In *Orchids* (1992; Fig. 6-20, Appendix) this sense of disentangling and simplifying the plant's form is seen in the intricate outline silhouetted against the rich, but neutral background.

Additionally, around this time Cotton was recovering from a serious illness and she related to the orchid as a living thing, for her they "seemed to speak to me about regeneration, recovery, possibility, survival" (Cotton, 2007). In the mid-1990s, Cotton, who has explored a wide range of subjects over the course of her career, discovered the work of Martin Johnson Heade. This encounter led to another direction in her work and a refocus on orchids. Using Heade's compositions as a point of departure, she took details of his work and integrated them into her own innovative conceptions. In *Fugue for MJH* (2003; Fig. 6-21, Appendix), the flowers and birds from *Two Fighting Hummingbirds with Two Orchids* (1875; Fig. 6-22, Appendix) are merged with an abstract pattern of swimming pool water reflections, as if seen below the surface. At the upper left, Cotton added a vignette from *Thunder Storm on Narragansett Bay* (1868) in completely contrasting colors, juxtaposing two of the subjects – still life and seascape – for which Heade was known, but which are now seen in a completely different and original way.

For the most part, the orchid has made infrequent appearances in Western art. Artists may have avoided them for a variety of reasons: early on, their lack of a religious association, later their sexual symbolism, or possibly overall, even the challenge of capturing the range and subtlety of their color and form. Discovering orchids in an artist's oeuvre is like encountering them in the wild – an unexpected, yet beautiful surprise.

Literature cited

Barnet, P. and N. Wu. 2005. *The Cloisters: Medieval Art and Architecture*. Yale University Press, New Haven, CT.

Cotton, J. 2007. Letter to K. Quinn. May 22, 2007.

Davies, J. 1991. *The Victorian Flower Garden*. BBC Books, London.

"Didymus" [M. J. Heade]. 1892. Taming Hummingbirds. *In* Forest and Stream 38, 15 (April 14, 1892), p. 348.

Distel, A., et al. 1994. *Gustave Caillebotte the Unknown Impressionist*. Royal Academy of Arts, London.

Farnham, E. 1971. *Charles Demuth: Behind a Laughing Mask*. University of Oklahoma Press, Norman, Oklahoma.

Foshay, E. 1980. Charles Darwin and the Development of American Flower Imagery. *In* Winterthur Portfolio, winter, 1980, pp. 299–314.

Ferry, R. J. 2006. Reichenbachia. *The McAllen International Orchid Society Journal* 7: 5–10.

Foshay, E. 1981. *Nineteenth Century American Flower Painting and the Botanical Sciences*. University Microfilms International, Ann Arbor, MI.

Foshay, E. 1984. *Reflections of Nature: Flowers in American Art*. Whitney Museum of American Art, New York.

Fulton, E., et al. 2002. The Methods and Materials of Martin Johnson Heade. *Journal of the American Institute for Conservation* 41: 155–184.

Gerdts, W. 1981. *Painters of the Humble Truth. Masterpieces of American Still Life 1801–1939*. Philbrook Art Center with University of Missouri Press, Columbia, Missouri.

Hall, J. 1979. Dictionary of Subjects and Symbols in Art. Harper & Row, New York.

Jacquet, P. 1994. History of Orchids in Europe, from Antiquity to the 17th Century. *In* J. Arditti (ed.), *Orchid Biology: Reviews and Perspectives VI*. Wiley, New York, pp. 33–102.

Keyes, D. D. 1984. The Orchid Studies of William Jacob Hays. *In* W. J. Hays (ed.), *Orchid and Wildflower Studies*, Berry-Hill Galleries, New York, pp. 4–10.

Loring, J. 1987. *Tiffany's 150 Years*. Doubleday & Co., Garden City, New York.

Lynes, B.B. 1999. *Georgia O'Keeffe Catalogue Raisonné. Volume Two*. Yale University Press, New Haven, CT.

Martin Johnson Heade Papers. no date. Archives of American Art, Smithsonian Institution, Washington, DC.

Novak, B. 1995. *Nature and Culture: American Landscape and Painting, 1825–1875*. Revised Edition, Oxford University Press, New York.

Pridgeon, A., (ed.) 1992. *The Illustrated Encyclopedia of Orchids*. Timber Press, Portland, Oregon.

Quinn, K. 1999. Orchids. *In* T. Stebbins, et al. (eds.), Martin Johnson Heade. Museum of Fine Arts, Boston, Boston, MA, pp. 109–113.

Quinn, K., with S. Kelberlau and J. Woodward. 2006. Rediscovering Fitz Henry Lane's *View of Coffin's Beach* on Cape Ann. *In* The Magazine Antiques, CLXX, no. 1, July 2006, pp. 66–69.

Reinikka, M. A. 1995. *A History of the Orchid*. Timber Press, Portland, OR.

Robinson, F. 1999. *The Country Flowers of a Victorian Lady*. Harper Collins, New York.

Stebbins, T. E. 2000. *The Life and Work of Martin Johnson Heade*. Yale University Press, New Haven, CT.

Weston, E. 1973. *The Daybooks of Edward Weston*. Volume II. California. Aperture, Millerton, New York.

Appendix: List of Original Illustrations

Since *Orchid Biology Reviews and Perspectives* cannot publish color illustrations and because black and white photographs will not do justice to the original works we have chosen not to include illustrations in this chapter. We are listing the locations of the works for those who may have an opportunity to visit the venues where they are displayed. Ownership, copyright, permissions and availability of illustrations are also reasons why illustrations could not be published.

Fig. 6-1. South Netherlandish
The Unicorn in Captivity, 1495–1505
Wool warp, wool, silk, silver, and gilt wefts
Metropolitan Museum of Art, Gift of John D. Rockefeller Jr., 1937 (37.80.6)

Fig. 6-2. Jan Brueghel the Younger
Lilies, Irises, Tulips, Roses, Orchids, Primroses, Peonies and other Flowers in a Sculpted Vase Decorated with the Figures of Amphitrite and Ceres, with a Branch of Flowers, a Stag Beetle and other Insects, 1620
Oil on panel
With Christie's, New York, January 27, 2000, lot 79

Fig. 6-3. George Henry Hall
Still Life with Flowers, 1868
Oil on canvas
Private collection

Fig. 6-4. William Jacob Hays
Four Orchids, 1870
Oil on canvas
Courtesy of Berry Hill Galleries, New York.

Fig. 6-5. William Jacob Hays
Bouquet of Orchids, 1871
Oil on fabric
Private collection

Fig. 6-6. Martin Johnson Heade
Cattleya Orchid and Three Brazilian Hummingbirds, 1871
Oil on wood
National Gallery of Art, Washington, D.C., Gift of The Morris and Gwendolyn Cafritz Foundation, 1982.73.1

Fig. 6-7. Martin Johnson Heade
The Pink Orchid, 1875–1890
Oil on canvas
Private collection

Fig. 6-8. Martin Johnson Heade
Orchids and Hummingbirds, 1875–1883
Oil on canvas
Museum of Fine Arts, Boston, Gift of Maxim Karolik for the M. and M. Karolik Collection of American Paintings, 1815–1865, 47.1164.

Fig. 6-9. Martin Johnson Heade
Two Hummingbirds above a White Orchid, 1875–1890
Oil on canvas
Amon Carter Museum, Fort Worth, Texas, 1981.33

Fig. 6-10. Martin Johnson Heade
Study of Laelia Purpurata and Another Flower, about 1870
Oil on canvas (unstretched)
Saint Augustine Historical Society, Saint Augustine, Florida.

Fig. 6-11. Gustave Caillebotte
Four paintings designed to decorate a door in the dining room, Petit Gennevilliers, about 1893
Clockwise:

a. *Orchids: Cattleya and Anthurium*
b. *Cattleya and Anthurium*
c. *White-Flowered Orchids*
d. *Cattleya and Red-Flowered Plants*

Oil on canvas
Private collection

Fig. 6-12. Charles Storer
Still Life of Orchids in an Urn, 1900
Oil on canvas
Spanierman Gallery, New York

Fig. 6-13. Ernst Papf
Fleurs Exotiques, 1895
Oil on canvas
With Sotheby's, New York, November 20, 2002, lot 70

Fig. 6-14. Emil Nolde
Orchids and Anemones with Bronze Idol, undated
Watercolor on paper
Museum of Fine Arts, Boston, Gift of John S. Newberry, Jr., 62.210

Fig. 6-15. Henri Matisse
Orchidée, 1943
Pen and Indian ink
With Sotheby's, London, June 25, 1997, lot 27

Fig. 6-16. Edward Weston
Orchid, 1931
Photograph, gelatin silver print
The Lane Collection, courtesy of the Museum of Fine Arts, Boston

Fig. 6-17. Georgia O'Keeffe
Narcissa's Last Orchid, 1941
Pastel on paper
The Art Museum, Princeton University, Gift of David McAlpin, Class of 1920 (x1982–357)

Fig. 6-18. Georgia O'Keeffe
An Orchid, 1941
Pastel on paper
The Museum of Modern Art, New York, Bequest of Georgia O'Keeffe [1987] (556.90)

Fig. 6-19. Charles Demuth
Wild Orchids, about 1920
Watercolor and graphite on paper
Kemper Museum of Contemporary Art, Bebe and Crosby Kemper Collection, Gift of the William T. Kemper Charitable Trust, 1995.24

Fig. 6-20. Judy Cotton
Orchids, 1992
Encaustic
Private collection

Fig. 6-21. Judy Cotton
Fugue for MJH, 2003
Oil on board
Private collection

Fig. 6-22. Martin Johnson Heade
Two Fighting Hummingbirds with Two Orchids, 1875
Oil on canvas
Private collection

HISTORY-POLLINATION

7

Hand-Pollination of *Vanilla*: How Many Discoverers?

JOSEPH ARDITTI, ADISHESHAPPA NAGARAJA RAO, AND HELEN NAIR

Contents

Introduction 234
Vanilla 235
Hand-Pollination of *Vanilla* 239
Belgium 239
France 242
Indonesia 243
Reunion 244
Padua Botanical Garden 245
Guadeloupe 246
Final Thoughts 246
Dedication 246
Literature Cited 247

Kull, Arditti and Wong (eds.), *Orchid Biology: Reviews and Perspectives, X*,

Introduction

Vanilla planifolia, "the ice cream orchid" (Ecott, 2004), is the only orchid (Fig. 7-1A–I) grown as an edible plantation crop. The plant is a vine which produces long green fruits (they are capsules, not beans as they are usually referred to) that change color to brown (Fig. 7-2) and develop their well known aroma following curing. To produce fruits *Vanilla* flowers (Fig. 7-3) must be pollinated. In the natural habitat this is accomplished by male and female euglossine bees (Hagsater et al., 2005). When *Vanilla* plants were taken to Europe and cultivated in greenhouses they did not produce fruits because their pollinators were not present.

Vanilla seems to have been introduced into Europe "prior to 1739" and "probably in 1510" (Childers, Cibes, and Medina, 1959). By 1807 it was cultivated in

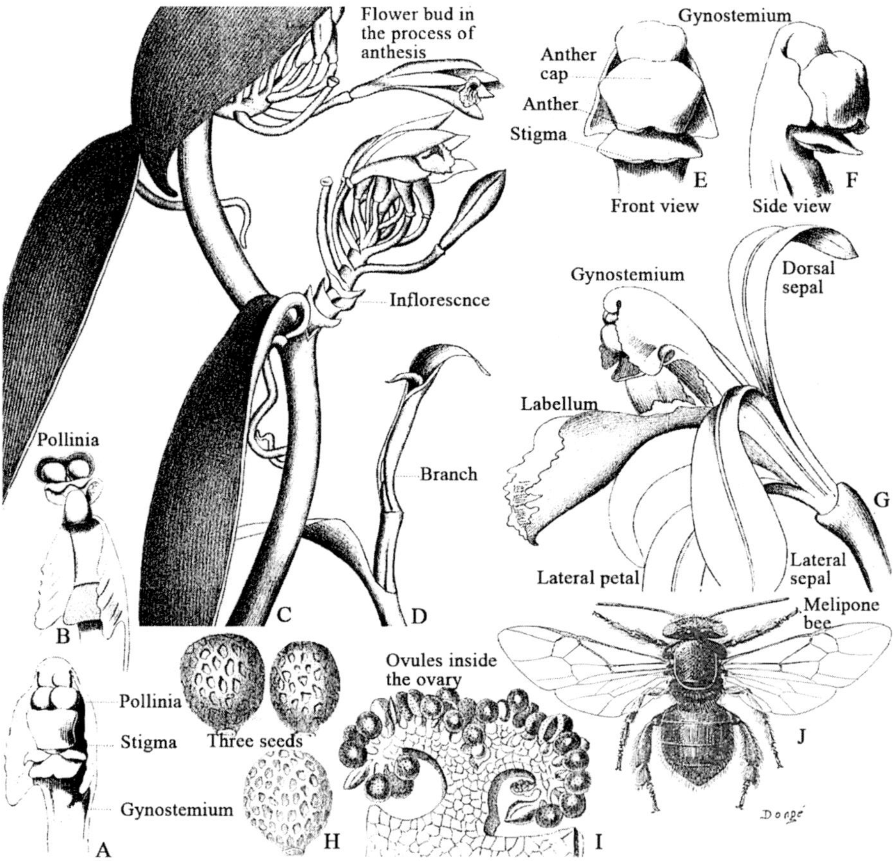

Fig. 7-1. *Vanilla*. **A, B**. Front view of upper part of gynostemium (column). **C, D**. Portion of stem. **E, F**. Front and side view of gynostemium tip. **G**. Flower. **H**. Seeds. **I**. Ovules. J. Natural pollinator, Mexican *Melipona* bee (**A–D**, Delteil, 1897; **E, F**, Delteil, 1874; **G**, Deleil, 1897; **H, I**, de Vriese, 1856).

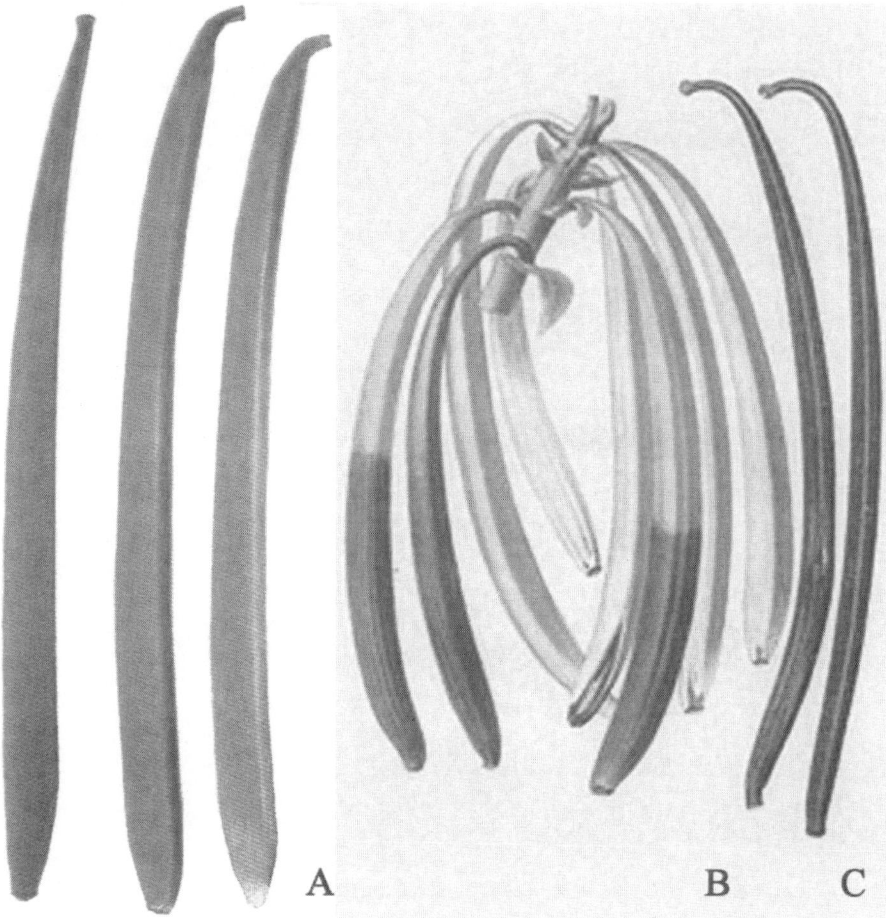

Fig. 7-2. *Vanilla* capsules. **A**. Immature (green). **B**. Various stages of ripening (light green, yellow, brown). **C**. Mature (ripe) and brown (Computer generated after Bouriquet, 1954).

Paddington, England (Childers, Cibes, and Medina 1959). After that cultivation became more widespread, but large scale commercial production of *Vanilla*, though of interest (Anonymous, 1855b), was not possible, the major reason being the failure of the plants to set fruit (Nolan, 1942; Childers et al., 1959). Introduction of *Melipona* bees (Fig. 7-1J), the natural pollinator (Beck, 1912) from Mexico was to no avail.

Vanilla

The genus *Vanilla* consists of 107 species, 54 of which are found in tropical America (Pridgeon et al., 2003). Best known of these is *Vanilla planifolia*, the source of the spice vanilla (Anonymous, 1894; Hart, 1894; Wigman, 1931; HWJ,

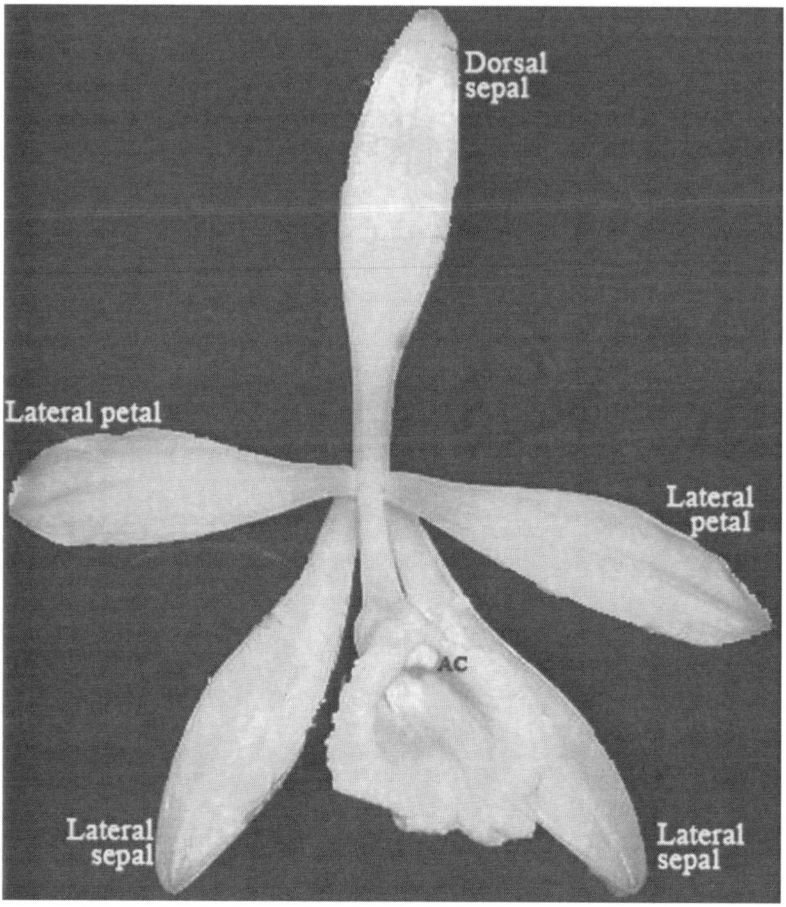

Fig. 7-3. *Vanilla* flower (photograph by J. Arditti and T. W. Yam).

1940; Nolan, 1942; Narodny, 1947; Childers and Cibes, 1948; Correll, 1953). One of the first written references to *Vanilla* is by Bernardino de Sahagún, a Franciscan friar who went to Mexico in 1529 and wrote *Historia General de las Cosas de Nueva España* in the Aztec language. It was translated into Spanish in 1829–1830 (Correll, 1953). Another early reference seems to be in the Badianus Codex (Fig. 7-4), a herbal written by Juan Badianus and an Indian convert named Martinus de la Cruz. The herbal was written in 1552 and refers to *Vanilla* by its Aztec name, *tlilxochitl* in describing its medicinal properties (Ecott, 2004).

Charles l'Écluse (Clusius, 1526–1609; Fig. 7-5A), Professor at the University of Leiden who received plants from Hugh Morgan (pharmacist to Queen Elizabeth I of England) was the first to write botanically about *Vanilla* calling it *Lobus oblongus aromaticus* in his *Exoticorum Libri Decem* which was published in 1605 (De Vriese, 1856; Rolfe, 1895, 1896; Lawler, 1984; Jacquet, 1994; Pridgeon et al., 2003). Father

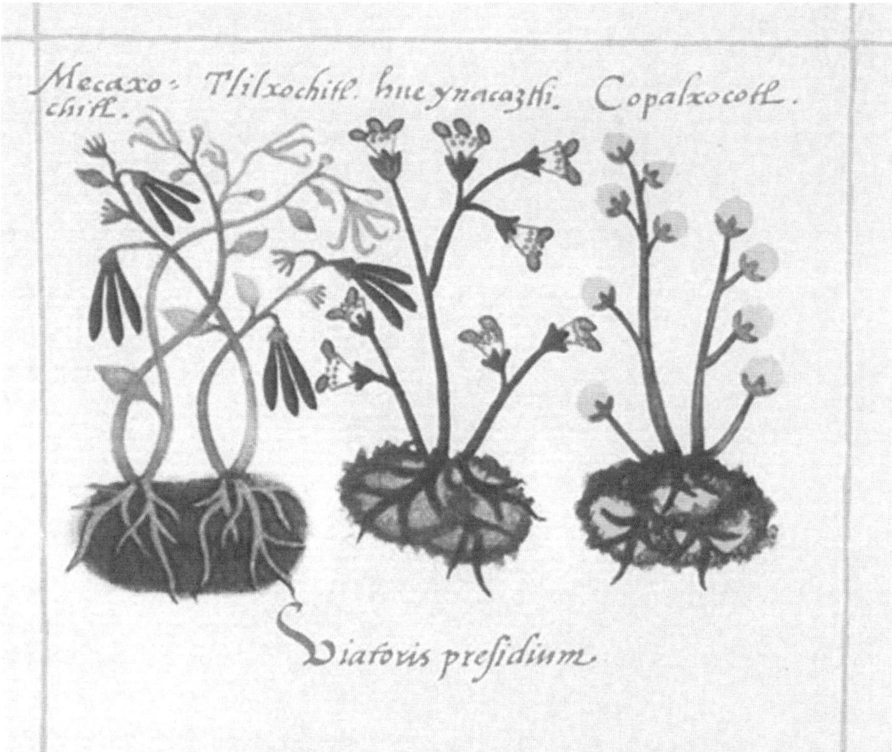

Fig. 7-4. *Vanilla* illustration in the *Badianus Codex* which dates back to 1552.

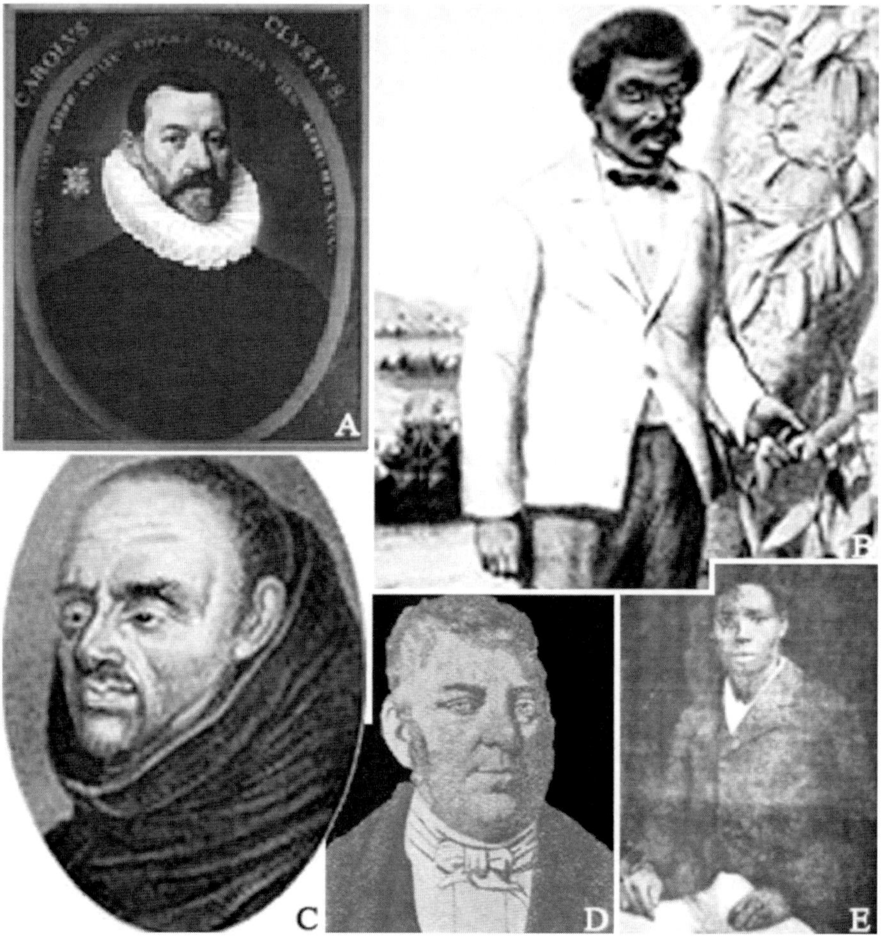

Fig. 7-5. Persons associates with *Vanilla*. **A**. Charles l'Écluse (Clusius). **B**. Edmond Albius as adult. **C**. Father Charles Plumier. **D**. Joseph Neumann. **E**. Young Edmond Albius (**A, C, D, J.** Arditti collection; **B**, Lithograph by Roussin, ca 1863 from Wikipedia; **E**, computer enhanced after Ecott, 2004).

Charles Plumier (1646 or 1648–1704; Fig. 7-5B), a Franciscan monk and a botanist was the first to use the generic name *Vanilla*. The name is a Latinization of *vainilla* the dimunitive for *vaina* and means "little pod" in Spanish (Delteil, 1874, 1884, 1897, 1902).

Vanilla planifolia is found in Oaxaca, Quintana Roo, and Yucatán in Mexico from sea level to 1.2 km altitude in mixed forests and swampy thickets (Wiard, 1987). It is also found in Florida, the West Indies, Central America, and South America as far south as Paraguay (Wiard, 1987). The plants are vines which climb on trees and can be up to about 25 m long. Leaves are succulent and elliptical, 9–23 cm in length and 2–8 cm in width. In Mexico the plant flowers from mid-February until the first part of

June. The inflorescences are 15–20 long racemes which arise from leaf axils. They bear approximately 20 yellowish green flowers which measure as much as 5 cm across. Their labella are 1.5 cm wide. Gynostemia are ca 3 cm long. The fruit, a capsule not a bean, reaches a length of 25 cm and a diameter of 1.5 cm. They are green, but become dark brown after curing (Anonymous, 1855b; Correll, 1953; Wiard, 1987).

Hand-Pollination of *Vanilla*

In nature orchids are pollinated by vectors, mostly insects, which are usually highly specific. There are very few, if any, common or non specific pollinators. Wind pollination is not known to occur. Parthenocarpy is known, but not common. Unpollinated flowers live for various periods, but eventually wither and abscise. When *Vanilla* plants were taken from their native habitats to other areas their pollinators remained behind and the plants did not set fruit. Therefore, large scale cultivation of *Vanilla* as a plantation crop became possible only after the discovery of hand-pollination (Fig. 7-6). The discovery seems to have been made, perhaps independently, several times.

Belgium

Vanilla was first taken to Belgium by Mr Parmentier d'Enghien (Morren, 1838–1839). Its flowers were successfully hand-pollinated for the first time at the Liege Botanical Gardens in 1836 by Charles Morren (Figs. 7-7 and 7-8; Morren, 1838; Poiteau, 1838; De Visiani, 1845; Anonymous, 1855a, b; van Gorkom, 1884; Delteil, 1884, 1902; Anonymous, no date-c) who was generally interested in hybridization (Mn, 1845a, 1945b) of orchids (Morren, 1827, 1829, 1848) and other plants (Morren, 1839b). There is no mention of orchid pollination, germination and hybridization in an earlier paper by him (Morren, 1829). Six years later he reported pollinating a flower which opened on 16 February 1836; the fruits ripened 1 year later (Morren, 1837a, b, 1838, 1838–1839, 1839a, 1850; Berlése, 1837). During a lecture to the Institute de France, Morren showed a fragment of a vine bearing three fruits (Childers et al., 1959).

Scion of a family which left Ireland during the reign of Henry the Eighth (Morren, 1860; Anonymous, 1869), Charles François Antoine Morren (05 March 1807, Gand-17 December 1858, Liége) was the son of Charles Ignace François Morren (1773–1837), a medical officer in the army and Marie Catherine Pole (?-1810). His mother died when he was 3 years old and he was raised by his grandparents (Anonymous, no date-a), or uncle (Morren, 1860; Anonymous, 1869; Crépin, 1899). He graduated from the l'Athénée Royal de Bruxelles on 14 August 1825 summa cum laude and proceeded to the University of Gand where on 1 August 1826 he was promoted to Candidate in Mathematics and Natural History. Morren received his doctorate in 1829. In 1831 he was appointed Professor of Physics at the University of Gand.

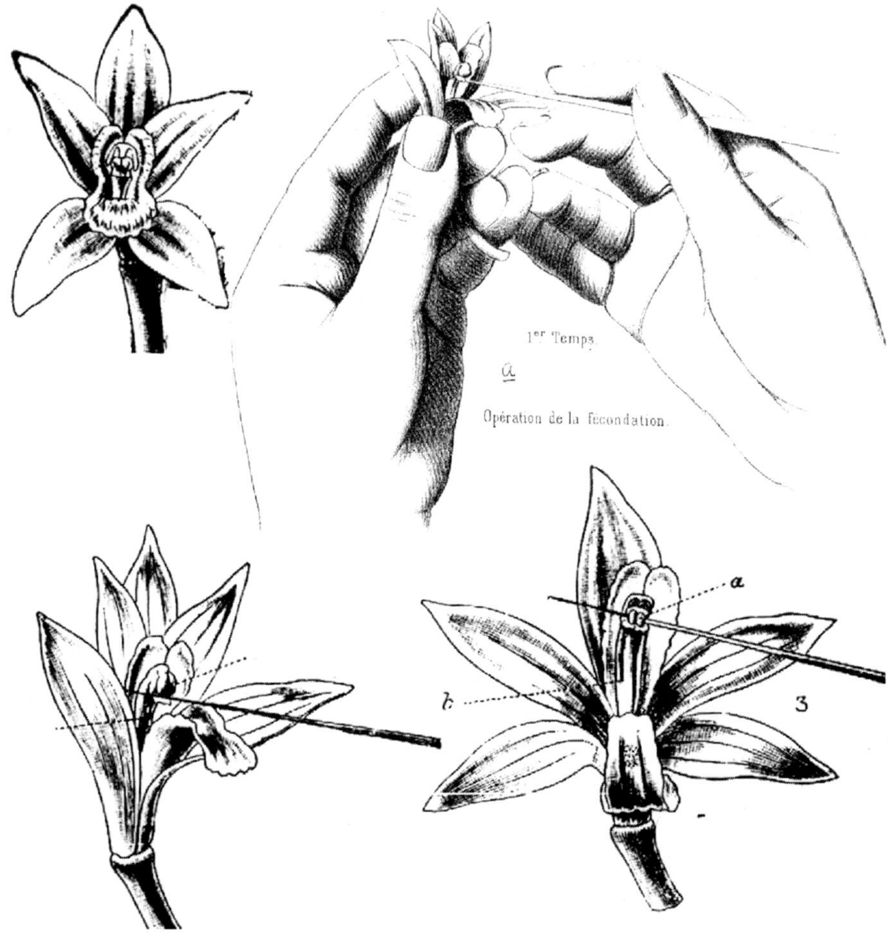

Fig. 7-6. Hand pollination of *Vanilla* (Delteil, 1874).

He married Marie Verrasselt in 1833. They had five children, one of whom Edouard Morren (Fig. 7-7G) followed his father as Professor of Botany at the University of Liége (Morren, 1856; Lawalrée, 1974; Anonymous, no date-f). Charles Morren was appointed as extraordinary professor (roughly equivalent to Associate Professor in the American system) at the University of Liége in 1835 and advanced to the ordinariat (i.e., Full Professor) in 1837. He was elected to the Royal Academy of Belgium in 1938.

Morren was very active and published a large number of scholarly papers (for lists see Morren, 1860; Anonymous, 1869). He was also the editor-in chief of l'*Horticulteur Belge* from 1833 until 1836, *Annales de la Société d'Agriculture et de Botanique de Gand* between 1845 and 1849, *Journal d'Agriculture Pratique* starting in 1848 and ending in 1855, and *Belgique Horticole* for the period of 1851–1855.

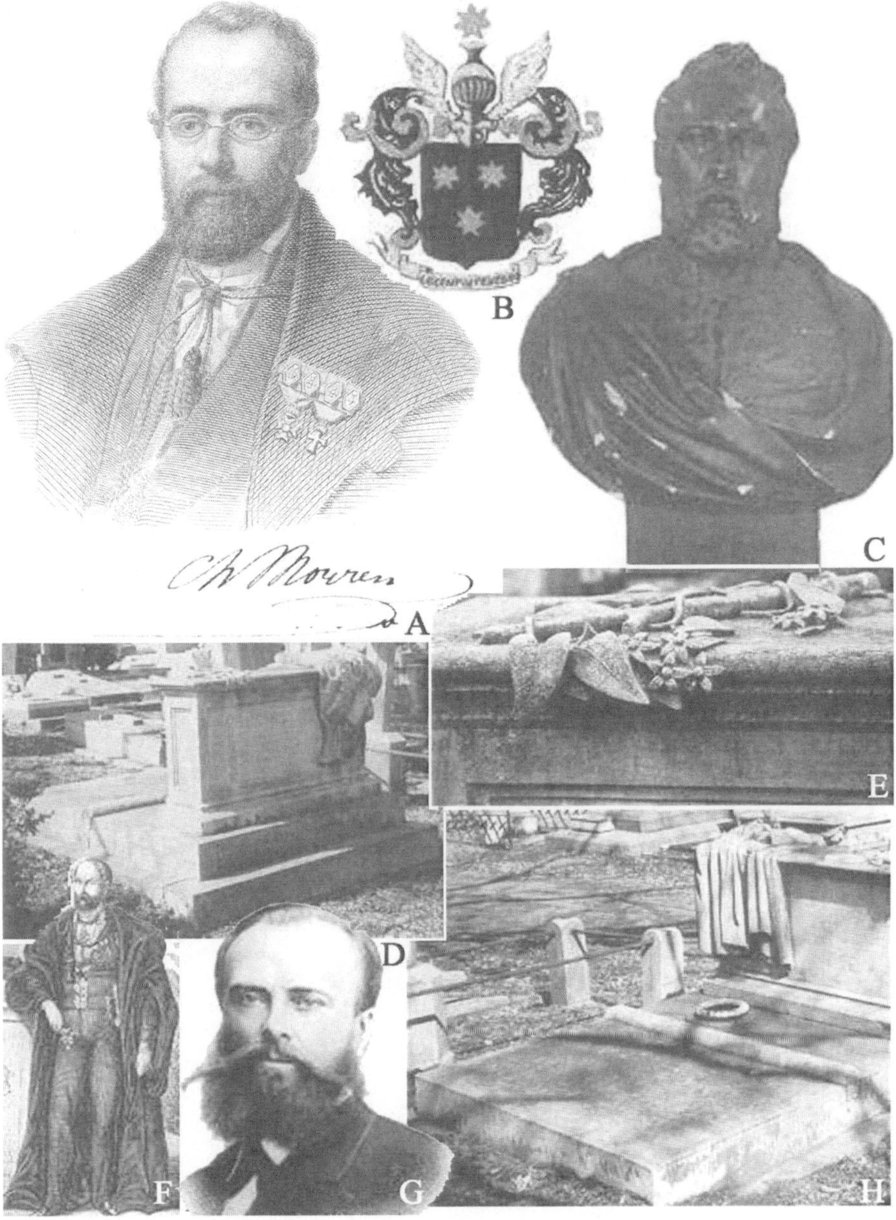

Fig. 7-7. The Morren family. **A**. Professor Charles Mores and his signature. **B**. The Morren family crest. **C**. Bust of Charles Morren. **D, E, H**. Grave of Charles and Edouard Morren. **F**. Charles Ignace François Morren, Professor Charles Morren's father. **G**. Professor Edouard Morren (**A-E, F**, No date-a; **G**, No date-f).

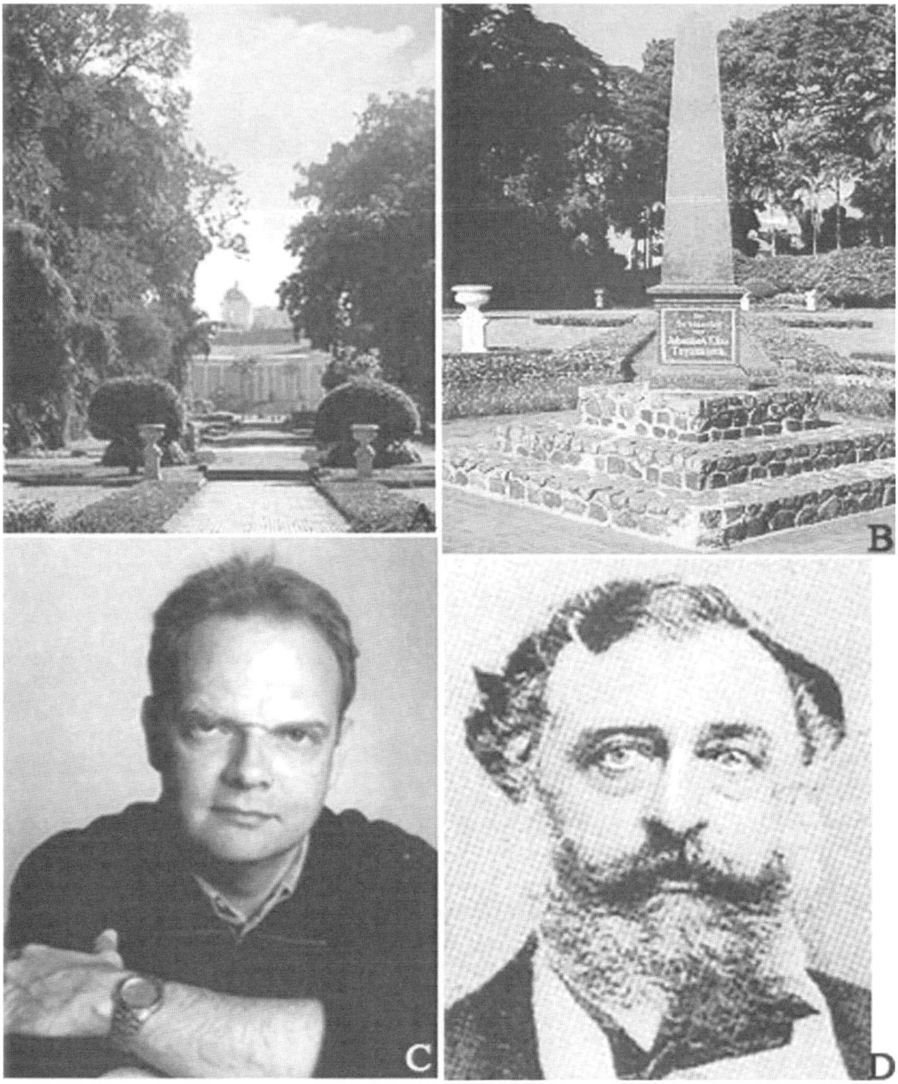

Fig. 7-8. Students of *Vanilla*. **A, B**. Johannes Elias Teijsman memorial garden at the Bogor Botanical Gardens in Indonesia. **C**. Tim Ecott. **D**. Johannes Elias Teijsman (**A, B**, photographs by Joseph Arditti; **C**, dust jacket of Ecott, 2004; **D**, Archives of the Bogor Botanical Garden, courtesy of the director, Dr. Irawati).

France

Joseph Neumann (1800–1858) at the Muséum National d'Histoire Naturelle Jardin des Plantes is credited with discovering the process in 1830 (Anonymous, 1855a; Delteil, 1874, 1884, 1897, 1902; Beck, 1912). He reported that of 11 flowers produced (apparently in 1837) at the Museum National d'Histoire Naturelle in Paris, three set fruit without pollination (Neumann, 1838). The plants flowered again in 1838 and 1839. They were hand-pollinated at that time and produced fruits (Neumann, 1838, 1841a, b; De Visiani, 1845). According to some reports, Neumann first pollinated *Vanilla* in France in 1830 (Delteil, 1884, 1902; van Gorkom, 1884). These reports have been questioned on two very reasonable grounds (Busse, 1899). First, this would have been a great discovery at the time and it is unlikely that Neumann would have waited 8 years to publish it. Secondly, the French were trying to pollinate *Vanilla* in Reunion, and it seems unlikely that it would have taken 10 years for this important discovery to reach their Indian Ocean dependency. Another point to consider is that Neumann seems to have made a habit of seeking to establish priority for discoveries others made and reported (Neumann, 1838, 1841a, 1841b, 1844), by describing earlier successes by himself which cannot be verified (for reviews see Arditti, 1984, 1992).

Indonesia

Vanilla was first taken to Java in1812 (de Vriese, 1856) and for a second time in 1840 by a Dr.[no initial given] Pierot (Teijsman, 1850, 1858). The plant flowered several years after being introduced, but set no fruits following the first bloom. This was attributed to the fact that the flowers dropped because of rain. However, it is also possible that they simply senesced and abscised because they were not pollinated. Interestingly, fruits of *Vanilla albida* Bl. and *Vanilla aphylla* Bl. from Java were described and illustrated in 1832 (Blume, 1832). However, in the case of *Vanilla planifolia* Andr. only the flowers were described and illustrated. This suggests that either the pollinators of *Vanilla albida* and *Vanilla aphylla,* even if present near the *Vanilla planifolia* plants, were not attracted to their flowers or could not pollinate them, or that in 1835 artificial pollination of *Vanilla planifolia* was unknown.

Subsequently, *Vanilla planifolia* flowers were pollinated successfully on 2 July 1852 by Graaf (initial not given) van den Bosch (Anonymous, 1852). They were also hand-pollinated by Johannes Elias Teijsman (1898–1882) and Simon Binnendijk (1821–1883) in Buitenzorg (Bogor) and produced fruits in or after 1850 (Teijsman, 1850, 1858; van Gorkom, 1884; Busse, 1899) and 1852 (van der Pant, 1852). These papers give no details regarding the method of hand-pollination or its origins. Therefore, it is not clear whether hand-pollination was discovered independently in Indonesia or if Morren's or another method discovered elsewhere was employed. However, according to another report (de Candolle, 1847), M. Falk, the Dutch ambassador in Bruxelles, provided details regarding the method used in Liege not long after a naturalist from the Royal Museum in Leyden named Pierot left for Java in 1840 for the purpose of introducing *Vanilla* there. If so, Pierot may have introduced Morren's method into Indonesia.

Reunion

Pierre-Henri Philibert, captain of the *Le Rhône* took *Vanill*a to Reunion (or Ile de Bourbon) from the French colony Cayenne in French Guiana in 1817, 1818 or 1819 (Focard, 1862; Legras, 1867–1868; Delteil, 1884, 1902; Bouriquet, 1954; Ecott, 2004). Captain Philibert (1774, Saint-Denis, Reunion-1824, France) was the fourth son of Pierre Nicholas Jacquin Philibert a marine commissioner. His wife was Augustine Thérèse Lheureux. In 1819 he became captain of the *Le Rhône* with the task of bringing useful plants from Cayenne. He was accompanied by a botanist named Perotet (Legras, 1867–1868 lists him as M. Perrotet, but the M. stands for "Monsieur" which is the French word for Mister and is not an initial). And, according to the *Feuille Habdomadaire* of 7 July 1819 *Vanilla* was one of the most interesting plants Philibert brought from Cayenne to Reunion (Legras, 1867–1868). The plants were probably *Vanilla pompona* (Ecott, 2004). Philibert and Perotet went on a second mission to find *Vanilla* the following year, this time to the Philippines (Ecott, 2004). They found *Vanilla* vines "… in the middle of a virgin forest, … climbing among the branches of an enormous bamboo thicket" (Ecott, 2004), collected many plants and brought them to Reunion. These plants may have been *Vanilla planifolia*. They were planted in two estates but all of them died (Ecott, 2004). *Vanilla planifolia* plants which finally thrived came to Reunion in 1822 from the *Jardin des Plantes* in Paris. These plants were descendants of those that were first grown in Britan (Ecott, 2004).

According to the British journalist Tim Ecott (Ecott, 2004) a planter named Ferréol Bellier-Beaumont had a 20-year-old *Vanilla* vine which failed to bear fruit in a plantation named Belle-vue near the city of Sainte-Suzanne. According to Ecott, early one morning in 1841, when Bellier-Beaumont was walking in his plantation with a young slave named Edmond (who was later given the surname Albius), he saw two fruits on the vine. Edmond told his master that he pollinated the flowers. Bellier-Beaumont did not believe Edmond at first. When a second flower showed signs of being pollinated and Edmond again claimed to have done the pollinating Bellier-Beaumont demanded to be shown how it was done. Edmond did. This story is romantic and there are suggestions that it may or may not be entirely accurate (Delteil, 1874; Busse, 1899; Arditti, 1971, 1984).

An early edition of a book on *Vanilla* by A. Delteil (page 12 of the 1874 edition according to Busse, 1899) reports that a Mr Perottet arrived in Reunion in 1839 and described Morren's pollination method which was published in Belgium in 1837 (Morren, 1837a) and in Paris in 1837 (Morren, 1837b) and 1838 (Morren, 1838). The account of Edmond Albius and his discovery appeared only in later editions of Delteil's book (Busse, 1899; Delteil, 1902), and there were questions whether Albius, an uneducated slave, could have discovered how to pollinate *Vanilla*. Consequently, some (Busse, 1899) found it easier to believe that Perottet introduced *Vanilla* pollination to Reunion. They may be wrong.

Tim Ecott devoted considerable time and effort to tracing Edmond's origins and history. He unearthed many details which lay forgotten in the island's *Archives Départementales* and reported them in his excellent book on *Vanilla* (Ecott, 2004) which is a fascinating read. The only logical conclusion that can be drawn from

Ecott's report and findings is that Edmond Albius discovered *Vanilla* pollination independently and on his own.

There is absolutely no reason to believe that Albius could not have figured out all by himself how to pollinate the flowers. And, since Ecott's search through the archives certainly seems to have been thorough his report should be accepted as definitive. If so there can be very little doubt that Albius discovered the pollination process on his own. On the other hand, in absolute terms and in the mind of scientists who are skeptical and questioning by nature a possibility still remains that Albius somehow became aware of Perotet's description of Morren's method without telling anyone and used or adapted it. The Albius story is fascinating, but the mystery may never be fully solved and some degree of uncertainty will remain despite Tim Ecott's interesting and largely convincing findings. However, even with that two facts are certain:

1. Edmond Albius made possible the *Vanilla* industry in Reunion and elsewhere.
2. The statement, "The reluctance of some experts to concede that lesser mortals were capable of research and hybridisation [sic, the correct spelling at least in the U. S. is hybridization] is not new. When Edmond Albion [sic, the correct spelling is Albius], a black slave on Reunion Island, developed his simple but very effective way of pollinating the Vanilla [sic, it should have been *Vanilla*] orchid in 1841, the experts were incredulous…" (Wright, 2000), bears no relation to reality and is only indicative of an agenda driven anti-expert, combative, chip-on-the-shoulder attitude by its ill informed author. Experts were not "incredulous." They only sought proof, as they always do, that a report is accurate and based on facts. Once proof became available the story was accepted even if questions may have remained in the minds of some (for reviews see Arditti, 1971, 1984, 1992; Ecott, 2004) simply because scientists tend be cautiously skeptical (good examples of this skepticism are the use, even at present, of terms like "cell theory" and "theory of evolution" notwithstanding the overwhelming evidence in favor of referring to each as an established fact or even a law). If scientists were not skeptical by nature the Piltdown man fraud would have been accepted as fact, Trofim Lysenko's charlatan genetics may have gone unchallenged, Paul Kaemmerer's rediscovery of Lamarckian inheritance with the midwife toad would not have been found to be fraudulent, John Heslop-Harrison's introduction of alien sedges into the Hebridean Island of Rum would have remained undiscovered and Martin Fleishmann's and Stanley Pons's cold fusion claims would not have been found as irreproducible and retracted (these are only a few examples, there are many more).

Padua Botanical Garden

A *Vanilla* plant flowered for the first time at Padua Botanical Garden in 1833. This plant continued to produce flowers over the years. Pollination was accomplished and a mature fruit was obtained in 1842 by Professor Roberto de Visiani (1800–1878), Dr. Giuseppe Clementi and Chief Gardener Carlo Caslini (Busse, 1899;

De Visiani, 1845). Despite being reported within a relatively short time their method did not attract much attention and was not used practically.

Guadeloupe

A pharmacist "*de 1re classe de la marine*" named Dupois discovered a method of *Vanilla* pollination on the French island of Guadeloupe (Delteil, 1884, 1902 does not give an initial). His method utilized scissors to perform a cut which allowed the pollen to drop into the stigma. He reported his method to the president of the Agricultural Chamber in Reunion. This method was considered to be cumbersome and of no practical value (Delteil, 1884, 1902).

Final Thoughts

The question of who discovered hand-pollination of *Vanilla* may never be answered to everyone's satisfaction and competing claims could be debated for a long time. But, this does not really matter. The possibility that several persons in different parts of the world discovered the process independently of each other is very real and not hard to accept. Regardless of who made the discovery there can be no question that Albius's method made the *Vanilla* industry possible in Reunion and other French dependencies or Francophone islands like Madagascar, Seychelles, and Tahiti (still the major sources of vanilla). Interestingly, vanilla production never became equally important in British dependencies and colonies and English speaking countries, despite several efforts to establish it. This is an intriguing question which may well be more difficult to answer than who discovered hand-pollination of *Vanilla* flowers.

Acknowledgments We thanks Joseph Beajean, Chief Technician, Botanical Garden, Bruxelles; Dr. Vincent Demoulin, Chargé de Cours, Department of Botany, University of Liege; Hans Janssens, Langelede 21, B-9185 Wachtebeke, Belgium and André Lawarée, Chef de Département, Jardin Botanique National, Bruxelles for copies of old and rare illustrations, papers, photographs and translations. Also we thank Jean Miller (formerly with the University of California, Irvine library and now retired) for obtaining rare and hard to find papers through interlibrary loan and Kathryn Kjaer for maintaining Joseph Arditti's current publications notifications.

Dedication

To the memory of my late friend of 30 years, Soediono (1931–2006) of Jakarta and Solo, Indonesia and Horizon Towers West in Singapore. He (an Indonesian who used only one name as is common in Indonesia) liked to describe himself as a country boy (which suited me just fine because I think of myself as a peasant), but was never out of place in even the most sophisticated establishments, among the superb orchids he grew with his wife, Noes and in the modern laboratory he designed, built and operated with her.– Joseph Arditti

Literature Cited

Some of the entries are annotated. Web sites may be altered, change names and even go off-line in time. Those cited here were consulted between 2005 and 2007.

Anonymous. 1852. Kungsmatige bevfruchting de vanille te Buitenzorg. *Natuurkundig Tijdschrift voor Nederlandsch Indie* 3: 484–485.

Anonymous. 1855a. Fécondation artificielle des orchidées. *L'Illustration Horticole* 2: 43–44. The editor of this journal at the time was Ch[arles] Lemaire. In those days editors wrote unsigned articles for their journals.

Anonymous. 1855b. Possibilité de la culture de vanillier en Europe. *L'Illustration Horticole* 2: 45–48. The editor of this journal at the time was Ch[arles] Lemaire. In those days editors wrote unsigned articles for their journals.

Anonymous. 1869. Morren (Charles-François-Antoine). *Liber Memorialis l'Universite de Liége*: 446–480.

Anonymous. 1894. *Vanilla* at Fiji. *Bulletin of Miscellaneous Information Royal Botanic Gardens, Kew* 103: 208–211.

Anonymous. No date-a. Morren, Charles (1807–1858). http://users.evl.net/~gpmoran/chasmorren.htm. Downloaded on 30 September 2006.

Anonymous. No date-b. *Vanilla*, the spice that is not a spice. http://www.botgard.ucla.edu/html/botanytextbooks/economicbotany/Vanilla/index.html. Downloaded on 30 September 2006.

Anonymous. No date-c. Charles Morren. http://fr.wikipedia.org/wiki/Charles_Morren. Downloaded on 30 September 2006.

Anonymous No date-d. Morren (Charles François Antoine). http://www.csmovisions.com/Morren.com. Downloaded on 30 September 2006.

Anonymous. No date-e. C'est le Liégeois Charles Morren qui a inventé en 1838 le moyen de dultiver la vanille! Un descendant du vanillier qui lui a permis de faire cette découverte vit toujours dans les serres de Liège. Two sides single page note published by *Comité de défense des Serres et du Jardin Botanique de Liège, 3, rue Fusch 4000 Liège*.

Anonymous. Nodate-f. Morren Edouard (1833–1886). http://users.evl.net/~gpmoran/EdouardMorren.htm.

Arditti, J. 1971. Vanilla: an historical vignette. *American Orchid Society Bulletin* 40. 610–613.

Arditti, J. 1984. An history of orchid hybridization, seed germination and tissue culture. *Botanical Journal of the Linnaean Society* 89: 359–381.

Arditti, J. 1992. *Fundamentals of orchid biology*. Wiley, New York.

Beck, G. 1912. Die Futterschuppen der Blüten von *Vanilla planifolia* Andr. *Mathematisch-Naturwissenschaftliche Klasse Akademie der Wissenschaften, Wien* 121: 509–524.

Berlése [no initials]. 1837. Note sur la première fructification du Vanillier en Europe. *Annales de la Société Royale d'Horticulture de Paris* 20: 331–334. Berlése is describe as "l'abbé" who presented the note on behalf of Charles Morren whose name appears at the end of the note.

Blume, C. L. 1832. *Vanilla*. *Rumphia* 1: 196–199, figs. 67, 68.

Bouriquet, G. (ed.) 1954. *Le vanillier et la vanille dans le monde*. Editions Paul Lechevalier, Paris.

Busse, W. 1899. Vanille. *Arbeiten aus dem Kaiserlichen Gesundheitsamte* 15: 1–113.

Childers, N.F. and H.R. Cibes. 1948. *Vanilla cultire in Puerto Rico*. Circular No. 28. Federal Experiment Station in Puerto Rico, U.S. Department of Agriculture, Mayaguez, Puerto Rico.

Childers, N. F., H. R. Cibes, and E. H. Medina. 1959. *Vanilla* – the orchid of commerce. In: C. L. Withner (ed.), *The orchids, a scientific survey*. The Ronald Press, New York, pp. 477–508.
Correll, D. S. 1953. *Vanilla*–its botany, history, cultivation and economic import. *Economic Botany* 7: 291–358.
Crépin, F. 1899. Morren (Charles-François-Antoine). *Biographie Nationale* (Belgium) 15: 275–279.
de Candole, A. 1847. Sur le musée botanique de M. Benjamin Deleseert de Paris, et surle martyreloge contemporain de la botanique et de l'horticulture. *Anales de la Société Royale de Agriculture et Botanique de Gand. Journal du Horticulture et Sciences Accessories* 3: 141–151.
Dclteil, A. 1874. *Étude sur la vanille*. Challamerl, Ainé, Libraire-Éditeur, Paris.
Delteil, A. 1884. *La vanille, sa culture et sa préparation*. Challamel, Ainé, Éditeur, Librairie Algériene et Coloniale, Paris.
Delteil, A. 1897. *La vanille, sa culture & sa préparation*, 4th ed. Augustin Challamel, Éditeur, Librairie Maritime et Coloniale, Paris.
Delteil, A. 1902. *La vanille, sa culture & sa préparation*, 5th ed. Augustin Challamel, Éditeur, Librairie Maritime et Coloniale, Paris.
De Visiani, R. 1845. Del metodo e delle avvertenz che si usano nell'orto botanico di Padova per la cultura, fecondazione e fruttificazione della vaniglia. *Memorie dell' I. R. Instituto Veneto di Scienze, Lettere ed Arti* 2: 3–18.
De Vriese, W. H. 1856. *De vanielje*. A. W. Sythoff, Leyden.
Ecott, T. 2004. *Travels in search of the ice cream orchid*, Vanilla. Grove Press, New York.
Focard, V. 1862. Introduction et fécondation du vanillier. *Bulletin de la Société des Sciences et Arts de l'Isle de la Réunion* [no volume number]: 223–235.
Hagsater, E., M. Soto, G. Salazar, R. Jiménez, M. López, and R. Dressler. 2005. Productos Phamacèuticos, S. A. De CV, 18 Lago Tangañica, 11520 Mexico City (www.chonoìn.com.mx).
Hart, J. H. 1894. *Vanilla planifolia. Bulletin of miscellaneous information*, Botanical Department, Trinidad and Tobago, pp. 240–241. (The author's name is given at the bottom of the first page only as J. H. H.)
Hemus, H. H. 1940. *Vanilla. Agricultural Journal* 11: 22–24. This journal was published by the Department of Agriculture, Colony of Fiji, Suva, Fiji.
H. W. J. 1940. Notes on curing Tahiti *Vanilla* beans. *Agricultural Journal Department of Agriculture, Fiji* 11: 22–23.
Jacquet, P. 1994. History of orchids in Europe, from antoquity to the 17th century. In: J. Arditti (ed.), *Orchid biology, reviews and perspectives*, vol. VI. Wiley, New York, pp. 33–102.
Lawalrée, A. 1974. La création d'*Oncidium limminghei* E. Morren. *Lejeunia Rev. Bot.* 72: 1–4.
Lawler, L. J. 1984. Ethnobotany of the Orchidaceae. In: J. Arditti (ed.), *Orchid biology, reviews and perspectives*, vol. III. Cornell University Press, Ithaca, NY, pp. 27–149.
Legras, A. 1867–1868. Philibert, introducteur de la vanille a Bourbon. *Bulletin de la Société des Sciences et Arts de l'Ile de la Reunion* [no volume number]: 201–224.
Meade, R. H., J. B. Thurston, D. Morris, and H. Arnold. 1894. *Vanilla* in Fiji. *Bulletin of Miscellaneous Information Royal Botanic Gardens, Kew* 7: 208–211.
Mn. 1845a. De la fécondation naturelle et artificialle des végétaux et de l'hybridation. *Anales de la Société Royale de Agriculture et Botanique de Gand, Journal du Horticulture et Sciences Accessories* 1: 235–239.
Mn. 1845b. Recherches et réflexions sur l'hybridation ches le acotylédones et les monocotylédones. *Anales de la Société Royale de Agriculture et Botanique de Gand, Journal du Horticulture et Sciences Accessories* 1: 323–332, 412–419.
Morren, [no initials given]. 1837a. Notice sur la vanille indigine et al fructification. *Bulletin de la Academie des Sciences, des Lettres et des Beaux-Arts de Belgique* 4: 225–237.
Morren, C. 1837b. Note sur le premiére fructification du vanillier en Europe. *Annales de le Société Royale d'Horticulture de Paris* 20: 331–334.
Morren, C. 1837c. Première notice sur la vanille indigène, don't il a réussi à produire la fructification. *Bull. Acad. Roy. Sci. Lett. Beaux Artes Belgique* Ser I 4: 225–237.
Morren, C. 1838. Sur la fructification de la vanille obtenue au moyen de la fécondation artificielle. *Comptes Rendus Hebdomadaires Academie des Sciences, Paris* 6: 489–492.

Morren, C. 1838–1839. Fructification de la vanille obtenue au moyen de la fécondation artificielle. *Agric. Pratique* Ser I 2: 114–116.
Morren, C. 1839a. On the production of vanilla in Europe. *Annals of Natural History or Magazine of Zoology, Botany and Geology* 3: 1–9.
Morren, C. 1839b. Déplacement et bouturage gigantesque d'un dragonier. *Annnales de la Société Royale d'Agriculture et de Botanique de Gand* 5: 234–237, plate 235.
Morren C. 1848. *Dossinia marmorata* Morr. *Annnales de la Société Royale d'Agriculture et de Botanique de Gand* 4: 170–172.
Morren, C. 1850. Memorandum sur la vanille, son histoire et sa culture. *Bull. Acad. Roy. Sci. Lett. Beaux Artes Belgique* 17: 108–133.
Morren, C. F. A. 1826. Disquisito de orchide latifolia. *Annales Academiæ Gandavensis* [no volume number]:1–92, plates I–VI.
Morren, C. F. A. 1829. Verhandelingen over de ware wijze, waarop de voortgang derstandelkruiden (orchides) met tweeknollige wortels plaats heeft, en bepoling van den natuurlijken vorm, dien zij zauden aannemen, Indien zijniet aan ein standvastige miswassing (avortement) onderwerpen waren; voorafgegaan van eenije aanmerkingen over de waardij van de kenmerk, genomen uit de verandering van plaats, tot onderscheiding van de beide rijken der levende wezens. *Bijdragen tot de Natuurkundige Wetenschappen* 4: 358–384.
Morren, C. F. A. 1927. Quœritur Orchidis latifoliœ descriptio botanica et anatomica. *Annales Academæ Gandavensis* [volume number not listed in the cover page]: 1–92, plates I–VI.
Morren, E. 1856. Histoire des plantes utiles. Histoire de la vanille. *La Beklgique Horticole, Journal des Jardins, des Serres et des Vergers* 6: 313–374, plate 76.
Morren, E. 1860. *Charles Morren sa vie et ses œuvres*, 2nd ed. Impremerie et Lithographie de C. Annoot-Braeckman, Gand, Belgium.
Narodny, L. H. 1947. *Vanilla*-growing on Dominica. *Agricultural Journal* 18: 88–89. This journal was published by the Department of Agriculture, Colony of Fiji, Suva, Fiji.
Neumann. 1838. Vanille. *Annales de Flore et de Pomone* 6: 316–320.
Neumann. 1841a. Vanille a feuiles planes, *Vanilla planifolia. Hervier Generale de l'Amateur* Série 2, 2: 1–5.
Neumann. 1841a. Vanille a feuiles planes, *Vanilla planifolia. L'Horticulteur Universel* Série 2, 2: 169–175.
Neumann. 1844. Nota. *Revue Horticole* 6: 38.
Nolan, L. C. 1942. The story of *Vanilla. Agriculture in the Americas* 2: 32–34.
Poiteau, [no initials given]. 1838. Note sur la vanille. *Annales de la Société Royale d'Horticulture de Paris* 23: 99–102.
Pridgeon, A. M., P. J. Cribb, M. W. Chase, and F. N. Rasmussen. 2003. *Genera Orchidacearum*, Vol. 3. *Orchidoideae (Part two) Vanilloideae*. Oxford University Press, Oxford.
Rolfe, R. A. 1895. Vanillas of commerce. *Bulletin of Miscellaneous Information Royal Botanic Gardens, Kew* 104: 169–178.
Rolfe, R. A. 1896. A revision of the genus *Vanilla. Journal of the Linnaean Society (Botany)* 32: 439–478.
Teijsman, J. E. 1850. Lands Plantentuin te Buitenzorg. *Natuurkundig Tijdschrift vor Neerlandsche Indie* 1: 431–440.
Teijsman, J. E. 1858. *Verslag ontent den staat vant's Lands Plantentuin in het Jaar 1850. Versllag van het beheer en den Staat der Nederlandsche besittingen en kolonien in Oest and West Indie enter kust von Guinea over 1850*. Bejlage A. Verschenen, Utrecht, pp. 93–97.
van der Pant, F. D. S. 1852. Kunsmatige bevfruchtung der vanille te Buitenzorg. *Natuurkundig Tijdschrift voor Nederlandsch Indie* 3: 486–487.
van Gorkom, K. W. 1884. *De Oost-Indische cultures, in betrekking tot handel en nijverhed.* J. H. De Bussy, Amsterdam.
Wiard, L. A. 1987. *An introduction to the orchids of Mexico*. Comstock Publishing Associates, Cornell University Press, Ithaca, NY.
Wigman, H. J. 1931.*Vanilla* Sw. *De Indische Culturen (Teysmannia)* 4: 127–132.
Wright, N. H. 2000. The origins of *Vanda* Miss Joaquim. *Malayan Orchid Review* 34: 70–73.

VIROLOGY

8

Molecular Biology of Two Orchid Infecting Viruses: Cymbidium Mosaic Potexvirus and Odontoglossum Ringspot Tobamovirus

PRABHA AJJIKUTTIRA AND SEK-MAN WONG

Contents

Distribution ... 252
Economic Significance and Incidence of CymMV and ORSV ... 252
Host Range and Symptomatology ... 252
Mode of Transmission ... 253
Molecular Structure and Composition ... 254
Important Molecular Aspects of CymMV and ORSV ... 257
Synthesis of Full-Length cDNA Clones ... 257
Molecular Methods Applied to the Detection of CymMV and ORSV ... 257
Studies of Mutations in the RdRp ... 258
Sequence Variability in the CP Genes ... 259
Regeneration of Transgenic Orchids ... 261
Synergism ... 262
Complementation of MP and/or CP genes ... 265
Future Challenges ... 268
Literature Cited ... 270

Kull, Arditti and Wong (eds.), *Orchid Biology: Reviews and Perspectives, X*,

Distribution

Cymbidium mosaic virus (CymMV) and *Odontoglossum ringspot virus* (ORSV)

Economic Significance and Incidence of CymMV and ORSV

Many members of the orchid family produce flowers in diverse shapes and exotic colours. These valuable flowers are a major product of the cut flower industry in Thailand, Singapore and other South-east Asian countries. The world market for tropical orchid cut flowers is estimated to be worth about S$180 million. Singapore is the second largest exporter of tropical orchid cut flowers (Yearbook of Statistics, Singapore, 1996). About 80 different hybrids of the major genera of orchids are cultivated in Singapore.

The orchid industry is affected by reduction in flower yield and quality caused by various pests and diseases. Of more than 50 viruses infecting orchids, *Cymbidium mosaic virus* (CymMV) and *Odontoglossum ringspot virus* (ORSV) are reported to be the most prevalent and economically important (Zettler et al., 1990). Both these viruses have been known for more than 50 years (Jensen and Gold, 1951). CymMV was first found in the *Cymbidium* species of orchids (Jensen, 1950), and was also known as orchid mosaic virus or *Cymbidium black streak virus* (Francki, 1970). ORSV was first found infecting *Odontoglossum grande* (Jensen and Gold, 1951), and was known as the orchid strain of *Tobacco mosaic virus* (TMV) (Kado, van Regenmortel, and Knight, 1968). Both these viruses have attained a world-wide distribution. In Singapore, the occurrence of CymMV infection in orchids is higher than that of ORSV (Wong et al., 1994). Prevalence of these two orchid viruses results in significant economic losses to the orchid industry caused by stunted growth and reduction in flower size and quality. Studies of these two viruses at the molecular level will help alleviate the impact of these viruses on the orchid industry.

Host Range and Symptomatology

The natural hosts for CymMV and ORSV are the monocotyledonous orchids. Among the orchids, some of the cultivars susceptible to CymMV are those belonging to the species of *Cymbidium*, *Cattleya*, *Epidendrum*, *Laelia*, *Laeliocattleya*, *Oncidium*, *Phalaenopsis*, *Vanda*, *Vanilla fragrans* and *Zygopetalum*. CymMV causes sunken chlorotic or necrotic patches on the leaves. In infected plants, the flowers become deformed and exhibit colour breaking symptoms. *Cymbidium* species (sp.), *Datura stramonium* and *Nicotiana* sp. are commonly used to propagate CymMV. Some orchids infected by ORSV are *Odontoglossum grande*, *Cymbidium*, *Cattleya*, *Dendrobium*, *Epidendrum* and *Zygopetalum*. ORSV also infects *Beta vulgaris*, *Cassia occidentalis*, *Chenopodium amaranticolor*, *C. quinoa*, *Datura stramonium*, *Gomphrena globosa* and *Nicotiana* sp. ORSV causes mottles, streaks,

Table 8-1. Species of plants susceptible to CymMV and ORSV infection.

CymMV	ORSV
Cassia occidentalis	*Beta vulgaris*
Cassia tora	*Cassia occidentalis*
Cattleya	*Chenopodium amaranticolor*
Chenopodium amaranticolor	*Chenopodium quinoa*
Chenopodium quinoa	*Cymbidium alexanderi*
Cucumis sativus	*Gomphrena globosa*
Cymbidium	*Nicotiana clevelandii*
Datura stramonium	*Nicotiana glutinosa*
Epidendrum	*Nicotiana benthamiana*
Gomphrena globosa	*Nicotiana tabacum*
Laelia	*Odontoglossum grande*
Laeliocattleya	*Tetragonia tetragonioides*
Nicotiana benthamiana	*Zinnia elegans*
Oncidium	
Oryza sativa	
Phalaenopsis	
Tropaeolum majus	
Vanda	
Vanda fragrans	
Zinnia elegans	
Zygopetalum	

stripes, mosaics or ringspots on the leaves. Infected flowers show ringspots and colour breaking. However, these viruses can infect plants without showing obvious foliar and floral symptoms.

In addition to orchids and other genera mentioned above, CymMV and ORSV are able to systemically infect a number of other plant species. Some of the plants they infect are indicated in Table 8-1. Of these, a common systemic host is the Solanaceous plant, *Nicotiana benthamiana*. Intermittent white lines on the leaves are typical symptoms of CymMV infection in these plants. Mild mosaics on leaves and distortion of emerging leaves are usual symptoms of ORSV in *N. benthamiana*.

Mode of Transmission

CymMV and ORSV are transmitted mechanically by inoculation of infected sap and contaminated cutting tools, equipment and potting media. These relatively heat stable viruses are able to retain infectivity for long periods in plant sap (Francki, 1970; Wisler, Zettler, and Sheehan, 1986). They are not transmitted by vectors or seeds (Namba and Iishi, 1971). CymMV and ORSV may be transmitted by contaminated pollen (Hamilton and Valentine, 1984; Hu et al., 1994).

Molecular Structure and Composition

CymMV belongs to the potexvirus group of viruses. *Potato virus X* (PVX) is the type member of this group. Viruses of this group are typically flexouous and filamentous, and the particles are 450–550 nm long (Francki, 1970). CymMV particles measure 480 nm in length and 13 nm in width. CymMV has a positive sense single-stranded (ss) RNA genome that is capped at the 5'-terminus and polyadenylated at the 3'-terminus. CymMV genomic RNA is 6227 nucleotides (nt) in length, excluding the poly (A) tail at the 3'-terminus (Wong et al., 1997). As is with all potexviruses, the CymMV genome consists of five open reading frames. They are the RNA-dependent RNA polymerase (RdRp) gene, the triple gene block (TGB) comprising of three overlapping genes and a coat/capsid protein (CP) gene (Fig. 8-1).

The RdRp gene (nt 73–4326) produces a 160 kilo Dalton (kDa) protein with three conserved domains: the methyltransferase domain (nt 73–975), the putative NTP-binding domain (nt 2049–2583) and the core binding domain (nt 3355–4101) (Wong et al., 1997). There is a six-nt intergenic region between the ORFs 1 and 2. ORFs 2, 3 and 4 overlap each other, are constituted by the TGB (nt 4333–5478) that is considered to be the movement protein (MP) gene (Fig. 8-3). The MP gene encodes three proteins of 26, 13 and 10 kDa respectively (Wong et al., 1997). TGB 1 (nt 4333–5022) contains the NTP-binding helicase motif (nt 4420–4446). A consensus sequence **PXXGDXXHXXPSGGXYXDGTKXXXY** is seen in the TGB 2 gene (nt 5115–5189). This sequence is also seen in other potexviruses and the carlaviruses. TGB 3 contains a high level of variability (Wong et al., 1997). The CP gene constitutes ORF 5 (nt 5480–6152) and is responsible for encapsidation of the viral RNA. N-terminus of the CP displays high level of variability and often results in low serological cross-reactivity in potexviruses (Chia, Chan, and Chua, 1992a).

ORSV belongs to the tobamovirus group. This virus produces particles that are rigid and rod-shaped, and are approximately 18 × 300 nm with a central hollow core that is 4 nm in diameter (Webster and Granoff, 1994). This virus has a ss positive sense RNA genome, which has a cap structure at the 5'-terminus but is not polyadenylated at the 3'-terminus. The Singapore isolate of ORSV (ORSV-S1) has a genomic RNA that is 6609 nt long (Chng et al., 1996) and encodes four genes: the 126/183 kDa RdRp at nt 63–3401/4901, the 33 kDa MP gene at 4,807–5,718 and

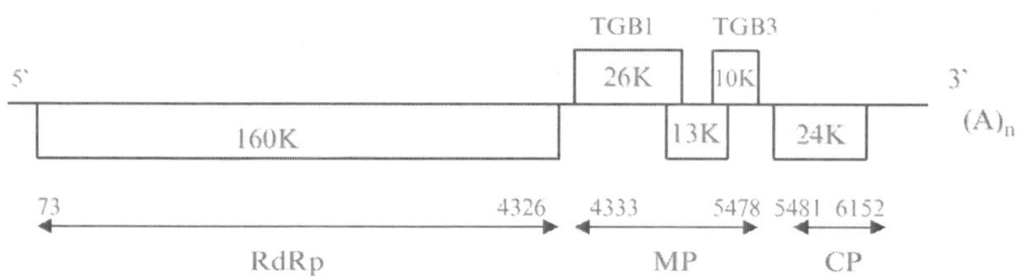

Fig. 8-1. Schematic representation of the genome of CymMV.

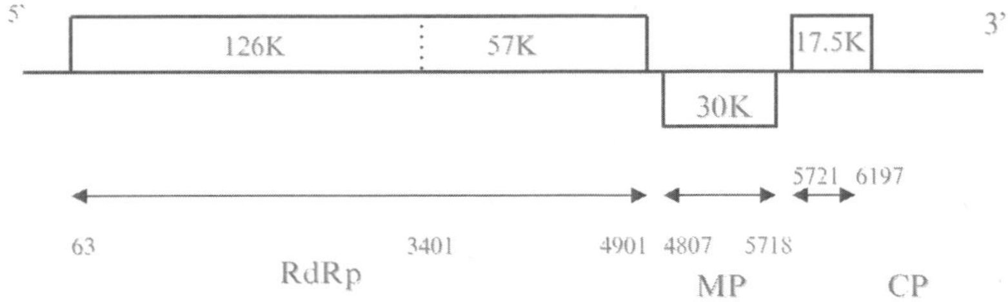

Fig. 8-2. Schematic representation of the genome of ORSV.

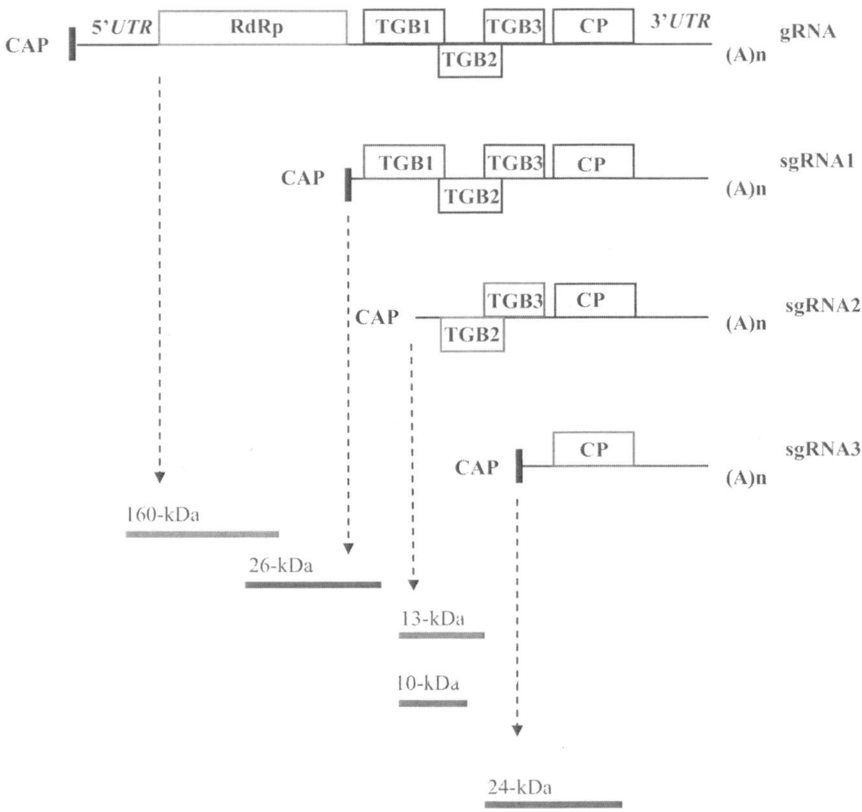

Fig. 8-3. Genome organization and translational strategy of CymMV. Open rectangles represent ORFs. *UTR* denotes untranslated region. Colored bars indicate proteins synthesized from the respective ORFs with their molecular weights. Drawing is not to scale.

the 18 kDa CP gene at nt 5721–6197 (Fig. 8-2). The 126 and 183 kDa proteins of the RdRp are translated from the same ORF, with the latter being produced by the readthrough of a leaky amber stop codon of the 126 kDa protein at nt 3399 (Fig. 8-4). Three functional domains have been identified in the RdRp of ORSV-S1. The first

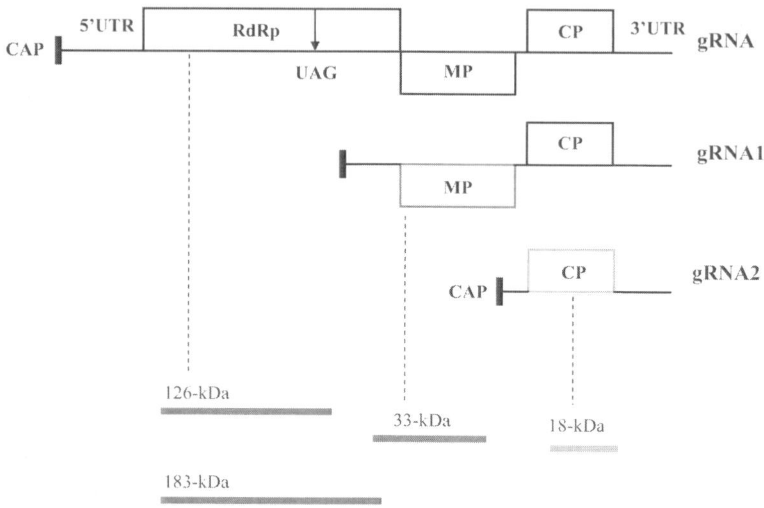

Fig. 8-4. Genome organization and replication strategy of ORSV. Open rectangles represent ORFs. UTR denotes untranslated regions. Colored bars represent proteins synthesized. Drawing is not to scale.

domain, the putative methyltransferase (MTase, Habili and Symons, 1989), has four distinct conserved motifs (I-IV) (Alonso et al., 1991) located at aa 72–287 (Chng et al., 1996) and may be responsible for the MTase activity that is required for cap formation. The second is the helicase domain at aa 820–1074 (Chng et al., 1996) with six conserved motifs (I-VI) (Habili and Symons, 1989; Evans, Haley, and Roth, 1985; Gorbalenya and Koonin, 1989). The third is the polymerase domain defined by a **GDD** consensus sequence and contains four conserved motifs (A-D) from aa 1372–1503 (Chng et al., 1996).

The 33 kDa MP gene (nt 4807–5718) overlaps with the 3'-terminus of the RdRp by 94 nt and is required for cell-to-cell movement of the virus (Ryu and Park, 1995a). A putative origin of assembly (Oa) of ORSV-S1 is located within the MP gene. The secondary structure of the Oa has been determined to possess two loops and a XXG repeat motif which are necessary for binding and initiation of assembly of the coat protein (Turner, Joyce, and Butler, 1988; Chng et al., 1996).

CP gene (18 kDa) of ORSV begins only two nt downstream of the MP and extends from nt 5721–6197. In ORSV-S1, this gene reveals three highly conserved RNA-binding motifs (Chng et al., 1996).

The 5'-untranslated region (UTR) is 62 nt long and contains three copies of ACAATTAC direct repeats and eight copies of CAA or ACA triplets in the Ω region. Containing 412 nt, the 3'-UTR is characterized by a tRNA-like structure and three consecutive homologous regions (Chng et al., 1996).

Research on CymMV and ORSV includes extensive studies on molecular biology of these two viruses. This chapter focuses on various previously published molecular biology work and where available, comparisons in similar contexts to other plant viruses.

Important Molecular Aspects of CymMV and ORSV

- Synthesis of full-length cDNA clones
- Molecualar methods applied to the detection of CymMV and ORSV
- Studies of mutations in the RdRp
- Sequence variability in the CP genes
- Regeneration of trangenic orchids harbouring the genes of CymMV and ORSV
- Complementation of cell-to-cell and long-distance movement
- Synergism between CymMV and ORSV

Synthesis of Full-Length cDNA Clones

Biologically active cDNA clones help us in the understanding of the molecular biology of RNA viruses, since modifications at the DNA level can be transcribed into viral RNA *in vitro* or *in vivo*. A complete viral sequence is required to produce biologically active cDNA clones. Usually, a cDNA is constructed by reverse transcribing the viral RNA into a ss DNA using a primer hybridizing to the 3' end of the viral sequence. The viral RNA is then eliminated, and the single-stranded DNA is converted to a double-stranded form by cDNA synthesis with a second primer that hybridizes to the 5' end of the viral RNA. The second primer usually carries the sequence for an RNA polymerase promoter fused to the viral sequence. A full-length viral cDNA clone may be inserted between the CaMV 35S promoter and the NOS gene (Weber, Haeckel, and Pfitzner, 1992).

A full-length cDNA clone encoding the genome of ORSV was syntheized and placed adjacent to a bacteriophage T7 RNA polymerase promoter (Yu and Wong, 1998a).

Capped *in vitro* transcripts proved highly infectious when inoculated on seedlings of *Oncidium* Gower Ramsey and *N. benthamiana*. Infectivity of a representative clone, designated pOT2 caused a disease phenotype identical to that produced by parental viral RNA. Full-length cDNA clones of CymMV were synthesized using the population cloning strategy (Yu and Wong, 1998b). Capped *in vitro* transcripts synthesized from the full-length cDNA fused to T7 RNA polymerase promoter were infectious.

Molecular Methods Applied to the Detection of CymMV and ORSV

Diagnosis of virus infections based on visual observation is unreliable as many infected orchids may not display symptoms. The application of serological methods like enzyme-linked immunosorbent assay (ELISA) is cumbersome and less sensitive when compared to molecular methods. The earliest molecular method applied to the detection of CymMV and ORSV was carried out using DNA probes and tissue prints (Chia et al., 1992b), the polymerase chain reaction (PCR) (Lim et al., 1993; Ryu and Park, 1995b; Barry et al., 1996). The method was used to detect CymMV and ORSV individually, and had to be therefore done twice to detect both CymMV and ORSV.

To be able to detect both viruses simultaneously, a single pair of common primers that could amplify both CymMV and ORSV viral sequences with equal efficiency was devised, based on overlapping alignments, secondary structures, possible priming sites and annealing temperatures (Seoh, Wong, and Zhang, 1998). These primers were used to detect CymMV and ORSV simultaneously by touch-down reverse-transcription PCR (TD/RT-PCR), making routine indexing of these two viruses more cost-effective in tissue culture before mass propagation. TD-PCR reduces the chances of mispriming and prevents the formation of spurious bands, therefore increasing the yield of the desired product since the annealing temperature starts several degrees above the calculated optimal annealing temperature, and is gradually lowered in subsequent cycles (Roux, 1995; Don et al., 1991).

Simultaneous quantitation of CymMV and ORSV was also carried out by TaqMan® real-time RT-PCR, a detection technique whose novelty lies in the fact that it combines RT-PCR with fluorescent detection. By this method, a quantity as low as 10^4 copies or 5 fg each of CymMV and ORSV could be detected simultaneously with either the RdRp or CP gene as the target (Eun, Seoh, and Wong, 2000). Taqman® probes were synthesized, each targeting the RdRp and CP genes of CymMV and ORSV. The reporter dye FAM (6-carboxyfluorescein) was used to label the 5' terminus of probes specific to CymMV, while TET (tetrachloro-6-carboxyfluorescein) was used for ORSV probes. In a more cumbersome method, digoxigenin (DIG)-labelled cRNA probes were used for the detection of CymMV and ORSV in orchids (Hu and Wong, 1998). This assay could detect 50 and 250 pg of purified CymMV and ORSV RNA, respectively. As little as 30 mg of *N. benthamiana* infected leaves was sufficient to provide detection. CymMV and ORSV were detected at 3,125 and 625 times dilution of leaf extracts, respectively. A novel fluorescence-based nucleic acid detection technique that consists of a probe termed a molecular beacon, was also used to detect CymMV and ORSV simultaneously (Tyagi and Kramer, 1996; Eun and Wong, 2000). Molecular beacons are ss nucleic acid molecules with a stem-loop conformation. Four molecular beacons specific to the RdRp and CP genes of CymMV and ORSV were designed. This technology was successfully applied to detect 0.5 ng of viral RNA of both orchid viruses simultaneously in 100 mg of coinfected *Oncidium* leaves.

Studies of Mutations in the RdRp

There is a high degree of similarity in the amino acid (aa) sequences of the putative RdRp of tobamoviruses and the putative RdRp of other animal and plant RNA viruses, with three distinct domains being conserved (Habili and Symons, 1989; Kramer and Argos, 1984; Rozanov, Koonin, and Gorbalenya, 1992).

Mutations in the RdRp domains that disable virus replication have been studied in *Brome mosaic virus* (BMV) (Kroner et al., 1989), PVX (Longstaff et al., 1993) and TMV (Ogawa et al., 1991). The helicase domain of the TMV RdRp has been shown to induce necrosis in N-gene-carrying tobacco even in the absence of virus replication (Abbink et al., 1998). Changes in different regions of the RdRp can alter symptom expression (Bao et al., 1996; Goregaoker, Lewandowski, and Culver, 2001; Shintaku et al., 1996), synthesis of subgenomic RNA (Watanabe et al., 1987), cell-to-cell movement (Hirashima and Watanabe, 2001) and host resistance responses (Hamamoto et al., 1997a, b).

In the ORSV, of six full-length clones of ORSV-S1 tested for infectivity during generation of full-length clones, one designated as pOT1 was not infectious (Yu, 1999). This mutant was found to contain a single nucleotide difference in the RdRp at nt 211 that led to an amino acid (aa) substitution at aa 50, resulting in the aa being changed from the normal phenylalanine in infectious clones to serine in the non-infectious clone. Nucleotide 211 is located outside the three conserved RdRp domains of the methyltransferase, helicase and polymerase.

The first 101 aa residues of the RdRp of 12 tobamoviruses were compared with those of ORSV-S1. It was found that the phenylalanine at aa 50 was conserved among all the viruses (Wang et al., 2004). Upon further characterization, it was found that it was not the nucleotide, but the amino acid change that was responsible for the inability of pOT1 to replicate. Minus-strand RNAs were not detected in mutants with a substitution for phenylalanine at aa 50. Identical results in TMV suggested that phenylalanine at aa 50 may play an important role in replication in all tobamoviruses. Complementation of a full-length mutant OT1 was demonstrated in a co-infected local lesion host, a systemic host and protoplasts by replication-competent mutants, suggesting that ORSV contains no RNA sequence inhibitory to replication *in trans*.

Sequence Variability in the CP Genes

Viruses retain their genetic structure during replication, altering only to a very minor degree, thus giving rise to variants. Lack of a proof-reading mechanism in DNA viruses cause the variation during replication (Steinhauer, Domingo, and Holland, 1992). Mutant forms may carry random mutations as compared to the parent strain, or they may be restricted to particular regions of the genome. These variations provide the basis for virus evolution. Viruses have extremely high evolutionary capacities, which have enabled them to parasitize all known groups of organisms and constantly broaden their host range. Mutation and recombination produces variants that can be distinguished on morphological, biological and serological lines. When the variants can be grouped based on their differences in epidemiology, serology and the host range, they are referred to as distinct strains (Matthews, 1991). A single virus particle gives rise to a local lesion, from which new mutants can appear. It is therefore highly likely that a virus culture actually consists of more than one strain. The concept is described by the term quasispecies (Domingo et al., 1995; Eigen, 1993, 1996; Holland, DeLaTorre. and Steinhauer, 1992; Moya and Garcia-Arenal, 1995) which was introduced to reflect the nature of RNA virus populations. The quasispecies concept predicts that a virus isolate rather than being a single RNA sequence is a mixture of mutant sequences that average around a consensus sequence. In any population, biological selection acts on the quasispecies to allow variants with improved fitness to arise, survive and dominate. These variations make it possible to classify the viruses. In some viruses such as *Banana bunchy top virus* (Wanitachacorn, Harding, and Dale, 2000), *Prunus necrotic ringspot virus* (Vaskova, Petrizk, and Karesova, 2000), *Ryegrass mosaic virus* (Webster et al., 1999) and *Citrus tristeza virus* (Ayllon et al., 2001), it has been possible to classify the isolates into different groups based on geographic origin and pathogenicity. In *Carnation mottle virus* (Canizares, Marcos, and Pallas, 2001) and *Alfalfa mosaic virus* (AlMV) (Parrella et al., 2000), definite co-variations at specific amino acids (aa) in the CP genes allowed the classification of geographically distinct virus isolates. Similarly, *Pea seed-borne mosaic virus* isolates from Pakistan could be placed into three different subgroups based on the differences in aa at 12 positions in the CP gene (Ali and Randles, 2001). Geographically distinct isolates of *Rice yellow mottle virus* showed differences in pathogenicity that could be associated with changes of aa in CP gene (Pinel et al., 2000).

There have been previous reports of variability studies in the sequences of potexviruses and tobamoviruses. In a comparative study of the CP sequences of eight potexviruses infecting different host ranges, an overall similarity in aa composition was observed, with variation of structurally important aa such as lysine, arginine, leucine and proline (Short and Davies, 1987). This could not lead to classification of the viruses. Members of the genus Tobamovirus have been shown to be genetically stable. A highly stable population maintaining its diversity through time has been reported in *Pepper mild mottle virus* (Rodriguez-Cerezo, Moya, and Garcia-Arenal, 1989). In a comparison of *Tobacco mild green mottle virus* (TMGMV) isolates infecting *Nicotiana glauca* from Australia, California and Spain, sequence similarity was reported, and no variable regions could be identified. (Fraile et al., 1996). An Australian isolate of TMGMV infecting *Nicotiana glauca* showed no increase in genetic diversity over a 90-year span (Fraile et al., 1997).

Genetic heterogeneity of the capsid protein genes of CymMV and ORSV and the possible occurrence of variability in isolates from different geographical locations was investigated in orchid samples of different genera infected with CymMV and ORSV (Ajjikuttira et al., 2002). Isolates were obtained from Korea, Singapore and Taiwan. All CymMV isolates from Korea and Singapore possessed a CP comprised of 223 aa in contrast to three previously published sequences (CyK3, SI and KI) that possessed a coat protein gene comprised of only 220 aa. These three sequences produced markedly different aa at the C-terminal of the CP. Among the Singapore isolates studied, 89.1–99.7% homology was exhibited at the nt level, and 93.2–100% homology was observed at the aa level. This meant that the isolates were very similar, and no particular pattern of variability was seen. Although co-variations were observed, it was not possible to separate the isolates studied into definitive groups due to the presence of additional random variations in the aa sequence throughout the CP gene of CymMV. The Singapore isolates showed a homology of 88.3–98.9% when compared with previously published aa sequences of CymMV infecting orchids in Thailand (Srifah, Loprasert, and Rungroj, 1996).

All the CP sequences of ORSV possessed a length of 158 aa, similar to that of previously published sequences. The N-terminal sequence of the CP was more conserved than the C-terminal. There were no distinct regions of variability in any of the sequences. Sequence identity matrices showed 97.4–99.3% homology between the Singapore isolates at the nt level and 96.2–99.3% at the aa level. At the nt level, the Korean isolates showed 96.6–99.7% homology. At the aa level, homology varied between 96.2–100%. The isolates from Taiwan showed 99.5–99.7% homology at the nt level and 98.7–99.3% homolo gy at the aa level. The percentage nt and aa sequence homology among all the isolates varied from 95.5–100% and 93–100%, respectively. These results indicated low genetic variability in the CP gene sequences of ORSV infecting orchids from different geographical areas. The isolates did not cluster according to hosts or geographic origin.

Regeneration of Transgenic Orchids

The success of biolistic techniques has made gene delivery into intact plant tissues a reliable process with many significant applications in plant biology. For success with this method, several parameters such as material and size of particles used in delivery, various means used to adsorb DNA to the particles, coating procedures and velocity of the particles need to be carefully considered. In addition, the efficiency of gene transfer is also associated with the target tissue. All these factors cumulatively make biolistic transformation cumbersome. Nevertheless, this method has been widely used to transform both monocotyledonous and dicotyledonous plant species. Transformation of orchid cultivars by this method has also been successful (Chia et al., 1994; Nan and Kuehnle, 1995; Kuehnle and Sugii, 1992; Yang et al., 1999).

Genetic transformation mediated by *Agrobacterium* has been successful among dicotyledonous plants and is gradually being extended successfully to monocotyledonous plants. Some economically important monocotyledonous species successfully transformed by this method are wheat (Cheng et al., 1997) and rice (Ranineri et al., 1990). Few reports are available on the successful *Agrobacterium*-mediated transformation of orchids. In the first of such reports, *Dendrobium* protocorms harbouring the reporter gene, GUS were produced (Nan and Kuehnle, 1998). Transformation of *Dendrobium* with the orchid DOH1 antisense gene was achieved by *Agrobacterium*-mediated transformation (Yu, Yang, and Goh, 2001).

Since this widely used gene transfer method could have a useful impact, we explored the possibility of producing transgenic *Dendrobium* Sonia harbouring the capsid protein genes of CymMV and ORSV. We followed the previously published protocol used for *Dendrobium* Madame Thong-In, but were largely unsuccessful in obtaining transformed lines. We attribute this to the continued necrosis of the cultured explants that led to bacterial infection that prevented the healthy growth of the explants. We therefore conclude that for *Dendrobium* Sonia, minimal necrosis of the explant must be achieved alongside maximal co-cultivation to improve the probability of generating successful transformants with this gene transfer method. Future genetic transformation studies by this method when successfully achieved in the orchid *Dendrobium* Sonia could have a tremendous economic potential.

Synergism

Multiple virus infections are commonly seen in the plant kingdom. Doubly or multiply infected plants show symptom intensification in the host plants. The phenomenon of severity of symptoms and higher amounts of virus accumulation of one or both viruses involved is referred to as synergism. Many synergistic interactions involving a potyvirus with other unrelated viruses have been described. Potyvirus associated synergisms include the *Cauliflower mosaic virus* (Khan and Demski, 1982), PVX (Rochow and Ross, 1955) and *Cowpea mosaic virus* (Anjos, Jarlfors, and Ghabrial, 1992). In these synergisms, the intensification of disease symptoms is due to the increased accumulation of the non-potyvirus component, with the level of the potyvirus remaining unchanged (Rochow and Ross, 1955; Calvery and Ghabrial, 1983; Goldberg and Brakke, 1987; Vance, 1991). Not all potyvirus related synergisms follow this pattern. In Peanut Mottle Virus, mixed infections with either *Tomato spotted wilt virus* (Hoffmann, Geske, and Moyer, 1998) or *Bean pod mottle virus* (Anjos et al., 1992), no synergism was observed. Enhanced potyviral accumulation was observed (Karyeija et al., 2000) while reduction of potyvirus has been reported (Poolpol and Inouye, 1986) in mixed infections with another virus. Of the potyviral synergisms, the PVX- *Potato virus Y* (PVY) synergism in tobacco has been well characterized (Rochow and Ross, 1955; Damirdagh and Ross, 1967; Goodman and Ross, 1974b; Vance et al., 1995). Doubly infected plants show severe vein clearing, necrosis of the first systemic leaf and increased accumulation of PVX. This synergism occurs by expression of the 5' proximal sequence encoding P1, helper component-proteinase and a fraction of P3 (the P1/HC-Pro sequence) of the potyviral genome (Vance et al., 1995). Mutational studies on the P1/HC-Pro sequence have shown that the amino-terminal 'zinc-finger' domain of HC-Pro is dispensable for induction of synergistic disease and transactivation of PVX multiplication, while regions within the central domain of HC-Pro are essential for both these responses (Shi et al., 1997). The central domain of HC-Pro mediates the suppression of posttranscriptional gene silencing (PTGS) (Anandalakshmi et al., 1998; Brigneti et al., 1998; Kasschau and Carrington, 1998). These results suggest that the two phenomena may be linked. The P1/HC-Pro sequence of potyviruses also enhances the pathogenicity and accumulation of TMV and CMV (Pruss et al., 1997). In a molecular characterization of the synergism between PVX and either PVY or TEV in *N. benthamiana* and *N. tabaccum*, neither PVX coat protein nor PVX (+) strand RNA accumulation increased in double infections although severe synergism was observed in symptom expression in *N. benthamiana* (Gonzalez-Jara et al., 2004). PVX CP levels in tobacco co-infected with PVY or *Tobacco etch virus* (TEV) were much higher than in plants infected with PVX alone. Tobacco plants co-infected with PVX and TEV did not develop synergistic symptoms as severe as the PVX and PVY combination, although a slight stunting was observed when compared with plants singly infected with either TEV or PVX. These findings therefore suggested that the relationship between viral titre enhancement and synergism in the PVX/potyvirus infection is host dependent. In the synergism between *Cucumber mosaic virus* (CMV) and the potyviruses, *Zucchini yellow mosaic virus* (ZYMV) and *Watermelon mosaic virus* (WMV), enhanced accumulation of CMV occurred (Wang et al., 2002). Co-infection of zucchini plants with the M strain of CMV (M-CMV) and the potyvirus *Zucchini yellow mosaic virus* strain A (ZYMV-A) allowed M-CMV to exhibit severe synergism in pathology (Choi et al., 2002). Infection of zucchini squash by M-CMV and an attenuated strain of ZYMV (ZYMV-AG) showed a milder symptom in pathology. Both ZYMV-A and ZYMV-AG facilitated the long-distance movement of M-CMV that is inhibited in some cultivars of zucchini squash. Synergistic interactions between a potyvirus and cucumovirus led to resistance breakage in double infections (Wang et al., 2004). Resistance to CMV in cucumber cv. Delila was broken by co-infection with the potyvirus ZYMV. The resistance breakage was accompanied by an increase in CMV RNA (+ and –) and capsid protein, with no increase in the level of accumulation of ZYMV. Resistance breakage was not accompanied by an increase in symptoms beyond those induced by ZYMV itself. Since increase of viral RNA and capsid protein accumulation were shown to occur without a corresponding increase in disease development, it has been suggested that virus accumulation and disease development are regulated by different host genes in CMV-resistant cucumber plants. In the synergistic interaction of *Maize cholorotic virus* (MCMV) and *Wheat streak mosaic virus* (WSMV), a potyvirus, the RNA concentrations of both the viruses were increased in mixed infections (Scheets, 1998). Synergism of geminiviruses infecting cassava cause symptoms more severe than in plants inoculated with either of *African cassava mosaic virus* or *East African cassava mosaic virus* alone (Fondong et al., 2000; Pita et al., 2001).

Synergism has been observed between potexvirus and tobamovirus in double infections (Goodman and Ross, 1974a; Taliansky et al., 1982a). Symptom severity in plants doubly infected with CymMV and ORSV has been reported (Lawson and Brannigan, 1986; Hadley, Arditti, and Arditti, 1987). Double infections of CymMV and ORSV resulted in severe mosaic symptoms with necrotic streaks (Pearson and Cole, 1991). These co-infections cause several diseases in orchids, showing more pronounced symptoms than single infections of either virus alone. Symptoms in orchids include mottling, colour breaking, ridging, thickening, curling and distortion of flowers, and abnormal growth and stunting of plants in *Cattleya*, *Cymbidium*, *Odontoglossum*, *Phalaenopsis* and *Oncidium*, and *Renanthera*. The molecular details of RNA accumulation in this double infection have not been determined. The increase in symptoms may be related to accumulation of either or both viruses.

In an orchid protoplast system, co-electroporation of CymMV and ORSV RNA resulted in enhancement of replication of both viruses when compared to singly electroporated protoplasts (Hu et al., 1998). This indicates that CymMV-ORSV synergism is caused by an enhancement of RNA replication.

In preliminary experiments to investigate this synergism, we merely mixed crude sap extracts of CymMV infected plants with those of ORSV infected plants and inoculated the mixture onto healthy *N. benthamiana* plants. Plants inoculated with crude sap extracts from single infections of CymMV or ORSV alone served as controls. It was observed that plants with mixed viral infections displayed intensified synergistic symptoms when compared to the milder symptoms seen in single infections of CymMV or ORSV alone. In the ORSV RNA accumulation experiment, the genomic RNA of ORSV was clearly detected in both single and double infections at 9 dpi. However, at 7 dpi, very small amounts of ORSV genomic RNA were detected in double infections. This result pointed to the fact that an acceleration of ORSV RNA accumulation was a likely feature of this synergism. A similar phenomenon was noticed in the synergism between *Panicum mosaic virus* (PMV) and *Satellite Panicum mosaic virus* (SPMV) where, in double infections of these viruses SPMV showed an acceleration of systemic transport than in single infections of SPMV alone (Scholthof, 1999). The study of RNA accumulation of CymMV in singly versus doubly infected tissues revealed that in both infections, the CymMV RNAs were detected *in vivo* at 12 dpi. Densitometric scans of the blots revealed that in single infections, CymMV genomic RNA was accumulated to a greater extent than in double infections on the first detection day itself (12 dpi). Subsequently, at 18 and 21 dpi the titer of CymMV genomic RNA in single infections were higher than in double infections. In virus preparations from equalized quantities of plant tissues subjected to transmission electron microscopy, there were far more ORSV viral particles in doubly infected preparations than in singly infected ones. Further, it became evident that the coat proteins of either CymMV or ORSV or both are involved in determining the synergism. In vitro transcripts of the infectious cDNA clone of ORSV (pOT2) were inoculated on transgenic plants carrying the CymMV CP transgene and observed against non-transgenic plants inoculated with the same *in vitro* transcripts. Transgenic ORSV CP plants inoculated with *in vitro* transcripts of the infectious cDNA clone of CymMV (p18Cy13) showed no symptoms of synergism. Transgenic CymMV CP plants inoculated with *in vitro* transcripts of pOT2 showed symptoms of synergism similar to those of double inoculations. Starting at 12 dpi, the uppermost leaves exhibited curling of the leaves, accompanied by dark-green islands that progressed over time to the lower leaves. In due course, the systemic leaves of the plants began showing intensified symptoms of synergism with dark-green islands and leaf shape distortion as observed on most of the systemic leaves. Hence the genomic RNA accumulation profiles of the CymMV CP transgenic plants that showed synergistic symptoms when inoculated with pOT2 were studied. The genomic RNA accumulation profile was similar to that of ORSV accumulation in double inoculations. More ORSV RNA was detected in CymMV CP transgenic plants than in non-transgenic plants inoculated with *in vitro* transcripts of pOT2. When plotted on a histogram, the average peak density of ORSV in test plants was very similar to the ORSV titer in doubly inoculated tissues. Since the values on the histogram were very similar, it is highly indicative that the CymMV CP plays a crucial role in the synergism, and could be responsible for the enhancement in the levels of ORSV RNA and the intensification of symptoms. In the synergism between PMV and SPMV, the SPMV CP was shown to be inducer of severe chlorosis characteristic of the synergism (Qiu and Scholthof, 2001). To understand plant viral synergisms, two mechanisms have been proposed (Goldberg and Brakke, 1987; Rochow and Ross, 1955). A perturbation of normal cellular regulation may cause a synergism by inducing a higher concentration of virus in each cell, or alternatively an increase in the number of infected cells may occur during the synergism due to the relaxation of cell-type specificity facilitating cell-to-cell movement. It would be of interest to assess the mechanism underlying the synergism between CymMV and ORSV.

Complementation of MP and/or CP genes

The structural organizations of CymMV and ORSV differ considerably, since they belong to unrelated taxonomic groups. The main distinction in the gene organization lies in the difference in the MP. In both CymMV and ORSV, the MP gene product performs the function of cell-to-cell movement of the virus, while the product of the CP gene allows for long-distance movement. While the ORSV MP is expressed from a single ORF, and produces a single product, the MP of CymMV produces three gene products expressed from the TGB. Despite structural differences in the organization of MPs, complementation of MP function has been noticed between unrelated groups of plant viruses (Atabekov and Taliansky, 1990; Ziegler-Graff, Guilford, and Baulcombe, 1991; Taliansky et al., 1993; Fuentes and Hamilton, 1991; Richins et al., 1993). The best studied examples of heterologous virus complementation are those involving the TMV MP in relation to other viral MPs. The MP of TMV is known to support the cell-to-cell movement of *Red clover necrotic mosaic virus* (RCNMV) (Giesman-Cookmeyer et al., 1995), *Cucumber mosaic virus* (CMV) (Cooper et al., 1996), *Barley stripe mosaic virus* (BSMV) (Solovyev et. al., 1996), PVX (Morozov et al., 1997) and *Bean necrotic yellow vein virus* (BNYVV) (Lauber et al., 1998). The functional equivalence of the MPs of TMV and RCNMV was studied using several approaches- creation of a chimeric virus in which the TMV MP gene was replaced by the RCNMV MP gene, complementation of movement-defective viruses by MP genes expressed in transgenic plants and helper virus complementation of movement-defective viruses. In all these experiments, the MPs of both TMV and RCNMV were able to provide cell-to-cell movement function to the heterologous movement-defective virus (Giesman-Cookmeyer et al., 1995). The movement protein of TMV supported the cell-to-cell spread, but not systemic transport of a movement defective CMV (Cooper et al., 1996). In a different set of experiments, it was later concluded that the TMV MP had the ability to support the cell-to-cell as well as long-distance movement of a movement defective CMV (Rao, Cooper, and Deom, 1998). In the same work, it was reported that neither the TMV MP nor RCNMV MP was able to rescue the defective cell-to-cell and long-distance movement of a movement-defective variant of BMV, but the RCNMV MP was able to complement the cell-to-cell, but not long-distance movement of *Cowpea chlorotic mottle virus* (CCMV). When the BSMV TGB coded MP was replaced with the 30-KDa MP of TMV, the hybrid virus was able to produce cell-to-cell infection in a host-dependent manner, but was however, unable to produce systemic infection (Solovyev et al., 1996). The cell-to-cell movement of a 25 kDa MP-defective full-length cloned PVX genome was restored by co-inoculation with 35S constructs containing the MP cDNAs of either of the tobamoviruses, TMV or *Tomato mosaic virus* (ToMV) (Morozov et al., 1997). The TMV MP could substitute for all the three TGB proteins of BNYVV, but the transcomplemented movement was less than that observed in the natural situation (Lauber et al., 1998).

The hordeivirus γb protein that is cysteine rich, has a putative zinc finger motif, possesses RNA-binding activity *in vitro*, and may influence hordeivirus gene expression, (Agranovsky et al., 1992; Donald and Jackson, 1994; Gustafson et al., 1987; Petty et al., 1990a; Petty, Donald, and Jackson, 1994) is dispensible for virus replication and affects infection phenotypes in barley (*Hordeum vulgare*) and *Chenopodium amaranticolor* (Petty et al., 1989; Petty, Edwards, and Jackson, 1990b). This protein is implicated in seed transmissibility of hordeiviruses (Edwards MC, 1995) and constitutes a virulence determinant (Donald and Jackson 1994; Petty et al., 1994). The potyviral HCpro is a protein known to suppress RNA silencing and stimulates viral genome amplification and long-distance movement (Kasschau and Carrington, 2001). Transgenically expressed PVA HCpro rescued defective long-distance movement of a γb gene deletion mutant (BSMVdel γb) in *N. benthamiana*. The data indicated that the long-distance movement functions of potyviral HCpro are specifically complementary to the hordeivirus γb protein and that the complementary functions can be provided in *trans* (Yelina et al., 2002).

Functional complementation of Potexvirus-Tobamovirus genes has been reported. Cell-to-cell movement of PVX is complemented efficiently by tobamoviruses (Taliansky, 1982a, b, c; Morozov et al., 1997; Atabekov et al., 1999). Cobombardment of plant tissues with MP deficient, GUS gene tagged PVX and cloned TMV MP gene showed that the TMV MP was functionally able to substitute the PVX MP (Morozov et al., 1997). The MP of *Sunn hemp mosaic virus* (SHMV) can substitute functionally for the PVX MP and CP (Atabekov et al., 1999).

Whilst most studies aim to dissect the movement protein complementarity between unrelated plant viruses, the coat protein genes have also been the focus of several complementation studies. In a majority of plant viruses, the movement of viral RNA between cells is supported by the movement protein(s), but in some viruses the coat protein is a primary determinant of cell-to-cell movement. Cell-to-cell movement is successfully accomplished by viruses when the movement proteins interact with the plasmodesmata and transfer the viral RNA into the neighbouring cells. For successful long-distance movement, the virus must move through the vascular system of the plant and into the uninoculated leaves and multiply, thus causing a systemic infection. Since functional coat proteins are essential for long-distance movement in several plant viruses, complementation studies focused on coat protein genes play a major role in understanding the long-distance movement function and possible enhanced movement abilities caused by complementation.

Groundnut rosette virus (GRV), like all other umbraviruses, does not code for a coat protein, but moves efficiently from cell to cell and long distance. In this virus, the protein encoded by ORF3 can functionally replace the coat protein of TMV for long-distance movement (Ryabov, Robinson, and Taliansky, 1999). A chimeric TMV with a replacement of the CP gene by GRV ORF3 was able to move rapidly through the phloem, and complement long-distance movement of another CP-deficient TMV derivative expressing the gene encoding the gene encoding the green fluorescent protein. It was therefore concluded that the GRV ORF3 protein represents a class of trans-acting long-distance RNA movement factors, and is a non-structural protein that can complement long-distance movement of an unrelated viral RNA.

A TMV-based vector Av., deficient in long-distance movement and limited to locally inoculated leaves because of the lack of native TMV coat protein was able to move systemically when *Alfalfa mosaic virus* (AlMV) coat protein was cloned into it (Spitsin et al., 1999). *N. benthamiana* and *N. tabacum* MD609 are both systemic hosts for TMV and AlMV, while *Spinacia oleracea* is systemically infected by only AlMV. The expression of AlMV CP in Av. directed by the subgenomic promoter of TMV CP permitted the systemic infection of *S. oleracea* with chimeric TMV, extending the host range of TMV. This report showed that complementation studies can lead to alteration of host range.

In transgenic plants expressing PVY CP, a PVX CP-deficient mutant (C-terminal truncation) was able to move cell-to-cell but not long-distance, suggesting that the CPs of potex- and potyviruses display complementarity in the viral movement process (Fedorkin et al., 2000). However, a deletion mutant of PVX CP was not able to move from cell to cell. Since trans-encapsidation of the PVX RNA by the potyviral CP was not observed here, the movement of the PVX CP by the PVY CP was thought to occur by a hitherto unexplained mechanism other than CP substitution. In transient co-expression experiments, the cell-to-cell movement of the PVX CP mutant lacking the C-terminal amino acid residues was rescued by potyvirus CPs and CPs of beet yellows closterovirus. These viruses could not however rescue the movement of a deletion mutant of PVX CP (Fedorkin et al., 2001). This indicated that a movement determinant within the PVX CP outside the C-terminal part could not be complemented by the heterologous CPs. These two mutants of PVX were also rescued by the CP of spherical cocksfoot mottle sobemovirus and TMV MP. The C-terminal mutant PVX CP could not be complemented by a TMV MP with an internal deletion, suggesting that the excluded region of the TMV MP (which includes a number of overlapping functional domains important for cell-to-cell transport) provides and activity complementing movement functions required by the C-terminal region of the PVX CP.

CMV and BMV require the CP and 3a movement protein for cell-to-cell movement, while CCMV does not. In bombardment-mediated transcomplementation studies, coexpression of CMV 3a and CP, but neither protein alone, complemented the defective movement of *Tomato mosaic virus* (ToMV) and PVX (Tamai et al., 2003).

Various approaches have been used to study the phenomenon of heterologous complementation of virus movement. The MP function was retained in a chimeric virus in which the native MP was replaced by the MP of a related virus (De Jong and Ahlquist, 1992; Solovycv et al., 1996). Co-bombardment of plant tissues with expression vectors carrying the movement-deficient virus and the MPs also produced complementation (Morozov et al., 1997). Co-inoculation of movement-deficient virus with the complementing virus (Taliansky et al., 1982a, b, c) or with replicons expressing the heterologous viral MP (Lauber et al., 1998) has been a successful strategy to study complementation between unrelated viruses.

How do the orchid-infecting CymMV and ORSV compare with the above situations? In contrast to most of the complementation cited above, CymMV and ORSV are known to naturally coinfect a common group of host plants- the monocotyledonous orchids. CymMV and ORSV show reciprocal complementation in cell-to-cell movement of movement protein genes (Ajjikuttira, Loh, and Wong, 2005). In the study which was carried out using the transgenic plant approach, complementation between the coat protein genes were also examined. ORSV CP could support the long-distance movement of CymMV RNA. Encapsidation of CymMV RNA with ORSV CP nor CymMV CP expression was detected, indicating this CP complementation was occurring by a mechnism other than encapsidation. Systemic movement of an ORSV CP-deficient mutant was not supported by CymMV CP, indicating that this phenomenon of CP complementation was not reciprocal. These studies confirmed that complementation of MPs and CPs of CymMV and ORSV facilitates movement of these viruses in plants.

Future Challenges

Development of viral vectors to express foreign genes: To express foreign genes in plants, vectors with additional viral subgenomic promoters have been constructed (Dawson et al., 1989). Attempts to express foreign genes in plants were largely unsuccessful due to either failure to infect plants (Joshi, Joshi, and Ow, 1990) or loss of long distance movement (Ahlquist, French, and Sacher, 1988; Takamatsu et al., 1987, 1990). Failure to infect plants resulted from the deletion of the added sequences or failure of systemic transport of these vectors. It was hypothesized that the recombination between two repeated subgenomic promoter sequences in the viral constructs caused the failure (Dawson et al., 1989). Therefore a new approach using duplication of a heterologous subgenomic mRNA promoter was adopted, which proved successful (Donson et al., 1991). Tobamoviruses had been used earlier to produce heterologous proteins. TMV has been used to express Leu-enkephlin in tobacco protoplasts (Takamatsu et al., 1990) and the bacterial chloramphenicol acetyltransferase gene in inoculated tobacco leaves (Takamatsu et al., 1987; Dawson et al., 1989). A hybrid viral RNA, designated TB2 constructed from two tobamoviruses (TMV-U1 and odontoglossum ringspot virus) was able to move systemically with two bacterial sequences- neomycin phosphotransferase and dihydrofolate reductase inserted independently (Donson et al., 1991). Hybrid RNAs containing both TMV-U1 and the inserted bacterial gene sequences were encapsidated by the ORSV coat protein, and were transmitted and amplified on passaging to subsequent plants. Using a similar approach, in our lab, we aim to characterize the subgenomic promoter of CymMV by adopting the PVX duplicated heterologous promoter approach.

Expression of non-viral genes in plants: Using plant viruses as vehicles, expression of non-viral genes in plants has been achieved (Scholthof, Scholthof, and Jackson, 1996; Yusibov et al., 1999; Gopinath et al., 2000). The ease with which RNA viruses can be manipulated *in vitro* through the reverse genetics approach and the high levels of expressed proteins obtained makes this method a more preferred one when compared to the use of the transgenic plant approach. Viruses that have been manipulated and used as viral vectors include the TMV (Yusibov et al., 1999), *Cowpea mosaic virus* (Dalsgaard et al., 1997), tomato bushy stunt virus (Joelson et al., 1997) and AlMV (Yusibov et al., 1997). Using TMV, efforts have been made to express a Leuenkephalin peptide (Takamatsu et al., 1990), an angiotensis-I- converting enzyme inhibitor peptide (Hamamoto et al., 1993), a 13 aa sequence of the murine zona pellucida ZP3 epitope (Fitchen, Beachy, and Hein, 1995) and a malarial epitope (Turpen et al., 1995). Recently, Nocistatin, a heptadecapeptide isolated from the bovine brain that appears to be a potential candidate for blocking 'neuropathic pain' was expressed from ORSV (Lim et al., 2002). A chimeric ORSV was constructed by fusing 17 aa of mouse nocistatin (mNST) to the C-terminal of the CP gene to yield ORSV-mNST. Expression of the mNST peptide was demonstrated by immuno-transmission electron microscopy, western blot, mass spectrometry and radioimmunoassay. The mNST was maintained in the chimera through six passages. The mNST peptide could be effectively cleaved and purified from the chimeric ORSV CP. This is a pioneering report of a successful attempt in obtaining a complete peptide with no additional aa sequence after expression and purification through the use of ORSV as a vector. Currently this is limited to very small peptides, and future challenges lie in successfully obtaining expression of larger peptides.

Development of virus-resistant transgenic orchids resistant to CymMV and ORSV: We have discussed earlier our efforts to generate transgenic *Dendrobium* Sonia harbouring the CP genes of CymMV and ORSV. We have also made efforts to find the conserved regions of the CP genes that would maximize our chances of producing transgenic orchid cultivars resistant to CymMV and ORSV of different geographical locations. Although there are reports of transgenic orchids being produced on a laboratory scale, there still remains the need for development of a protocol for large scale production of transgenic plants that would be resistant to infection of CymMV and ORSV from different geographical distributions of the world.

Literature Cited

Abbink, T. E. M., P. A. Tjernberg, J. F. Bol, and H. J. M. Linthorst. 1998. Tobacco mosaic virus helicase domain induces necrosis in *N* gene-carrying tobacco in the absence of virus replication. *Mol. Plant Microbe. Interact.* 12: 1242–1246.

Agranovsky, A. A., A. V. Karasev, V. K. Novikov, N. A. Lunina, S. Loginov, and L. G. Tyulkina. 1992. Poa semilatent virus, a hordeivirus having no internal polydisperse poly A in the 3' non-coding region of the RNA genome. *J. Gen. Virol.* 73: 2085–2092.

Ahlquist, P., R. French, and R. Sacher. 1988. *Current Communications in Molecular Biology: Viral Vectors.* Gluzman Y. and Hughes S. H. (eds.). Cold Spring Harbor Lab., Cold Spring Harbor, NY, pp. 183–189.

Ajjikuttira, P., C. S. Loh, and S. M. Wong. 2005. Reciprocal function of movement proteins and complementation of long-distance movement of Cymbidium mosaic virus RNA by Odontoglossum ringspot virus coat protein. *J. Gen. Virol.* 86: 1543–1553.

Ajjikuttira, P. A., C. L. Lim-Ho, M. H. Woon, K. H. Ryu, C. A. Chang, C. S. Loh, and S. M. Wong. 2002. Genetic variability in the coat protein genes of two orchid viruses: Cymbidium mosaic virus and Odontoglossum ringspot virus. *Arch. Virol.* 147: 1943–1954.

Anandalakshmi, R., G. A. Pruss, X. Ge, R. Marathe, A. C. Mallory, T. H. Smith, and V. B. Vance. 1998. A viral suppressor of gene silencing in plants. *Proc. Natl. Acad. Sci. USA.* 95: 13079–13084.

Anjos, J. R., U. Jarlfors, and S. A. Ghabrial. 1992. Soy bean mosaic potyvirus enhances the titer of two comoviruses in dually infected soy plants. *Phytopathology* 82: 1022–1027.

Atabekov, J. G. and M. E. Taliansky. 1990. Expression of a plant virus-coding function different viral genomes. *Adv. Virus. Res.* 38: 201–248.

Atabekov, J. G., S. I. Malyshenko, S. Yu Morozov, M. E. Taliansky, A. G. Solovyev, A. A. Agranovsky, and N. A. Shapka. 1999. Identification and study of tobacco mosaic virus movement function by complementation tests. *Phil. Trans. R. Soc. Lond. B.* 354: 629–635.

Ayllon, M. A., C. Lopez, J. Navas-Castillo, S. M Garnsey, J. Guerri, R. Flores, and P. Moreno. 2001. Polymorphism of the 5' terminal region of citrus tristeza virus. CTV RNA: Incidence of three sequence types in isolates of different origin and pathogenecity. *Arch. Virol.* 146: 27–40.

Ali, A. and J. W. Randles. 2001. Genomic heterogeneity in Pea seed-borne mosaic virus isolates from Pakistan, the centre of diversity of the host species, Pisum sativum. *Arch. Virol.* 146: 1855–1870.

Alonso, E., I. Garcia-Luqus, A. de la Cruz, B. Wicke, M. J. Avila-Rincom, M. T Serra, C. Castresana, and J. R. Diaz-Ruiz. 1991. Nucleotide sequence of the genomic RNA of pepper mild mottle virus, a resistance-breaking tobamovirus in pepper. *J. Gen. Virol.* 72: 2875–2884.

Bao, Y., S. A. Carter, and R. S. Nelson. 1996. The 126- and 183-kilodalton proteins of tobacco mosaic virus, and not their common nucleotide sequence, control mosaic symptom formation in tobacco. *J. Virol.* 70: 6378–6383.

Barry, K., J. S. Hu, A. R. Kuehnle, and N. Sugii. 1996. Sequence analysis and detection using immunocapture-PCR of cymbidium mosaic virus and odontoglossum ringspot virus in Hawaiian orchids. *J. Phytopathol.* 144: 179–186.

Brigneti, G., O. Voinett, W. X. Li, and L. H. Ji. 1998. Viral pathogenicity determinants are suppressors of transgene silencing in *Nicotiana benthamiana. EMBO J.* 17: 6739–6746.

Calvery, L. A. and S. A. Ghabrial. 1983. Enhancement by soybean mosaic virus of bean pod mottle virus titer in doubly infected soybean. *Phytopathology* 73: 992–997.

Canizares, M. C., J. F. Marcos, and V. Pallas. 2001. Molecular variability of twenty-one geographically distinct isolates of Carnation mottle virus CarMV and phylogenetic relationships within the Tombusviridae family. *Arch. Virol.* 146: 2039–2051.

Chapman, S. N., G. Hills, J. Watts, and D. C. Baulcombe. 1992b. Potato Virus X as a vector for gene expression in plants. *Plant J.* 549–557.

Cheng, M., J. E. Fry, S.-. Pang, H.-P. Zhou, C. M. Hironaka, D. R. Duncan, T. W. Conner, and Y.-C. Wan. 1997. Genetic transformation of wheat mediated by Agrobacterium tumefaciens. *Plant Physiol.* 115: 971–980.

Chia, T. F., Y. S. Chan, and N. H. Chua. 1992a. Characterisation of cymbidium mosaic virus coat protein gene and its expression in transgenic tobacco plants. *Plant Mol. Biol.* 18: 1091–1099.

Chia, T. F., Y. S. Chan, and N. H. Chua. 1992b. Detection and localization of viruses in orchids by tissue-print hybridization. *Plant Pathology* 41: 355–361.

Chia, T. F., Y. S. Chan, and N. H. Chua. 1994. The firefly luciferase gene as a non-invasive reporter for *Dendrobium* transformation. *Plant J.* 6: 441–446.

Chng, C. G., S. M. Wong, P. H. Mahtani, C. S. Loh, C. J. Goh, M. C. C. Kao, M. C. M. Chung, and Y. Watanabe. 1996. The complete sequence of a Singapore isolate of odontoglossum ringspot virus and comparison with other tobamoviruses. *Gene* 171: 155–161.

Choi, S. K., J. Y. Yoon, K. H. Ryu, J. K. Choi, P. Palukaitis, and W. M. Park. 2002. Systemic movement of a movement-deficient strain of cucumber mosaic virus in zucchini squash is facilitated by a cucurbit-infecting potyvirus. *J. Gen. Virol.* 83: 3173–3178.

Cooper, B., I. Schmitz, A. L. N. Rao, R. N. Beachy, and J. A. Dodds. 1996. Cell-to-Cell transport of movement-defective cucumber mosaic and tobacco mosaic viruses in transgenic plants expressing heterologous movement protein genes. *Virology* 216: 208–213.

Dalsgaard, K., A. Uttenthal, T. D. Jones, F. Xu, A. Meeryweather, W. D. O. Hamilton, J. P. M. Langeveld, R. S. Boshuizen, S. Kamstrup, G. P. Lomonosoff, C. Porta, C. Vela, J. L. Casal, R. H. Meloen, and P. B. Rodgers. 1997. Plant-derived vaccine protects target animals against a viral disease. *Nat Biotechnol.* 15: 248–252.

Damirdagh, I. S. and A. F. Ross. 1967. A marked synergistic interaction of potato viruses X and Y in inoculated leaves of tobacco. *Virology* 31: 296–307.

De Jong, W. and P. Ahlquist. 1992. A Hybrid plant virus made by transferring the noncapsid movement protein from a rod-shaped to an icosahedral virus is competent for systemic infection. *Proc. Natl. Acad. Sci. USA* 89: 6808–6812.

Dawson, W. O., D. J. Lewandowski, M. E. Hilf, P. Bubrick, A. J. Raffo, J. J. Shaw, G. L. Grantham, and P. R. Desjardins. 1989. A tobacco mosaic virus-hybrid expresses and loses an added gene. *Virology* 172: 285–292.

Derrick, P. M., S. A. Carter, and R. S. Nelson. 1997. Mutation of the tobacco mosaic tobamovirus 126- and 183-kDa proteins: effects on phloem-dependent virus accumulation and synthesis of viral proteins. *Mol. Plant-Microbe Interact.* 10: 589–596.

Domingo, E., J. Holland, C. Biebreicher, and M. Eigen. 1995. Quasi-species: the concept and the word. *In* A. J. Gibbs, C. H. Calisher, and F. Garcia-Arenal, (eds.), *Molecular Basis of Virus Evolution*, Cambridge University Press, Cambridge, pp. 181–191.

Don, R. H., P. T. Cox, B. J. Wainwright, K. Baker, and J. S. Mattick. 1991. 'Touchdown' PCR to circumvent spurious priming during gene amplification. *Nucl. Acids Res.* 19, 4008.

Donald, R. G. and A. O. Jackson. 1994. The barley stripe mosaic virus γb gene encodes a multifunctional cysteine-rich protein that affects pathogenesis. *Plant Cell* 6: 1593–1606.

Donson, J., C. M. Kearney, M. E Hilf, and W. O. Dawson. 1991. Systemic expression of a bacterial gene by a tobacco mosaic virus-based vector. *Proc. Natl. Acad. Sci. USA* 88: 7204–7208.

Edwards, M. C. 1995. Mapping of the seed transmission determinants of barley stripe mosaic virus. *Mol-Plant Microbe Interact.* 8: 906–915.

Eigen, M. 1993. Viral quasispecies. *Sci. Am.* July, 42–49.

Eun, A. J. C. and S. M. Wong. 2000. Molecular beacons: a new approach to plant virus detection. *Phytopathology* 90: 269–275.

Eun, A. J. C., M. L. Seoh, and S. M. Wong. 2000. Simultaneous quantitation of two orchid viruses by the TaqMan® real-time RT-PCR. *J. Virol. Meth.* 87: 151–160.

Evans, R. K., B. E. Haley, and D. A. Roth. 1985. Photoaffinity labelling of a viral induced protein from tobacco. *J. Biol. Chem.* 260: 7800–7804.

Fedorkin, O. N., A. Merits, J. Lucchesi, A. G. Solovyev, M. Saarma, S. Yu Morozov, and K. Makinen. 2000. Complementation of the movement-deficient mutations in Potato Virus X:

Potyvirus coat protein mediates cell-to-cell trafficking of C-terminal truncation but not deletion mutant of potexvirus coat protein. *Virology* 270: 31–42.
Fedorkin, O. N., A. G. Solovyev, N. E. Yelina, A. A. Zamyatnin Jr., R. A. zinovkin, K. Makinen, J. Schiemann, and S. Yu. Morozov. 2001. Cell-to-cell movement of potato virus X involves distinct functions of the coat protein. *J. Gen. Virol.* 82: 449–458.
Fitchen, J., R. N. Beachy, and M. B. Hein. 1995. Plant virus expressing hybrid coat protein with added murine epitope elicits autoimmunity response. *Vaccine* 13: 1051–1057.
Fondong, V. N., J. S. Pita, M. E. C. Rey, A. de Kochko, R. N. Beachy, and C. M. Fauquet. 2000. Evidence of synergism between African cassava mosaic virus and a new double-recombinant geminivirus infecting cassava in Cameroon. *J. Gen. Virol.* 81: 287–297
Fraile, A., J. M. Malpica, M. A. Aranda, E. Rodriguez-Cerezo, and F. Garcia-Arenal. 1996. Genetic diversity in tobacco mild green mosaic tobamovirus infecting the wild plant Nicotiana glauca. *Virology* 223: 148–155.
Fraile, A., F. Escriu, M. A. Aranda, J. M. Malpica, J. M. Gibbs, and F. Garcia-Arenal. 1997. A century of tobamovirus evolution in an Australian population of Nicotiana glauca. *J. Virol.* 71: 8316–8320.
Francki, R. I. B. 1970. Cymbidium mosaic virus. *In Descriptions of Plant Viruses No. 27*. CMI/AAB, Wellesbourne, Warwick.
Fuentes, A. L. and R. I. Hamilton. 1991. Sunn-hemp mosaic virus facilitates cell-to-cell spread of southern bean mosaic virus in a nonpermissive host. *Phytopathology* 81: 1302–1305.
Giesman-Cookmeyer, D., S. Silver, A. A. Vaewhongs, S. A. Lommel and C. M. Deom. 1995. Tobamovirus and dianthovirus movement proteins are functionally homologous. *Virology* 213: 38–45.
Goldberg, K. B. and M. K. Brakke. 1987. Concentration of maize chlorotic mottle virus increased in mixed infections with maize dwarf mosaic virus, strain B. *Phytopathology* 77: 162–167.
Gonzalez-Jara, P., B. Tenllado, B. Martinez-Garcia, F. A. Atencio, D. Barajas, M. Vargas, J. Diaz-Ruiz, and J. R. Diaz-Ruiz. 2004. Host-dependent differences during synergistic infection by potyviruses with potato virus X. *Mol. Plant Pathol.* 5: 29–35.
Goodman, R. M. and A. F. Ross. 1974a. Independent assembly of virions in tobacco doubly infected by potato virus X and potato virus Y or tobacco mosaic virus. *Virology* 59: 314–318.
Goodman, R. M. and A. F. Ross. 1974b. Enhancement by potato virus Y of potato virus X synthesis in doubly infected tobacco depends on the timing of invasion by the viruses. *Virology* 58: 263–271.
Gopinath, K., J. Wellink, C. Porta, K. M. Taylor, G. P. Lomonosoff, and A. van Kanmen. 2000. Engineering cowpea mosaic virus RNA-2 into a vector to express heterologous proteins in plants. *Virology* 267: 159–173.
Gorbalenya, A. E., and R. H. Koonin. 1989. Viral proteins containing the purine NTP-binding sequence pattern. *Nucleic Acids Res.* 17: 7735–7762.
Goregaoker, S. P., D. J. Lewandowski, and J. N. Culver. 2001. Identification and functional analysis of an interaction between domains of the 126/183-kDa replicase-associated proteins of Tobacco mosaic virus. *Virology* 282: 320–328.
Gustafson, G. B., R. Hunter, R. Hanau, S. L. Armour, and A. O. Jackson. 1987. Nucleotide sequence and genetic organization of barley stripe mosaic virus RNA gamma. *Virology* 158: 394–406.
Habili, N. and R. H. Symons. 1989. Evolutionary relationship between luteoviruses and other RNA plant viruses based on sequence motifs in their putative RNA polymerases and nucleic acid helicases. *Nucleic Acids Res.* 17: 9543–9555.
Hadley, G., M. Arditti, and J. Arditti. 1987. Orchids diseases-a compedium. *In* J. Arditti (ed.), *Orchid Biology, Reviews and Perspectives IV*. Cornell University Press, Ithaca, pp. 263–322.
Hamamoto, H., Y. Sugiyama, N. Nakagawa, E. Hashida, Y. Matsunaga, S. Takemoto, Y. Watanabe, and Y. Okada. 1993. A new tobacco mosaic virus vector and its use for the systemic production of angiotensin converting enzyme inhibitor in transgenic tobacco and tomato. *Biotechnology* 11: 930–932.

Hamamoto, H., Y. Watanabe, H. Kamada and Y. Okada. 1997a. Amino acid changes in the putative replicase of tomato mosaic tobamovirus that overcome resistance in Tm-1 tomato. *J. Gen. Virol.* 78: 461–464.

Hamamoto, H., Y. Watanabe, H. Kamada, and Y. Okada. 1997b. A single amino acid substitution in the virus-encoded replicase of tomato mosaic tobamovirus alters host specificity. *Mol. Plant Microbe Interact.* 10: 1015–1018.

Hamilton, R. I. and B. Valentine. 1984. Infection of orchid pollen by odontoglossum ringspot and cymbidium mosaic viruses. *Can. J. Plant Pathol.* 6: 185–190.

Hirashima, K. and Y. Watanabe. 2001. Tobamovirus replicase coding region is involved in cell-to-cell movement. *J. Virol.* 75: 8831–8836.

Hoffmann, K., S. M. Geske, and J. W. Moyer. 1998. Pathogenesis of tomato spotted wilt virus in peanut plants dually infected with peanut mottle virus. *Plant Dis.* 82: 610–614

Holland, J. J., J. C. DeLaTorre, and D. A. Steinhauer. 1992. RNA virus populations as quasispecies. *In* J. J. Holland (ed.), *Genetic Diversity of RNA Viruses*. Springer Verlag, Berlin, pp. 1–20.

Hu, J. S., S. Ferreira, M. Q. Xu, M. Lu, M. Iha, E. Pflum, and M. Wang. 1994. Transmission, movement and inactivation of cymbidium mosaic and odontoglossum ringspot viruses. *Plant Dis.* 78: 633–636.

Hu, W. W. and S. M. Wong. 1998. The use of DIG-labelled cRNA probes for the detection of cymbidium mosaic potexvirus and odontoglossum ringspot tobamovirus in orchids. *J. Virol. Meth.* 70: 193–199.

Hu, W. W., S. M. Wong, C. S. Loh, and C. J. Goh. 1998. Synergism in replication of cymbidium mosaic potexvirus CymMV and odontoglossum ringspot tobamovirus ORSV RNA in orchid protoplasts. *Arch. Virol.* 143: 1265–1275.

Jensen, D. D. 1950. Mosaic of Cymbidium orchids. *Phytopathology* 40: 966–967.

Jensen, D. D. and A. H. Gold. 1951. A virus ringspot of odontoglossum orchid: symptoms, transmission and electron microscopy. *Phytopathology* 41: 648–653.

Joelson, T., L. Akerblom, L. Oxefelt, B. Strandberg, K. Tomenius, and T. J. Morris. 1997. Presentation of a foreign peptide on the surface tomato bushy stunt virus. *J. Gen. Virol.* 78: 1213–1217.

Joshi, R. L., V. Joshi and D. W. Ow. 1990. Barley stripe mosaic virus genome mediated expression of a foreign gene in dicot and monocot plant cells. *EMBO J.* 9: 2663–2669.

Kado, C. I., M. H. V. van Regenmortel, and C. A. Knight. 1968. Some studies on strains of tobacco mosaic virus in orchids. I. Biological, chemical and serological studies. *Virology* 34: 17–24.

Karyeija, R. F., J. F. Kreuze, R. W. Gibson, and J. P. T Valkonen. 2000. Synergistic Interactions of a potyvirus and a phloem-limited crinivirus in sweet potato plants. *Virology* 269: 26–36.

Kasschau, K. D. and J. C. Carrington. 1998. A counter-defensive strategy of plant viruses: Suppression of posttranscriptional gene silencing. *Cell* 95: 461–470.

Kramer, G. and P. Argos. 1984. Primary structural comparison of RNA-dependent polymerases from plant, animal and bacterial viruses. *Nucleic Acids Res.* 12: 7269–7282.

Kroner, P., D. Richards, P. Traynor, and P. Ahlquist. 1989. Defined mutations in a small region of the brome mosaic virus 2a gene cause diverse temperature-sensitive RNA replication phenotypes. *J. Virol.* 63: 5302–5309.

Lauber, E., C. Bleykasten-Grosshans, M. Erhardt, S. Bouzoubaa, G. Jonard, K. E. Richards, and H. Guilley. 1998. Cell-to-cell movement of beet necrotic yellow vein virus: I. Heterologous complementation experiments provide evidence for specific interactions among the triple gene block proteins. *Mol. Plant-Microbe Interact.* 11: 618–625.

Lawson, R. H. and M. Brannigan. 1986. Virus diseases of orchids. In: *Handbook of Orchid Pests and Diseases*. American Orchid Society, West Palm Beach, FL.

Lim, A. L., S. Tachibana, Y. Watanabe, and S. M. Wong. 2002. Expression and purification of a neuropeptide nocistatin using two related plant viral vectors. *Gene* 289: 69–79

Lim, S. T., S. M. Wong, C. Y. Yeong, S. C. Lee, and C. J. Goh. 1993. Rapid detection of cymbidium mosaic virus by the polymerase chain reaction. *J. Virol. Meth.* 41: 37–46.

Longstaff, M., G. Brigneti, F. Boccard, S. N. Chapman, and D. C. Baulcombe. 1993. Extreme resistance to potato virus X infection in plants expressing a modified component of the putative viral replicase. *EMBO J.* 12: 379–386.

Matthews, R. E. F. 1991. *Plant Virology*, 3rd ed. Academic Press, San Diego

Moya, A. and F. Garcia-Arenal. 1995. Population genetics of viruses: An introduction. *In* A. J. Gibbs, C. H. Calisher, and F.Garcia-Arenal, (eds.), *Molecular Basis of Virus Evolution*. Cambridge University Press, Cambridge, pp. 213–223.

Morozov, S. Yu, N. A. Miroshnichenko, A. G. Solovyev, O. N. Fedorkin, D. A. Zelenina, L. I. Lukasheva, A.V. Karasev, V. V. Dolja, and J. G. Atabekov. 1991. Expression strategy of the potato virus X triple gene block. *J. Gen. Virol.* 72: 2039–2042.

Morozov, S Yu, O. N. Fedorkin, G. Juttner, J. Schiemann, D. C. Baulcombe, and J. G. Atabekov. 1997. Complementation of potato virus X mutant mediated by bombardment of plant tissues with cloned viral movement protein genes. *J. Gen. Virol.* 78: 2077–2081.

Namba, R. and M. Ishii. 1971. Failure of aphids to transmit the ORSV and CymMV to orchid plantlets derived from meristem cultures. *Phytopathology* 61: 582–583.

Nan, G. L. and A. R. Kuehnle. 1995. Genetic transformation of dendrobium orchid. Biotechnology in agriculture and forestry, Vol. 34. *In* Y.P.S. Bajaj (ed.), *Plant Protoplasts and Genetic Engineering VI*. Springer-Verlag, Berlin/Heidelberg.

Nan, G. L. and A. R. Kuehnle. 1998. Transgenic Dendrobium orchid through Agrobacterium-mediated transformation. *Malayan Orchid Rev*. 32: 93–96.

Kasschau, K. D. and J. C. Carrington. 2001. Long-distance movement and replication maintainence functions correlate with silencing suppression activity of potyviral HC-Pro. *Virology* 285: 71–81.

Khan, M. A. and J. W. Demski. 1982. Identification of turnip mosaic and cauliflower mosaic viruses naturally infecting collards. *Plant Dis*. 66: 253–256.

Kuehnle, A. R. and N. Sugii. 1992. Transformation of *Dendrobium* orchid using particle bombardment of protocorms. *Plant Cell Rep*. 11: 484–488.

Ogawa, T., Y. Watanabe, T. Meshi, and Y. Okada. 1991. Trans complementation of virus-encoded replicase components of tobacco mosaic virus. *Virology* 185: 580–584.

Parrella, G., C. Lanave, G. Marchoux, M. M. Finetti Sialer, A. Di Franco, and D. Gallitelli. 2000. Evidence for two distinct subgroups of Alfalfa mosaic virus AMV from France and Italy and their relationships with other AMV strains. *Arch. Virol.* 145: 2659–2667.

Pearson, M. N. and J. S. Cole. 1991. Further observations on the effects of cymbidium mosaic virus and odontoglossum ringspot virus on the growth of Cymbidium orchids. *J. Pathol.* 131: 193–198.

Petty, I. T., B. G. Hunter, N. Wei, and A. O. Jackson. 1989. Infectious barley stripe mosaic virus RNA transcribed in vitro from full-length genomic cDNA clones. *Virology* 171: 342–349.

Petty, I. T., R. French, R. W. Jones, and A. O. Jackson. 1990a. Identification of barley stripe mosaic virus genes involved in viral RNA replication and systemic movement. *EMBO J*. 9: 3453–3457.

Petty, I. T., M. C. Edwards, and A. O. Jackson. 1990b. Systemic movement of an RNA plant virus determined by a point substitution in a 5' leader sequence. *Proc. Natl. Acad. Sci. USA* 87: 8894–8897.

Petty, I. T., R. G. Donald, and A. O. Jackson. 1994. Multiple genetic determinants of barley stripe mosaic virus influence lesion phenotype on *Chenopodium amaranticolor. Virology* 198: 218–226.

Pinel, A., P. N. Guessan, M. Bousalem, and D. Fargette. 2000. Molecular variability of geographically distinct isolates of Rice yellow mottle virus in Africa. *Arch. Virol.* 145: 1621–1638.

Pita, J. S., V. N. Fondong, A. Sangare, G. W. Otim-Nape, S. Ogwal, and C. M. Fauquet. 2001. Recombination, pseudorecombination and synergism of geminiviruses are determinant keys to the epidemic of severe cassava mosaic disease in Uganda. *J. Gen. Virol.* 82: 655–665.

Poolpol, P. and T. Inouye. 1986. Enhancement of cucumber mosaic virus multiplication by zuchinni yellow mosaic virus in doubly infected cucumber plants. *Ann. Phytopathol. Soc. Jpn.* 52: 22–30.

Pruss, G., X. Ge, X. M. Shi, J. C. Carrington, and V. B. Vance. 1997. Plant viral synergism: the potyviral genome encodes a broad-range pathogenicity enchancer that transactivates replication of heterologous viruses. *Plant Cell* 9: 859–868.

Qiu, W. P. and K. B. G. Scholthof. 2001. Genetic identification of multiple biological roles associated with the capsid protein of satellite panicum mosaic virus. *Mol. Plant-Microbe Interact.* 14: 21–30.

Ranineri, D. M., P. Bottino, M. P. Gordon, and E. W. Nester. 1990. Agrobacterium-mediated transformation of rice Oryza sative L. *Biotechnology* 8: 33–39.

Rao, A. L. N., B. Cooper, and C. M. Deom. 1998. Defective Movement of Viruses in the Family Bromoviridae is differentially complemented in *Nicotiana benthamiana* expressing Tobamovirus or Dianthovirus Movement Proteins. *Phytopathology* 88: 666–672.

Richins, R. D., J. Donson, D. J. Lewandowsky, D. A. Ducasse, W. O. Dawson, and R. J. Shepherd. 1993. Temperature-dependent complementation of movement-deficient mutants of tobacco mosaic virus by gene I of peanut chlorotic streak caulimovirus. *Phytopathology* 83: 1349.

Rochow, W. F. and A. F. Ross. 1955. Virus multiplication in plants doubly infected by potato viruses X and Y. *Virology* 1: 10–27.

Rodriguez-Cerezo, E., A. Moya, and F. Garcia-Arenal. 1989. Variability and evolution of the plant RNA virus Pepper mild mottle virus. *J. Virol.* 63: 2198–2203.

Roux, K. H. 1995. Optimization and troubleshooting in PCR. PCR Methods Appl. 4, S185–S194.

Rozanov, M. N., E. V. Koonin, and A. E. Gorbalenya. 1992. Conservation of the putative methyltransferase domain: a hallmark of the 'Sindbis-like' supergroup of positive-strand RNA viruses. *J. Gen Virol.* 73: 2129–2134.

Ryabov, E. V., D. J. Robinson, and M. E. Taliansky. 1999. A plant virus-encoded protein facilitates long-distance movementof heterologous viral RNA. *Proc. Natl. Acad. Sci. USA*. 96: 1212–1217.

Ryu, K. H. and W. M. Park. 1995a. The complete nucleotide sequence and genome organisation of odontoglossum ringspot tobamovirus RNA. *Arch. Virol.* 140: 1577–1587.

Ryu, K. H. and W. M. Park. 1995b. Rapid detection and identification of odontoglossum ringspot virus by polymerase chain reaction amplification. *FEMS Microbiol. Lett.* 133: 265–269.

Scholthof, H. B., K. B. G. Scholthof, and A. O. Jackson. 1996. Plant virus gene vectors for transient expression of foreign proteins in plants. Ann. Rev. *Phytopathology* 34: 299–323.

Seoh, M. L., S. M. Wong, and L. Zhang. 1998. Simultaneous TD/RT-PCR detection of cymbidium mosaic potexvirus and odontoglossum ringspot tobamovirus with a single pair of primers. *J. Virol. Meth.* 72: 197–204.

Scheets, K. 1998. Maize chlorotic mottle machlomovirus and wheat streak mosaic rymovirus concentrations increase in the synergistic disease corn lethal necrosis. *Virology* 242: 28–38.

Scholthof, K-B. G. 1999. A synergism induced by satellite panicum mosaic virus. *Mol. Plant-Microbe Interact.* 12: 163–166.

Shi, X. M., H. Miller, J. Verchot, J. C. Carrington, and V. B. Vance. 1997. Mutations in the region encoding the central domain of helper component-proteinase HC-Pro eliminate potato virus X/potyviral synergism. *Virology* 231: 35–42.

Shintaku, M. H., S. A. Carter, Y. Bao, and R. S. Nelson. 1996. Mapping nucleotides in the 126-kDa protein gene that control the differential symptoms induced by two strains of tobacco mosaic virus. *Virology* 221: 218–225.

Short, M. N. and J. W. Davies. 1987. Host ranges, symptoms and amino acid compositions of eight potexviruses. *Ann. Appl. Biol.* 110: 213–219.

Skyrabin, K. G., A. S. Kraev, S. Y. Morozov, M. N. Rozanov, B. K. Chernov, L. I. Lukasheva, and J. G. Atabekov. 1988. The nucleotide sequence of potato virus X RNA. *Nucl. Acids Res.* 16: 10929–10930.

Solovyev, A. G., D. A. Zelenina, E. I. Savenkov, V. Z. Grdzelishvili, S. Y Morozov, D-E. Lesemann, E. Maiss, R. Casper, and J. G. Atabekov. 1996. Movement of a barley stripe mosaic virus chimera with a tobacco mosaic virus movement protein. *Virology* 217: 435–441.

Spitsin, S., K. Steplewski, N. Fleysh, H. Belanger, T. Mikheeva, S. Shivprasad, W. Dawson, H. Koprowski, and V. Yusibov. 1999. Expression of alfalfa mosaic virus coat protein in tobacco mosaic virus TMV deficient in the production of its native coat protein supports long-distance movement of a chimeric TMV. *Proc. Natl. Acad. Sci. USA* 96: 2549–2553.

Srifah, P., S. Loprasert, N. Rungroj. 1996. Use of reverse transcription-polymerase chain reaction for cloning of coat protein-encoding genes of cymbidium mosaic virus. *Gene* 179: 105–107.

Steinhauer, D. A., E. Domingo, and J. J. Holland. 1992. Lack of evidence for proofreading mechanisms associated with an RNA virus polymerase. *Gene* 122: 4673–4680.

Takamatsu, N., M. Ishikawa, T. Meshi, and Y. Okada. 1987. Expression of bacterial chloramphenicol acetyltransferase gene in tobacco plants mediated by TMV-RNA. *EMBO J.* 6: 307–311.

Takamatsu, N., Y. Watanabe, H. Yanagi, T. Meshi, T. Shiba, Y. Okada. 1990. Production of enkephalin in tobacco protoplasts using tobacco mosaic virus RNA vector. *FEBS Lett* 269: 73–76.

Taliansky, M. E., T. I. Atabekova, I. B. Kaplan, S. Yu, S. I. Malyshenko, and J. G. Atabekov. 1982a. A study of TMV ts mutant Ni2519. I. Complementation experiments. *Virology* 118: 301–308.

Taliansky, M. E., S. I. Malyshenko, E. S. Pshennikova, I. B. Kaplan, E. F. Ulanova, and J. G. Atabekov. 1982b. Plant virus-specific transport function. I. Virus genetic control required for systemic spread. *Virology* 122: 318–326.

Taliansky, M. E., S. I. Malyshenko, E. S. Pshennikova, I. B. Kaplan, E. F. Ulanova, and J. G. Atabekov. 1982c. Plant virus-specific transport function. II. A factor controlling virus host range. *Virology* 122: 327–331.

Taliansky, M. E., C. P. de Jager, J. Wellink, J. W. M. van Lent, and R. W. Goldbach. 1993. Defective cell-to-cell movement of cowpea mosaic virus mutant N123 is efficiently complemented by sunn-hemp mosaic virus. *J. Gen. Virol.* 74: 1895–1901.

Tamai, A., K. Kubota, H. Nagano, M. Yoshii, M. Ishikawa, K. Mise, and T. Meshi. 2003. Cucumovirus- and bromovirus-encoded movement functions potentiate cell-to-cell movement of tobamo- and potexviruses. *Virology* 315: 56–67.

Turner, D. R., L. E. Joyce, and P. J. G. Butler. 1988. The tobacco mosaic virus assembly origin RNA: functional characteristics defined by directed mutagenesis. *J. Mol. Biol.* 203: 531–547.

Turpen, Th., S. Reinl, Y. Charoenvit, S. L. Hoffman, V. Fallarme, and L. K. Gill. 1995. Malarial epitopes expressed on the surface of chimeric tobacco mosaic virus. *Biotechnology* 13: 53–57.

Tyagi, S. and S. R. Kramer. 1996. Molecular beacons: Probes that fluoresce upon hybridization. *Nature Biotechnol.* 14: 303–308.

Vance, V. B. 1991. Replication of potato virus X RNA is altered in coinfections with potato virus Y. *Virology* 182: 486–494.

Vance, V. B., P. H. Berger, J. C. Carrington, A. G. Hunt, and X. M. Shi. 1995. 5' Proximal potyviral sequences mediate potato virus X/potyviral synergistic disease in transgencic tobacco. *Virology* 206: 583–590.

Vaskova, D., K. Petrizk, R. Karesova. 2000. Variability and molecular typing of the woody-tree infecting prunus necrotic ringspot ilarvirus. *Arch. Virol.* 145: 699–709.

Wang, Y.-Z., V. Gaba, Yang J., Palukaitis, P., and A. Gal-On. 2002. Characterization of synergy between cucumber mosaic virus and potyviruses in Cucurbit hosts. *Phytopathology* 92: 51–58.

Wang, Y., K. C. Lee, V. Gaba, S. M. Wong, P. Palukaitis, and A. Gal-On. 2004. Breakage of resistance to cucumber mosaic virus by co-infection with Zucchini yellow mosaic virus: enhancement of CMV accumulation independent of symptom expression. *Arch. Virol.* 149: 379–396.

Wanitachacorn, R., R. M. Harding, J. L. Dale. 2000. Sequence variability in the coat protein gene of two groups of banana bunchy top isolates. *Arch. Virol.* 145: 593–602.

Watanabe, Y., N. Morita, M. Nishiguchi, and Y. Okada. 1987. Attenuated strains of tobacco mosaic virus. Reduced synthesis of a viral protein with a cell-to-cell movement function. *J. Mol. Biol.* 194: 699–704.

Weber, H., P. Haeckel, and A. J. Pfitzner. 1992. A cDNA clone of tomato mosaic virus is infectious in plants. *J. Virol.* 66: 3909–3912.

Webster, R. D. and A. Granoff. 1994. Tobamoviruses. *In Encyclopedia of Virology*. Academic Press, Harcourt Brace, New York, pp. 1436–1441.

Webster, D. E., P. L. Guy, D. L. Beck, R. L. S. Forster. 1999. Distribution and diversity of New Zealand isolates of ryegrass mosaic virus. *Arch. Virol.* 144: 2059–2064.

Wisler, G. C., F. W. Zettler, T. J. Sheehan. 1986. Common questions and misconceptions concerning orchid viruses. *Am. Orchid Soc. Bull.* 55: 472–479.

Wong, S. M., C. G. Chng, Y. H. Lee, Y. Tan, F. W. Zettler. 1994. Incidence of cymbidium mosaic and odontoglossum ringspot viruses and their significance in orchid cultivation in Singapore. *Crop Protect.* 13: 235–239

Wong, S. M., P. H. Mahtani, K. C. Lee, H. H. Yu, Y. Tan, K. K. Neo, Y. Chan, M. Wu, and C. G. Chng. 1997. Cymbidium mosaic potexvirus RNA: complete nucleotide sequence and phylogenetic analysis. *Arch. Virol.* 142: 383–391.

Yang, J., H. J. Lee, D. H. Shin, S. K. Oh, J. H. Seon, K. Y. Paek, and K. H. Han. 1999. Genetic transformation of Cymbidium orchid by particle bombardment. *Plant Cell Rep.* 18: 978–984.

Yelina, N. E., E. I. Savenkov, A. G. Solovyev, S. Y. Morozov, and J. P. Valkonen. 2002. Long-distance movement, virulence, and RNA silencing suppression controlled by a single protein in hordei- and potyviruses: complementary functions between virus families. *J. Virol.* 76: 12981–12991.

Yu, H. H. 1999. *Molecular manipulation of two orchid viruses*. Ph. D. Thesis. National University of Singapore, Singapore.

Yu, H. H. and S. M. Wong. 1998a. A DNA clone encoding the full-length infectious genome of odontoglossum ringspot tobamovirus and mutagenesis of its coat protein gene. *Arch. Virol.* 143: 163–171.

Yu, H. H. and S. M. Wong. 1998b. Synthesis of biologically active clones of cymbidium mosaic potexvirus using a population cloning strategy. *Arch. Virol.* 143: 1617–1620.

Yu, H., S. H. Yang, C. J. Goh. 2001. *Agrobacterium*-mediated transformation of a *Dendrobium* orchid with the class 1 *knox* gene DOH1. *Plant Cell Reports* 20: 301–305.

Yusibov, V., A. Modelska, K. Steplewski, M. Agadjanyan, D. Weiner, C. Hooper, H. Koprowski. 1997. Antigens produced in plants by infection with chimeric plant viruses immunize against rabies virus and HIV-1. *Proc. Natl. Acad. Sci. USA* 94: 5784–5788.

Yusibov, V., S. Shivprasad, T. H. Turpen, W. Dawson, and H. Koprowski. 1999. Plant viral vectors based on tobamoviruses. *Curr. Top Microbiol. Immunol.* 240: 181–194.

Zettler, F. W., N. J. Ko, G. C. Wisler, M. S. Elliot, and S. M. Wong. 1990. Viruses of orchids and their control. *Plant Dis.* 74: 621–626.

Ziegler-Graff, V., P. J. Guilford, and D. C. Baulcombe. 1991. Tobacco rattle virus RNA-1 29 K gene product potentiates viral movement and also affects symptom induction in tobacco. *Virology* 182: 145–155.

APPENDIX

A list of Orchid Books

TIM WING YAM, BENJAMIN SINGER,
CHOY SIN HEW, TIIU KULL, IRINA TATARENKO,
AND JOSEPH ARDITTI

Two private libraries, Benjamin Singer's (which he donated to the American Orchid Society) and Joseph Arditti's (its future is yet to be decided, it may be donated to an academic or scientific institutions or sold), served as primary sources for this list. However other sources were also used. The use of multiple sources increased the number of books which are listed but may have introduced errors or imperfections for following reasons.

One and the same book may have been listed under different names erroneously.
Names of authors may have been misspelled.
When books have more than one author, the order of authors may have been presented differently in different lists and/or one or more names may have been omitted, added or misspelled.
A book may have been published under different names in more than one country.
Books are sometimes published by one publisher in one country and another in a different one.
Spelling errors in different lists
Translations
Different editions
Lack of information
Conventions used in spelling names like "de" and "van."
Erroneous assumptions regarding Chinese surnames. The Chinese traditions is to list surname first, as for example, Yam Tim Wing which may end up incorrectly as Wing, Y. T. in some lists compiled in the West and correctly as T. W. Yam in others.
Only the last names of some authors are listed.
Some entries listed as books may in fact be no more than reprints.
Several lists did not provide all relevant information about a book (dates of publication and names of publishers, for example).
In some cases we could check the accuracy or a listing and make appropriate corrections or find missing information and add it. However in many instances this was not possible and listings are incomplete.
When what seemed to be the same book appeared under several "identities" in different sources all are included in this list because we deemed multiple listings preferable to omission.

On the whole this list is far from perfect, but it we hope that it will prove to be useful.

Author(s)	Date	Title and publisher
Abraham, A., and P. Vatsala	1981	Introduction to orchids. 533 pp., numerous color and B&W photographs and line drawings. Tropical Botanic Gardens and Research Institute, Trivandrum, India.
Ackerman, J. D.	1995	An orchid flora of Puerto Rico and the Virgin Islands. Memoirs of The New York Botanical Garden Volume 73, New York Botanical Garden, Bronx, New York. 203 pp., 96 line drawing plates.
Acuna, G. J. A	1938	Catalogo Descriptivo de las Orquideas Cuba.
Adams, P. B. (ed.)	1988	Reproductive biology of species orchids, principles and practice. School of Botany, University of Melbourne, Parkville, Australia and the Orchid Species Society of Victoria. 93 pp., line drawings, maps, color photographs.
Adams, P. B., and S. D. Lowson	1995	*Dendrobium kingianum*: A unique Australian orchid. Central Queensland University Press, Rockhampton, Queensland, Australia. 197 pp., 199 color photographs, numerous maps.
Addison, H.	1961	Malayan orchid hybrids, 1st supplement. Government Printing Office, Singapore. 93 pp., numerous B&W photographs, one color photograph. See also Henderson and Addison, 1959.
Ageenko, V.	1887	Notes about some Crimean orchids. *Trudi Saint-Peterburgskogo obschestva estestvoispitateley*. 18(1). (In Russian.)
Agnes, R.	1994	Brazilian orchids in focus
Ahundov, G. F., J. M. Isaev, L. I. Prilipko, and R. J. Rzazade	1952	Orchidaceae. Pages 240–271. In: Flora of Azerbaijan. 2. Baku. (In Russian.)
Aiton, W.	1813	Gynandria. Pages 185–227. In: A catalogue of plants cultivates in the Royal Botanic Garden, Kew or Hortus Kewensis Volume 5. Longman, Hurst, Rees, Orme and Brown, Paternoster Row, London.
Akhalkatsi, M., M. Kimeridze, S. Künkele, R. Lorenz, and M. Mosulishvili	2003	Diversity and conservation of Georgian orchids. 40 pp., with color photographs. Tbilisi. (In Georgian and English.)
Albertis, C. and A. Albertis	1985	Wild orchids of Crete.
Alcorn, A. R. W. (F.) and M. S. C. Hallett	No date	*Lycaste* orchids, cultivation and hybridization. No publisher listed. Published in Australia. 59 pp., numerous color photographs.

(continued)

(continued)

Author(s)	Date	Title and publisher
Alekhin, A. A., N. B. Gaponenko, and V. G. Sobko	2003	Orchids of Far East. Kiev. pp. 1–203, 65 line drawings. (In Ukrainian.)
Alessandrini, A., and A. Busetto	1985	Le orchidee spontanee dell'Emilia-Romana. Grafis Edizioni, Casalecchio di Reno, Italy, 149 pp., 58 plates of color paintings and B&W drawings each.
Alkimos, A.	1988	The orchids of Greece (in Greek).
Allan, B., P. Woods, and S. Clarke	1993	Wild orchids of Scotland. HMSO, Edinburgh. 135 pp., with numerous color photographs, line drawings and maps.
Allen, C.	1996	North American native terrestrial orchids propagation and production, conference proceedings March 16 & 17, 1996. North American Native Orchid Conference, 14320 Poplar Hill Rd., Germantown, MD 20874. 116 pp., B&W photographs.
Allen, P. H., and D. O. Allen	1953	The orchids of Panama. Reprinted a softcover book from The Orchid Journal, Volume II, pp. 20–67, 104–108, 181–185, 212–216, 259–263,319–323, 362–368, with numerous B&W photographs and 41 plates of line drawings.
Allikas, G., and N. Nash	2000	Orchids. Thunder Bay Press, San Diego, CA. 176 pp., numerous color photographs.
Allikas, G., and N. Nash	2004	A pocket guide to orchids. Book Sales. 256 pp.
Allikas, G., and N. Nash	2005	The world's most beautiful orchids. Thunder Bay Press, San Diego, CA. 448 pp., numerous color photographs.
Allorge, F., and L. Blaringhem	1937	Volume publié a la mémorie de Julien Costantin 1857–1936. Masson et Cie, Éditeurs, Paris. 469 pp., numerous B&W photographs and line drawings.
Alocock, J.	2006	An enthusiasm for orchids. Oxford University Press, Oxford.
Alrich, P.	1991	Orchids on stamps
American Orchid Society	1990–1991	Commercial orchid growers directory. 44 pp.
Ames, B.	1959	Drawings of Florida orchids. Botanical Museum of Harvard University, Cambridge, MA. 190 pp., numerous line drawings.
Ames, O.	No date	Orchidaceae in Flora of Costa Rica. Pages 197–306, Field Museum of Natural History–Botany Volume 18.
Ames, O.	1910	The genus *Habenaria* in North America. 1979 reprint by Earl M. Coleman, Stanfordville, NY with an introduction by L. A. Garay. 288 pp., 79 line drawings by Blanche Ames.

(continued)

(continued)

Author(s)	Date	Title and publisher
Ames, O.	1948	Orchids and retrospect. Botanical Museum of Harvard University, Cambridge, MA. 172 pp., numerous line drawings.
Ames, O.	1905–1922	Illustrations and studies of the family Orchidaceae. Houghton, Mifflin and Company, Boston, MA (Vols. I and II), Merrymount Press, Boston, MA (Vols. III and VII).
Ames, O.	1905–1922	Studies in the Family Orchidaceae I–VII
Ames, O.	No date	Orchidaceae in Flora of Costa Rica. Pages 197–306 Field Museum of Natural History–Botany Volume 18.
Ames, O.	1924	An enumeration of the orchids of the United States and Canada. American Orchid Society, Boston, MA.
Ames, O.	1936	The genus *Epidendrum* in the Unites States and middle America. Botanical Museum, Harvard University, Cambridge, MA. 233 pp.
Ames, O., and B. Ames	1975	Orchids at Christmas. Botanical Museum of Harvard University. 50 pp., ca. 13 line drawings and several B&W photographs.
Ames, O., and D. S. Correll	1952, 1953, 1965	Orchids of Guatemala, volumes 1 (Fieldiana: Botany, Vol. 26, No. 1, pp. xiii+1–395, line drawing Figs. 1–107), 2 (Fieldiana: Botany, Vol. 26, No. 2, pp. 399–72, line drawing Fig. 108–198) and supplement by D. S. Correll (Fieldiana: Botany, Vol. 31, No. 7, pp. 177–221, line drawing Fig. 53).
Ames, O., and E. Quisumbing	1931–1936	New or Noteworthy Philippine Orchids I–VI
Ames, O., and E. Quisumbing	1931, 1932	New or noteworthy Philippine orchids, I. The Philippine Journal of Science 44 (4): 369–383, plates 1, 2 (color)-16 (B&W) and New or noteworthy Philippine orchids, I. The Philippine Journal of Science 47 (2): 197–219, plates 1, 32 (color) 29 (B&W).
Ames, O., and C. Schweinfurth	1923–1930	Schedulae Orchidianae volumes 1–10. Lawrence Press, Boston, MA.
Anderson, F. J.	1979	An illustrated treasury of cultivated flowers. Crown Publishers, New York. 104 pp., 50 color plates.
Anderson, F. J.	1979	An illustrated treasury of orchids. Abbeville Press Publishers, New York. 157 pp., 73 colored plates.
Anderson, F. J.	1979	An illustrated treasury of orchids. Crown Publishers, New York. 104 pp., 50 color plates.

(continued)

(continued)

Author(s)	Date	Title and publisher
Anderson, J. F.	1981	Orchids. Aberville Press, New York. 112 pp., numerous color photographs.
Andreev, L. N., A. S. Demidov, B. N. Golovkin, E. S. Smirnova, G. V. Porubinovskaya, T. K. Kriulinov, M. M. Serebryanii, V. G. Shatko, and M. G. Vahrameeva (eds.)	1988–1987	Ohrana I kultivirovanie orhidei. Abstracts of the All-Union Symposium, Main Botanical Garden, USSR Academy of Sciences, February 1987. Moscow. (In Russian.) 84 pp.
Angel, H.	1977	British wild orchids. Jarrold & Sons, Ltd., Norwich, UK.
Anonymous	1856	Russian flora. Russian Orchids. *Zhurnal Sadovodstva*. 1: 155–159. (In Russian.) Probably the first publication about orchids in Russia.
Anonymous	No date	Meristerm tissue culture–A selection of articles from the American Orchid Society Bulletin. 72 pp., numerous line drawings and B&W photographs.
Anonymous	1965 or earlier	Orchid culture in Victoria. Victorian Orchid Club of Australia, Glen Iris, Victoria, Australia. 28 pp.
Anonymous	1974	Modern methods in propagation, hybridizing and culture of orchids. Prague, Chechoslovakia.
Anonymous	1976	Orchids. Countryside books/A. B. Morse Company, Barrignton, IL, USA. 48 pp., 17 B&W photographs, one line drawing, 26 color photographs (including front and back covers).
Anonymous	1977	Cultivation and mericloning of Malaysian orchids. Mardi manual No: 1 (1975), KDN 19705. Malaysian Agricultural Research and Development Institute (MARDI) Serdang, Selangor. 75 pp., 24 color and 23 B&W photographs.
Anonymous	1977	The Philippines recommends for orchids. University of the Philippines, Los Baños. 99 pp., numerous line drawings and B&W and color photographs.
Anonymous	1978	Lanhua, a collection of color photographs of Chinese cymbidiums.

(continued)

(continued)

Author(s)	Date	Title and publisher
Anonymous	1978	Ninth world orchid conference activities booklet. No pagination, numerous B&W and color photographs.
Anonymous	1978	Proceedings of the Symposium of the Symposium on Orchidology, Singapore 1978. Published by the Orchid Society of South East Asia (Singapore). 101 pp., numerous illustrations.
Anonymous	1979	Export of orchids [from] Thailand, Singapore and Malaysia. An internal report for the Ceylon Tobacco Company, Sri Lanka.
Anonymous	1979	Hawaiian orchids. 14 pp., 23 color illustrations. A small book. Boom Books, Hilo Bay, Hawaii.
Anonymous	1981	Native orchids in Melbourne, revised edition. Australasian Native Orchid Society (Victorian Group). 56 pp., numerous line drawings.
Anonymous	1984	Australasian native orchid society, 21st anniversary workshop.
Anonymous	1984	Introduction to orchids. South Florida Orchid Society. Mimeograph. 51 pp.
Anonymous	1986	Handbook on orchid pests and diseases. American Orchid Society, Delray Beach, FL. 108 pp., numerous B&W and color photographs.
Anonymous	1990	First Australasian native orchid conference and show. Announcement.
Anonymous	1990	La Reunion, ile de vanille. Ocean Editions, Saint André, Reunion Island. 144 pp., B&W and color photographs.
Anonymous	1992	Iconography of wild and cultivated orchids in China. ISBN 957–531–185-X. 140 pp., numerous color photographs.
Anonymous	1995	Orchid pests and diseases. American Orchid Society, Delray Beach, FL. 118 pp., numerous B&W and color photographs.
Anonymous	1997	11th European orchid congress, Geneva. Lectures program.
Anonymous	2000	*Paphiopedilum* in Taiwan II. Taiwan Paphiopedilum Society. 144 pp., numerous color photographs.
Anonymous	2001	First International orchid conservation congress announcement.
Anonymous	2003	First international conference on neotropical orchidology. Announcement, program, etc.
Anonymous	2003	Protection and cultivation of orchids. Proceedings of International Conference, 6–8 October 2003. Kharkov. 89 pp. (In Russian and Ukrainian.)

(continued)

(continued)

Author(s)	Date	Title and publisher
Anonymous	No date	A folio of paintings of European orchids. Brücke-Verlag, Kurt Schmersow, Hildesheim.
Anonymous	No date	Cultivo de orquideas. Sociedad Colombiana de Orquideologia. 48 pp., B&W photographs in the text, color photographs on the cover.
Anonymous	No date	Enjoyment of orchids. American Orchid Society. 66 pp., Line drawingsn and B&W photographs.
Anonymous	No date	Meristem tissue culture. A collection of articles. American Orchid Society, Del Ray Beach, FL. 72 pp., numerous B&W photographs.
Anonymous	No date	Orchid publications on microfiche. Interdocumentation Company AG, Zurich Switzerland. 12 pp.
Anonymous	No date	Singapore orchids. Singapore Trade Development Board. 66 pp., numerous color photographs.
Anonymous	No date	List of hybrid cymbidiums. San Diego Orchid Society.
Anonymous	No date	Mitteleuropäische orchideen. Carl Zeiss, Oberkochen/Würt, Germany. 12 color slides.
Anonymous	1967	Your first orchids and how to grow them. Oregon Orchid Society. 96 pp., line drawings, color and B&W photographs.
Anonymous	1874	The cultivation of South Australian native orchids. Mimeograph. 19 pp.
Anonymous	1886	The report of the orchid conference held at South Kensington on May 12th and 13th, 1885. The Journal of the Royal Horticultural Society, Vol. 7, No. 1, 155 pp. Original.
Anonymous	1886	The report of the orchid conference held at South Kensington on May 12th and 13th, 1885. The Journal of the Royal Horticultural Society, Vol. 7, No. 1, 155 pp. Reprint 1985 in slip case with the report of the 1985 conference. See Napier, 1985.
Anonymous	1904	Hand list of orchids cultivated in the Royal Botanic Gardens, Kew, 2nd ed. Her Majesty's Stationary Office. 230 pp.
Anonymous	1904	Hand-list of orchids cultivated in the Royal Botanic Gardens. His Majestry's Stationary Office, London. 229 pp.
Anonymous	1960	Orchids. Oswald Wolff (Publishers) Ltd., London. 6 pp., 10 color plates. Book measure 13.3 by 18 cm.
Anonymous	1964	10th Western orchid congress. A collection of announcements and reports.
Anonymous	1964	Orchid culture. American Orchid Society. 63 pp., B&W photographs and line drawings.

(continued)

(continued)

Author(s)	Date	Title and publisher
Anonymous	1965	Orchids in your home. Pages 12–15, 5 color photographs in The Bridge, Lloyd Brothers, Inc., Hoechst Co., Cincinnati, OH.
Anonymous	1967	Special issue: *Calanthe*. Shinkashi Magazine. 62 pp., numerous B&W and color photographs.
Anonymous	1969	Orchids of Thailand. Thai Export Bulletin No. 7. Board of Export Promotion, Bangkon. 26 pp., color photographs..
Anonymous	1971	Orchid culture. American Orchid Society. 80 pp., color photographs.
Anonymous	1904	Hand list of orchids cultivated in the Royal Botanic Gardens, Kew, 2nd ed. Her Majesty's Stationary Office. 230 pp.
Anonymous	1972	Las orquideas y su conservacion. Publication No. 2, Comite Latinoamericano de orquideologia. 20 pp., bilingual English and Spanish.
Anonymous	1974	An orchidist's glossary. American Orchid Society. Botanical Museum Harvard University. 95 pp., numerous B&W photographs.
Anonymous	1986	Handbook on orchid culture. American Orchid Society. West Palm Beach, FL. 92 pp., color photographs.
Anonymous	1993	Growing orchids. American Orchid Society. 136 pp., numerous color photographs.
Anonymous	1996	Guide to the AOS. 41 pp. American Orchid Society, Palm Beach.
Anonymous	1997	1st BIMP EAGA Orchid Congress, Davao, Philippines. Orchid Society of Davao. 66 pp., numerous B&W and color photographs.
Anonymous	1997	The orchids of Bruce & Grey. Bruce and Grey Plant Committee (Owen Sound Field Naturalists), Stan Bown Printers Limited, Owen Sound Ontario. 106 pp., numerous line drawings, 48 color photographs.
Anonymous	1876 or so	Orchids and their cultivation. Pages 905–988 and full page color paintings from The Practical Gardener bound into a book.
Anonymous	1896–1897	Hand lists of orchids and tender monocotyledons excluding orchids cultivated in the Royal Botanical Gardens, Kew. Her Majesty's Stationary Office. Two volumes (225 and 347 pp.) in one.
Anonymous	1955–1956	Beginner's handbook – XIII. Building your orchid collection. American Orchid Society, pp. 67–203.

(continued)

(continued)

Author(s)	Date	Title and publisher
Anonymous	1955–1956	Beginner's handbook – XXV. Miscellaneous genera for the hobbyist. American Orchid Society, pp. 204–281.
Anonymous	1990–1991	American Orchid Society Yearbook. 36 pp. American Orchid Society, Palm Beach.
Anonymous	No date	Collection of color photographs of *Calanthe*. Variations in spring blooming Calanthe. Garden Life Magazine. Seibundo Synkosya. (In Japanese.) 200 pp., numerous color photographs.
Anonymous	No date	Western [Tropical] orchid dictionary. (In Japanese.) 332 pp., numerous B&W and color photographs and line drawings.
Anonymous	No date	Orchids. Ukrainian SSR Ministry of Municipal Services, Kiev. Folded poster.
Anonymous (ed.)	1985	Proceedings of the 4th ASEAN Orchid Conference. Philippine Council for Agriculture and Resources Research and Development. Ministry of Agriculture and Food and the Ministry of Foreign Affairs, Philippines. 225 pp.
Anonymous (ed.)	1930	Centenaire de Lamarck. Archives du Muséum National d'Histoire Naturelle Ser 6, Vol. 6, pp. 1–73, numerous black and illustrations.
Anonymous (ed.)	1968	Voprosyi biologyi i ecologyi dominatov i edificatorov rastitelnyih soobchtev (Questions of biology and ecology of dominant plant communities-in Russian). Orchids are discussed on page 186. Ministry of Education RSFCR, USSR.
Anonymous (ed.)	1974	Orchid evaluation course, 2nd ed. South Florida Orchid Society, Miami. 75+viii pp., few line drawings.
Anonymous (ed.)	No date	Handbook of orchid culture. Santa Barbara International Orchid Show, Inc. 52 pp., numerous line drawings and B&W illustrations, some color.
Anonymous (ed.)	1937	Monja Blanca (Lycaste skinneri alba). Flor Nacional de Guatemala. Publicaciones de la Secretaria de Educacion Publica, Guatemala. A collection of poems, songs and other outpourings about the National Flower of Guatemala. Xerocopy, 95 pp.
Anonymous (ed.)	1969	An orchidist's lexicon, 1st printing. Oregon Orchid Society, Inc. 73 pp., one map.
Anonymous (ed.)	1975	The First ASEAN Orchid Congress, Report of the Congress. Ministry of Agriculture, Ministry of Foreign Affairs and Kasetsart University, Bangkok, Thailand, January 17–21, 1975. 284 pp., mimeographed pages.

(continued)

(continued)

Author(s)	Date	Title and publisher
Anonymous (ed.)	1980	Proceedings of the Second ASEAN Orchid Congress, April 6–10, 1977. Published by the Indonesian Orchid Society and the Ministry of Agriculture, Republic of Indonesia. 127 pp., line drawings and color photographs.
Anonymous (ed.)	1986	21st annual conference, Tropical Queensland Orchid Council, Townsville. Townsville Orchid Society.
Anonymous (ed.)	1986	Brochures relative to the 6th ASEAN Orchid Congress in Bangkok.
Anonymous (ed.)	1987	Proceedings of the World Orchid Hiroshima Symposium, Post-conference of the 12th World Orchid Conference, Hiroshima, Japan, March 23–25, 1987. Hiroshima Botanical Garden168 pp., maps, line drawings, B&W and color photographs.
Anonymous (ed.)	1994	Handbook of the ninth ASEAN orchid congress, 25 February–2 March 1994, Manila, Philippines, 99 numbered pages and about as many unpaginated.
Anonymous (ed.)	2004	Proceedings of the 8th Asia Pacific orchid conference (APOC 8), Tainan, Taiwan, March 6–8, 2004. 640 pp., line drawings, B&W and color photographs with a 78 page abstracts publication and 32 page program.
Anonymous (ed.)	ca 1969	Meristem tissue culture. A collection of articles. American Orchid Society, Delray Beach, FL. $15.
Anonymous (ed.)	Date?	*Cymbidium kanran* in Tosa. Tosa Orchid Society, Kohchi Prefercture, Japan. 290 pp., numerous color photographs. (In Japanese.)
Anonymous, P. Vermeulen, H. W. Pugsley, A. J. Wilmot, G. C. Druce, P. M. Hall, T, Stephenson, J. A. Richardson, W. H. Pearsall, and R. B. Ullman	1930, 1947, 1956, 1957	A collection of articles on British orchids in from several journals.
Appleby, T.	1861	The orchid manual, for the cultivation of stove, greenhouse, and hardy orchids, 1st ed. Journal of Horticulture and Cottage Gardener, London. 92 pp., line drawings.
Arana, C. B.	1962	Las orquideas: Como propagarlas. Extension Service, University of Puerto Rico, Rio Piedras, Puerto Rico. 16 pp., photographs, line drawings.

(continued)

(continued)

Author(s)	Date	Title and publisher
Arditti, J. (ed.)	1977	Orchid Biology: Reviews and Perspectives, Volume I. Cornell University Press, Ithaca, NY
Arditti, J. (ed.)	1982	Orchid Biology: Reviews and Perspectives, Volume II. Cornell University Press, Ithaca, NY
Arditti, J. (ed.)	1984	Orchid Biology: Reviews and Perspectives, Volume III. Cornell University Press, Ithaca, NY
Arditti, J. (ed.)	1987	Orchid Biology: Reviews and Perspectives, Volume IV. Cornell University Press, Ithaca, NY
Arditti, J. (ed.)	1990	Orchid Biology: Reviews and Perspectives, Volume V. Timber Press, Portland, OR
Arditti, J.	1992	Fundamentals of Orchid Biology, Wiley-Interscience, New York.
Arditti, J. (ed.)	1994	Orchid Biology: Reviews and Perspectives, Volume VI. Wiley-Interscience, New York.
Arditti, J. and R. Ernst	1993	Micropropagation of Orchids. Wiley-Interscience, New York
Arditti, J., and Alec M. Pridgeon (eds.)	1997	Orchid Biology: Reviews and Perspectives, Volume VII. Kluwer Academic Publishers, Dordrecht/Boston, MA/London
Argus, G. W., Hess, W. J., and J. L. Strothers	2002	Flora of North America north of Mexico. Oxford University Press, New York and Oxford. 723 pp., numerous maps.
Arnold, P.	1994	Orchids. Rizzoli International Publications, Inc., New York. Numerous line drawings and color photographs.
Arosemena, A., C. de Jurado, R. Estrada, and M. Konanz	1988	Orchids from the Coast of Equador. Associación Ecuatoriana de Orquideologia, P. O. Box 1033, Guaquil, Ecuador. Bilingual Spanish and English.
Asociacion Risaraldense de Orquideologia	1993	Las orquideas, guia practica para su cultivo.
Assavaphiches, C.	No date	Beautiful orchids.
Attwood, J. T.	1989	Orchids of Costa Rica, part 1. Icones Plantarum Tropicarum Fascile 14, Plates 1351–1400. Marie Selby Botanical Garden, Sarasota, FL.
Atwood, J. T., and D. E. Mora de Retana	1999	Flora Costaricensis, Family #39 Orchidaceae: Tribe Maxillarieae: Subtribes Maxillariinae and Oncidiinae. Fieldiana, Botany New Series No. 40. 183 pp., 51 line drawing plates.
Au Yong Nang Yip	No date	Notable orchids of Borneo. 3 page mimeograph.
Aulisi, A. C.	1997	*Cattleya luedemanniana* y sus variedades.

(continued)

(continued)

Author(s)	Date	Title and publisher
Aulisi A. C, and E. Foldats	1989	Monography of the Venezuelan cattlleyas and its varieties. Published by Carlo Aluisi, Venzuela.191 pp., numerous line drawings and color photographs.
Author ?	1976	The native orchids of Japan. S+S Publishing Co., Tokyo. 195 pp., numerous color photographs. Book measures 37.5 × 26 cm.
Author ?	1980	Orchid, part II. Geibun Mooks, Japan. 198 pp., numerous B&W and color illustrations, some line drawings.
Author ?	1981	Excellent orchids. Geibun Mooks, Japan. 150 pp., numerous B&W and color illustrations, some line drawings.
Author ?	1982	Beautiful orchids. Geibun Mooks, Japan. 178 pp., numerous B&W and color illustrations, some line drawings.
Author ?	Date?	Oriental orchids. How to grow and enjoy them. (In Japanese.)
Author(s) ?	1986	An introduction to the cultivation of 476 popular orchids. Shufunotomo Co., Ltd. 176 pp., numerous color line drawings and photographs. (In Japanese.)
Author(s) ?	?	Face of wild orchids. (In Japanese.) 173 pp., 140 colored and B&W photographs.
Author?	1960	Spring flowering cymbidiums of Japan and China. 114 pp., numerous B&W photographs.
Averyanov, L. V.	1994	Opredelitel orhidyi (Orchidaceae Juss.) Vietnama. Russian Academy of Sciences, Botanical Institute named after V. L. Komarov, and the National Institute of Natural Sciences and Technology, Published by Mir I Cemya, St. Petersburg. 433 pp., 4 plates of line drawings, 21 line drawing figures in the text. (In Russian.)
Averyanov, L. V.	1994	Opredelitel orhideih (Orchidaceae Juss.) Vietnama [Identification guide to Vietnamese orchids (Orchidaceae Juss.). Mir I Semya Publishing House, St. Petersburg, Russia. 432 pp., line drawings.
Averyanov, L. V. et al. (eds.)	1990	Vascular plants synopsis of Vietnamese flora. Nauka Publishing House, Leningrad (St. Petersburg). 199 pp.
Averyanov, L., P. Cribb, K. L. Phan, and T. H. Nguyen	2003	Slipper orchid of Vietnam. Timber Press, Portland, OR. 308 pp., 218 figures which include maps, line drawings and color photographs and paintings.

(continued)

(continued)

Author(s)	Date	Title and publisher
Averyanov, L. V., and A. L. Averyanova	2003	Updated checklist of the orchids of Vietnam. Vietname National University Publishing House, Hanoi. 101 pp., 11 line drawing figures.
Backer, C. A.	1952	Beknopte flora van Java 3 volumes
Backhouse, G., and J. Jeanes	1995	The orchids of Victoria
Bailes, C.	1986	The Eric Young Orchid Foundation souvenir guide. Eric Young Orchid Foundation, Jersey. 40 pp., line drawings and color photographs.
Bailey, F. M.	1883	Orchid pages (506–533) from A synopsis of the Queensland flora. James C. Beal, Government Printer, Brisbane.
Bailey, F. M.	1902	The Queensland flora: Orchidaceae. Printed for the Queensland Government by H. J. Diddams & Co, Brisbane. pp. 1514–1592.
Baker, M. L., and C. O. Baker	1991	Orchid species culture. *Pescatorea, Phaius, Phalaenopsis, Pholidota, Phragmipedium, Pleione*. Timber Press, Portland, OR. 250 pp.
Baker, M. L., and C. O. Baker	1996	Orchid species culture, *Dendrobium*. Timber Press, Portland, OR. 852 pp.
Baldwin, H.	1884	The orchids of New England. John Wiley & Sons, New York. 192 pp., 40 B&W line drawing figures.
Balis, J., and A. Lawalrée	1961	L'Orchidée en Belgique. Bibliothéque Albert I, Buxelles. 80 pp., line drawings, B&W drawings and photographs, color paintings.
Ball, A. V.	1965	Studies of the South African species of *Eulophia*. Journal of South African Botany, Supplementary Volume V. 248 pp., 43 maps.
Ball, J. S.	1978	South Africa epiphytic orchids.
Banerji, M. L.	1978	Orchids of Nepal. Today & Tomorrow's Printers & Publishers, New Delhi. 130 pp., 54 line drawing figures.
Banerji, M. L., and P. Pradhan	1984	The orchids of Nepal-Himalaya. J. Cramer, Vaduz, Germany. 534 pp., 247 line drawing plates, some of them foldouts. The book measures 34.3 × 25 cm.
Banks, D.	2003	Handy pocket guide to orchids
Banks, D. P.	2005	Orchid grower's companion. Cultivation, propagation, and varieties. Timber Press, Portland, OR.
Banks, D. P.	1999	Tropical orchids of Southeast Asia. Periplus Editions (HK) Ltd. 64 pp., numerous color photographs.

(continued)

(continued)

Author(s)	Date	Title and publisher
Barbosa Rodrigues, J.	1882	Genera et species Orchidearum novarum. Sebastianapolis. Vol I: 302+XVI pp.; Vol II: 302+XVI pp. (In French.)
Bark, D.	2006	Wildwachsende orchideen.
Barla, J. B.	1968	Flore Illustree de Nice et des Alpes-Maritimes-Iconographie des Orchidees.
Barretto, G. D'A., and J. L. Y. Sate	1980	Hong Kong orchids. Urban Council, Hong Kong. 109 pp., 2 line drawings, numerous color photographs.
Barringer, K.	1986	Typification of Schlecter's Costa Rican Orchidaceae. I. Types collected by A. Brenes. Fieldiana Botany, New Series, No. 17, Publication 1365. iii + 24 pp., line drawings.
Bartolini, L.	1996	Le orchidee della Valle Camonica
Batchelor, S. R.	No date	Yoir first orchid: Guide for the beginner. American Orchid Society, Delray Beach, FL
Bateman, J.	1867	A second century of orchidaceous plants. L. Reeve & Co., London. One hundred plates (numbered 101–200) from Curtis' Botanical Magazine.
Bateman, J.	1874	Monograph of *Odontoglossum*
Bateman, J.	1979	Hundert orchdeen. Harenberg Kommunikation, Dortmund, Germany. 224 pp., 100 color paintings by Walter Hood Fitch, selected and annotated by Edmund Launert.
Bateman, J.	1837–1842	The Orchidaceae of Mexico and Guatemala. James Ridgway & Sons. Facsimile edition 1974 by Johnson Reprint Corporation, New York.
Batygina T. B. (ed.)	1997	Embryology of flowering plants. Terminology and conceptions. II. Saint-Petersburg – chapters about early stages of orchid ontogenesis. (In Russian.)
Batygina T. B. (ed.)	2000	Embryology of flowering plants. Terminology and conceptions. III. Saint-Petersburg – chapters about early stages of orchid ontogenesis. (In Russian.)
Batygina, T. B. and Yakovlev (eds.)	1979	Comparative embryology of flowering plants. The monocotylodinae. Nauka. Leningrad – chapters about orchid embryology. (In Russian.)
Bauer, F., and J. Lindley	1830–1838	Illustrations of orchidaceous plants. James Ridgway and Sons, London. No pagination.
Bauman, H. et al.	2006	Orchideen Europas mis angrezenden gebieten
Baumann, H., and S. Kunkele	1982	Die Wildwachsenden Orchideen Europas
Baumann, H., and S. Kunkele	1983	Thieme's orchideeegids

(continued)

(continued)

Author(s)	Date	Title and publisher
Baumann, H., and S. Kunkele	1988	Die orchideen Europas
Bechtel, H.	1969	Orchideen im heim. Landbuch-Verlag, GMBH. Hannover, Germany. 180 pp., numerous line drawings and color photographs.
Bechtel, H.	1971	Exotische orchideen, Kosmos. Gesellschaft der Naturfreunde Franckh'sche Verlaghandlung, Stuttgart. 71 pp., 120 color photographs.
Bechtel, H.	1977	Wundebare und geheimnisvolle Welt der exotischen Orchideen. Bertelsmann Lexikon-Verlag, Berlin. 179 pp., numerous color photographs.
Bechtel, H.	1982	Orchideen mein hobby, 2nd ed. Hallwag Verlag, Bern, Switzerland. 128 pp., numerous color photographs.
Bechtel, H., P. Cribb, and E. Launert	1980	Orchideenatlas. Verlag Eugen Ulmer, Stuttgart. 475 pp., numerous line drawings and color photographs. (In German.)
Bechtel, H., P. Cribb, and E. Launert	1981	The manual of cultivated orchid species, 1st ed. MIT Press. 444 pp., numerous line drawings and color photographs.
Bechtel, H., P. Cribb, and E. Launert	1981	The manual of cultivated orchid species, 1st ed. Blandford Press, Poole, Dorset, UK. 444 pp., numerous line drawings and color photographs.
Bechtel, H., P. Cribb, and E. Launert	1986	The manual of cultivated orchid species, revised edition. MIT Press. 444 pp., numerous line drawings and 720 color photographs.
Bechtel, H., P. Cribb, and E. Launert	1992	The manual of cultivated orchid species, 3rd ed. MIT Press. 585 pp., numerous line drawings and 864 color photographs.
Bedford, R. B.	1969	A guide to native Australasian orchids. Angus and Robertson. Sydney. 132 pp., 7 plates of color paintings, 58 plates of line drawings.
Beekman, E. M.	1999	The Ambonese curiosity cabinet. Yale University Press, New Haven, Connecticut, 567 pp., with all illustrations from the original which was published in the 1600s or 1700s.
Beekman, E. M.	2003	Rumphius' orchids. Yale University Press. 172. pp., numerous line drawings.
Beer, J. G.	1863	Beiträge zur morphologie und biologie der familie der orchideen. Druck and Verlag von Carl Gerold's Sohn, Vienna.
Bégaud, J.	1995	Orchidées indigenes de Nouvelle-Caledonie. Société Néo-Calédonienne d'orchidophilie, Noumea, New Caledonia. 180 pp., numerous color photographs.
Béguin, J.-J.	1993	A. P. O. orchid source book. Jaysquare Associates S, A. R. L. Publication, Geneva, Switzerland. 131 pp.

(continued)

(continued)

Author(s)	Date	Title and publisher
Behar, M.	1993	Orquídeas de Guatemala. Iconos Publishers, Guatemala.
Behar, M., and O. Tinschert	1998	Guatemala y sus orquídeas/Guatemala and its orchids. Bancafe, Guatemala City. Bilingual, 240 pp., numerous color and B&W photographs. Includes a CD.
Belitskii, I. V.	2001	Orchids (Orchidei). Moscow: Astrel. 175 pp., numerous color photographs. (In Russian.)
Bender, G. A., and R. A. Thom	1961	Great moments in medicine. Parke-Davis, Detroit. 277 pp., about 60 color plates. $30
Bennett, D. E., and E. A. Christenson	1993	Icones Orchidacearum Peruvianum. A. Pastorelly de Bennett, Lima. No page numbers. Plates 1–600, line drawings.
Bennett, K. S.	1984	The tropical Asiatic slipper orchids. Angus and Robertson Publishers, North Ryde, NSW, Australia. 91 pp., numerous color photographs.
Bentham, G.	1881	Notes on orchideae. Journal of the Linnean Society (London), Volume 18, pp. 281–360.
Bentley, S. L.	2000	Native orchids of the Southern Appalachian mountains. The University of North Carolina Press, Chapel Hill, NC. 235 pp., numerous color photographs and maps.
Berliocchi, L.	1996	The orchid in lore and legend. Translated by Lenore Rosenberg and Anita Weston, edited by Mark Griffiths. Timber Press, Portland, OR. 184 pp., B&W and color illustrations.
Berliocchi, L.	No date	Il fiore degli dei. Stampe Alternativa. Italy. 199 pp., 16 color plates, numerous line drawings.
Bernadi, L., and P. A. Robert	1966	Fleurs tropicales. Éditions Delachaux & Niestle, Neuchatel, Switzerland. 326 pp., 20 B&W and 30 color illustrations.
Bernard, N.	1916	L'Évolution des plantes. Librairie de Félix Alcan, Paris. 314 pp., 29 line drawings.
Bernard, N.	1921	Principles de biologie végétale. Librairie de Félix Alcan, Paris. 212 pp., 18 line drawings. Bernard died in 1911. This book was edited/written by his mentor Julien Costantin and his friend and cousin Joseph Magrou from Bernard's notes. The preface is by Bernard's widow Marie-Louise. In a separate note she thanked Costantin and Magrou. The book was published posthumously.
Bernhardt, P.	1989	Wily violets & underground orchids. Revelations of a botanist. William Morrow and Company, Inc., New York. xii+255 pp., line drawings, B&W and color illustrations.

(continued)

(continued)

Author(s)	Date	Title and publisher
Bernhardt, P.	2003	Wily violets and underground orchids. The University of Chicago Press, Chicago, IL. 256 pp., some line drawings and B&W photographs.
Bevilacqua, F	2000	Le orchidee spontanee del Parco Regionale Roccamonfino-Foce del Garigliano
Bicalho, H. D.	1969	Subsidios à orquidocultura Paulista, Secretaria da Agricultura, Instituto de Botânica, São Paulo, Brasil. 121 pp., and excellent B&W and color illustrations of orchid pests.
bin Mohamad, M.	1970	The Malay dilemma. Donald Moore for Asia Pacific Press, Singapore. Xerocopy.
Bingham, M. T.	1939	Orchids of Michigan. Cranbrook Institute of Science, Bloomfield, MI. Bulletin No. 15. 88 pp., 21 plates (most B&W photographs, a few are color paintings).
Birk, L.	1983	The *Paphiopedilum* grower's manual, 1st ed. Pisang Press, Santa Barbara, CA 93101. 207 pp., numerous color and B&W illustrations.
Birk, L.	1975	Growing *Cymbidium* orchids at home. Published by the author. 47 pp., numerous B&W and color photographs.
Birk, L. A.	2004	The *Paphiopedilum* grower's manual, 2nd ed. Pisang Press, Santa Barbara, CA 93101. 284 pp., numerous color and B&W illustrations.
Bishop, T.	1996	Field guide to the orchids of New South Wales and Victoria. University of New South Wales Press, Sydney. 257 pp., 5004 color photographs.
Bisset, J.	ca. 1950	A handbook on orchids. Published by John Bisset, Abbotsford, NSW, Australia. 114 pp., 35 B&W plates. NSW.
Bisset, J., A. B. Porter and L. J. Lawler	1957	The Orchid Society of New South Wales, Ltd. 77 pp.
Bisset, J., and H. W. Wilson	No date	*Cymbidium* culture and virus diseases. Published by John Bisset, Abbotsford, Sydney, NSW, Australia. 93 pp., B&W photographs.
Black, M. McK	1988	The complete book of orchid growing, 2nd ed. Trafalgar Square Publishing, North Pomfret, VT. 160 pp., line drawings and 100 color photographs.
Black, M. McK	1992	The Ward Lock book of orchid growing. Ward Lock, Ltd., London. 160 pp., line drawings and 100 color photographs.
Black, P. M.	1998	Orchid growing. Ward Lock, Wellington House, 125 Strand, London. 160 pp., numerous color photographs, some line drawings.

(continued)

(continued)

Author(s)	Date	Title and publisher
Black, P. McK	1973	Beautiful orchids. Hamlyn Publishing Group, London. 128 pp., numerous B&W and color photographs. Some of the photographs are of orchid history.
Black, P. McK	1992	Orchid growing, 2nd ed. revised by W. Rittershausen and with a contribution by W. Rysy. Ward Lock, London. 160 pp., numerous color photographs, several line drawings.
Black, P., and C. Dixon	No date	Orchids. Evening Standard and Orchid Growers Association. 6 pp., one line drawing.
Blackmore, S.	1985	Bee orchids. Shire Publications, Ltd., Aylesbury, Bucks, UK. 24 pp., Line drawings and color photographs.
Blake, M.	1992	Wild Orchid Echo. Volume 1, Winter 1992. No other information is available.
Blaxell, D. F.	1973	The status os Schlechter's specimens of Orchidaceae held at the National Herbarium of New South Wales. I. New Guinea. Contributions from the New South Wales National Herbarium 4 (7): 457–470.
Blombery, A. M., and B. Maloney	1998	Growing Australian orchids. Kangaroo Press/ Simon and Schuster Australia, East Roseville, Australia 2069. 72 pp., numerous color photographs and B&W paintings.
Blowers, J. W.	1957	Orchid growing, revised edition. Dover Publications, Inc., New York. 94 pp., 8 B&W photographs.
Blowers, J. W.	1962	Orchids. Blandford Press, London. 129 pp., numerous B&W and color photographs.
Blowers, J. W.	1962	Orchids. The garden Book Club, London. 121 pp., line drawings and B&W illustrations.
Blowers, J. W.	1966	Pictorial orchid growing. Published by John W. Blowers, 96 Marion Crescent, Maidstone, Kent, UK. 127 pp., 266 B&W photographs.
Blowers, J. W.	1971	Pictorial orchid growing, 2nd ed. Wyld Court Orchids, Hampstead-Norris, Newbury, Berks, UK. 137 pp., 307 B&W photographs.
Blowers, J. W.	No date	Introducing orchids. British Orchid Growers' Association, London. 15 pp.
Blowers, J., and A. Moon	1993	Your first orchids, where and how to grow them. The Eric Young Orchid Foundation, Jersey. 80 pp., numerous color and B&W photographs.
Blume, C. L.	1857	Flora of Java: Orchidaceae
Boardman, R. C.	1906	Lilies and orchids. Robert Grier Cooke, Inc., New York. 24 pp., 29 color plates of color paintings.

(continued)

(continued)

Author(s)	Date	Title and publisher
Bockemül, L.	1969	*Odontoglossum*. Brücke-Verlag Kurt Schmersow, Hildesheim, Germany. 344 pp., numerous line drawings and color and B&W photographs.
Bodegom, J., van	1973	Einge orchideeen van West Nieuw Guinea
Bohnhof, E.	1893	Dictionnaire des orchidées hybrides. Octave Doin, Éditeur, Paris. 139 pp.
Bois, D.	1893	Les orchidées manuel de l'amateur. Librairie J.-B. Baillière et Fils, Paris. 323 pp., 119 line drawings.
Bois, D.	1893	Les orchidées, manuel de l'amateur. Librairie J.-B. Baillière et Fils, Paris. 323 pp., 119 line drawings.
Bolus, H.	1918	The orchids of the Cape Peninsula, 2nd ed. Darter Bros. & Co., Cape Town. 142 pp., 119 color paintings.
Bolus, H.	1896–1913	Orchids of South Africa, 3 volumes
Bond, R.	1988	All about growing orchids. Ortho Books, San Francisco, CA.96 pp., numerous color photographs.
Bone, G	1928	The hidden orchids. The Medici Society, London & Boston, MA. 49 pp., line drawings.
Bonham, N., and D. Bonham	1990	A history of orchid growers in New Zealand. Published by the Orchid Council of New Zealand. 146 pp., several B&W illustrations.
Borg-Karlson, A.-K.	1985	Chemical and behavioral studies of pollination in the genus *Ophrys* L. (Orchidaceae).
Bose, T. K., and S. K. Bhattacharjee	1980	Orchids of India. Naya Prokash, Calcutta. 538+xxiii pp., numerous line drawings and color photographs.
Boulard, B.	1985a	Un biologiste d'exception: Noel Bernard 1874–1911. Published either by the author or the University of Rouen. An excellent and very rare biography of Noel Bernard.
Bouriquet, G.	1954	Le vanillier et la vanille dans le monde. Èditions Paul Lechevalier, Paris. 748 pp., color plates, numerous line drawings, maps and B&W photographs.
Bowden, C.	2002	Blood orchid. North Point Press, New York.
Bowen, L.	1976	The art and craft of growing orchids. G. P. Putnam's Sons, New York. 126 pp., color and B&W photographs and line drawings.
Boyd, J.	1978	The pollinators of Eden. A novel. Penguin Books, New York. 212 pp.
Boyle, F.	1893	About orchids, a chat. Chapman & Hall, London. 250 pp., eight plates of color paintings.
Boyle, F.	1896	Uber orchideen.

(continued)

(continued)

Author(s)	Date	Title and publisher
Boyle, F.	1902	The culture of greenhouse orchids. Chapman and Hall, Ltd., London. 231 pp., 3 color plates, 50 B&W illustrations.
Boyle, F.	1901	The woodland orchids. Macmillan and Co., Ltd. London. 274 pp., 16 color plates.
Boyle, L. M.	1946	El Rancho Rinconada, Orchid town. Published by the author at Rancho Rinconada, Ojain, CA. No pagination (about 50–60 pp.), bound, numerous B&W and color paintings, drawings and photographs.
Boyle, L. M.	1947	My observations on growing *Cymbidium* orchids out of doors. *Cymbidium* Orchid Town, El Rancho Rinconada, Ojai, CA. 47 pp., B&W photographs.
Boyle, L. M.	1950	*Cymbidium* orchids and you.
Boyle, L. M.	1952	Out west growing *Cymbidium* orchids and other flowers. The story of rancho Rinconada. Published by the author. 526 pp., numerous B&W photographs.
Bracey, B. O. (?)	1927	Commercial orchids and their production. Armacost and Royston, Inc., Los Angeles, CA. 22 pp., 10 B&W photographs.
Brackley, F. E.	1985	The orchids of New Hampshire. Rhodora 87, No. 849, pp. 1–117. Numerous maps.
Braecklein, A.	1904	Die orchideen und ihre kultur im zimmer, 1st ed. Verlag von Trowitzsch & Sohn, Frankfurt an den Oder. 100 pp., 50 line drawings and B&W photographs and paintings.
Braecklein, U.	1918	Die orchideen und ihre kultur im zimmer, 2nd ed. Verlag von Trowitzsch & Sohn, Frankfurt an den Oder. 120 pp., 77 line drawings and B&W photographs and paintings.
Braem, G. J.	1988	*Paphiopedilum*. Brücke-Verlag Kurt Schmersow, D-3200 Hildesheim, Germany. 249 pp., 110 color photographs, bilingual English and German
Braem, G. J.	No date	Cattleya, The Brazilian bifoliate cattleyas. Brücke-Verlag Kurt Schmersow, Hildesheim. 94 pp., 47 color 3 B&W plates and 2 line drawings.
Braem, G. J.	No date	*Cattleya*, The unifoliate cattleyas. Brücke-Verlag Kurt Schmersow, Hildesheim. 96 pp., 43 color plates.
Braem, G. J., C. O. Baker, and M. L. Baker	1998–1999	The genus *Paphiopedilum*, natural history and cultivation. Botanical Publishers, Inc., 2501 Old Lake Wilson Road, Kassimmee, FL 34747. Two volumes, 362 pp. of a projected three volume work. The third volume was never published. Numerous color photographs, some line drawings.

(continued)

(continued)

Author(s)	Date	Title and publisher
Braem, G. J., and G. R. Chiron	2003	*Paphiopedilum*. Tropicalia, F-69230 Saint-Genis Laval, France. 440 pp., numerous line drawings, water colors, and color photographs.
Bransilver, C.	2004	Wild love affair. Essence of Florida's native orchids. Westcliffe Publishers, Inc., Englewood, CO. 126 pp., numerous color photographs.
Bransilver, C.	2004	Wild love affair-essence of Florida's native orchids.
Brieger, F. G., R. Maatsch, and K. Senghas (eds.)	1970–2003	Rudolf Schlechter Die Orchideen, 3rd ed. Parts IA (pp. 1–944, 739 figures), IB (pp. 945–1976), IC (1977–2897) and II (727 pp.) plus a separate volume which lists literature, has an index, includes a biography of Schlechter and provides addenda, 268 pp). Parey Buchverlag, Berlin.
Briscoe, T. W.	1939	Orchids for amateurs, 2nd ed. W. H. & L. Collingridge, Ltd., London. 128 pp., line drawings and B&W photographs.
Briscoe, T. W.	1948	Orchids for amateurs. W. H. & L. Collingridge, Ltd., Covent Garden, UK. 128 pp., line drawings and B&W photographs.
Briscoe, T. W.	ca 1938	Orchids for amateurs. W. H. & L. Collingridge, Ltd., London. 137 pp., numerous line drawings and color photographs.
Bristow, A.	1982	Orchids. Wisley Handbook 42. The Royal Horticultural Society. London. 49 pp., 8 color photographs and some line drawings.
Bristow, A.	1992	Orchids. The Royal Horticultural Society. 64 pp., line drawings and color photographs.
Britten, J.	1878	Orchids for amateurs.
Brooke, J.	1950	The wild orchids of Britain. The Bodley Head Publishers, London. 139 pp., 40 color painting by Gavin Bone, Muirhead Bone and Stephen Bone.
Brooke, J. & Co.	1872	The Fairfield orchids. Bradbury, Evans & Co., London. 128 pp.
Brown, P. M.	2004	Wild orchids of the southeastern United States north of peninsular Florida. University Press of Florida, Gainesville, FL. 396 pp., numerous maps and color painting and photographs.
Brown, P. M. and S. Folsom	1993	Filed and study guide to the orchids of New England and New York. Orchis Press, 15 Dresden Street, Jamaica Plains, MA. 247 pp., numerous line drawings.
Brown, P. M. and S. Folsom	1997	Wild orchids of the Northeastern United States. A field guide. Comstock Publishing Associates, Cornell University Press, Ithaca, NY. 236 pp., numerous line drawings and color photographs.

(continued)

(continued)

Author(s)	Date	Title and publisher
Brown, P. M., and S. Folsom	2002	Wild orchids of Florida. University Press of Florida, Gainesville, numerous maps, line drawings and color photographs
Brown, R.	1831	Observations on the organs and mode of fecundation of Orchidaceae and Asclepiadaceae. No publisher listed. 36+2 pp.
Brühl, P.	1926	A guide to the orchids of Sikkim. Thacker, Spink & Co., Calcutta and Simla, India. 208 pp.
Bulat, T. J., and M. Bulat	1995	Hidden orchids. Rudi Publishing, P. O. Box 1066, Iowas City, IA 52244. 112 pp., numerous color photographs.
Burberry, H. A.	1894	The amateur orchid cultivators' guide book, 1st ed. Blake & MacKenzie Printers and Publishers, Liverpool. 144 pp., line drawings and B&W photographs.
Burberry, H. A.	1895	The amateur orchid cultivators' guide book, 2nd ed. Blake & MacKenzie Printers and Publishers, Liverpool. 200 pp., line drawings, B&W photographs and three color paintings.
Burberry, H. A.	1899	The amateur orchid cultivators' guide book, 1st ed. Blake & MacKenzie Printers and Publishers, Liverpool. 188 pp., line drawing, B&W photographs and color frontispiece and plates.
Burbery, H. A.	1900	The amateur orchid grower's guide book, 3rd ed.
Burbidge, F. W.	1874	Cool orchids and how to grow them. Robert Harwicke Publishing, London. 180 pp., line drawings and color paintings.
Burbidge, F. W.	1875	Die Orchideen des temperirten und kalten hauses.
Burbidge, F. W.	1882	Die Orchideen des temperirten und kalten hauses.
Burddett, F. D.	1930	The odyssey of an orchid hunter. Herbert Jenkins, Ltd., London. 317 pp., frontispiece, 15 B&W plates of photographs, maps.
Burgeff, H.	1909	Die wurzelpilze der orchideen ihre kultur und ihr leben in der pflanze. Verlag von Gustav Fischer in Jena. 220 pp., 2 fold out plates, 38 line drawings and B&W photographs in the text.
Burgeff, H.	1932	Saprophytismus und symbiose. Verlag von Gustav Fischer in Jena. 249 pp., 176 line drawings and B&W photographs.
Burgeff, H.	1936	Samenkeimung der orchideen. Verlag von Gustav Fischer in Jena. 312 pp., 186 line drawings and B&W photographs.
Burgeff, H.	1954	Samekeimung und kultur Europäischer erdorchideen. Verlag von Gustav Fischer in Jena. 48 pp., 10 line drawings and B&W photographs.

(continued)

(continued)

Author(s)	Date	Title and publisher
Burgeff, H.	1911	Die anzucht tropischer orchideen aus samen. Verlag von Gustav Fischer, Jena. 90 pp.
Burger, A.	1992	Orchideen fur den garten
Burian, R., and W. Bluhm	2000	Native orchids of Oregon. Oregon Orchid Society, Inc., P. O. Box 14182, Portland, OR. 40 pp., 25 color photographs.
Burnett, H. C.	1974	Orchid diseases. Bulletin 10, Florida Department of Agriculture and Consumer Services. 66 pp., B&W and color photographs.
Burnnett, H. C.	1965	Orchid diseases. Publication volume I, Number 3, State of Florida Department of Agriculture. 56 pp., 13 color pictures.
Burnnett, H. C.	1974	Orchid diseases. Bulletin 10, State of Florida Department of Agriculture. 66 pp., 13 color pictures.
Burrage, A. C.	1930	The orchidvale collection, Orchidvale, Beverly Farms, Massachusetts. The Plimpton Press, Norwood, MA, USA. 188 pp.
Butterfield, H. M.	1956	Orchids for the California amateur, second printing. California Agricultural Experiment Station Extension Service Manual 18.
Butterfield, I.	No date	How to grow *Pleione* orchids.
Buttler [no initials]	1996	Steinbachs naturführer orchideen.
Buttler, K. P.	1986	Orchideen. Mosaik Verlag. Munich. 288 pp., 562 color photographs, 197 line drawings, 16 maps.
Buttler, K. P.	1991	Field guide to orchids of Britain and Europe. The Crowood Press, Swindon, UK. 288 pp. 750+ line drawings and color photographs.
Buysson, F. du	1878	L'Orchidophile
Byron, Camus, Dostoevski, Kierkegaard, Tagore and Wilde and others	1970	Springs of friendship. A booklet with quotes from various authors and orchid paintings. Herder and Herder, New York. No pagination.
Cabalero, R.	1986	Orquideas de Costa Rica
Cabrera, A. L.	1868	Flora de la provincia de Buenos Aires. Coleccion Cientifica Del I. N. T. A., Instituto Nacional de Tecnoloia Agropecuaria, Buenos Aires, Argentina. 627 pp., 129 line drawings.
Cady, L., and E. R. Rotherham	1970	Australian native orchids in color. A. H. & A. W. Reed, Sydney. 112 pp., 100 color photographs.
Cameron, J. W.	1976	The orchids of Maine.
Cameron, K. M., J. Arditti, and T. Kull (eds.)	2007	Orchid Biology: Reviews and Perspectives. Volume IX. The New York Botanical Garden Press.
Camus, E. G.	1894	Monographie des orchidées de France.

(continued)

(continued)

Author(s)	Date	Title and publisher
Camus, E. G.	1885–1891	Iconographie des orchidées des environs de Paris.
Camus, E. G., P. Bergon, and A. Camus	1908	Monographie des orchidées d l'Europe, de l'Afrique septentrionale, de l'Asie Mineure et les Provinces Russes transcaspienenes. Jaques Lechevalier, Paris. 484 mimeographed pages, 32 plates with 1,100 figures. In some copies of this book several of the plates were hand colored. Only 175 copies were published. Some were numbered.
Camus, E. G., and A. Camus	1885	Notes sure les orchidées.
Camus, E. G., and A. Camus	1921	Iconographie des orchidées d'Europe et du diagr méditerranéen. Paul Lechevalier, Paris. Text and figure atlas. Figures 1–110 are colored. Figures 111–122 are not in color.
Capeder, E.	1898	Orchid morphology
Capps, A. L.	1965	The source handbook of orchid species. Published by the author at Route 4, Box 376, Vienna, Virginia 22180.
Carbone, G. et al.	2002	The little book of orchids.
Carlson, E. J.	ca 1988	*Miltoniopsis* culture
Carr, C. E.	1928	Orchid pollination notes
Carr, G.	1981	Native orchids in Melbourne. Australasian Native Orchid Society (Victorian Branch). 36 pp., line drawings.
Carr, G. F., Jr.	1995	*Cychnoches*, 165 years of confusion.
Carr, G. W.	1980	Mellblom's spider-orchid conservation. Environmental design report No. 4. Alcoa Portland aluminium smelter. Kinhill Planners Pty., Ltd., 530 Little Collins Street, Melbourne, Australia.
Carroll, C., and C. H. Fischer	2002	Orchid of the Bayou. A novel. Gallaudedt University Press, Washington, DC. 253 pp.
Case, F. W., Jr.	1964	Orchids of the western Great Lakes region. Cranbrook Institute of Science, Bloomfield Hilla, Michigan 48013. 147 pp., numerous B&W photographs and maps.
Cash, C.	1991	The slipper orchids. Timber Press, Portland, OR. 228 pp., 225 color photographs, a few line drawings.
Castle, L.	1887	Orchids, their structure, history and culture, 3rd ed. Journal of Horticulture Publishers, London.
Castle, L.	1889	Les orchidées, structure, histoire et culture. Translated by A. de Meulenaere. Imprimerie F & R. Buyck, Gand, Belgium.
Castle, L.	ca 1886	Orchids, their structure, history and culture, 2nd ed. Journal of Horticulture Publishers. London. 106 pp., 21 line drawing illustrations.

(continued)

(continued)

Author(s)	Date	Title and publisher
Castle, L.	1886	Orchids, their structure, history and culture, 1st ed. Journal of Horticulture Office, 171 Fleet Street. 105 pp., black and white illustrations.
Cavallo, O., R. Cavallo, G. Dellapiana	1993	Guida alle orchidee spontanee delle Langhe. Amici Del Museo "F. Eusebio," Alba, Italy. 263 pp., 170 figures (color and B&W photographs, line drawing, maps).
Ceibb, P. J., J. Greatwood, and P. F. Hunt	1985	Handbook of orchid nomenclature and registration, 3rd ed. International Orchid Commission, Royal Horticultural Society, London. 143 pp.
Ceulemans, N.	2006	Lindedn, explorer master of the orchids
Ceutericik, A.	1908	Le Comte Oswald de Kerkhove de Denterghem. Ad. Hoste, Gand, Belgium. 165 pp. He was a well known orchid grower in his time.
Chadwick, A. A. and A. Chadwick	2006	The classic cattleyas.
Chan, C. L., A. Lamb, P. S. Shim, and J. J. Wood	1994	Orchids of Borneo, Vol 1. The Sabah Society and Royal Botanic Gardens, Kew. 401 pp., numerous line drawings in 100 figures and color photographs in 22 plates, a few maps.
Chandler, R.	1976	The big sleep. A novel. Vintage Books, New York. 217.
Chang, Z. Q.	1988	Handbook of orchid propagation diagrams&$$$;. Breeding and tissue culture of the Butterfly Orchid [meaning Phalaenopsis]. Published inn Taiwan in Chinese. 98 pp., numerous line drawings, color photographs and media recipes.
Chapman, W. K.	1997	Orchids of the Northeast, a field guide. Syracuse University Press, Syracuse, NY. 200 pp., line drawings and color photographs.
Chardard, R.	1962	Recherches sur les cellules-meres des microspores des orchidees. Etude au microscope electronique.
Chase, F. W. Jr.	1987	Orchids of the western Great Lakes region. Revised edition. Cranbrool Institute of Science Bulletin 48. 251 pp., numerous line drawings, maps and color photographs.
Chase, J. H.	1980	No orchids for Miss Blandish. A novel. Penguin Books, New York. 176 pp.
Chase, M. (ed.)	1997	The pictorial encyclopedia of *Oncidium*. ZAI Publications, New York City. 164 pp., numerous color paintings by H. Zelenko, B. D. Zelenko and J. Warshaw.
Chase, M. W	1993	*Oncidium* alliance. Part Three of Thesaurus Woolwardiae. Missouri Botanical Garden in association with the The Royal Botanic Gardens, Kew. No pagination, 15 color plates. See also Cribb, 1993; Stewart, 1993; Wood, 1993.

(continued)

(continued)

Author(s)	Date	Title and publisher
Chatin, M. A.	1850	Anatomie des plantes aeriennes de l'ordre des orchidees
Chávez Salazar, G. A., and M. A. Soto Arenas	1996	El genero Lepanthes SW. en Mexico. Orquidea (Mex.) Volumen 14, Numero Unico. Association Mexicana de Orquidelogia, A. C., Mexico, D. F. 230 pp., 77 line drawing plates, 74 color photographs, one B&W photograph.
Chen S. C., Z. Tsi, and Y. Luo	1999	Native orchids of China in color
Chen, Y. C.	1989	The new and champion of *Cattleya* (Japanese and English).
Cheng, C	No date	Formosan orchids. Chow Cheng Orchids. 103 pp., 170 color photographs.
Cheng, C.	1970	Chinese cymbidiums of Taiwan. No publisher listed. 41 pp., numerous color and B&W photographs. (In Chinese.) Some illustrations have Latin name captions.
Cheng, C.	1968 (?)	Taiwan native orchids. Published by Chow Cheng Orchids, Taichung, Taiwan. No pagination, numerous B&W and color photographs.
Cherevchenko, T. M.	1993	Tropitcheskie I subtropitcheskie orhidei. Naukova Dymka Publishing House, Kiev, Ukraine. 253 pp., 12 line drawings, numerous color photographs. (In Russian.)
Cherevchenko, T. M., L. I. Byun, L. A. Kovalska, and V. S. Vahrushkin	2001	Orhidei. (In Russian or Ukrainian.) Prosvita, Kiev, Ukraine, 224 pp., numerous color photographs.
Cherevchenko, T. M., and G. P. Kushnir	1986	Orhidei v kulture. Naukova Dumka Publishing House, Kiev, Ukraine. 200 pp., 85 line drawing and B&W photograph figures in the text, 16 pages of color photographs. (In Russian.)
Chernyakovskaya, E. G.	1932	Orchidaceae. Pages 330–338. In: Flora of Turkmenia. Leningrad. (In Russian.)
Chernyakovskaya, E. G.	1941	Orchidaceae. Pages 522–533. In: Flora of Uzbekistan. 1. (In Russian.)
Chiron, G., and W. Cavestro	1999	*Paphiopedilum*. Tropicalia, 69230 Saint-Genis-Laval, France. 157 pp., numerous maps, 60 line drawings and 80 color photographs.
Chitty, F. D.	1976	Flora de la Hacienda El Limon, Distrito Federal. Memoria de la Sociedad de Ciencias Naturales La Salle. Vol. 36, No. 105, pp. 7–110, several photographs.
Chitty, F. D.	1978	Aportes al Conocimiento de la etnobotanica del estado Cojedes. Fundacion La Sale de Ciencia Naturales, Caracas. 128 pp., 41 line drawing plates.

(continued)

(continued)

Author(s)	Date	Title and publisher
Chowdhery, H. A., and G. D. Pal	1997	Orchidaceae of Arunachal Pradesh.
Chowdhery, H. J.	1998	Orchid Flora of Arunachal Pradesh.
Christenson, E.	2003	Machu Picchu orchids.
Christenson, E. A.	2001	*Phalaenopsis*, a monograph. Timber Press, Portland, OR. 330 pp., numerous line drawings and color photographs.
Christie, C. B., and D. Cohen	1980	Massey University proceedings of orchid culture workshop, 18–20 November 1980. 89 pp.
Christie, C. B., and D. Cohen (ed.)	1980	Proceedings of orchid culture workshop. Department and Horticulture and Plant Health, Massey University, New Zealand. 89 pp., line drawings.
Christofides, Y.	2001	Orchids of Cyprus.
Chtanondh, H.	1987	A color picture book of orchids, Vol. 1.
Chuo, S. K.	1979	Common pests of orchids in Singapore. Primary Production Department, Ministry of National Development, Singapore. 30 pp., 24 color pictures.
Cingel, N. A. van der	1995	An atlas of orchid pollination. European orchids Rotterdam; Brookfield.
Cingel, N. A. van der	2001	An atlas of orchid pollination. America, Africa, Asia and Australia. CRC Press, Boca Raton, FL.
Clark, J. (ed.)	1999	16th World Orchid Conference (April 22–May 3, 1999) registrants' handbook and abstracts (44 pp.) and 26 pp., general announcement.
Clarke, A. C.	1970	Tales from the White Hart. A collection of stories one of which deals with an orchid. Harcourt, Brace & World, Inc., New York. 179 pp.
Clayton, D.	2002	The genus *Coelogyne*: A synopsis.
Clement, J. L.	1978	Connaissance des orchidees sauvages.
Clements, M. A. (compiler and editor)	1982	Preliminary checklist of Australian orchids. National Botanic Gardens, Canberra, Australia. 216 pp.
Clements, M. A.	1989	Catalogue of Australian Orchidaceae. Volume 1 of Australian Orchid Research (D. L. Jones, editor). Australian Orchid Foundation, 107 Roberts Street, Essendon, Victoria, Australia. 160 pp., 6 plates of line drawings.
Clifton, T.	1911	Pilgrims to the Isles of Penance. Orchid Gathering in the West. John Long, Ltd., London. 320 pp., 54 illustration, 1 map.
Clyne, D.	1970	Australian ground orchids. Periwinkle Books/ Landsdowne Press Pty., Ltd., Melbourne, Victoria, Australia. 112 pp., numerous B&W and color illustrations, several line drawings and one map.

(continued)

(continued)

Author(s)	Date	Title and publisher
Clyne, D.	1972	Australian rock and tree orchids.
Cogniaux, A.	1909–1910	Orchidaceae from volume 6 of Symbolae Antillanae (Flora of the West Indies, Venezuela, etc) edited by Urban. Reprint 1964 by A. Asher & Co., Amsterdam. 721 pp.
Cogniaux, A.	1893–1896	Flora Brasiliensis, Volume III, part IV. Orchidaceae. Editor: Martius, G. F. P., and A. G. Eichler. Monachii, Lipsiae.
Cogniaux, A.	1893–1906	Orchidaceae parts I–III. In: C. P. P. de Martius, A. G. Eichler and I. Urban (eds.), Flora Brasiliensis, Volumes III parts 4–6 (three volumes plus figure atlas). Reprint 1975 by Otto Koeltz Science Publishers, D-624 Koenigstein, Germany.
Cogniaux, A. and A. Goosens		Chronique Orchidienne.
Cogniaux, A., and A. Goosens	1896–1903	Dictionnaire iconographique des orchidees.
Cohen, B., and E. Roberts	1975	Growing orchids in the home. Hodder & Stoughton, London. 129 pp., 23 line drawings, 8 color painting plates.
Coleman, R. A.	1995	The wild orchids of California, 1st ed. Comstock Publishing Associates (Cornell University Press), Ithaca, NY. xvi+201 pp., numerous line drawings, maps and color photographs.
Coleman, R.	2002	The wild orchid of Arizona and New Mexico. Comstock Publishing Associates/Cornell University Press, Ithaca, NY. 248 pp., numerous color photographs.
Colman, J.	1932 or 1933	Hybridization of orchids. Gatton Park Orchids. Published for private circulation by the author. In original binding. 103 pp.
Comber, J.	1981	Wayside orchids of Southeast Asia. Heinemann Educational Books (Asia), Ltd., Kuala Lumpur, Malaysia. 28 pp., 32 B&W and color plates, 14 line drawings.
Comber, J. B.	1990	Orchids of Java. Bentham-Moxon Trust, Royal Botanic Gardens, Kew. 407 pp., numerous color photographs (about 700).
Comber, J. B.	2001	Orchids of Sumatra. Royal Botanic Gardens, Kew. 1026 pp., numerous color photographs.
Contoni, M.	1992	Orchidee spontanee del Monte Amiata.
Coomans de Ruiter, L.	1935	Borneo-orchideeën. Joh. Enschede en Zonen Grafische Inrichting N. V., Haarlem. 80 pp., 62 B&W photographs.
Coomans de Ruiter, L.	1955	Indonesiche orchideeen en hun beschering.

(continued)

(continued)

Author(s)	Date	Title and publisher
Cooper, D.	1981	A field guide to New Zealand orchids. Price Milburn, Wellington for the Wellington Orchid Society. 103 pp., numerous line drawings and 16 color photographs.
Cootes, J.	1991	The orchids of the Philippines. Timber Press, Portland, OR. 232 pp., numerous color photographs.
Copley, G. C., E. M. Tweedie, and E. W. Carroll	1964	A key and check list to Kenya orchids, Part I. Journal of the East Africa Natural History Society and Coryndon museum, Vol. 24, No. 4, pp. 1–59. Together with: Moreau, W. M., and R. E. Moreau. 1943. An introduction to the epiphytic orchids of East Africa. Journal of the East Africa Natural History Society and Coryndon museum, Vol. 17, No. 1 & 2, pp. 1–32 and. Copley, G. C., E. M. Tweedie, and E. W. Carroll. 1964. A key and check list to Kenya orchids, part II. Journal of the East Africa Natural History Society and Coryndon museum, Vol. 24, No. 5, pp. 85–91.
Correa, M. N.	1969	Flora patagonica. Collection Cientifica Del INTA, Buenos Aires. 221 pp., 184 line drawings full page figures.
Correll, D. S.	1950	Native orchids of North America north of Mexico. Chronica Botanica Company. 399 pp., 146 black and white paintings and line drawings.
Correvon, H.	1893	Les orchidées rustiques. Published by the author in Geneva and by O. Doin in Paris. 242 pp., 39 line drawings and many handwritten notations in the text.
Correvon, H.	1899	Album des orchidées d'Europe, 1st ed. Librairie Gerog & Cie, Geneva. 90 pp., 60 color plates (actually plate 60 is a B&W line drawing).
Correvon, H.	1923	Album des orchidées d'Europe, 2nd revised and augmented edition. H. Correvon "Floraire", Geneva. 70 pp., 66 color plates (actually plate 66 is a B&W line drawing).
Corrigan, M. J. G. (ed.)	1971	Proceedings of the sixth world orchid conference, September, 12–18, 1969, Sixth World Orchid Conference, Sydney, New South Wales, Australia. 276 pp., maps, line drawings, B&W photographs, color plates, a 16-page program brochure, 20 page information brochure and a 56 page magazine-like brochure..
Costantin, J.	1920–1926 (?)	Atlas des orchidées cultivées. L. Orlhac, Éditeur, Paris. 91 pp., about 1,000 paintings in 30 plates.

(continued)

(continued)

Author(s)	Date	Title and publisher
Costantin, J.	1917	Le vie des orchidées. Ernest Flammarion, Éditeur, Paris. 187 pp., line drawings.
Costerus [?]	1894	*Grammatophyllum speciosum.*
Couret, P.	1982	Jewels of Venezuelan orchids.
Craighead, F. C.	1963	Orchids and other air plants of the Everglades National Park. University of Miami Press. 127 pp., 8 pages of color and numerous B&W photographs and line drawings..
Cribb, P.	1984	Flora of tropical east Africa, Orchidaceae (part 2). A. A. Balkema, Rotterdam. 412 pp., 46 line drawing plates, one map.
Cribb, P.	1987	The genus *Cypripedium*. The Royal Botanic Gardens, Kew in association with Timber Press, Portland, OR. 301 pp., numerous maps, line drawings, color photographs and color paintings by a number of illustrations.
Cribb, P.	1987	The genus *Paphiopedilum,* 1st ed. The Royal Botanic Gardens, Kew in association with Collingridge. 222 pp., numerous maps, line drawings and color paintings by Pandora Sellars.
Cribb, P.	1987	The genus *Paphiopedilum*, 2nd ed. Natural History Publication (Borneo) in association with The Royal Botanic Gardens, Kew. 427 pp., numerous maps, line drawings, color photographs and color paintings by Pandora Sellars and Carol Woodin.
Cribb, P.	1997	Slipper orchids of Borneo. Natural History Publications (Borneo) Sdn. Bhd., Kota Kinabalu, Sabah, Malaysia. 118 pp., 82 color photographs.
Cribb, P.	1986	The 'Antelope' dendrobiums. Booklet reprinted from Kew Bulletin Vol. 41, No. 3, pp. 615–692. plates 6–13, Figures 1–20, maps 1–7.
Cribb, P.	1992	The forgotten orchids of Alexander Brun. Grove Press, New York. 159 pp., numerous B&W illustrations, line drawings and color paintings.
Cribb, P. J.	1983	A revision of *Dendrobium* sect. Latourea (Orchidaceae). Booklet reprinted from Kew Bulletin Vol. 38, No. 2, pp. 229–306, plates 10–17, Fig. 1–18, maps 1–10.
Cribb, P. J.	1993	The slipper orchids. Part One of Thesaurus Woolwardiae. Missouri Botanical Garden in association with the The Royal Botanic Gardens, Kew. No pagination, 15 color plates. Book measures 42 × 30 cm. See also Chase 1993; Stewart, 1993; Wood, 1993.

(continued)

(continued)

Author(s)	Date	Title and publisher
Cribb, P. J.	1983 and 1986	A revision of the "Antelope" and "Latourea" dendrobiums. Book reprinted from Kew Bulletin Vol. 38, No. 2, pp. 229–306, plates 10–17, Fig. 1–18, maps 1–10 and Vol. 41, No. 3, pp. 615–692., plates 6–13, Fig. 1–20, maps 1–7.
Cribb, P., and C. Bailes	1992	Orchids, a romantic history with a guide to cultivation. Running Press, Philadelphia. 64 pp., numerous B&W line drawings and photographs and color paintings and photographs.
Cribb, P., and I. Butterfield	1999	The genus 2nd ed. Botanical Magazine Monographs, Royal Botanic Garden, Kew. 166 pp., 74 color pictures, 18 plates of color paintings, 19 maps.
Cribb, P., I. Butterfield and C. Z. Tang	1988	The genus *Pleione*, 1st ed. Illustrated with line drawings and color paintings by Christabel King, Rodella Purves and Margaret Stones. The Royal Botanical Gardens, Kew in association with Christopher Helm and Timber Press. 94 pp., color photographs.
Cribb, P. J., and J. M. Fay	1982	The mountain flowers of southern Tanzania. A. A. Balkema, Rotterdam, The Netherlands. 244 pp., line drawings, color photographs.
Cribb, P. J., and J. M. Fay	1987	Orchids of the Central African Republic – A provisional checklist. Booklet reprinted from Kew Bulletin Vol. 42, No. 3, pp. 711–737.
Cribb, P. J., C. Z. Tang, and I. Butterfield	Ca 1983	The genus *Pleione*. Curtis Botanical Magazine Vol. 184, part 3, New Series, pp. 93–147, 12 color plates and a number of drawings and maps.
Cribb, P., and M. Tibbs	2004	A very Victorian passion. The orchid paintings of John Day 1863 to 1888. Blacker Publishing & Royal Botanic Gardens, Kew in association with Thames & Hudson. 464 pp., numerous color paintings and some B&W photographs.
Cribb, P., and M. Tibbs	2004	A very Victorian passion – the orchid paintings of John Day.
Cribb, P., and W. A. Whistler	1996	Orchids of Samoa. Royal Botanic Gardens, Kew. 141 pp., 24 plates of color photographs, numerous line drawings.
Cullen, J. (ed.)	1992	The orchid book. Cambridge University Press. 529 pp., 16 color photographs, numerous line drawings.
Cullina, W.	2004	Understanding orchids. Houghton Mifflin Co., Boston, MA. 260 pp., numerous color photographs.
Curtis, C. H.	1910	Orchids for everyone. J. M. Dent & Sons, Ltd., London and E. P. Dutton & Co., New York. 234 pp., 51 color and 43 B&W illustrations.

(continued)

(continued)

Author(s)	Date	Title and publisher
Curtis, C. H.	1950	Orchids, their culture and cultivation. Putnam & Company, Ltd. 274 pp., numerous B&W photographs and colored paintings.
Curtis, W. M.	1979	Angiospermae: Orchidaceae, Part 4A of The student's flora of Tasmania. T. J. Hughes, Government Printer, Hobart, Tasmania. 138 pp., 27 line drawings, 28 color pictures.
Curtis, W. M.	1979	The student's flora of Tasmania, Part 4A, Angiospermae: Orchidaceae. Government Printer, Hobart, Tasmania. 138 pp., numerous line drawings and color photographs.
Da Silva Ramos, M. S.	1969	A orquídea e sua reprodução pela seemente. Published by the author in Campinas, São Paulo, Brazil. 103 pp., color photographs.
D'A. Guimaraes, J.	1887	Orchidographia Portugueza.
Dacy, M.	No date	Victorian orchids in habitat.
Dafni, A.	1981	Orchids in Israel. Massad, Ltd., Israel. (In Hebrew.) Numerous line drawings and B&W and color photographs.
Dakkus, P. M. W.	1935	Orchideeën welke in Nde.-Indië gekweekt kunnen worde, 3rd ed. Algemeen Landbouwweekblad voor Nederlansch-Indië, A. C. Nix & Co., Bandoeng. 367 pp., line drawings and B&W photographs.
Dakkus. P. M. W.	1930	Beschrijving van Orchideeën die in Ned.-Indië gekweekt kunnen worden. Algemeen Landbouwweekblad v. Ned.-Indië, Buitenzorg. 265 pp., numerous line drawings and B&W photographs and paintings.
Danesch, E., and O. Danesch	1966	Atudies in nature photography. Leben Bild, Aalen, Stuttgart. 46 pp., numerous B&W and color photographs including orchids.
Danesch, E., and O. Danesch	1977	Orchideen nitteleuropa, 2nd ed. Hallwag Verlag, Bern, Switzerland. 128 pp., numerous color photographs
Danesch, O., and E. Danesch	1962	Orchideen Europas. Mitteleuropa. Verlag Hallwag, Bern and Stuttgart. 264 pp., 165 B&W and color photographs, a few line drawings.
Danesch, O., and E. Danesch	1963	Nos orchidées. Europe occidentale et centrale. Editions Payot, Lausanne, Switzerland/Société du Livre, Paris. 264 pp., 165 pp., B&W and color photographs, some line drawings.
Danesch, O., and E. Danesch	1969	Orchideen Europas. Südeuropa. Verlag Hallwag, Bern and Stuttgart. 256 pp., 184 color photographs, a few line drawings.
Danesch, O., and E. Danesch	1972	Orchideen Europas. Ophrys-hybriden. Verlag Hallwag, Bern and Stuttgart. 268 pp., 220 color photographs, a few line drawings.

(continued)

(continued)

Author(s)	Date	Title and publisher
Danesch, O., and E. Danesch	1977	Tiroler orchideen. Verlagsanstalt, Athesia, Austria. 151 pp., numerous color photographs.
Darnell, A. W.	1930	Orchids for the outdoor garden. L. Reeve & Co., Ltd., Ashford, Kent, UK. 467 pp., colored frontispiece, numerous B&W paintings. See also I 1976 reprint by Dover Publications, Inc., New York. 467 pp., numerous line drawings.
Darnell, A. W.	1930	Orchids for the outdoor garden. Reprint 1976 by Dover Publications, Inc., New York. 467 pp. numerous line drawings.
Darwin, C.	1862	Einrichtungen zur befruchtung Britischer und auslanderdischer orchideen.
Darwin, C.	1877	Die verschiedenen Einrichtungen durch welche Orchideen von Insekten befruchtet werden. E. Schweizerbart'sche Verlagshandlung (E. Koch), Stuttgart. 259 pp., line drawings.
Darwin, C.	1891	De la fécondation des orchdées par les diagram. C. Reinwald & Cie Libraires-Éditeurs, Paris. 352 pp., numerous line drawings. Paperback.
Darwin, C.	1904	On the various contrivances by which orchids are diagrams &$$$; by insects, 7th printing of the 2nd ed. John Murray, London. 300 pp., 38 line drawings.
Das, S., and S. K. Jain	1980	Orchidaceae-genus *Coelogyne*. Fascicle 5. Fasciles of Flora of India. Botanical Survey of India, P. O. Botanic Garden, Howrah 711103. 33 pp., line drawings.
Dassanayake, M. D., and F. R. Fosberg	1981	A revised handbook to the flora of Ceylon, A. A. Balkema. Rotterdam. 511 pp., numerous line drawings. Orchidaceae by D. M. A. Jayaweera on pp. 4–386.
Davies, P., and A.Huxley	1883	Wild orchids.
Davis, R. S., and M. L. Steiner	1952	Philippine orchids. The William-Frederick Press, New York. 270 pp., map, some line drawings, numerous B&W photographs.
De Blume, C. L.	1858	Collection des orchidees les plus remarquables de larchipel Indien et du Japon.
De Britto Pereira, C. E. (ed.)	1998	Proceedings of the 15th world orchid conference, September 14–23, 1996 Rio de Janeiro, Brasil. Naturalia Publications, Transfaire SA, F-04250 Turriers, France. 500 copies printed 496 pp., maps, line drawings, B&W and color photographs, an unpaginated registrant's handbook and a 31 page abstracts book.

(continued)

(continued)

Author(s)	Date	Title and publisher
De Candole, A.	1884	Origin of cultivated plants. Kegan Paul, Trench & Co., London. 468 pp.
de Garmo, L. R. (ed.)	1966	Proceedings of the fifth world orchid conference, April 11–21, 1966, Fifth World Orchid Conference, Inc., Long Beach, CA, USA. 326 pp., maps, line drawings, B&W photographs, color plates. There is separately published index by Robert M. Hamilton.
De Giorgio, M.	1982	Le orchidee Italiane.
De Kerchove de Denterghem, O., Le Comte	1898	Le Livre des orchidées. Ad. Hoste Éditeur, Gand Belgium and G. Masson, Éditeur, Paris. 601 pp., 310 line drawings, 31 color plates.
De la Bathie, H. P.	1939	Les orchudées de la diagra Malgache (variation, biologie et distribution). Mémoires de Muséum National d'Histoire Naturelle, N. S., Vol. 10, 5th fascicle, pp. 237–297. Bound as a booklet.
De Lapiner, J. M.	No date	Orquìdeas Michoacanas. CFEM. Mexico. 63 pp., numerous color photographs.
De Leon Flores, F.	1976	Philippine orchid culture. Published by the author. 60 pp., numerous B&W and color photographs.
De Maria, G.	1982	Le orchidee Italiane. Sagep Editrice, Genoa. Numerous line drawings, maps and color photographs.
De Martino, E.	1998	Le orchidee spontanee del Montovolo. Alesa Editirice. Bologna. 88 pp., numerous color photographs and maps, B&W photographs.
De Martino, E.	2000	Le orchidee spontanee dell' Emilia Romagna
De Melville, R., and T. B. Shresta	2004	Nepal orchids in pictures.
De Ospina, B. H.	1970	My orchid gardens, 1st English ed. Published by the author in Medellín, Colombia. 109 pp., color and B&W photographs.
De Puydt, E.	1880	Les orchidées. J. Rothschild, Editeur, Paris. viii and 348 pages, 244 text figures and 50 hand colored plates.
De Rosa, P. A., and P. E. Trastulli	1980	Orchidee Romane, paintings by Enrico Coleman. Newton Compton Editori, Rome. 222 pp., 83 color plates of paintings, some B&W photographs and line drawings.
De Sousa da Camara	1966	Catalgus systematicus fungorum omnium Lusitaniae. Part II Fungi imperfecti Part I. Ministry of Economics, Lisbon, Portugal. 378 pp.
de Vogel, E. F.	1986–1997	Orchid monographs Volumes 1–8. Rijksherbarium, Leiden, The Netherlands.
Decker, J. S.	1946	Cultura das orcquideas do Brasil.

(continued)

(continued)

Author(s)	Date	Title and publisher
Deinum, D.	1942 (?)	Die orchideeën van Nederland. CV. Allert de Lange, Amsterdam. 174 pp., numerous line drawings, 27 B&W photographic plates.
Del Prete, C., and G. Tosi	1981	Orchidee spontanee dell'Argentario. A Cura Della Comunità Montana No. 24, Monte Argentario (Grosseto), Italy. 38 pp., 5 B&W plates (copies from an old book).
Del Prete, C., H. Tichy, G. Tosi	1993	Le orchidee spontanee della Maremma Grossetana. Libreria Massimi, Porto Ercole, Italy. 143 pp., numerous line drawings, maps, color photographs.
Delchevalerie, G.	1869	Les orchidées culture, propagation, nomenclature. Librairie Agricole de la Maison Rustique. 133 pp., 32 line drawings.
Delfino, P.	1994	Orchidee spontanee della Sardegna.
Delforge, P.	1994	Guide des orchidées d'Europe, d'Afrique du nord et du Proche-Orient. Delachaux et Niestlé, Lausanne, Switzerland. 481 pp., numerous color photographs.
Delforge, P.	1995	Orchids of Britain and Europe.
Delforge, P.	2006	Orchids of Europe, North Africa and the Middle East. Edited by Simon Harrap. A&C Black Publishers, Ltd., London. $60.00. 640 pp., numerous color photographs.
Delforge, P., and D. Tyteca	1984	Europese orchideeen.
Denissova, L. V. and M. G. Vakhrameeva	1978	Genus Cypripedium L. Pages 62–71. In: T. A. Rabotnov (ed.) Biological Flora of Moscow Province. Vyp. 4. MGU, Moscow. (In Russian.)
Demares, M.	1997	Atlas des orchidees de haute-Normandie.
Deorani, S. C., and H. B. Naithani	1995	Orchids os Nagaland.
Department of Agriculture, Ceylon	1938	Notes on orchid cultivation and on some orchids cultivated in Ceylon.
Der Star, J. A., van	1938	Het kweeken van orchideeen door den liefhebber.
Derx, H. G.	1937	De betekenis van het orchidologische werk van dr. J. J. Smith voor den Neederlandschen kweeker.
Desbois, F.	1898	*Cypripedium, Selenipedium & Uropedium.* Imprimerie F. Meyer-van Loo, Gand. 544 pp., numerous line drawings.
Deva, S., and H. B. Naithani	1986	The orchid flora of north west Himalaya. Print & Media Associates, New Delhi, India. 459 pp., numerous line drawings.
Diaz, M. A.	1988	Las orquideas nativaes de Cuba.
Diekman, R.	2001	Die gattung Cattleya.

(continued)

(continued)

Author(s)	Date	Title and publisher
Diels, L.	1908	Die orchideen.
Dietrich, H.	1979	Bibliographia Orchidacearum. Aussereuropäische arten. Volume 2/1. Friedrich Schiller University, Jena, Germany. 159 pp.
Dietrich, H.	1979	Bibliographia Orchidacearum. Aussereuropäische arten. Volume 2/2. Friedrich Schiller University, Jena, Germany. 320 pp.
Dietrich, H.	1984	Orchideenmosaik. Bibliographische Mitteilungen der Universitätsbibliothek Jena No. 30, Beiheft zu Band 3, 52 pp., 15 color photographs and paintings.
Dietrich, H.	1985	Bibliographia Orchidacearum. Aussereuropäische arten. Volume 3. Friedrich Schiller University, Jena, Germany. 220 pp.
Dietrich, H.	1986	Orchideenkaleidoscop.
Dietrich, H.	1988	Bibliographia Orchidacearum. Aussereuropäische arten. Volume 4. Friedrich Schiller University, Jena, Germany. 116 pp.
Digby, G.	1982	Slipper orchids.
Dillon, G. (ed.)	1958	Proceedings of the second world orchid conference, September 19–23, 1957, Honolulu, HI. Harvard University Press Printing Office, Cambridge, MA. 253 pp., numerous line drawings.
Dillon, G. W. et al. (ed.)	1969	Handbook of orchid nomenclature and registration, 1st ed. International Orchid Commission, Royal Horticultural Society, London. 143 pp.
Dix, M. A., and M. W. Dix	2000	Orchids of Guatemala. A revised annotated checklist. Missouri Botanical Garden Press, St. Louis, MO.
Dixon, K. W., B. Biurchell, and M. T. Collins	1989	Orchids of Western Australia, 2nd ed. Western Australian Native Orchid Study and Conservation Group (Inc.), Victoria Park, Western Australia. 68 pp., line drawings and color photographs.
Dixon, K., and B. Buirchell (eds.)	No date	Orchids of Western Australia, cultivation and natural history. Western Australian Native Orchid Study & Conservation Group, Inc. 102 pp., B&W illustrations.
Dockrill, A. W.	1967	Australasian Sarcanthinae. The Australian Native Orchid Society. 41 pp., 43 plates of line drawings.
Dockrill, A. W.	1969	Australian indigenous orchids. The Society for Growing Australian Plants. Australia. 825 pp., 30 color plates, numerous line drawings. Attached a reprint: Schlechter, R. 1914. Caladenien. Orchis 8 (1):3–7+color plate.

(continued)

(continued)

Author(s)	Date	Title and publisher
Dockrill, A. W.	1992	Australian indigenous orchids volumes I and II. Surrey Beatty & Sons Pty., Ltd., Chipping Norton, NSW 2170. liv+1062 pages, 157 color plates, numerous plates of line drawings. There are two second editions. The first was published in a single volume in 1969.
Dodds, J. H., and L. W. Roberts	1995	Experiments in plant tissue culture, 3rd ed. Cambridge University Press, New York. 256 pp., few illustrations.
Dodson, C. H.	1965	Agentes de polinizacion y su influencia sobre la familia Orquidaceae. Instituto General de Investigacion, Universidad Nacional de la Amazonia Peruana. 128 pp., 8 line drawing figures, 36 B&W photographs.
Dodson, C. H.	2002	Native Ecuadorian orchids, volume III: Lepanthopsis-Oliveriana. Published by the Dodson Trust, 1550 Eastbrook Drive, Sarasota, FL., pp. 420–651, Figs. 859–1450 (color photographs and line drawings).
Dodson, C. H.	2003	Native Ecuadorian orchids, volume IV: Oncidium-Restrepiopsis. Published by the Dodson Trust, 1550 Eastbrook Drive, Sarasota, FL., pp. 652–883, Figs.1451–2034 (color photographs and line drawings).
Dodson, C. H.	2004	Native Ecuadorian orchids, volume V: Rodriguezia-Zygosepalum. Published by the Dodson Trust, 1550 Eastbrook Drive, Sarasota, FL, pp. 884–1187, Figs. 2035–2572 (color photographs and line drawings).
Dodson, C. H.	No date	Native Ecuadorian orchids, volume II: Dresslerella-Lepanthes. Published by the Dodson Trust, 1550 Eastbrook Drive, Sarasota, FL, pp. 209–419, Figs. 373–858 (color photographs and line drawings).
Dodson, C. H., and D. Bennett, Jr.	1989	Icones plantarum tropicarun. Series II. 6 Fascicles.
Dodson, C. H., and P. M. Dodson	1980	Icones plantarum tropicarum, Fascicles 1–4, plates 1–400. Marie Selby Botanical Garden, Sarasota, FL.
Dodson, C. H., and P. M. Marmol de Dodson	1982, 1984	Icones platarum tropicarum. Orchids of Ecuador plates 401–500, 901–1000. Marie Selby Botanical Garden, Sarasota, FL.
Dodson, C. H., and R. Escobar R.	No date	Native Ecuadorian orchids, volume I: AA-Dracula. Compañia Lithografica Nacional S. A., Colombia. pp. 1–208, Figs. 1–372 (color photographs and line drawings).
Dodson, C. H., and R. J. Gillespie	1967	The biology of the orchids. The Mid America Orchid Congress, Inc. 158 pp., line drawings, B&W and color photographs.

(continued)

(continued)

Author(s)	Date	Title and publisher
Doin, O.	1895	Dictionnaire des orchidees diagram.
Domokos, M.	1972	Orchideák. Mezögazdaasági Kiadó, Budapest. 149 pp., 4 color photographs, numerous B&W photographs.
Dransfield, J., J. B. Comber, and G. Smith	1986	*Corybas* west of the Wallace line. Royal Botanic Gardens, Kew. Kew Bulletin 41 (3): 575–578, 15 line drawings, 5 color plates.
Drar, M.	1970	A botanic expedition to the Suan in 1938. Publication No. 4 of the Cairo University Herbarium, Cairo University Press. 114 pp.
Dressler, R. L.	1981	The orchids, natural history and classification. Harvard University Press. 332 pp., numerous line drawings and B&W photographs and some color photographs.
Dressler, R. L.	1981	The orchids, natural history and classification. Harvard University Press. 332 pp., numerous line drawings and B&W photographs and some color photographs. Paperback reprint.
Dressler, R. L.	1993	Field guide to the orchids of Costa Rica and Panama. Comstock Publishing Associates, Cornell University Press, Ithaca, NY. 374 pp., numerous B&W and color photographs.
Dressler, R. L.	1993	Phylogeny and classification of the orchid family. Dioscorides Press (an imprint of Timber Press), Portland, OR.
Dressler, R. L., and C. H. Dodson	1960	Classification and phylogeny in the Orchidaceae. Reprinted from the Annals of the Missouri Botanical Garden 47: 125–168.
Dressler, R. L., and G. E. Pollard	1974	The genus *Encyclia* in Mexico. Associación Mexicana de Orquideología, A. C., Mexico City. 151 pp., 153 pp., line drawings, maps, 51 color photographs.
Drude, O.	1873	Die biologie von *Monotropa hypopitys* L. und *Neottia nidus* avis L.
Du Buysson, F., Le Comte	1878	L'Orchidphile, traité théorique et pratique sur la culture des orchidées. Auguste Goin Èditeur, Paris. 536 pp.
du Petit Thouars, L.-M., A.-A.	1822	Histoire particuliere des plantes orchdées recueilles sure les isles de France, de Bourbon et de Madagascar.
Du Petit-Thouars, A.-A.	1822 (reprint 1979)	Histoire perticulière des plantes orchidées recueillies sur les &$$$;iagr iles australesd'Afrique. Earl M. Coleman, Publisher, Stanfordville, New York. 110 pp., 110 line drawings.
Du Puy, D., and P. Cribb	1988	The genus *Cymbidium*. Timber Press, Portland, OR. 204 pp., 143 color photographs, 28 paintings, 31 plates of line drawings, 11 maps.

(continued)

(continued)

Author(s)	Date	Title and publisher
Du Puy, D., P. Cribb, J. Bosser, J. & C. Hermans	1990	The orchids of Madagascar. Royal Botanic Gardens, Kew. 346 pp., 48 color plates of 1–4 pictures each.
Dulson, R.	1993	Orchids.
Dunmire, J. R.	1998	Orchids.
Dunsterville, G. C. K.	1957	Orquideas de Caracas. Boletin de la Sociedad Venezuelana de Ciencias Naturales, Vol. 18, No. 89, pp. 1–14. Line drawings and color photographs.
Dunsterville, G. C. K.	1960	Orquideas de Venezuella, Compañia Shell de Venezuela. Unpaginated book 43.5 by 33 cm in Spanish and 18 paintings 28.8 by 24 cm by the author.
Dunsterville, G. C. K.	1963	Variation within species. Proceedings of the World Orchid Conference, Singapore (Y. B. Choon, ed.), pp. 48–60, line drawings.
Dunsterville, G. C. K.	1964	Introduction to the world of orchids. Doubleday & Company, Inc. Garden City, New York. 104 pp., 40 color plates, numerous line drawings.
Dunsterville, G. C. K.	1965	Auyantepui. Boletin de la Sociedad Venezuelana de Ciencias Naturales, Vol. 26, No. 100, pp. 163–171. Two foldout maps.
Dunsterville, G. C. K.	1972	Some orchids of Brazil's highest highlands (bilingual: English-Portuguiese). Bradea, Vol. 1, No. 12, pp. 83–121.
Dunsterville, G. C. K.	1975	Orquideas en la cime del cero Autana. Acta Botanica Venezuelica, Vol. 10, Nos. 1–4, pp. 251–262, line drawings.
Dunsterville, G. C. K.	1987	Venezuelan orchids. E. Armitano, Publisher (probably in Venezuela). Only 1,000 copies of the English edition were published. 125 pp., 247 line drawings and 50 color photographs.
Dunsterville, G. C. K., and L. A. Garay	1959–1976	Venezuelan Orchid Illustrated, Vols 1–6. A total of 1,000 line drawings. Andre Deutsch, London
Dunsterville, G. C. K., and L. A. Garay	1979	Orchids of Venezuela, an illustrated field guide. Botanical Museum, Harvard University, Cambridge, MA. 3 volumes. X+694 pages, 694 line drawings plates (total 1050) by Mr. G. C. K. Dunstervile.
Duperrex, A.	1955	Orchidees d'Europe.
Duperrex, A.	1961	Orchids of Europe. Translated by A. J. Huxley. Blandford Press, London. 235 pp., line drawings and color photographs.
Dusak, F.	2002	Orchidees sauvages de l'Ile de France.
Dusek, J., and J. Krístek	1986	Orchideje. Academia. Prague.

(continued)

(continued)

Author(s)	Date	Title and publisher
Dusterville, G. C. K. and Dusterville	1988	Orchid hunting in the lost world (and elsewhere in Venezuela). American Orchid Society, Inc., West Palm Beach, FL.
Duthie, J. F.	1906	The orchids of north-western Himalaya. Annals of the Royal Botanic Garden, Calcutta 9: i–ii, 81–21, 58 plates of B&W drawings. Bound with Hooker, 1895 and King and Pantling, 1898. Reprints 1967.
Duval, L.	1900	Les *Odontoglossum*.
Duval, L.	1905	Les orchidées, 3rd ed. Librairie Horticole, Paris. 183 pp., 72 line drawings.
Duval, L.	1907	Traite de culture pratique des *Cattleya*.
Ebel, F., and O. Birnbaum	1971	Schoene und seltsame welt der orchideen. Verlag Ernst Wasmuth, Tübingen. 207 pp. numerous B&W and color photographs, line drawings.
Ebel, F., and O. Birnbaum	1971	The strange and beautiful world of orchids.
Ebel, F., and O. Birnbaum	1983	Orchideen – Juwelen im pflanzenreich. Martin-Luther-Universität Halle-Wittenberg. 112 pp., 64 B&W and 64color illustrations.
Eberle, G.	1961	Die orchideen der Deutschen heimat, 2nd ed. Verlag von Waldemar Kramer, Freanfurt am Main, 128 pp., 105 B&W photographs.
Ecarius [?]	1983	Die orchideen des kreises Eisenbach.
Ecott, T.	2004	*Vanilla*. Travels in search of the ice cream orchid. Grove Press, New York. 278 pp., illustrations.
Ede, A., and J. Ede	1985	Living with orchids. MPH, Singapore. 91 pp., numerous color photographs.
Edinger, P.	1970	How to grow orchids, first printing. A Sunset Book. Lane Books, Menlo Park, CA. 64 pp., numerous B&W photographs.
Edinger, P.	1971	How to grow orchids, second printing. A Sunset Book. Lane Books, Menlo Park, CA. 64 pp., numerous B&W photographs
Editors of Consumer Guide	1976	Bromeliads and orchids. Simin and Schuster, New York. 66 pp., numerous color illustrations and a few line drawings.
Eigeldinger, O.	1957	Orchids for everyone. John Gifford, Ltd., London. 144 pp., numerous B&W and a few color photographs.
Eigeldinger, O., and L. S. Murphy	1972	Orchids, a complete guide to cultivation.
Eiko, I.	1980	Japanese wild orchids and summer *Calanthe*. 235 pp., 112 color photographs.
Elbert, V. F.	1983	Orchids of the world coloring book. Dover Publications, Inc., New York. 46 pp., about 40 line drawings.

(continued)

(continued)

Author(s)	Date	Title and publisher
Ellenberger, D.	1984	Orchid lovers sales directory & guide to regional sources. Orchid Eden, P. O. Box 17125, Rochester, NY 14617, USA, 84+2 unnumbered pp.
Elliott, J. (ed.)	1993	Orchid growing in the tropics. Times Editions, Singapore. 207 pp., numerous color photographs.
Elliott, J., P. Tan, S. Y. Alsagoff, and W. Chew (eds.)	2005	Orchid hybrids of Singapore 1893–2003. Orchid Society of South East Asia, Singapore. 304 pp., numerous B&W and color photographs.
Endlicher, S.	1838	Iconographia generum plantaru. Several pages of text and plates.
Engler, A., and K. Prantl	1889	Die natürlichen pflanzenfamilien, part II, No. 6 (diagrams the orchids). Verlag von Wilhelm Engelmann, Leipzig.
Erdman, D. J.	1982	Notes on the subgenus Brachypetalum of the genus *Paphiopedilum*. Published by the author in Rochester, NY. 95 pp., B&W photographs and line drawings.
Erdmann, D. J.	1980	Notes on the *Brachypetalum* group of *Paphiopedilum*. Rochester, NY. 53 pp., line drawings and a map.
Erickson, R.	1951	Orchids of the west, 1st ed. Paterson Brokensha Pty., Ltd., Perth, Western Australia. 109 pp., line drawings, color paintings.
Erickson, R.	1965	Orchids of the west, 2nd ed. Paterson Brokensha Pty., Ltd., Perth, Western Australia. 109 pp., line drawings, color paintings.
Ernst, R.	1979	Biological effects of surfactants. Ph.D. dissertation, University of California, Irvine. This work was done with orchids.
Escobar R. R. (ed.), and J. M. Múnera B.	1990–1998	Native Colombian orchids, Volumes 1–6. Published for the Colombian Orchid Society by Compañia Litografica Nacional S. A., Medellín.
Escobar R. R., and C. H. Dodson	ca 1996	Native Ecuadorian orchids.
Estacio de Yeiga, S. P. M.	1886	Orchideas de Portugal. Academia Real das Sciencias de Lisboa, Portugal. 52 pp., 36 B&W drawings.
Ettlinger, D. M. T.	1976	British and Irish orchids, a field guide.
Eun, J. C. A., W. Y. Lee and R. D. Samynadan	1999	First colloquium on orchid research and industry in Singapore, abstracts. 27 pp.
Evans, R. E. (ed.)	1984	The botanical museum of Harvard University. Botanical Museum Leaflets, Harvard University. 62 pp., B&W photographs.
Faber, F. CF., von	1904	Beitrag zur vergleichenden anatomie der Cypripedilinae. Inaugural-Dissertation.

(continued)

(continued)

Author(s)	Date	Title and publisher
Fabre, J. H.	1855	These de botanique-recherches sur les tubercules de l'Hmatoglossum hircinum. Inaugural-Dissertation.
Fandani, A., W. Rossi	1988	Simon & Schuster guide to orchids. Simon & Schuster, New York. Translated from the Italian by John Gilbert. 256 pp., numerous line drawings, some line drawings.
Fanfani, a.	1989	The MacDonald encyclopedia of orchids.
Fantova, M. C.	1977	Introducción a las orquídeas Españolas. Fundación Juan March. Serie Universitaria 122. 54 pp., includes presentation letter (form letter) from the publisher.
Fast, G.	1980	Orchideen kultur. Verlag Eugen Ulmer, Stuttgart. Line drawings, graphs, B&W and color photographs.
Fawcett, W., and A. B. Rendle	1910	Flora of Jamaica, Vol. 1 Orchidaceae. Longmans & Co., London. Reprint 1963 by George W. Hart, Kingston, Jamaica. 121 pp., numerous line drawings.
Fayé, C.	No date	Orchids from Hawaii in counted cross stitch. Needlepoint & Cross Stitch Designs. 7 pp., 7 line drawings.
Fedtschenko, B. A.	1927	Orchids of the Middle Asia. In: Anniversary book devoted to I. P. Borodin. Leningrad. (In Russian.)
Fedtschenko, B. A.	1898	On some plants of Moscow Gubernia. P. 278 In: Dairy of the X meeting of nature researchers and medical doctors in Kiev. 8. Kiev. (In Russian).
Feldmann [?]	2001	Orchidees sauvages de uadeloupe.
Feldmann, R.	1987	Orchideen as zimmepflanzen.
Feldmann, R.	1988	Orchideen as zimmepflanzen. Slightly different edition from the one above.
Fennell, Jr., T. A.	1956	Orchids for home and garden, revised edition. Rinehart & Company, New York. 160 pp., 27 line drawings.
Fennell, Jr., T. A.	1959	Orchids for home and garden, revised edition. Rinehart & Company, New York. 160 pp., 27 line drawings. Slightly different edition from the one above.
Fennell, T. A.	1959	Orchids for home and garden, 2nd ed. Rinehart & Company, New York. 160 pp., 27 line drawings, B&W and color photographs.
Fennell's Orchid Jungle	ca. 1984	The joy of orchids. Fennell's Orchid Jungle, Homestead, FL. 80 pp., numerous line drawings, color and B&W photographs.
Ferlinghetti, R.	2001	Orchidee spontanee della provincia di Bergamo.

(continued)

(continued)

Author(s)	Date	Title and publisher
Fernandez Perez, A.	1985	La Real expedicion botanica del Nuevo Reino de Granada, volume IX. Microspermae: Orchidaceae, III. Ediciones Cultura Hispanica, Madrid. 45 pp., approx. 50 color plates, maps, B&W illustrations. Book measures 54.7 × 37 cm. Perez-Albelaez et al., 1954; Schweinfurth et al., 1963, 1969; Ortiz y Valdivieso, 2000.
Fernandez Perez, A., and C. Schweinfurth	1995	La Real expedicion botanica del Nuevo Reino de Granada, volume X. Microspermae: Orchidaceae, IV. Ediciones Cultura Hispanica, Madrid. 96 pp., 53 color and 4 monochrome plates, maps, B&W illustrations. Book measures 54.7 × 37 cm. Perez-Albelaez et al., 1954; Schweinfurth et al., 1963, 1969; Ortiz y Valdivieso, 2000 are part of the same series.
Fernández, C.	2003 or 2004	Orquídeas iagram del Táchira. www.Loteriadeltachira.com.ve. 246 pp., numerous color photographs.
Ferry Townsend, R. J.	1999	An anatomical study of the epidermis of leaves of the genus *Stanhopea* (Orchidaceae) and their taxonomic implications. Ph.D. dissertation. The Autonomous University of Nuevo Leon, Monterey, N. L., Mexico.
Fillion-Delamain, C.	1992	Les orchidees d'uncoteau Charentais.
Finet, M. E. A.	1907	Classification et enumeration des orchidees Africaines de la tribu des Sarcanthees.
Finnis, J (ed.)	1992	The little book of orchids. Color Library Books, Godalming, Surrey, UK. 60 pp., line drawings and color paintings and photographs.
Finnis, J (ed.)	1992	The little book of orchids. Mallard Press, an imprint of BDD Promotional Book Company Inc., New York. 60 pp., line drawings and color paintings and photographs.
Firth, M. J.	1965	Native orchids of Tasmania. Printed by C. L. Richmond and Sons Pty., Ltd., Davenport, Tasmania. 90 pp., line drawings and B&W photographs. 1st ed.
Fischer, C. E. C.	No original date	Orchidaceae in Flora of the Presidency of Madras edited by J. S. Gamble. Reprint 1984 by Bishen Singh Mahendra Pal Singh, 23-A Connaught Place, Dehra Dun-248001, India, pp. 1399–1478.
Fisher, J. R.	1981	Bishop's interim list of orchid hybrids.
Fisher, R. M.	1980	Guide to the orchids of the Cypress Hills. Published by the author or O. Pall 1612 – 46 Street NW, Calgary, Alberta, Canada T3B 1A9. 44 pp., numerous line drawings and color photographs.

(continued)

(continued)

Author(s)	Date	Title and publisher
Fitch, C. M.	1981	All about orchids. Double Day & Co., Inc., Garden City, NY. 276 pp., numerous B&W and color photographs.
Fitch, C. M.	No date	Growing orchids under lights. American Orchid Society. 76 pp., numerous color photographs.
Fitch, C. M. (ed.)	2004	The best orchids for indoors. Brooklyn Botanic Garden, Brooklyn, NY. 128 pp., numerous color photographs.
Fitch, C. M. (ed.)	2004	The gardener's guide to growing orchids.
Fitch, C. M., and B. Pesch	1985	Orchids for the home and garden. Plants and Gardens, Vol. 41, No. 2, #107, Brooklyn Botanical Garden. 88 pp., B&W and color photographs.
Fitch, N.	1882	Orchids, a selection of 40 paintings by this famous illustrators first published in the Orchid Album. Wadsworth Editions, Ware, Hertforshire, UK, 92 pp., 40 color plates. Book measures 37 × 27.5 cm.
Fitzgerald, R. D.	1977, 1979	Australian orchids. Vols I and II, 48.9 × 33 cm. Volume I has 360 pages, 65 full page and 4 double page color plates of paintings. Volume II has 320 pages, 46 full page and 5 double page color plates of paintings. Lansdowne Editions, East Melbourne, Australia. This fascimile edition was limited to 350 copies. The authenticity of each volume is attested to with the signature of Robert D. Fitzgerald, grandson of the author. The book was originally published by the Government printer in Sydney between 1874 and 1892.
Fiveash, R.	1974	Rosa Fiveash's Australian orchids, text by N. Lothian. Robert Hale & Company, London and Rigby, Ltd., Adelaide. 149 pp., 99 color paintings.
Fiveash, R.	1981	Rosa Fiveash's Australian orchids, text by N. Lothian. Rigby Publishers, Ltd., Adelaide. 149 pp., 99 color paintings.
Fowler, J. A.	2005	Wild orchids of South Carolina. University of South Carolina, Columbia, South Carolina.
Fowlie, J.	1970	The genus *Lycaste*. Azul Quinta Press, La Canada, CA. 90 pp., numerous B&W paintings. Includes diagrams and color pictures of the species illustrated in the book. Book is 36.5 × 29. This is a self published book.
Fowlie, J.	1977	The Brazilian bifoliate cattleyas and their color varieties. Azul Quinta Press, La Canada, CA. 132 pp., numerous color photographs and paintings. This is a self published book.
Franchet, A.	1894	Les *Cypripedium*. Journal de Botanique, Vol. 8, pp. 225–233, 249–256, 265–271.

(continued)

(continued)

Author(s)	Date	Title and publisher
Franke, H. W.	1973	The orchid cage. A novel. Daw Books, New York. 174 pp.
Franssem, C. J. H, and L. M. J. Tiggelovend	1935 or 1936	De vijanden en de ziekten der Orchideeën op Java en hunne bestrijding. G. Kolff & Co. Batavia, Dutch East Indies. 86 pp., numerous line drawings and one color plate of orchid pests.
Freed, H.	1970	Orchids and serendipity. Prentice Hall, Englewood Cliffs, NJ. 184 pp., color and B&W photographs.
Freed, H.	1979	New horizons in orchid breeding. Malibu, CA. 148 pp., numerous B&W photographs.
Friend, R. G. M.	2004	Growing orchids in your garden.
Fritzen, J.	1978?	Orchideen, maybe 2n ed. Albrecht Philler Verlag., Minden. 61 pp., 5 line drawings, 31 B&W photographs and 4 color plates.
Fritzen, J.	No date	Orchideen, maybe 1st ed. Albrecht Philler Verlag., Minden. 61 pp., 5 line drawings, 31 B&W photographs and 4 color plates.
Frowine, S. A.	2005	Fragrant orchids. Timber Press, Portland, OR. 200 pp., numerous color photographs.
Frowine, S. A.	2005	Orchid for dummies. Willey Publishing, Hoboken, NJ. 266 pp., numerous B&W and some color photographs.
Fuchs, L.	1542	De historia stirpium diagrams&$$$; insignes. Published as "The Great Herbal of Leonhart Fuchs in two volumes (commentary by F. G. Meyer, E. E. Trueblood, and J. L. Heller with a foreword by J. Ewan. Facsimile edition.
Fuhrmann, E.	1924	Die welt der pflanze. Band I. Orchideen. Auriga-Verlag, Berlin. 15 pp., text followed by 40 pages of B&W photographs by Albert Renger-Patsch.
Fukuhara, Y.	1991	100 orchids. A book in two volumes. One 175 pp. is in Japanese and has 100 color photographs of orchids. The other has 87 pages, small B&W pictures of orchids and English text.
Fulcher, M.	1999	Enchanted orchid. New Holland Publishing (Australia) Pty., Ltd., Sydney. 144 pp.
Fuller, A. M.	1933	Studies of the flora of Wisconsin. Part I: The orchids; Orchidaceae. Bulletin of the Public Museum of the City of Milwaukee. Vol. 14, No. 1, pp. 1–284, text figs 1, 2, maps, 1–24, plates, 1–54.
Fuller, F.	1967	*Limodorum, Epipogium, Neottia, Corallorhiza.* Die Neue Brehm-Bücherei, A. Ziemsen Verlag, Wittenberg Lutherstadt, 54 pp., 62 B&W and line drawing illustrations, 4 color plates.
Fuller, F.	1969	*Ophrys*, 2nd ed. Die Neue Brehm-Bücherei, A. Ziemsen Verlag, Wittenberg Lutherstadt. 56 pp., 55 B&W and line drawing illustrations, 2 color plates.

(continued)

(continued)

Author(s)	Date	Title and publisher
Fuller, F.	1969	*Platanthera, Gymnadenia, Leucorchis*. Die Neue Brehm-Bücherei, A. Ziemsen Verlag, Wittenberg Lutherstadt, 56 pp.59 B&W and line drawing illustrations, 1 color plate.
Fuller, F.	1970	*Aceras* und *Anacamptis*. Die Neue Brehm-Bücherei, A. Ziemsen Verlag, Wittenberg Lutherstadt, 50 pp.43 B&W and line drawing illustrations, 1 color plate.
Fuller, F.	1970	Frauenschuh und riemenzunge, 2nd ed. Die Neue Brehm-Bücherei, A. Ziemsen Verlag, Wittenberg Lutherstadt. 64 pp., 49 B&W and line drawing illustrations, 2 color plates.
Fuller, F.	1972	Alpine und nordish-Alpine orchideen. Die Neue Brehm-Bücherei, A. Ziemsen Verlag, Wittenberg Lutherstadt, 80 pp.76 B&W and line drawing illustrations, 2color plates.
Fuller, F.	1972	*Orchis* und *Dactylorhiza*, 2nd ed. Die Neue Brehm-Bücherei, A. Ziemsen Verlag, Wittenberg Lutherstadt. 127 pp., 104 B&W and line drawing illustrations, 3 color plates.
Fuller, F.	1974	*Epipactis* und *Cephalanthera*, 2nd ed. Die Neue Brehm-Bücherei, A. Ziemsen Verlag, Wittenberg Lutherstadt, 96 pp.,81 B&W and line drawing illustrations,10 color plates.
Fuller, F.	1975	*Goodyera* und *Spiranthes*, 2nd ed. Die Neue Brehm-Bücherei, A. Ziemsen Verlag, Wittenberg Lutherstadt.52 pp.,45 B&W and line drawing illustrations,1 color plate.
Fuller, F.	1976	*Malaxis, Hammarbya, Liparis*, 2nd ed. Die Neue Brehm-Bücherei, A. Ziemsen Verlag, Wittenberg Lutherstadt,48 pp., 47 B&W and line drawing illustrations.
Fuller, F.	1977	*Limodorum, Epipogium, Neottia, Corallorhiza,* 2nd ed. Die Neue Brehm-Bücherei, A. Ziemsen Verlag, Wittenberg Lutherstadt, 55 pp., 63 B&W and line drawing illustrations, 4 color plates.
Fuller, F.	1981	Frauenschuh und riemenzunge, 3rd ed. Die Neue Brehm-Bücherei, A. Ziemsen Verlag, Wittenberg Lutherstadt. 64 pp., 56 B&W and line drawing illustrations, 2 color plates.
Fuller, F.	1986	*Epipactis und Cephalanthera*, 3rd ed. Die Neue Brehm-Bücherei, A. Ziemsen Verlag, Wittenberg Lutherstadt, 92 pp., 87 B&W and line drawing illustrations,14 color plates.
Fultz, F. M.	1928	Lily, Iris and orchid of Southern California. Spanish American Institute Press, Gardena, CA. 135 pp., 64 B&W photographs.
Gade, H.	1967	Orchideer.

(continued)

(continued)

Author(s)	Date	Title and publisher
Gagnepain, F., and A. Guillaumin	1932–1934	Flore générale de l'Indo-chine: Orchidacées. Volime 6, Fascicle 2, pp. 142–288, Fig. 17–23; Fascicle 3, pp. 289–432, Fig. 24–39; Fascicle 4, pp. 433–576, Fig. 40–54; Fascicle 5, pp. 577–720, Fig. 55–71. Masson et Cie, Paris.
Galbraith, S. J.	1898	*Vanilla* culture as practiced in the Seychelles islands. US Department of Agriculture, Washington, DC. 24 pp.
Galé, J. A.	1987	Catalogo descriptivo de las orquideas Cubanas. Bishen Singh Mahendra Pal Singh, Dehra Sun, India, 221 pp.
Gallagher, D., J. Harris, M. Hewitt, and C. Jennings	1987	Australian orchid growing. Volume 1. Cymbidiums. Australian Orchid Council, 36 East Avenue, Black Forest, South Australia. B&W and color illustrations.
Gammie, G. A.	1905–1912	The orchids of Bombay Presidency. 13 parts from the Journal of the Bombay Natural History Society volumes 16–21, 11 color plates accompanied by text.
Gandhad, H. U.	1931	Doe tropischen orchideen.
Garay, L., and H. R. Sweet	1974	Orchids of the Southern Ryuku Islsands. Botanical Museum, Harvard University. 180 pp., 18 line drawings.
Garay, L. A.	1978	Orchidaceae (Cypripedioideae, Orchidoideae and Neottiodeae), No. 9 in Flora of Ecuador edited by G. Harling and B. Sparre. Department of Systematic Botany and Section of Botany, Riksmuseum, Stockholm. 305 pp., 1 color figure, 86 line drawing plates.
Garay, L. A.	1980	A generic revision of the Spiranthinae. Botanical Museum Leaflets Harvard University, 278–425.
Garay, L. A., F. Hamer and E. S. Siegerist	1994	The genus *Cirrhopetalum* and the genera of the *Bulbophyllum* alliance. Nordic Journal of Botany 14: 609–646.
Garay, L. A., and H. R. Sweet	1966, 1969	Natural and artificial hybrid generic names of orchids 1887–1965, and Natural and artificial hybrid generic names of orchids, Supplement I: 1966–1969. Botanical Museum Leaflets, Harvard University, Vol. 21, No. 6, pp. 141–212 and 22, No. 8, pp. 273–296.
Garay, L., and H. Sweet	1974	Flora of the Lesser Antilles. Orchidaceae.
García Castro, J. B. (ed.)	1995	Orquídeas de Costa Rica y su cultivo. Vol I. *Cattleya* y géneros relacionados. 2 Asociación Costarricense de Orquidiologia, San José. 59 pp., color painting and color photographs.

(continued)

(continued)

Author(s)	Date	Title and publisher
Gardner, C. A.	1947	West Australian wild flowers, 6th ed. 116 pp., numerous color illustrations. Note on title page: "Every good wish from [unreadable] Sunday Times, Perth, Western Australia."
Garrardin, J.	1966	Growing orchids for pleasure.
Geerinck, D.	1984	Flore d'Afrique Centrale, Vol I (Zaire, Rwanda, Burundi). Jardin Botanique Belgique. 296 pp., 16 color photographs, 48 plates of line drawings.
Gellert, M.	1923	Anatomische studien uber den bau der orchideenblute.
Gentry, E., and P. Foreman	1979	Native orchids of South Australia. Published by the authors, no city given. 96 pp., many line drawings.
George, A.	No date	Orchids of Western Australia. Westview Pty., Ltd., Probably in Perth, Western Australia. No pagination, 30 pp., numerous line drawings and color photographs.
George, A. S.	1971	A collection of articles on orchids, Nuytsia, Vol. 1m, No. 2, pp. 158–216.
Gerard, J.	1663	The herbal or General history of plants. The complete 1633 edition as revised and enlarged by Thomas Johnson. Reprint 1975 by Dover Publications, Inc, New York. 1631 pp.+index. Contains sections on orchids.
Gerasimov, S. O. and I. M. Zhuravlev	1988	Orchids. Moscow, Rosagropromizdat. (In Russian.)
Gerasimov, S. O., and I. M. Zhuravlev	1988	Orhidei. Rosagropromizdat, Moscow. 208 pp., Line drawings and color photographs.
Gerlach, G., and R. Schill	1993	Die gattung *Coryanthes* Hook (Orchidaceae). Akademie der Wissenschaften und der Literatur, Mainz and Franz Steiner Verlag, Stuttgart. 205 pp., numerous line drawings and B&W photographs.
Gerritsen, M. E., and R. Parsons	2005	Masdevallias. Timber Press, Portland, OR. 299 pp., numerous line drawings and colot photographs and several maps.
Ghose, B. N.	1959	Beautiful Indian orchids. Published by G. Ghose & Co., Town-End, Darjeeling, India. 155 pp., line drawings, color plates and B&W photographs.
Ghose, B. N.	1968	Beautiful Indian orchids, revised edition. Published by G. Ghose & Co., Town-End, Darjeeling, India. 200 pp., line drawings, color plates and B&W photographs.
Gibson, W. H., and H. L. Jelliffe	1905	Our native orchids. Doubleday, Page & Company, New York. 158 pp., numerous line drawings.

(continued)

(continued)

Author(s)	Date	Title and publisher
Gilbert, E. L. and P. H. Raven	1975	Coevolution of animals and plants. University of Texas Press. Chapter on bees and orchids by C. H. Dodson on pages 91–99. $30. 246 pp
Gilbert, L.	1992	Orchid man, the life, work and memoirs of the Rev H. M. R. Rupp 1872–1956. Kangaroo Press, Kenthurst, NSW. 248 pp., numerous B&W photographs and line drawings, colored line drawings.
Gilbert, P. A.	1945	The charm of growing orchids, 1st ed. John Bisset, Publisher. Abbotsford, NSW. 232 pp., 7 B&W photographs, a few line drawings.
Gilbert, P. A.	1951	Orchids, their culture and classification. The Shepherd Press, Sydney. 252 pp., one color and severak B&W photographs.
Gilbert, P. A.	1952	The charm of growing orchids, 2nd ed. The Shepherd Press, Sydney, Australia. 242 pp., 7 B&W photographs, 11 drawings.
Gilespie, R. J.	1961	Orchid culture. T. F. H. Publications, Inc., Neptune City, NJ. 32 pp., numerous B&W photographs.
Gilland, T.	2000	Orchids are easy. Guild of Master Craftsman Publications, Ltd., Lewes, East Sussex, UK. 120 pp., numerous color photographs.
Gillespie, R. J.	No date	Orchid culture. Missouri Botanical Garden. 23 pp., B&W photographs and line drawings.
Gilmour, J.	1946	British botanists. Collins, London. 48 pp., color and B&W illustrations.
Ginadi, T.	1979a	Anggrek dari bibit hinga berbunga. Perhimpunan Anggrek Indonesia. Cabang Bandung Priangen, Indonesia. 327 pp., numerous line drawings.
Girelli, E.	1987	Le orchidee della val d'Astico e della val Leogra Nel Vicentino. 152 pp., numerous color photographs.
Gitsham, A. T.	1089	Hedley home of orchids.
Gleason, H. A.	1952	The new Britton and Brown illustrated flora of the northeastern United States and adjacent Canada. Published for the New York Botanical Garden by Hafner Press, a division of Macmillan Publishing Co., New York. 482 pp., numerous line drawings.
Gloudon, A., and C. Tobish	1995	Orchids of Jamaica. The Press, The University of the West Indies, Kingston, Jamaica. 222 pp., numerous color photographs and line drawings.
Godfery, M. J., and H. M. Godfery	1933	Monograph & iconograph of native British Orchidaceae, 259 pp., 57 colored plates. Cambridge University Press, Cambridge. 57 colored plates.

(continued)

(continued)

Author(s)	Date	Title and publisher
Golamco, A.	1933	Philippines' book on orchids.
Goldsack, H.	1965	Orchid of the national park and wild life reserves. Field Naturalists' Society of South Australia, Box 1594M, G. P. O., Adelaide, South Australia. Pages 46–64, Fig. 22–27.
Golovanov, V. D. et al. (ed.)	1988	Krasnaya kniga RSFCR rastenyi (Red Book of Russian plants). Rosagropromisdat, Moscow. 392 pp., numerous color painting and photographs. The book has a good chapter on orchids.
Gordon, B.	1990	Culture of the *Phalaenopsis* orchid, 1st revision. Laid Back Publications, 276 East Shamrock, Rialto, California. 188 pp., few illustrations.
Gordon, B.	1985	Culture of the *Phalaenopsis* orchid. Laid-Back Publications, Rialto, CA. 118 pp., numerous B&W photographs, some line drawings.
Gordon, B.	1991	Orchid seedling care. Laid Back Publications, 276 East Shamrock, Rialto, California. 162 pp., few illustrations.
Gordon, B. (ed.)	1988	*Phalaenopsis* culture: A worldwide survey of growers. Laid Back Publications, Rialto, CA. 311 pp.
Gotwald, F. G.	1983	The Nero Wolfe handbook. Published by the author (a clergyman at St. John's Lutheran Church) in Salisbury, NC. 48 pp. This privately published book diagrams information on Nero Wolfe's orchids.
Gould, J.	2001	Time to say good-bye. A novel. Signet Books. 402 pp.
Graebner, K. E.	1982	Freude an orchideen. Verlag J. Berg, Munich and Prisma Verlag Gmbh, Gütersloh, 128 pp., 80 color photographs, 8 line drawings.
Graham, R., and R. Roy	1983	Slipper orchids. The art of Digby Graham. A. H. & A. W. Reed, Wellington, NSW, Australia. 109 pp., numerous plates of color paintings.
Grant, B.	1893	The orchids of Burma (Including the Andaman Islands). The Hanthawaddy Press, Rangoon, errata page+ii+424 pp.+ 8 pp. separately paginated index+memorandum page (a form provided by the author for readers to send him information about Burmese orchids).
Grant, B.	1895	The orchids of Burma. Reprint 1966 by Central Press, Rangoon.
Gratiot, J.	1934	Les orchdées leur culture. Librairie Agricole de la Maison Rustique/Librairie de l'Académie d'Agriculture, Paris. 170 pp., 49 B&W photographs and line drawings.
Gravendeel, B.	2000	Reorganising the orchid genus *Coelogyne*. A phylogenetic classification based on morphology and molecules. National Herbarium Nederland, Universiteit Leiden branch. 208 pp., 28 color and numerous B&W photographs.

(continued)

(continued)

Author(s)	Date	Title and publisher
Gray, A.	1868	Gray's Lessons in botany. Ivison, Blakeman, Taylor & Co., New York. 703 pp., 20 plates of line drawings, 800+ line drawings in the text. Has sections on orchids.
Greatwood, J., P. F. Hunt, P. J. Cribb, and J. Stewart	1993	The handbook of orchid nomenclature and registration, 4th ed. International Orchid Commission, Royal Horticultural Society, Wisley, Woking, Surrey, UK
Green, J. R.	1909	A history of botany 1860–1900. The Clarendon Press, Oxford. 543 pp.
Greer, B.	1998	The astonishing stanhopeas. Published by the author at 102 Anzac Ave., Collaroy Beach, NSW, Australia. 80 pp., numerous B&W and color photographs and line drawings by Walter T. Upton.
Greisen, K. S.	2002	Commercial propagation of orchids in tissue culture: Seed-flasking methods. An illustrated laboratory manual. K. S. Griese Specialties, P. O. Box 5818, Oakland, CA. An awfully bad self published book.
Grezaffi, J., III.	No date	Orchid culture and techniques. Grezaffi Orchids, W. Melbourne, FL.
Griffiths, M.	2002	Orchids from the archives of the Royal Horticultural Society. Harry N. Abrams, Inc., Publishers, New York. 336 pp., 300 color plates a few in B&W.
Grisebach, A. H. R.	1963	Flora of the British West Indian islands, (orchid pages are 606–644). Reprint 1963 by J. Cramer, Weinheim, Germany; Wheldon & Wesley, Ltd., Codicote, Herts; and Hafner Publishing Co., New York.
Grontved, J.	1947	Orchidernes udbredelse Denmark.
Gross [?]	1966	Blühende kleinoden heimische Orchideen.
Grove, D. L.	1995	Vandas and Ascocendas and their combinations with other genera. Timber Press, Portland, OR. 241 pp., numerous color photographs and paintings.
Grubb, R.	1961–1970	Select orchidaceous plants. Volume One, Parts 1–4 (1450 copies published): 147 pp., 50 colored plates of paintings and 26 B&W illustrations. Volume Two (1,000 copies published): 77 pp., 13 colored plates and several B&W illustrations Printed and published by Roy & Ann Grubb Partnership, 62 Chaldon Common Road, Caterham, Surrey.
Gruss, O., and M. Wolff	1995	Phalaenopsis.
Gsell, R.	1935/1951	Orchideen.
Gudest, S., and S. Guest	1997	*Cymbidium* mania.

(continued)

(continued)

Author(s)	Date	Title and publisher
Gudžinskas, Z., and M. Ryla	2006	Lietuvos gegužraibiniai (Orchidaceae). Orchids (Orchidaceae) of Lithuania. Botaniskos institutio leidykla, Vilnius. 104 pp., numerous color photographs and distribution maps.
Guest S., and G. Guest	1999	I've got spikes! Guest orchids. Business Centre Post Office, Box 183, Salisbury South, South Australia 5106. 80 pp., numerous line drawings and B&W and color photographs.
Guest, G. and Guest, S.	1992	Cymbidiums, your guide to a rewarding pastime.
Guest, G. and Guest, S.	1996	More cymbidiums. Guest Orchids, Burton, South Australia. 88pp., numerous line drawings and B&W and color illustrations.
Guillaumin, A.	1929	Les *Phalaenopsis* diagram issus du *P. amabilis* Bl. *Archives du Muéum National d'Histoire Naturelle* 4: 35–36, 2 colored plates of paintings.
Guillaumin, A.	1930	Les *Phalaenopsis* diagram autres que ceux issus du *P. amabilis* Bl. *Archives du Muéum National d'Histoire Naturelle* Ser 6, Vol. 6, pp. 75–80, 2 colored plates of paintings.
Guillaumin, A., and R. Lami	1935	Les diagram intergénériques d'orchidées-Sarcanthées et le X Vandacostylis Bernardii. Archives du Muséum, 6th Series, Vol. 12, pp. 607–613, two B&W photographs, one color painting, three line drawings.
Guillochon, L.	1896	Calendrier mensuel du cultivateur d'orchidées. Librairie Horticole du "Jardin," Paris. 88 pp.
Guillot, G.	1986	Les orchidées sauvages de France. Hatier Publishers, Paris. 120 pp., numeros color photographs.
Gunadi, T.	1977	Mengali Anggrek. Perhimpunan Anggrek Indonesia, Cabang Bandung/Priangan. 128 pp., numerous line drawings.
Gunadi, T.	1979	Anggrek Indonesia. Published by Perhimpunan Anggrek Indonesia (Indonesian Orchid Society), Jakarta. 96 pp., color and B&W photographs.
Gunther, R. T.	1968	The Greek herbal of Dioscorides. Fcsimile of the 1934 edition. Hafner Publishing Co., London and New York. 701 pp., 396 illustrations.
Gupton, O. W., and F. C. Swope	1986	Wild orchids of the middle Atlantic states. The University of Tennessee Press, Knoxville. 112 pp., numerous color photographs.
Haber, O.	No date	Orchideeen in huis. Dutch translation of Haber, 1966.
Haber, W.	1966	Orchideen inm haus. C. Bertelsmann Verlag. 191 pp., numerous line drawings, B&W and color illustrations.

(continued)

(continued)

Author(s)	Date	Title and publisher
Hackney, C. T.	2004	American cattleyas. No publisher listed. 142 pages of text and 76 pages of color photographs.
Hagasater, E., V. Dumont, and A. M. Pridgeon	1996	Orchids, status survey and conservation action plan. IUCN/SSC Orchid Specialist Group, Gland Switzerland and Cambridge, UK. 152 pp., numerous B&W photographs.
Hágsater, E. et al.	1993	Icones Orchidacearum, Fascicles 2–4. The Genus *Epidendrum*. Parts 1–3. Plates 100–200, 301–500. A century of new species in Epidendrum. Associacion Mexicana de Orquideologia A. C. Apartado Postal 53–123, 11320 México D. F. English and Spanish. No pagination, 200 line drawing plates, numerous maps.
Hágsater, E. et al. (ed.)	2005	Orchids of Mexico
Hágsater, E. et al. (ed.)	2002 on	Icones Orchidacearum
Hágsater, E., and G. A. Salazar.	1990	Icones Orchidacearum, Fascicle 1. Plates 1–100. Associacion Mexicana de Orquideologia A. C. Apartado Postal 53–123, 11320 México D. F. English and Spanish. No pagination, 200 line drawing plates.
Halbinger, F.	1993	Laelias de Mexico. Associación Mexicana de Orquideologia, A. C., Mexico City. 71 pp., 49 color photographs, numerous line drawings and maps.
Halbinger, F., and M. Soto	1997	Laelias of Mexico. Orquidea (Méx), Herbario AMO, Mexico City. 160 pp., numerous color photographs, line drawings and maps.
Halcrow, M, ands M. L. Halcrow	1967 or 1968	Orchids of Belize. Government Printer, Belize. 151 pages, about 75 line drawings, map. Not on sale on the WWW in 2005.
Hall, A. V.	1965	Studies of the South African species of *Eulophia*
Hallé, N.	1977	Flore de la Nouvelle Caledonie et Dépendances. Orchidacées. Muséum Nationale d'Histoire Naturelle, Paris. 565 pp., 218 line drawing and color plates, 172 maps. Includes reprint of addenda.
Haller, A.	1736	Methodico studio botanices. (In Latin.)
Hamer, F.	1988, 1990	Orchids of Central America. An illustrated field guide. Selbyana Volumes 10 (Supplement; pp. 1–430, illustrations 1–422, A-L) and 11 (Supplement; pp. 423–860, illustrations 423–860, M-Z).
Hamer, F.	1982–1985	Orchids of Nicaragua parts 1–6, Icones Plantarum Tropicarum fascicles 7–13, Marie Selby Botanical Garden, Sarasota, FL. 1,300 line drawings plates and the same number of maps.

(continued)

(continued)

Author(s)	Date	Title and publisher
Hamer, F. or Hammer?	1974	Las orquideas de El Salvador/The orchids of El Salvador/Die Orchideen El Salvador, volumes I (374 pp., numerous color photographs and line drawings), II (427 pp., numerous color photographs and line drawings; Ministerio de Educacion, Direccion de Publicaciones, San Salvador, El Salvador), III (supplement, 304 pp., numerous color photographs and line drawings; Marie Selby Botanical Garden, Sarasota FL). Tri lingual, Spanish-English-German.
Hamilton, R. M.	1967	Orchid flower index. A world list of reproductions in color in books and periodicals 1736 to 1966. Compiled and published by the author at 921 Beckwith Road, Richmond, British Columbia, Canada. 124 pp., one color plate as frontispiece.
Hamilton, R. M.	1977	When does it flower (Orchids in the USA). Published by Robert M. Hamilton, 9211 Beckwith Road, Richmond, BC, Canada V6X 1V7, 76 pp.
Hamilton, R. M.	1979	Orchid flower index – Volume 2. A world list of reproductions in color in books and periodicals 1966–1979. Compiled and published by the author at 921 Beckwith Road, Richmond, British Columbia, Canada V6X 1V7. 107 pp.
Hamilton, R. M.	1980	The orchid doctor. Published by the author. Vi plus 152 pp.
Hamilton, R. M.	1986	When does it flower? Orchid species in the greenhouse since 1881. Compiled and published by the author at 921 Beckwith Road, Richmond, British Columbia, Canada V6X 1V7. 120 pp.
Hamilton, R. M.	1988	The new orchid doctor. Compiled and published by the author at 921 Beckwith Road, Richmond, British Columbia, Canada V6X 1V7. 130 pp.
Hamilton, R. M.	1992	Address by James Bateman (1867) and other papers in the memory of George Ure Skinner. Orchid History Reference Paper No. 8. Published by the author (a Canadian from Richmond, BC) at P. O. Box 3232, Blaine, Washington. 25 pp.
Hamilton, R. M.	1992	Index to orchid orchid history in Paxton's Horticultural Register 1831–1836. Orchid History Reference Paper No. 8. Published by the author (a Canadian from Richmond, BC) at P. O. Box 3232, Blaine, Washington. 10 pp.

(continued)

(continued)

Author(s)	Date	Title and publisher
Hamilton, R. M.	1992	James Bateman: On the cool treatment of orchids, three selections, 1862–1864 Orchid History Reference Paper No.10. Published by the author (a Canadian from Richmond, BC) at P. O. Box 3232, Blaine, Washington. 45 pp., sketches of the inside of a greenhouse and of Bateman.
Hamilton, R. M.	1992	The Gardener's Chronicle index to orchid history 1841–1846 when Lindley was editor, with new orchids described 1841–1878. Orchid History Reference Paper No. 5. Published by the author (a Canadian from Richmond, BC) at P. O. Box 3232, Blaine, Washington.
Hamilton, R. M.	1993	Orchidaceae Lindenianae, or, notes upon a collection of orchids formed in Colombia and Cuba by Mr. J. Linden. Brandbury and Evans, Whitefriars, 1846. Orchid History Reference Paper No. 14, 1993 – Reprint. 35 pp.
Hamilton, R. M.	1993	Orchids and ordeals in Guatemala and England, 1830–1867, 260 letters by George Ure Skinner and friends. Orchid History Reference Paper No.12. Published by the author (a Canadian from Richmond, BC) at P. O. Box 3232, Blaine, Washington. 215 pp., color map, sketches by Skinner.
Hamilton, R. M.	1994	John Lindley (1799–1865) "Father of Modern Orchidology". A gathering of his correspondence issued in installments. Part One – 117 letters with a chronology and index. Published (only 50 copies) by the author (a Canadian from Richmond, BC) at P. O. Box 3232, Blaine, Washington. x and 46 pp. $20.00. Also included is an article on the centenary of Lindley's death from the Journal of the Royal Horticultural Society, Volume 90, part 11, pp. 459–462 with a B&W photograph or painting.
Hamilton, R. M.	1994	Sertum Orchidaceum by J. Lindley. Text only with index. Orchid History Reference Paper No. 15. xviii pp.
Hamilton, R. M.	1995	John Lindley (1799–1865) "Father of Modern Orchidology". A gathering of his correspondence issued in installments. Part two – 136 letters with a portrait in color, sketch of his life by his son, several graphic al sketches and an index to letters. Published (only 50 copies) by the author (a Canadian from Richmond, BC) at P. O. Box 3232, Blaine, Washington. xviii and 71 pp.

(continued)

(continued)

Author(s)	Date	Title and publisher
Hamilton, R. M.	1996	John Lindley (1799–1865) "Father of Modern Orchidology". A gathering of his correspondence issued in installments. Part three – 173 letters by him 1819–1865, 45 letters by family and friends, illustrations in black and white, a view of his tombstone recently destroyed and an index. Published (only 50 copies) by the author (a Canadian from Richmond, BC) at P. O. Box 3232, Blaine, Washington. vii and 145 pp.
Hamilton, R. M.	1996	Orchid books prices digest. Old & in-print for buyers and sellers, 1990–1995.
Hamilton, R. M.	1996	Sarah Lindley's family letters to Henry Crease. Part One, July–December 1849. Orchid History Reference Paper No. 21. xii+80 pp.
Hamilton, R. M.	1997	Sarah Lindley's family letters to Henry Crease. Part four and last, 1856 (20 letters), 1858 (21 letters), 1859 (4 letters). Orchid History Reference Paper No.24. xvii+92 pp.
Hamilton, R. M.	1997	Sarah Lindley's family letters to Henry Crease. Part Two, January to September 1851. Orchid History Reference Paper No. 22. xvii+90 pp.
Hamilton, R. M.	1998	John Lindley (1799–1865) "Father of Modern Orchidology". A gathering of his correspondence issued in installments. Part four – 175 letters 1820–1860 mostly to W. J. Hooker with an appreciation of Lindley of 1846, extracts about him from his son's mss autobiography, a revised sketch of the life of Miss Drake, some illustrations by Lindley and others, and an index. Published (only 50 copies) by the author (a Canadian from Richmond, BC) at P. O. Box 3232, Blaine, Washington. xxvi and 129 pp.
Hamilton, R. M.	No date	Gardener's Magazine index to orchid history 1826–1844. Orchid History Reference Paper No. 6. Published by the author (a Canadian from Richmond, BC) at P. O. Box 3232, Blaine, Washington. 28 pp.
Hamilton, R. M.	No date	Index to Catalogue of Orchidaceaous plants in the collection of the Rev. John Clowes 1842. Orchid History Reference Paper No. 4. Published by the author (a Canadian from Richmond, BC) at P. O. Box 3232, Blaine, Washington. 27 pp.

(continued)

(continued)

Author(s)	Date	Title and publisher
Hamilton, R. M.	No date	Index to orchid species (1787–1950) and orchid history (1787–1880) in Curtis's Botanical Magazine. Orchid History Reference Paper No. 13. Published by the author (a Canadian from Richmond, BC) at P. O. Box 3232, Blaine, Washington.73 pp.
Hamilton, R. M.	No date	Index to Orchideae in the collection of Conrad Loddiges and Sons 1844. Orchid History Reference Paper No. 2. Published by the author (a Canadian from Richmond, BC) at P. O. Box 3232, Blaine, Washington. 30 pp.
Hamilton, R. M.	No date	James Bateman: The Orchidaceae of Mexico and Guatemala, a readable reprint of the text with one color plate only, numerous line drawings and an index. Orchid History Reference Paper No. 16. Published by the author (a Canadian from Richmond, BC) at P. O. Box 3232, Blaine, Washington. ix and 116 pp.
Hamilton, R. M.	No date	Orchid history in Edward's Botanical Register 1815–1847. Orchid History Reference Paper No.3. Published by the author (a Canadian from Richmond, BC) at P. O. Box 3232, Blaine, Washington. 22 pp.
Hammer, F.	1974–1981	Las Orquideas de El Salvador. Volumes 1–3.
Handcock, R., and M. Smith	1959	You too can grow orchids, 2nd ed. Dymock's Book Arcade, Ltd., Sydney. 58 pp., two color and many B&W photographs.
Handcock, R., and M. Smith	1995	You too can grow orchids.
Handcock, R., and M.Smith	1990	Orchids month by month a guide for Australian gardeners.
Handcock, R., and M. Smith	1991	Orchids month by month a guide for Australian gardeners.
Hansen, E.	2000	Orchid fever. Pantheon Books, Inc. New York. 273 pp., 17 line drawings.
Hansen, E.	2000	Orchid fever. Vintage Departures/Vintage Books/ Random House, Inc. New York. 272 pp., 17 line drawings.
Hansen, E.	2000	Orchideeënkoorts. Uitgeverij BZZTôH. 271 pp., 17 line drawings.
Hansen, E.	2002	Orchideenfieber. Klett-Cotta. 311 pp., 17 line drawings.
Hansen, E.	2001	Orchid fever. NHK. 288 pp., line drawings and color photographs. Japanese edition.
Hansen, G.	1895	The orchid hybrids.
Hansen, G.	1897	The orchid hybrids.

(continued)

(continued)

Author(s)	Date	Title and publisher
Hansson, S.	1985	Orckidéer I Svensk diagra. Wiken, Switzerland. 128 pp., numerous maps and color photographs.
Harch, E. D.	No date	Collection of articles on New Zealand orchids. Transaction of the Royal Society of New Zealand. Vol. 78 and 80, 1951–1953.
Harjadi, S. S., Irawati, Krisantini, N. M. A.Wiendi, and A. Munandar	1990	Proceedings of the seventh ASEAN orchid congress, November 5–7, 1988, Jakarta. Central Research Institute for Horticulture, Pasar Mingu, Jakarta and Directorate for Horticulture Production, Ministry of Agriculture, Jakarta, Indonesia. 183 pp., published with 15 page seminar schedule booklet, 32 page program brochure and 122 magazine like publication in Indonesian with numerous color photographs.
Harley, J. L.	1969	The biology of mycorrhiza. Leonard Hill, London. 334 pp., B&W photographs, graphs and line drawings.
Harley, J. L., and S. E. Smith	1983	Mycorrhizal symbiosis. Academic Press, New York. 483 pp. Photographs and line drawings. S. E. Smith is J. L. Harley's daughter which means that this book is a family effort.
Harris, W. K.	1987	Modern orchid growing for pleasure and profit. Orchid Club of South Australia, Adelaide. 152 pp., numerous B&W and color photographs.
Harrison, C. A.	1911	Orchids for amateurs. W. H. & L. Collingridge, London. 148 pp., 19 B&W illustrations.
Harrison, C. A.	1914	Commercial orchid growing.
Harrison, C. R.	1973	Physiology and ultrastructure of *Cattleya aurantiaca* (Orchidaceae) germination. Ph.D. dissertation, University of California, Irvine.
Harrison, E. R.	1972	Epiphytic orchids of southern Africa. Published by the Natal Branch of the Wildlife Protection and Conservation Society of South Africa. 107 pp., numerous maps and line drawings.
Harrori, Y., and H. Goda	1974	Orchids, Yoran Publishers, Japan. 232 pp., 247 color pictures, 506 numbered B&W pictures+some not numbered, some line drawings.
Hartman, W. L.	1971	Introduction to the cultivation of orchids. Editorial Fournier, Mexico, D. F. 106 pp., line drawings, map, numerous color photographs.
Hashimoto, T., and K. Kanda	1981	Japanese indigenous orchids in color. Ienohikari Association, Tokyo. 247 pp., numerous color photographs and several line drawings. (In Japanese.)

(continued)

(continued)

Author(s)	Date	Title and publisher
Hashimoto, T., Kanda, K. and H. Murakawa	1991	Japanese Indigenous Orchids in color. Revised and enlarged. Ienohikari Association, Tokyo. 295 pp., Numerous color photographs. (In Japanese.)
Hashimoto, T., and S. Kimura	1983	Exotic orchids in color. Ienohikari Association, Tokyo. 248 pp., numerous color photographs and several line drawings. (In Japanese.)
Hauschild, J.	1986	Orchids and exotic flowers, charted designs. Dover Publications, Inc. New York. 32 pp., 18 color illustrations inside covers, one of front and 3 on back covers..
Hauzeau, de Lehaie, J.	1914	Dissemination des orchidees indigenes
Hawkes, A. D.	1961	Orchids, their botany and culture. Harper & Brothers, New York. 297 pp., numerous line drawings and B&W and one color photographs.
Hawkes, A. D.	1965	Encyclopaedia of cultivated orchids. Faber and Faber, Ltd., London. 602 pp., numerous illustrations, some in color.
Hayashi, L. A., and K. W. Bishop	1998	Fables from the garden. University of Hawaii Press, Honolulu, HI.
Hayek, A., and F. Markgraf	1933 (reprint 1971)	Prodromus florae peninsulae Balcanicae. Verlag des Repertoriums, Fabeckstrasses 49, Dahlem bei Berlin. Reprinted in 1971 by Koelz Scientific Books, Loeningstein, Germany. 472 pp. Orchids are on pp.371–416.
He, Q. Z., and X. Q. Chen (eds.)	1990	Chinese orchid. Vol 1 (132 pp., numerous color photograph), and Vol 2 (132 pp., numerous color photograph). Sichuan Art Publishing House. (In Chinese with some English captions.)
Heaps, G.	1998	The new grower's guide to orchids. Gordon Heaps Horticultural Services, Spruce Grove, Alberta, Canada.
Hedge, S. N.	1984	Orchids of Arunachal Pradesh. Forest Department, Arunachal Pradesh, Itanagar, India, 99 pp., 120 color photographs.
Hefka, A.	1914	Catleyen und Laelien. Verlag von Wilhelm Frick, k. u. k. Hafbuchändler, Vienna and Leipzig. 82 pp., 20 B&W photographs.
Henderson, P.	1896	Practical Horticulture. Orange Judd Compnay, New York. 325 pp., numerous line drawings. New and expanded edition. Has information on orchids.
Henderson, M. R., and G. H. Addison	1959	Malayan orchid hybrids. Government Printing Office, Singapore. 191 pp., numerous B&W photographs, color frontispiece. See also Addison, 1961.

(continued)

(continued)

Author(s)	Date	Title and publisher
Hennessy, E. F., and T. A. Hedge	1989	The slipper orchids. Acorn Books CC., P. O. Box 4845, Randburg, 2125, Republic of South Africa, 263 pp., 104 plates of color paintings.
Henry, L. K., W. E. Buker, and D. L. Pearth	1975	Western Pennsylvania orchids. Castanea Vol. 40, No. 2, pp. 93–168, 43 line drawing figures.
Henshall, J.	1845	A practical treatise on the cultivation of Orchidaceous plants.
Henshall, J.	1845	The cultivation of Orchidaceus plants.
Henslow, G.	1888	The origin of floral structures through insect and other agencies. Kegan Paul, Trench & Co., London. 349 pp., 88 line drawings. Original.
Herjakob, G.	1974	Orchids of Greece. 1. The genus *Ophrys*. The Goulandris Natural History Museum, Kifissia, Greece. No pagination. Numerous color photographs of 33 Ophrys species and one line drawing.
Hermans, J., and C. Hermans	1997	An annotated checklist on the genus *Dracula*. Published by the Orchid Digest Corporation, Huntington Library, San Marino, CA.
Hernstrom, T. C.	1982	*Paphiopedilum*. TJD Co., 65 pp., numerous maps and line drawings.
Herter, G.	1943	Flora illustrada del Uruguay. Part V Cyperaceae-Orchidaceae. Staatliche Botanische Anstalten, Krakau. Pages 161–256. No text, only line drawings.
Hess, H. (poetry) and J. Thomas (paintings)	1962	Orchids and the orchid isle. Malama Arts, Inc., Honolulu, HI. 83 pp., line drawings and color paintings.
Hetherington, E.	No date	How to grow your cattleyas. Fred A. Stewart Orchids, 121 E. Las Tunas Drive, San Gabriel, CA 91778. No pagination, 7 pp.
Heusser, K.	1914	Die entwicklung der generativen organe von *Himatoglossum hircinu*m Spr. Inaugural-Dissertation.
Hew, C. S., and W. H. Yong	1997	The physiology of tropical orchids in relation to the industry, 1st ed. World Scientific Publishing Co., Pte., Ltd. 331 pp., line drawings, graphs, B&W photographs.
Hew, C. S., and W. H. Yong	2004	The physiology of tropical orchids in relation to the industry, 2nd ed. World Scientific Publishing Co., Pte., Ltd. 370 pp., line drawings, graphs, B&W photographs.
Hifrmer, M.	1974	Wunderwelt der orchideen.
Hill, G. L.	1971	White orchids. A novel. Spire Books, New York. 218 pp.
Hillerman, F.	1982	A cultural manual for Angrecoid orchid growers. Published by the author. 32 pp., color photographs on cover, line drawings and B&W photographs in text.

(continued)

(continued)

Author(s)	Date	Title and publisher
Hillerman, F.	No date	A culture guide for Aerangis orchid growers. Fred Hillerman, P. O. Box 976, Grass Valley, CA 95945. 32 pp.
Hillerman, F. E., and A. W. Holst	1986	An introduction to the cultivated Angrecoid orchids of Madagascar. Timber Press, Portland, OR. 302 pp., numerous line drawings and color photographs.
Hirmer, M.	1920	Beitrage zur organographie derorchideenbute.
Hiroe, M.	1971	Orchid flowers. Kyoto-Shoin Co., Ltd., Shijo-ogaru, Kawaramachi, Nagyo-ku, Kyoto, Japan.. Two volumes. Numerous full page color photographs. (In Japanese.)
Hitchinson, J., J. M. Dalzel and F. N. Hepper	1968	Flora of west tropical Africa, Vol. Three, Part One. 275 pp., numerous line drawings. Orchids by V. S. Summerhays on pp. 180–275.
Hoagland, D. R., and D. I. Arnon	1950	The water-culture method for growing plants without soil. California Agricultural Experiment Station Circular 347, revised January 1950 by D. I. Arnon. The College of Agriculture, University of California, Berkeley, CA. This is a classic publication in the area of nutrient solutions which support plant growth (even orchids) hydroponically and in vitro.
Hodene, F. C.	1930	Album de Orchidaceas Brasileiras.
Hodene, F. C.	1930	Orchidaceas Brasileiras.
Hodene, F. C.	1936	Orchidaceas dos herbarios.
Hodgson, M. and R. Paine	1989	Field guide to Australian orchids.
Hodgson, M., R. Payne, and N. Abderson	1991	A guide to orchids of the world. Angus & Robertso, Sydney. 236 pp., with many color photographs.
Hoehne, F. C.	1940	Flora Brasílica Orchidaceae Fasc. 1, Vol. XII, I (1–12 compl), 218 pp., numerous line drawings, some color paintings (in Portuguese). Secretaria Da Agricultura, Indústria e Comércio de São Paulo, Brasil.
Hoehne, F. C.	1942	Flora Brasílica Orchidaceae Fasc. 5, Vol. XII, VI (97–114), 254 pp., numerous line drawings. (In Portuguese.) Secretaria Da Agricultura, Indústria e Comércio de São Paulo, Brasil.
Hoehne, F. C.	1945	Flora Brasílica Orchidaceae Fasc. 8, Vol. XII, II (13–43), no text, diagrams line drawings. (In Portuguese.) Secretaria Da Agricultura, Indústria e Comércio de São Paulo, Brasil.
Hoehne, F. C.	1949	Iconografia de orchidaceas do Brasil, 301 pp., numerous line drawings, B&W and colored illustrations. S. A. Indúsrias Graphicars F. Lanzara, Sao Paolo, Brasil.

(continued)

(continued)

Author(s)	Date	Title and publisher
Hoehne, F. C.	1953	Flora Brasílica Orchidaceae Fasc. 10, Vol. XII, VII (115–148), 397 pp., numerous line drawings, B&W photographs and color paintings. (In Portuguese.) Secretaria Da Agricultura, Indústria e Comércio de São Paulo, Brasil).
Hoffman, N., and A. Brown	1984	Orchids of south-west Australia, 1st ed. University of Western Australia. 382 pp., numerous line drawings, maps and color photographs.
Hoffman, N., and A. Brown	1998	Orchids of south-west Australia, revised 2nd ed. with supplement. University of Western Australia. 470 pp., numerous line drawings, maps and color photographs. Two versions, a hadrcover and a paperback.
Hofmeister, G., and A. Springer	1967	Orchideen im zimmer und im garten, 4th ed. Verlag Paul Parey, Hamburg. 80 pp., 34 B&W photographs.
Hogg, B.	1857	Orchids: Their culture.
Holladay, K., W. Rodehamel, and C. A. Luer	1997	A *Dracula* cultural guide. Published by the Pleurathallid Alliance, 201 Lee Drive North, Middleburg, FL 32068. x+56 pp., line drawings and graphs.
Hollander, S.	1932	Ernahrungsphysiologische uuntersuchungen an wurzelpilzen saprophytisch lebender orchideen.
Holley, A. L.	2005	Orchids from the beginning. Alva L. Holley.
Holliman, J.	2002	Botanica's orchids. Laurel Glen Publishing, Advantage Publishers Group, San Diego, CA. 608 pp., over 1,200 color photographs.
Hollingsworth, A. P.	1989	Growing orchids is fun.
Holst, A. W.	1999	The world of catasetums. Timber Press, Portland, OR. 306 pp., numeros line drawings and B&W and color photographs.
Holzhausen, A.	1916	Orchideer.
Holttum, R. E.	1953	A revised flora of Malaya. Volume I. Orchids. 1st ed. Government Printing Office, Singapore. 753 pp., numerous line drawings, some color.
Holttum, R. E.	1953	Gardening in the lowlands of Malaya. The Straits Times Press, Ltd., Singapore. 323 pp., numerous line drawings, 18 pages of B&W and 16 pages of color photographs.
Holttum, R. E.	1954	A revised flora of Malaya. Volume I. Orchids. Government Printing Office, Singapore. 759 pp., numerous line drawings, some color.
Holttum, R. E.	1957	A revised flora of Malaya. Volume I. Orchids. 1st ed. Government Printing Office, Singapore. 759 pp., numerous line drawings, some color.

(continued)

(continued)

Author(s)	Date	Title and publisher
Holttum, R. E.	1969	Plant life in Malaya. Longman Group, Ltd., London. 254, B&W photograph frontispiece, 51 line drawings.
Homoya, M. A.	1993	Orchids of Indiana. Indiana Academy of Sciences at the Indiana University Press, Bloomington, IN. 276 pp., numerous maps and 92 color photographs.
Hoog, R. B.	No date	Orchids for everybody.
Hooker, J. D.	1894	Flora of British India, Vol. V. Chenopodiaceae to Orchideae. L. Reeve & Co., Ashford, Kent, UK. 1954 reprint. 910 pp.
Hooker, J. D.	1894	Flora of British India, Vol. VI. Orchideae to Cyperaceae. L. Reeve & Co., Ashford, Kent, UK. 1954 reprint. 793 pp.
Hooker, J. D.	1895	A century of Indian orchids. *Annals of the Royal Botanic Garden, Calcutta* 5: 1–68, 100 plates of B&W drawings. Bound with King and Pantling, 1898 and Duthie, 1906. Reprint.
Hooker, J. D.	1855	Himalayan journals. Reprint 1980 by Today and Tomorrow's Printers and Publishers, New Delhi. Two volumes in one. Volume I: xvi+34 pp., volume II: xii+345 pp., numerous line drawings.
Hopkinson, A. E.	1896	The orchid hybridists handbook
Hopp, W.	1957	Blütenzauber der orchideen. Safari-Verlag, Berlin. 236 pp., 16 color and 48 B&W photographs.
Horowitz, A.	1902	Uber den anatomischen Bau und der Aufspringen der Orchideen Fruchte. Inaugural Dissertation.
Houston Orchid Society	2001	Orchids of the Philippines.
Howard, R. A.	1974	Flora of the lesser diagrams, Leeward and Windward Islands. Orchidaceae. Arnold Arboretum, Harvard University, Jamaica Plain, MA. 235 pp.83 line drawings.
Hsiang, R.-C.	1984	*Cattleya* superior parentages. The Orchid Society of the Republic of China, Tainan, Taiwan. 364 pp., numerous color photographs.
Hsiang, R.-C.	No date	Phalaenopsis, superior parentages and awards (1962–1978). The Orchid Society of the Republic of China, 512 Pei Men Road, Tainan, Taiwan. Original price $45 plus postage.
Hsu, C. H.	1984	Quality cattleyas in color.
Hu, S. Y.	1971	The Orchidaceae of China parts 1 and 2. Quarterly Journal of the Taiwan Museum Volume 24, Nos. 1–5, pp. 67–103, 181–262, photographs, line drawings, maps.

(continued)

(continued)

Author(s)	Date	Title and publisher
Hu, S. Y.	1977	The genera of Orchidaceae in Hong Kong. The Chinese University Press. 160 pp., color painting frontispiece, 70 line drawing figures, 4 multi picture color plates.
Huber, G. (ed.)	1980	Anais do 1 encontro nacional de orquidófilos e orquidólogos. Expresäo e Cultura, Rio de Janeiro. 250 pp., line drawings, B&W and color photographs. Bilungual: Portoguese and English.
Humbert, H.	1939–1941	Flore de Madagascar. Orchideess. Parts I and II.
Humphris, T.	1970	Garden glory. The County Book Club, London. 212 pp., numerous line drawings.
Hungria, M. S. A., de	1996	Orchids from Brasil.
Hunt, D. R.	1981	Orchids from Curtis's Botanical Magazine. Curwen Books, London. 125 pp., 31 color plates, line drawings.
Hunt, P. F.	1979	The international book of orchids. Chartwell Books, Secaucus, NJ. 172 pp., numerous line drawings, color paintings and B&W and color photographs..
Hunt, P. F., and M. A. Grierson	1973	Orchidaceae. The Bourton Press. 144 pp., 1 table, 40 colored plates. No. 31 of 620 copies inscribed in print and calligraphy.
Hunt, P. F., and M. Grierson	1978	The country life book of orchids. The Bourton Press, Ltd., London. 128 pp., numerous colors and B&W illustrations.
Hunt, P. F., and T. Kajima	1978	The orchid. Octopus Books, Ltd., London. 207 pp., 159 color photographs.
Hurst, C. C.	1898	Notes on some curiosities of orchid breeding.
Hurst, C. C.	1925	Experiments in genetics. Cambridge University Press. Contains sections on orchids with is probably the first coverage of orchid genetics.
Hüssy, H. W.	1976	Orchideen der Schweiz. Verlag Fasler-Druck AG, Aarau, Switzerland. No pagination, numerous colot photographs. Book measures 35.5 × 26 cm.
Hyde, S. Mrs. (ed.)	1961	Your first orchids and how to grow them. Oregon Orchid Society. 80 pp., line drawings, color and B&W photographs.
Ichihashi, S.	1993	*Phalaenopsis*. S+S, Tokyo. 234 pp., numerous B&W photographs and line drawings. (In Japanese.)
Ichihashi, S.	2002	Flower dome 2002. Aichi Flower Festival. Nagoya International Orchid Show Proceedings. (In Japanese.) 105 pp., numerous color photographs.

(continued)

(continued)

Author(s)	Date	Title and publisher
Ichihashi, S. and H. Inoue	1995	Proceedings of the Nagoya International Orchid Show '95. The Organizing Committee of the Nagoya International Orchid Show '95, Secretariat of NIOS '95, c/o Meitetsu Agency, Inc., 2–4 Meieki 1-chome, Nakamura-ku, Nagoya, 450 Japan.174 pp., line drawings, B&W and color photographs.
Ichihashi, S. and H. Inoue	1996	Proceedings of the Nagoya International Orchid Show '96. The Organizing Committee of the Nagoya International Orchid Show '96, Secretariat of NIOS '96, c/o Meitetsu Agency, Inc., 2–4 Meieki 1-chome, Nakamura-ku, Nagoya, 450 Japan.128 pp., line drawings, B&W and color photographs.
Ichihashi, S., and H. Inoue	1997	Proceedings of the Nagoya International Orchid Show '97. The Organizing Committee of the Nagoya International Orchid Show '97, Secretariat of NIOS '97, c/o Meitetsu Agency, Inc., 2–4 Meieki 1-chome, Nakamura-ku, Nagoya, 450 Japan.74 pp., line drawings, B&W and color photographs.
Ichihashi, S., and H. Nagata	1992	Proceedings of the Nagoya International Orchid Show '92. The Organizing Committee of the Nagoya International Orchid Show '92, Secretariat of NIOS '92, 1–1 Kawaseyama, Igaya-cho, Kariya-shi, Aichi-ken 448, Japan. 230 pp., line drawings, B&W and color photographs.
Ichihashi, S., and H. Nagata	1993	Proceedings of the Nagoya International Orchid Show '93. The Organizing Committee of the Nagoya International Orchid Show '93, Secretariat of NIOS '93, c/o Meitetsu Advertising Inc, 2–4 Meieki 1-chome, Nakamura-ku, Nagoya, 450 Japan. 204 pp., line drawings, B&W and color photographs.
Ichihashi, S., and H. Nagata	1994	Proceedings of the Nagoya International Orchid Show '94. The Organizing Committee of the Nagoya International Orchid Show '94, Secretariat of NIOS '94, c/o Meitetsu Agency, Inc., 2–4 Meieki 1-chome, Nakamura-ku, Nagoya, 450 Japan. 220 pp., line drawings, B&W and color photographs.
Ichihashi, S., and H. Nagata	2001	Proceedings of the seventh Asia Pacific orchid conference (APOC), Nagoya, Japan, March 15–21, 2001. Announcement, CD, registration information, lectures, exhibits, sales and social programs brochure.

(continued)

(continued)

Author(s)	Date	Title and publisher
Ilsley, P.	No date	Great orchids. The Ilsley Orchids, Los Angeles, CA. 56 pp., color photographs.
Imes, R.	1993	Orchids.
Imes, R.	1993	The orchid identifier.
International Orchid Commission	1993	The handbook on orchid nomenclature and registration.
Irawati [no initials]	1987	*In vitro* culture and symbiotic relationships of *Paphiopedilum niveum, Taeniophyllum obtusum* and *Vanda hookerana* with their associated fungi. Faculty of Agriculture, University Pertanian Malaysia.
Irmish, T.	1853	Beiträge zur biologie und morphologie der orchideen. Verlag von Ambrosius Abel, Leipzig. 82 pp., 6. See also Irmisch, 1842–1877.
Irvine, K.	1982	Orchids for the Australian garden.
Isaac-Williams, M. L.	1988	An introduction to the orchids of Asia. Angus & Robertson, North Ryde, NSW, Australia. 261 pp., numerous color photographs and line drawings.
Izumi, E.	1980	Japanese wild orchids. Numerous color photographs. (In Japanese.)
Ito, I., and K Karasawa	1969	*Calanthe* in Japan and its kinds. Seibundo Shinkosha, Tokyo. 310 pp., line drawings, B&W and color photographs.
Ives, P.	1989	Catalogue of living plant collections, Part 2. Royal Botanic Garden, Kew. 163 pp.
Izumi, E.	1980	Japanese wild orchids. Numerous color photographs. (In Japanese.)
Jackson, I.	1976	Kangaroo Island orchids. Island Newspapers Publishers, Kangaroo Island, South Australia. 28 pp., numerous line drawings.
Jacquet, P.	1988	Une diagrams 'n des orchidées sauvages de France, 2nd ed. Societe Francaise d'Orchidophile, Paris. 75 pp., numerous color photographs and maps.
Jacquet, P.	1995	Une diagrams 'n des orchidées sauvages de France, 3rd ed. Societe Francaise d'Orchidophile, Paris. 100 pp., numerous color photographs and maps.
Jacquet, P.	1996	Les diagrams Lyonnais du XVIe siècle et 14 letters de Dalechamp à Camerarius (1582–1585). 85 pp., illustrations.
Jacquet, P.	2002	Une histoire de l'orchidologie française. Société Française d'Orchidologie. 198 pp., many illustrations of orchids and botanists.

(continued)

(continued)

Author(s)	Date	Title and publisher
Jaeger, P., N. Hallé, and J. G. Adam	1968	Contribution a l'Étude de orchidées des Monts Loma (Sierra Leone). Adansonia (Muséum National d'Histoire Naturelle) N. S., Vol. 8, Fascicle 3, pp. 265–310, 9 figures (photographs and line drawings), 2 maps.
Jafri, S. M. H., and A. El-Gadi (eds.)	1985	Flora of Libya. Orchidaceae by A. B. M. Enayet-Hossain and A. El-Gadi, 37 pp., and 13 line drawings and Cyperaceae by F. B. Erteeb and A. S. Sherif, 46 pp., and 14 line drawings. Al Faateh University Faculty of Science, Department of Botany, Tripoli, Libya.
Jain, S. K., and A. Mehrotra	1984	A preliminary inventory of Orchidaceae in India. Botanical Survey of India, Howrah, India.
James Veitch & Sons	1887–1894	A manual of Orchidaceous plants cultivated under glass in Great Britain, volumes 1 and 2. Every section in each volume is paginated separately. Numerous line drawings. Reprint 1963 by A. Asher & Co., Amsterdam.
James, I. D.	1988	Orchid growers handbook. Blandford Press, London. 112 pp., line drawings and color photographs.
James, I. D.	1993	Growing orchids. Cassell Publishers, Ltd., London. 96 pp., numerous line drawings and color photographs.
James, L. D., and D. Kendall	1985	The New Zealand orchid grower. Landsdowne Press, Aukland. 160 pp., numerous color photographs, some line drawings.
Japan Orchid Growers Association	1984	Quality stream cattleyas.
Japan Orchid Growers Association	1987, 1989, 1995, 1996	Register of awards.
Jasper, B.	No date	Mericlone manuscript, commercial.
Jaworski, H.	1992	Orchids simplified. Chapters Publishing, Ltd., Shelburne, VT, USA. 143 pp., numerous color photographs.
Jayaweera, D. M. A.	1981	Flora of Ceylon. Orchidaceae.
Jenkinson, M. N.	1991	Wild orchids of Dorset.
Jennings, S.	1875	Orchids and how to grow them.
Jenny, R.	1993	Monograph of the genus *Gongora* Ruiz&Pavon. Koeltz Scientific Books USA), Champaign, IL. 136 pp., 23 pp. of B&W and color photographs.
Jesurún, A.	No date	Las orquídeas y su propagación. Comite Latinoamericano de Orquideologia. 20 pp., 4 color photographs, 6 line drawings.
Ježek, Z.	2004	Geilustreede orchdeeën encyclopedie.

(continued)

(continued)

Author(s)	Date	Title and publisher
Johns, J., and B. Molloy	1983	Native orchids of New Zealand. A. H. & A. W. Reed, Wellington, New Zealand. 124 pp., 37 B&W and 120 color photographs.
Johnsen, B.	1994	Skandinaviens Orckideer-Orchids of Scandinavia. Rhodos International Science and Art Publishers, Copenhagen. 127 pp., 50 color plates.
Johnston, E. R.	No date	Orchid culture in silvabark. No publisher listed. 33 pp., numerous B&W photographs.
Jones, D. L.	1991	New Taxa of Australian Orchidaceae. Volume 1 of Australian Orchid Research (D. L. Jones, ed.). Australian Orchid Foundation, 107 Roberts Street, Essendon, Victoria, Australia. 208 pp., numerous color photographs, 110 plates of line drawings.
Jones, D. L.	1998	Exotic orchids in Australia. How to grow hundreds of species. New Holland Publishers, Pty., Ltd., 128 pp., numerous color photographs, few line drawings.
Jonsson, L.	1981	A monograph of the genus *Microcoelia* (Orchidaceae). Acta Universitatis Upsaliensis, Symbolae Botanicae Upsaliensis XXIII:4, 8 unnumbered pp., 151 pp., 89 figures on line drawings, maps, electron microscope photographs and one color picture.
Joseph, J.	1982	Orchids of Nilgiris. Records of the Botanical Survey of India Volume 22. Botanical Survey of India, P. O. Botanic Garden, Howrah-711 103.
Josst, F.	1851	Beschreibung und Kultur einer grossen unzahl tropischer, der Kultur werther und in eurpäischen Gärten eingefühter Orchideen. Prag (Prague) 1851.
Jouandoudet, F.	2004	A la découverte de orchidées sauvages d'Aquitaine.
Judging Committee	1979	The genus *Paphiopedilum*-A guide to the species. Sub Tropical Orchid Council, Queensland. 26 pp., two pages of line drawings.
Kaazik, K., M. Lyik, Yu. Martin, T. Relve, and V. Roost (eds.)	1980	Ohrana I kultivirovanie orhidei. Abstracts of the All-Union Symposium, March 18–20, 1980. The Academy of Sciences of the Estonian SSR, Tallin Botanical Garden. 168 pp. (In Russian, English and German.)
Kaga, M. S.	1946	A record of an orchid collection by M. S. Kaga (1888–1954).
Kaiser, R.	1993	The scent of orchids. Edtiones Riches, Basel. 259 pp., numerous colored photographs and chemical structures.
Kajima, T.	1987	Orchids, wonders of nature. Salamander Books, London and New York. 207 pp., numerous color photographs.

(continued)

(continued)

Author(s)	Date	Title and publisher
Kajima, T.	1988	Orchids, wonders of nature.
Kajima, T.	1988	The orchid.
Kajima, T.	1988	The orchid. Graphic-sha Publishing Co., Ltd., Tokyo. 68+XIX pp., 68 color photographs.
Kajima, T.	No date	The original orchids. (In Japanese.)
Kako, S. (ed.)	1988	Illustrated orchid biotechnology: Micropropagation, seedlings and crossbreeding. 138 pp., numerous line drawings and B&W and color illustrations.
Kallmeyer [], and [] Ziesche	1996	Die Orchideen Sachsen-Anhalts-Vorbereitungsatlas.
Kalman, B.	1999	Rare orchids. Bulfinch Press/Little, Brown and Company, Boston, MA. 144 pp., numerous color photographs and line drawings.
Kamata, K., and K. Uchida	1982	The orchid. (In Japanese.) 135 pp., 179 color photographs, maps, diagrams and numerous line drawings.
Kamemoto, H., T. D. Amore, and A. Kuehnle	1999	Breeding *Dendrobium* orchids in Hawaii. University of Hawaii Press, Honolulu, HI. 166 pp., numerous color photographs.
Kamemoto, H., and R. Sagarik	1975	Beautiful Thai orchid species. The Orchid Society of Thailand, G. P. O. Box 953, Bangkok, Thailand. 186 pp., numerous B&W and color photographs, map.
Kamemoto, H., R. Tanaka, and K. Kosaki	1961	Chromosome numbers of orchids in Hawaii. Hawaii Agricultural Experiment Station Bulletin 127. 28 pp., 9 B&W photographs. Contains procedure for chromosome counting.
Kamjaipai, K.	1984	Diseases and pests of orchids. Bangkok Flower Centre Co., Ltd. 116 pp., 35 color plates.
Kanda, K.	1984	Japanese wild orchids. 179 pp., numerous color photographs. (In Japanese with English caption.)
Kapteyn den Bourmeester, D. W., and P. Brederoo	1991	Veldwerk Europese orchideeen.
Karasawa, K.	1981	Species of Western [i. E., tropical] orchids. Seibundo SinkohsyA, Tokyo. 223 pp., numerous color photographs.
Karasawa, K.	1982	The genus *Paphiopedilum*. S+S, Tokyo. 256 pp., numerous line drawings, B&W and color photographs, chromosome pictures. (In Japanase with some English titles.)
Karasawa, K.	1996	Orchids. Yama-Kei Publishers, Tokyo. 672 pp., ca 2000 color photographs.
Karasawa, K.	1886–1997	Orchid atles, volumes I–VIII.
Karasawa, K., and G. Ishida	1983	My green deluxe. A book about orchids. 160 pp., numerous line drawings and color photographs. (In Japanese.)

(continued)

(continued)

Author(s)	Date	Title and publisher
Karasawa, K., and G. Ishida	1998	*Calanthe*. Yasaka Shobo, Inc., Tokyo. 327 pp., numerous, line drawings and B&W and color photographs. (In Japanese with English titles on illustrations.)
Karasawa, K., and I. Ito	1976	*Calanthe* in Japan and its breeding. Color books published by Hoikusya, Japan. 160 pp., numerous B&W and color photographs.
Kashemsanta, M. R. S.	1980	Proceedings of the 9th world orchid conference, January 18–25, 1978. Published for the 9th World Orchid Conference, Bangkok, Thailand. 313 pp., 33 color plates.
Kataki, S. K.	1986	Orchids of Meghalaya. Government of Meghalaya, Shilong, India. 238 pp., 82 plates of line drawings, 18 plates of color photographs, 4 plates of B&W photographs.
Katkov, N.	1984	Blood & orchids. A novel. New American Library, New York. 456 pp.
Kaushik, P.	1983	Ecological and anatomical marvels of the Himalayan orchids. Today and Tomorrow's Printers & Publishers, New Delhi.
Kavulak, J.	1986	The Australian native orchid hybrid guide. Jodi Enterprises Pty., Ltd., Jodi's Orchid Nursery, Fernleigh, Australia. 98 pp.
Kazaryan, V. O. (ed.)	1989	Red Data Book of the Armyanskaya SSR. Endangerous and rare plant species. Erevan, Aiastan. 270 p. (In Russian and Armenian.)
Kavulak, J.	1988	The Australian native orchid hybrid guide, 2nd ed.
Kavulak, J.	1993	The Australian native orchid hybrid guide.
Keenan, P. E.	1983	Orchids. A complete guide to Maine's orchids. DeLorme Publishing Company, Freeport, Maine 04032. 48 pp., 2 line drawings, 42 color photographs.
Keenan, P. E.	1999	Wild orchids across North America. Timber Press, Portland, OR. 321 pp., numerous color photographs.
Kelleher, J.	1984	Intriguing masdevallias. HGH Publications, Wokingham, Bershire, England. 76 pp., numerous color photographs and line drawings.
Keller, G., and R. Schlechter	1928	Monographie und iconographie der orchideen Europas und des Mittelmeergebietes. Volume I. Monographie der gattungen und arten (mit blütenanalysen). Repertorium Specierum Novarum Regni Vegetabilis. 304 pp., 37 plates of line drawings.
Keller, G., and R. V. Soó	1930–1940	Monographie und iconographie der orchideen Europas und des Mittelmeergebietes. Volume II in a series edited by Keller, G. and R. Schlechter. Repertorium Specierum Novarum Regni Vegetabilis, Sonderheft A. Reprint 1972 by Otto Koeltz, Koenigstein-Taunus, Germany.

(continued)

(continued)

Author(s)	Date	Title and publisher
Kemularia-Natadze, L. M. and D. I. Sosnowskii	1941	Orchidaceae. Pages 545–578. In: Flora of Georgia. 2. Tbilisi (in Georgian).
Kenkyusha, G. S. (ed.)	1987	Encyclopedia of Japanese native orchids. (Yasei-ran jiten). Kanuma City, Tochigi, Tochinoha Shobo. 304 pp. 2nd ed. Numerous color photographs. (In Japanese.)
Kenny, J.	1988	Native orchids of the eastern Caribbean. Macmillan Caribbean, London. 88 pp., 4 line drawings, 1 map, numerous color photographs.
Kerner, A.	1865	Die hybrider orchideen der Osterreichischen flora.
Kernohan, J., Bonham, N., Bonham, D., and L. Cobb (eds.)	1990	Proceedings of the 13th world orchid conference, September 6–17, Auckland. 1990 World Orchid Conference Trust. 336 pp., maps, line drawings, B&W and color photographs, 750 printed.
Kerr, M.	1966	Botanical orchids. The Blotter Press, Lamar State College, Beaumont, Texas. No pagination. Numerous color painting. Small book. Press run limited to 300 copies.
Kerr, R. (ed.)	1986	Orchids in Australia. Graphic World Publishers Pty., Ltd., Sydney. 207 pp., map, color and B&W photographs.
Khan, I.	1964	Preliminary investigations into the “Crassulacean Acid Metabolism” of Vanda Miss Joaquim. Honors Thessis supervised by Professor P. N. Avadhani, University of Singapore. 84 pp., graphs, pictures.
Khattab, A., and M. N. El-Hadidi	1971	Results of a botanical expedition to Arabia in 1944–1945. Publication No. 4 of the Cairo University Herbarium, Cairo University Press. 96 pp.
Kiew, R., and I. M. Turner	2001	Singapore botanic gardens, a souvenir guide. Landmark Books, Singapore. 48 pp.
Kijima, T.	1987	Orchids, wonders of nature. Salamander Books, London and New York. 207 pp., numerous color photographs.
Kijima, T.	1988	The orchid. Graphic-sha Publishing Co., Ltd., Tokyo. 68+XIX pp., 68 color photographs.
Killick, D. J. B.	1981	The Flowering Plants of Africa, Vol. 46, parts 3 and 4, plates 1821–1340. Text and color plates of paintings. 20 plates and a B&W photograph of Professor E. C. L. E Schelpe.
Kimber, S.	1964	A handbook of orchids. Sheila Kimber,, “Oakfield Lodge”, Conghusrst Lane, Hawkhurst, Kent, UK. 55 pp., 22 color paintings by the author.

(continued)

(continued)

Author(s)	Date	Title and publisher
Kimura, K., S. Ichihashi, and H. Nagata	1990	Proceedings of the Nagoya International Orchid Show '90. The Organizing Committee of the Nagoya International Orchid Show '90, Secretariat of NIOS '90, 1–1 Kawaseyama, Igaya-cho, Kariya-shi, Aichi-ken 448, Japan. 274 pp., line drawings, B&W and color photographs.
Kimura, K., S. Ichihashi, and H. Nagata	1991	Proceedings of the Nagoya International Orchid Show '91. The Organizing Committee of the Nagoya International Orchid Show '91, Secretariat of NIOS '91, 1–1 Kawaseyama, Igaya-cho, Kariya-shi, Aichi-ken 448, Japan. 230 pp., line drawings, B&W and color photographs.
King, G., and R. Pantling	1898	Orchids of the Sikkim-Himalaya. *Annals of the Royal Botanic Garden, Calcutta* 8: i–iii, then i–iv, 1–11, then 1–342, then 1–2, another 1–2, another 1–2, 448 plates of B&W drawings. Bound with Hooker, 1895 and Duthie, 1906. Reprint 1967.
Kindlmann, P. J. H. Willems, D. F. Whigham	2002	Trends and fluctuations and underlying mechanisms in terrestrial orchid populations. Backhuys Publishers, Leiden. 254 pp.
Kipping, J. L.	1971	Pollination studies of native orchids. M.S. thesis, San Fransico State College. 87 pp., illustrations.
Kirch, W.	No date	Sears orchid guide. Sears, Alamoana, HI. 26 pp.
Kirsch, O., M. Warne, E. Warne, and J. Lewis	1976	New and revised beginner's handbook for orchid growing in Hawaii. Pacific Orchid Society of Hawaii, P. O. Box 1091, Honolulu, HI. 39 pp., numerous line drawings.
Klaus, H.	1936	Die kulture der orchideen.
Klinge, J.	1898	Dactylorchidis. Acta Horti Ptropolitani 17, Fascicles 1, 3: 1–64.
Klinge, J.	1899	Die homo-und polyphyletischen formenkreise der Dactyorchis-arten. Acta Horti Ptropolitani 17, Fascicle 2, No. 6, pp. 69–146, 2 plates.
Klinge, J.	1899	Zur geographischen verbreitung und entstehung de Dactylorchis-arten. Acta Horti Ptropolitani 17, Fascicle 2, No. 7, pp. 149–267, one map.
Klinge, J.	1899	Zur orientitierung der Orchis-bastarde und zur polymorphie der Dactylorchis-arten. Acta Horti Ptropolitani 17, Fascicle 2, pp. 1–64.
Klopfenstein, E.	ca 1995	West European orchids.
Koay, S. H.	1984	Cultivation of orchids. Primary Productivity Department, Singapore. 45 pp., 25 B&W and 27 color photographs.
Koay, S. H., J. S. Loi, and L. T. Lam-Chan	1989	Cultivated tropical orchids.

(continued)

(continued)

Author(s)	Date	Title and publisher
Koerper, H.	2002	The disobedient genus *Stanhopea*.
Koester, A. R.	1979	The *Cymbidium* list (1799–1979), Vol. I. Arthur R. Koester Publisher, Burbank, CA. 192 pp.
Koester, A. R.	1982	The *Cymbidium* list, Vol. II. Arthur R. Koester Publisher, Burbank, CA.
Kohlhaupt, P.	1970	Bunte welt der gartenblumen. Kosmos, Stuttgart. 72 pp., numerous color photographs and line drawings.
Kohlhaupt, P.	1971	Bunte welt der orchideen. Kosmos, Stuttgart. 73 pp., numerous color photographs, one line drawing.
Kohlhaupt, P.	No date	Wilde orchideeën. B. V. W. J. Thieme & Cie, Zutphen, The Netherlands. 72 pp., 120 color photographs.
Kojyo, K.	1987	*Cymbidium* and *Dendrobium*. Shufu-To-Seikatsusha, Tokyo. 132 pp., line drawings, B&W photographs, numerous color photographs. (In Japanese.)
Kokusaki Nursey Co., Ltd.	1985	Catalog, 78. No. 1–10, Taishido 2-chome, Setagaya-ku, Tokyo. 160 pp., numerous color photographs.
Kokusaki Nursey Co., Ltd.	1986	Catalog, 79. No. 1–10, Taishido 2-chome, Setagaya-ku, Tokyo. 160 pp., numerous color photographs.
Kolakovskii, A. A.	1986	Orchidaceae. In: Flora of Abkhazia. Vol. 4. (In Russian.)
Komarov, V. L.	1935	Flora of the U. S. S. R. Volume IV Liliiflorae and Microspermae. Translated by N. Landau and published in 1968 by the Israel Program for Scientific Translations, Jerusalem. Xxxiv+586 pages, 44 line drawing plates, 2 folded maps.
Koopowitz, H.	2000	Paphiopedilums. Reprint from the Orchid Digest.
Koopowitz, H.	2001	Orchids and their conservation. Timber Press, Portland, OR. 176 pp., numerous color photographs.
Koopowitz, H., and N. Hasegawa	1989	Slipper orchids.
Korhonen, M.	1987	Suomen orkideat.
Krackowizer, F. J.	1950	Monographia da *Laelia purpurata*.
Kraenzlin, F and R. Schlechter	1907	Das pflanzenreich, Orchidaceae- Monandrae-Coelogyninae
Kraenzlin, F.	1925	Monographie der Gattungen *Masdevallia* Ruiz et Pavon, L*othiania* Kraenzl, *Scaphocephalum* Pfitzer, *Cryptophoranthus* Barb. Rodr. *Pseudoctomeria* Kraenzl. Repertorium Specierum Novarum Regni Vegetabilis. Dahlem bei Berlin, Beihefte, Band XXXIV, 240 pp. There is also a 1980 reprint by Otto Koeltz Antoquariat, Koeningstein, Germany.

(continued)

(continued)

Author(s)	Date	Title and publisher
Kraenzlin, F.	1926	Monographie der gattung *Polystachya* Hook. Repertorium Specierum Novarum Regni Vegetabilis Beihefte Vol. 39, 136 pp.
Kraenzlin, F.	1901, 1904	Orchidacearum genera et species. Vol 1 (pp. 1–986, 1901) and 2, part1 (pp. 1–140, 1904). Reprint 1969 by Verlag von J. Cramerl Lehre, Germany, Wheldon and Wesley, Ltd., Codicote, Herts, and S. H. Service Agency, Inc., New York.
Kraenzlin, F.	1931	Orchidacearum Sibiriae Enumeratio.
Kraenzlin. F.	1911	Beitrage zurorchideenflora Sudamerikas
Kramer, J.	1963	Growing orchids at your windows. D. Van Nostrand Company, Inc. 151 pp., 41 line drawings and 7 B&W photographs.
Kramer, J.	1974	Orchids for your home. Cornerstone Library (Simon and Schuster), New York. 144 pp., numerous line drawings and B&W photographs.
Kramer, J.	1975	Orchids, flowers of romance and mystery. Harry N. Abrams, Inc., publishers, New York. 309 pp., numerous color and B&W photographs.
Kramer, J.	1989	The World Wildlife Fund book of orchids. Abbeville Press, New York. 276 pp., 212 color illustrations, many line drawings and other illustrations.
Kramer, J.	1994	Orchids for the south. Taylor Publishing Co., Dallas, Texas. 163 pp., numerous color photographs, some B&W photographs and drawings.
Kramer, J.		Growing hybrid orchids indoors. Universe Books, New York. 208 pp., numerous color and B&W illustrations.
Kramer, J.	1968	Rare orchids everyone can grow. Doubleday & Company, Garden City, New York. 144 pp., numerous line drawings and B&W and color photographs.
Kramer, J.	1979	Orchids, flowers of romance and mystery. Harry N. Abrams, Inc., New York. 156 pp., 106 illustrations of which 61 are color photograph plates.
Kramer, J.	1988	Botanical orchids and how to grow them.
Kramer, J., and R. L. Crafton	1982	Miniature orchids and how to grow them. W. W. Norton, New York. 1997 pp., numerous B&W photographs.
Kranzlin, F.	1931	Orchidacearum Sibirae enumerative. *Repert. Sp. Nov. Fedde Beih.* 65: 1–103.
Kranzlin, F.	1913	Orchidaceae Sibiriae. *Russkiy Botan. Zhurn.* 1–4: 17–60. (In Russian.) Kreutz, C. A. J.

(continued)

(continued)

Author(s)	Date	Title and publisher
Kreutz, C. A. J.	1987	De verspreiding van de Inheemse orchideeen in Nederland.
Kreutz, C. A. J.	1994	Orchideeen in Zuid-Limburg.
Kreutz, C. A. J.	1998	Die orchideen der Turkei.
Kreutz, C. A. J.	1999	Die orchideeen van het Gerendal.
Kreutz, C. A. J.	2002	Feldfuhrer Deutsche orchideen.
Kreutz, C. A. J.	2004	Kompendium der Europäischer orchideen.
Kreutz, C. A. J. and H. Dekker	2000	De orchideeen van Nederland.
Kronenberg, H. G.	1993	De kennis en de kunst van het kweken van orchideeen.
Kronenberg, H. G. (ed.)	1976	Proceedings of the fourth European orchid congress, Amsterdam. 80 pp., line drawings.
Kropf, J., and W. Purzig. (eds.)	No date	Zeiss information, Vol. 11, No. 49. Article on Ophrys species on Majorca, pp. 103–106, B&W and color photographs.
Kränzlin, F.	1911	Beiträge zur orchideenflora Südamerikas. Kungl. Svenska Vetenskapsakademiens Handlingar 46 (10): 1–105., 13 plates of B&W line drawings.
Kränzlin, Fr.	1910	Das pflanzenreich, IV.50. II. B. 21. Orchidaceae-Monandrae-Dendrobiinae, Part I, genera 275–277. part 45. Reprint 1957 by Verlag von H. R. Engelmann (J. Cramer) Weinheim/ Begrstrasse. 169 pp., 327 line drawing figures in 35 plates.
Kränzlin, Fr.	1911	Beiträge zur orchideenflora Südamerikas. Kungl. Svenska Vetenskapsakademiens Handlingar 46 (10): 1–105, 13 plates of B&W and color and line drawings.
Kränzlin, Fr.	1919	Beiträge zur kenntnis der gattung Telipogon H. B. K. Reprint from Annalen des Naturhistorischen Hofmuseums Vienna, Volume 33: 9–38.
Kränzlin, Fr.	1922	Das pflanzenreich, IV.50. Orchidaceae-Monandrae, Tribus Oncidiinae-Odontoglosseae Part II, part 50. Reprint 1957 by Verlag von H. R. Engelmann (J. Cramer) Weinheim/Begrstrasse. 344 pp., 282 line drawing figures in 29 plates.
Kränzlin, Fr.	1923	Das pflanzenreich, IV.50. Orchidaceae-Monandrae, Pseudomonopodiales, part 83. Reprint 1957 by Verlag von H. R. Engelmann (J. Cramer) Weinheim/Begrstrasse. 66 pp., 101 line drawing figures in 5 plates.

(continued)

(continued)

Author(s)	Date	Title and publisher
Kränzlin, Fr.	1910, 1911	Das pflanzenreich, IV.50. II. B. 21. Orchidaceae-Monandrae-Dendrobiinae, Part II, genera 278–279, part 50. There is also a 1957 reprint 1957 by Verlag von H. R. Engelmann (J. Cramer) Weinheim/Begrstrasse. 182 pp., 240 line drawing figures in 35 plates. Bound with IV.50. II. B. 23. Orchidaceae-Monandrae-Thelasinae, genera 280–280a, Part 50, 46 pp., 103 line drawing figures in 5 plates.
Krishtofovich, A. N.	1907	Notes on Russian orchids. *Estestvoznanie I geographia*. 9: 1–6. (In Russian.)
Kronenberg, H. G. (ed.)	1976	Proceedings of the fourth European orchid congress, Amsterdam. 80 pp., line drawings.
Kropf, J., and W. Purzig (eds.)	No date	Zeiss information, Vol. 11, No. 49. Article on Ophrys species on Majorca, pp. 103–106, B&W and color photographs.
Ktamer, J.	1985	Growing hybrid orchids indoors. Unioverse Books, New York. 208 pp., numerous color and B&W illustrations.
Kucherov, E. V., A. A. Muldashev and A. Kh. Galeeva	1987	Protection of rare plants on the Middle Ural. Nauka, Moscow. 205 pp. (In Russian.)
Kueh, C. K.	1949	Some orchids and ferns in Sarawak.
Kull, T.	1987	Kuldking. Valgus, Tallinn. 80 pp. (summary: Lady's Slipper)
Kull, T. (ed.)	1994	Orchid ecology and protection is Estonia. ELF, Tartu, Estonia. 75 pp., line drawings, map.
Kull, T., and J. Arditti (eds.)	2002	Orchid Biology: Reviews and Perspectives, Volume VIII. Kluwer Academic Publishers, Dordrecht/Boston, MA/London
Kull, T., and T. Tuulik	2002	Kodumaa käpalised [Orchids in Estonia]. Eesti Orhideekaitse Klubi, Tallinn. 95 pp., numerous color photograps and distribution maps
Kullenberg, B.	1961	Studies in *Ophrys* pollination reprinted from Zoolıgska Bıdrag från Uppsala, Volume 34. Almqvist & Wiksells Boktryckeri AB, Uppsala.
Kullenberg, B, G. Bregström, B. G. Svenson, J. Tnegö, and L. Ågren	1984	The ecological station of Upsala University on Öland 1963–1973, Nova Acta Regiae Sociatatis Scientiarum Upsalensis, Ser. V: C, Vol 3. Selection of articles a number of which are on *Ophrys* pollination and the pollinators. 205 pp., numerous color and B&W photographs and maps.
Kullenberg, B, and E. Stenhagen	1973	The ecological station of Uppsala University on Öland 1963–1973, Zoon, Supplement 1. Selection of articles a number of which are on Ophrys pollination and the pollinators. 151 pp., numerous color and B&W photographs and maps.

(continued)

(continued)

Author(s)	Date	Title and publisher
Kumpel, H.	1996	Die wild-wachsenden orchideen der Rhon.
Kupper, W., and W. Linsenmaier	1961	Orchids. Translated by Jean W. Little. Thomas Nelson and Sons, London. 128 pp., 60 color plates.
Kurth, A. W. F., and G. L. Piper	1956	An introduction to orchid growing. Queensland Orchid Society, Brisbane. 16 pp., line drawings.
Kuznetsov, N. M. and N. V. Pavlov	1958	Orchidaceae. Pages 255–274. In: Flora of Kazakhstan. Alma-Ata. (In Russian.)
L. E. D. E. N.	1975	Atlas des orchidees.
L. Sherman Adams Co.	1943	Catalog. L. Sherman Adams Co., Wellesley, MA. 65 pp. Color and B&W illustrations.
la Croix, I.	2000	Orchid basics. Sterling Publishing Co., New York. 128 pp., numerous color illustrations.
La Croix, I. F., E. A. S. La Croix, and T. M. La Croix	1991	Orchids of Malawi. A. A. Balkema, Rotterdam. 358 pp., numerous lines drawings and color photographs and three maps.
la Croix, I., and E. la Croix	1997	African orchids in the wild and in cultivation. Timber Press, Portland, OR. 379 pp., numerous line drawings and color photographs.
la Croix, I., and P. J. Cribb	1995	Orchidaceae, pp. vi+1–320 in G. V. Pope (ed.), Flora Zambesiaca, Vol. 11, Part one. Numerous line drawings..
la Croix, I., and P. J. Cribb	1998	Orchidaceae (Part) 2 in volume 11 of Flora Zambesiaca. Royal Botanic Gardens, Kew, London. 569. pp., 24 color photographs, numerous line drawings.
la Croix, I., E. A. S. la Croix, T. M. la Croix, J. A. Hutson, N. G. B Johnson-Stewart	1983	Malawi orchids. Volume 1. Epiphytic orchids. National Fauna Preservation Society of Malawi. 150 pp., 100 line drawings, one map.
Laibach, F.	1929	Untersuchungen uber die postfloration der Orchideen.
Laimer, M., and W. Rücker	2003	Plant tissue culture, 100 years since Gottlieb Haberlandt. Springer Verlag, Vienna. 260 pp., color and B&W illustrations.
Lambert, F. H., K. Staples, and S. Monkhouse	1980	A book for orchid lovers. 56 pp., numerous B&W and color photographs and line drawings. The Orchid Club of South Australia, Torrens Park, South Australia.
Lambert, H.	1920	Les orchidees.
Lambert, H. F., K. Staples, and S. Monkhouse (eds.)	1976	A book for orchid lovers. The Orchid Club of South Australia, Inc., 38 Princes Rd., Torrens Park, South Australia 5062. 56 pp., numerous B&W and color illustrations.
Landau, D.	1996	Living with wild orchids in Sumatra.

(continued)

(continued)

Author(s)	Date	Title and publisher
Landwehr, J.	1977	Wilde orchideen van Europa. Verenging tot Behoud van Natuurmonumenten in Nederland's-Graveland. Two volumes, 574 pp., 258 color painting and numerous line drawings.
Lang, D.	1980	Orchids of Britain. Oxford University Press, Oxford, UK. 213 pp., numerous color photographs.
Lapiner, J. M. de	1974	Orquideas Michoacanas.
Lasser, T.	1969–1970	Flora de Venezuela. Vol. 15, parts 1–5 plus a separate issue which contains an index and keys. Every part is paginated separately, but the figures are numbered consecutively. There are 929 plates of line drawings.
Latham, A.	1977	Orchids for mother. A novel. Bantham Books, New York. 344 pp.
Latif, S. M.	1953	Bunga anggerik.
Latif, S. M.	1960	Bunga anggerik. Penerbitan "Sumur Bandung," d/h N. V. Mij Vorking-Van Hoeve, Bandung, Indonesia. (In Indonesian.) 446 pp., numerous line drawings, B&W and color paintings and severak maps. Soetan Mahmoed Latif.
Latif, S. M.	1972	Kembang angerrik. Penerbit N. V. Masa Baru, Bandung. 140 pp., 87 line drawings.
Latif, S. M.	No date	Nomenklatur anggerik merangkap siklopedia. P. T. Saksama Publishing House, Jakarta. 179 pp., 12 line drawings of orchid personalities. (In Indonesian.) Published ca 1960.
Lavarack, P. S., B. Gray and A. W. Dockrill	1985	Tropical orchids of Australia. Thomas Nelson Australia, Melbourne. 177 pp., numerous line drawings and color photographs.
Lavarck, B., and B. Gray	1992	Australian tropical orchids. Frith and Frith Books, Malanda, Queensland 4885, Australia.70 pp., line drawing maps and numerous color photographs.
Lavarack, B., W. Harris, and G. Stocker	2000	*Dendrobium* and its relatives. Timber Press, Portland, OR. 286 pp., numerous color photographs.
Lawler, L., and R. D. Kerr (eds.)	1981	Proceedings of the orchid symposium held as a satellite function of the 13th International Botanical Congress. Published by the Orchid Society of New South Wales.
le Moyne, J.	No date	Le Moyne's botanical watercolors. Pomegranate Calendars and Books, San Francisco. Actually a daybook. No orchids, No pagination. Color Illustrations.
Lecoufle, M.	1981	Orchidées exotiques. La Maison Rustique, Paris. 191 pp., numerous color photographs.

(continued)

(continued)

Author(s)	Date	Title and publisher
Lecoufle, Marcel	1992	Letter, catalog, information about J. Costantin, pictures.
Lecoufle, M. and H. Rose	1957	Orchids culture and descriptions. Crosby Lockwood & Son, Ltd., London. 112 pp., numerous B&W and color photographs.
Lecoufle, M., and H. Rose	1958	Orchidee. Antonion Vallardi Editor. 117 pp., B&W and color photographs. (In Italian.) $65, Has ex libris: "Piero Piani."
Lecoufle, M., and H. Rose	1964	Orchidées, 1st reprinting. La Maison Rustique, Paris. 168 pp., numerous line drawings, B&W and color photographs.
Ledin, F.	1904	Kultura krasivotsvetushih orhidei. (In Russian.) Published by A. F. Devriena, St. Petersburg before the communist revolution. 84 pp., 32 line drawing and B&W photographs.
Lee, C. K.	1983	Orchids, their cultivation and hybridization. Eastern University Press (M) Sdn. Bhd., Singapore, Kuala Lumpur and Hong Kong, 94 pp., 72 color photographs, 27 B&W photographs, 3 line drawings, 89 tables.
Lee, C. K.	1979	Orchids, their cultivation and hybridization. Eastern University Press (M) Sdn. Bhd., Singapore, Kuala Lumpur and Hong Kong, 120 pp., 72 color photographs, 27 B&W photographs, 3 line drawings, 89 tables.
Leigh, D.	1990	Orchids, their care and cultivation.
Leitger, H.	1864	Die luftwurzeln der orchideen.
Leon, H. (J. S. Sauget y Barbier)	1946	Flora de Cuba, Vol. 1. Cultural, S. A., La Habana. 441 pp., numerous line drawings and B&W photographs. Orchids on pp. 341–404.
Leonhardt, K., and K. Sewake	1999	Growing *Dendrobium* orchids in Hawaii. College of Tropical Agriculture and Human Resources, University of Hawaii, Manoa. 92 pp., numerous color photographs.
Leonhardt, K.	1977	Chromosome numbers and cross-compatibility in the genus *Cymbidium* and some related tropical genera (Orchidaceae) Inaugural-Dissertation.
Lepage, H. S., and E. R. de Figueiredo Jr.	1947	As pragas das orquidaceas.
Leroy-Terquiem, G., and J. Parisot	1993	Orchids care and cultivation. 200 pp., numerous B&W and color photographs and paintings. Cassell Publishers, Ltd., London.
Leroy-Terquem, G., and J. Parisot	1999	Orchids care and cultivation. 200 pp., numerous B&W and color photographs and paintings. Cassell Publishers, Ltd., London.

(continued)

(continued)

Author(s)	Date	Title and publisher
Leroy-Terquem, G., and J. Parisot	2004	Orchid care and cultivation. Cassell Illustrated. Octopus Publishing Group, 2–4 Heron Quays, London, UK.
Lewis, B. A., and P. J. Cribb	1991	Orchids of the Solomon Islands and Bougainville. Royal Botanic Gardens Kew. 335 pp., 16 color plates, 80 line drawing plates.
Lewis, B. A., and P. J. Cribb	1991	Orchids of Vanuatu. Royal Botanic Gardens, Kew. 335 pp., 16 color plates, 80 line drawing plates.
Lewis, J.	1992	Walter Hood Fitch. Royal Botanic Gardens, Kew/HMSO, London. 73 pp., several B&W illustrations and numerous color paintings by W. H. Fitch who painted the orchids in the Orchid Album and many other publications, some of which are on this list.
Levy, B., S. Berliner, and Z. Silberstein	1958	Mt. Carmel flowers. Department of Education, Municipality of Haifa, Israel. 32 color plates, XVI pp.
Liggio, J, and A. O. Liggio	1999	Wild orchids of Texas. University of Texas Press, Austin. 228 pp., numerous maps and color photographs.
Light, M. H. S.	1995	Growing orchids in the Caribbean. Macmillan Education, Ltd. vii+67 pp., 15 color pages.
Lin, T. P.	1976–1977	Native orchids of Taiwan. Volumes 1 (271 pp., 156 color photographs, numerous drawings) and 2 (355 pp., 170 color photographs, numerous B&W photographs and line drawings). Ji-Chyi Wang, Publisher, 236 Hsin Rong Road, Chiayi, Taiwan. Bilingual, Chinese-English.
Lin, W.-H., and Chou, G.	1986	*Phalaenopsis* kingdom from Formosa. Asia-Agribusiness, Taipei. 299 pp., numerous color illustrations.
Lindell, C. L.	1944	Floea of Texas-Orchidaceae
Linden, L.	1888–1894	Lindenia Iconographie des orchidees. 17 volumes. (In French.)
Linden, J.	1860	Pescatorea. English translation by G. Braem published in 1994 by Naturalia Publications, Turriers, France. 155 pp., 48 plates of color paintings. 1500 copies published.
Linden, L.	1894	Les orchidées exotiques et leur culture en Europe. Published by the author in Bruxelles, Belgium and by Octave Doin Publisher in Paris. 1020 pp., 141 line drawings.
Linder, H. P., and H. Kurzweil	1999	Orchids of southern Africa. A. A. Balkema, Rotterdam, The Netherlands. 492 pp., 458 maps, 82 line drawing figures, 500 color plates.

(continued)

(continued)

Author(s)	Date	Title and publisher
Lindley, J.	1838	Sertum Orchidaceum. James Ridgway & Sons. There is also a facsimile edition of 1,000 copies published in 1974 by Johnson Reprint Corporation, New York, 50 color plates. Both the original and facsimile measure 49 × 34.5 cm.
Lindley, J.	1840	The genera and species of orchidaceous plants. Riddgways, London, 553 pp. There are also a 1963 Reprint by A. Asher & Co., Amsterdam and a more recent reprint from India.
Lindley, J.	1852	Folia Orchidaceae. J. Matthews, London. There is also a 1964 reprint by A. Asher & Co., Amsterdam which has many parts, each paginated separately, about 330 pages followed by a reprint on green paper of the original wrappers and a more recent reprint from India.
Lindley, J.	1857	Contribution to the orchidology of India. There is also a 1982 reprint by Bishen Singh Mahendra Pal Singh, Dehra Dun, India.
Lindley, J.	1858	List of Orchidaceous plants collected in the East of Cuba by Mr. C. Wright with characters of the new species. *Annals and Magazine of Natural History* for May 1858., 11 pp.
Link, H. F.	1849	Bemerkungen uber den bau der orchideen.
Linn, J.	1966	Cypripedioidae.
Lipanovich, M.	1998	Orchids. Sunset Books, Menlo Park, CA. 112 pp., numerous color photographs.
Litvinskaya, S. A.	1993	Protection of Gene- and Coenofunds in the North-Western Caucasus. Rostov-na-Donu, SKNC VSh. 110 pp. Distribution of orchids in the area. (In Russian.)
Litvinskaya, S. A., A. P. Tilba, and R. G. Filimonova	1983	Rare and endangerous species of Kuban. Krasnodar. 160 p. (In Russian.)
Litzelmann, E.	1950	Heimische orchideen.
Litzelmann, E., and F. Bohne	1951	Heimische orchideen.
Litzelmann, E., G. Eberle, K. Herschel, and H. Schurhammer	1931	Heimische Orchideen. Brehm Verlag, Berlin. 32 pp., 24 B&W photographs.
Liverani, P.	1991	Orchidee species spontanee.
Llamacho, J., and J. Larramend	2005	The orchids of Cuba
Logan, H. B., and L. C. Cosper	1949	Orchids are easy to grow. Prentice-Hall, New York. 312 pp., line drawings and 21 color photographs.

(continued)

(continued)

Author(s)	Date	Title and publisher
Logan, H. B., and L. C. Cosper	1951	Orchids are easy to grow.
Loiseau, J.	1953	Chercheur d'orchidées de plaine et de montagne. Vigot Frères, Éditeurs, Paris. 86 pp., 85 plates of line drawings.
Long, J. C.	1970	Native orchids of Colorado, 2nd printing. Museum Pictorial No. 16. Denver Museum of Natural History, Denver. 17 pp., one line drawing, numerous color photographs.
Long, J. C.	1970	Native orchids of Colorado. Museum Pictorial No. 16, Denver Museum of Natural History, Denver. 37 pp., numerous color photographs
Lorenzini, C.	1985	Orchidee d'Italia.
Lucke, E.	No date	Orchideenkultur für alle. Lehrmeister-Bücherei Nr. 463/Albrecht Philler Verlag, Minden, Germany. 120 pp., 70 B&W illustrations, 4 color photographs.
Luckel, E.	1993	Orchid limericks. Published by the author at 2 Bornmannstrasse, Frankfurt. 24 pp.
Luebbermann, M., and F. Echtermeyer	1996	Easy orchids. Chronicle Books, San Francisco. 108 pp., 90 color photographs.
Luebbermann, M., and F. Echtmeyer	1996	Easy orchids. Raamcoast Books, 9050 Shaughnessy Street, Vancouver, BC, Canada.
Luer, C. A.	1972	The native orchids of Florida. The New York Botanical Garden, New York. 295 pp., 84 color plates, numerous maps and line drawings.
Luer, C. A.	1975	The native orchids of the United States and Canada. The New York Botanical Garden, New York. 363 pp., 95 color plates, numerous maps and line drawings.
Luer, C. A.	1986	Icones Pleurothallidinarum. I. Systematics of the Pleurothallidinae (Orchidaceae). Missouri Botanical Garden.
Luer, C. A.	1986	Icones Pleurothallidinarum. II. Systematics of *Masdevallia* (Orchidaceae). Missouri Botanical Garden.
Luer, C. A.	1986	Icones Pleurothallidinarum. III. Systematics of *Pleurothallis* (Orchidaceae). Missouri Botanical Garden.
Luer, C. A.	1987	Icones Pleurothallidinarum. IV. Systematics of *Acostea, Condylago* and *Porroglossum* (Orchidaceae). Missouri Botanical Garden.
Luer, C. A.	1988	Icones Pleurothallidinarum. V. Systematics of *Dresslerella* and *Scaphosepalum* (Orchidaceae). Missouri Botanical Garden.

(continued)

(continued)

Author(s)	Date	Title and publisher
Luer, C. A.	1989	Icones Pleurothallidinarum. VI. Systematics of Pleorothallis subgenus *Ancipitia,* subgenus *Scopula* and *Trisetella* (Orchidaceae). Missouri Botanical Garden.
Luer, C. A.	1990	Icones Pleurothallidinarum. VII. Systematics of *Platystele* (Orchidaceae). Missouri Botanical Garden.
Luer, C. A.	1991	Icones Pleurothallidinarum. VIII. Systematics of *Lepanthopsis, Octomeria* subgenus *Pleurothallopsis, Restrepiella, Restrepiopsis, Salpistele* and *Tegueia* (Orchidaceae). Missouri Botanical Garden.
Luer, C. A.	1992	Icones Pleurothallidinarum. IX. Systematics of *Myoxanthus* (Orchidaceae). Missouri Botanical Garden.
Luer, C. A.	1993	Icones Pleurothallidinarum. X. Systematics of *Dracula* (Orchidaceae). Missouri Botanical Garden.
Luer, C. A.	1994	Icones Pleurothallidinarum. XI. Systematics of *Lepanthes* subgenus *Brachycladium* and *Pleurothallis* subgenus *Aenigma* subgenus *Elongatia* subgenus *Kraenzlinella* addenda to *Dracula, Lepanthopsis, Myoxanthus, Platystele, Porogolossum* and *Trisetella* (Orchidaceae). Missouri Botanical Garden.
Luer, C. A.	1995	Icones Pleurothallidinarum. XII. Systematics of *Brachionidium* addenda to *Dresslerella, Platystele* and *Porroglossum* (Orchidaceae). Missouri Botanical Garden.
Luer, C. A.	1996	Icones Pleurothallidinarum. XIII. Systematics of *Restrepia* (Orchidaceae). Missouri Botanical Garden.
Luer, C. A.	1996	Icones Pleurothallidinarum. XIV. Systematics of *Draconanthes, Lepanthes* subgenus *Marsipanthes* and subgenus *Lepanthes* of Ecuador addenda to *Brachionidium Lepanthes* subgen. Aenigma and subgen. *Ancipitia* (Orchidaceae). Missouri Botanical Garden.
Luer, C. A.	1997	Icones Pleurothallidinarum. XV. Systematics of *Trichosalpinx* addenda to *Dracula, Masdevallia, Myoxanthus* and *Scaphosepalum* corrigenda to *Lepanthes* of Ecuador (Orchidaceae). Missouri Botanical Garden.
Luer, C. A.	1998	Icones Pleurothallidinarum. XVI. Systematics of *Pleurothallis, Pleurothallis* subgenera *Crocodeilanthe Rhynchopera, Talpinaria.* Missouri Botanical Garden.

(continued)

(continued)

Author(s)	Date	Title and publisher
Luer, C. A.	1998	Icones Pleurothallidinarum. XVII. Systematics of subgen. *Pleurothallis* sect. Abortivae sect. Truncatae sect. *Pleurothallis* subsect. Acrontiae subsect Pleurothallis subgen Dracontia subgen Unciferia. Missouri Botanical Garden.
Luer, C. A.	1999	Icones Pleurothallidinarum. XVIII. Systematics of *Pleurothallis* subgen. *Pleurothallis* sect. *Pleurothallis* subsect. Antenniferae subsect Longiracemosae subsect. Macrophyllae-Racemosae subsect Perplexae subgen. *Pseudostelis* subgen *Acuminata*. Missouri Botanical Garden.
Luer, C. A.	2000	Icones Pleurothallidinarum XIX. Systematics of *Masdevallia* Part One. Missouri Botanical Garden.
Luer, C. A.	2000	Icones Pleurothallidinarum XIX. Systematics of *Masdevallia* Part Two. Missouri Botanical Garden.
Luer, C. A.	2000	Icones Pleurothallidinarum XX. Systematics of *Jostia, Andinia, Barbosella, Barbrodria*, & *Pleurothallis* subgen. *Antilla,* subgen *Effusia* subgen. *Restrepioidia*. Missouri Botanical Garden.
Luer, C. A.	2001	Icones Pleurothallidinarum XIX. Systematics of *Masdevallia* Part Three. Missouri Botanical Garden.
Luer, C. A.	2002	Icones Pleurothallidinarum XIX. Systematics of *Masdevallia* Part Four. Missouri Botanical Garden.
Luer, C. A.	2003	Icones Pleurothallidinarum XIX. Systematics of *Masdevallia* Part Five. Missouri Botanical Garden.
Luer, C. A.	2004	Icones Pleurothallidinarum XXVI. *Pleurothallis* subgenus *Acianthera* and three allied sub-genera A second century of new species of *Stellis* of Ecuador *Epibator, Ophidion Zootrophion*. Missouri Botanical Garden.
Luer, C. A.	1986–1994	Icones Pleurothallidinarum I–VI, XI. Missouri Botanical Garden.
Luer, C. A.	2005	Ein schatz von *Masdevallia*
Luer, C. A., and S. Dalstrom	1996–2004	A treasure of *Masdevallia*. Missouri Botanical Garden, St. Louis, MO. Fascicle 21, no pagination, 17 plates; Fascicle 22, no pagination, 15 plates; Fascicle 23, no pagination, 18 plates; Fascicle 24, no pagination, 16 plates; Fascicle 25, no pagination, 16 plates; Fascicle 26, no pagination, 18 plates; Fascicle 27, no pagination, 15 plates; Fascicles measure 42 × 30cm.

(continued)

(continued)

Author(s)	Date	Title and publisher
Luer, C. A., R. Escobar R., and S. Dalstrom	1988–1992	Thesaurus Dracularum. Missouri Botanical Garden. No pagination in any of the volumes, 15 color plates per volume. Books measure 42 × 30cm.
Lugo, H. L.	1954	Important facts about *Vanilla.*
Lumbreras, E. L. (ed.)	2001	Orquideas silvestres de la comunidad Valenciana.
Löhndorf, E.	1956	The forest of fear. Souvenir Press, London. 182 pp., B&W photographs. This is a tale by an orchid hunter.
Lüttge, U. (ed.)	1989	Vascular plants as epiphytes. Springer-Verlag, Berlin. 270 pp., numerous B&W illustrations.
Lyons, J. C.	1845	A practical treatise on the management of orchidaceous plants. Hodges and Smith, Dublin and J. Ridgway, London. 234 pp.
Lyons, J. C.	1851	Praktische anweisung zur cultur der tropischen orchideen. E. Schweizerbart'sche Verlagshandlung und Drukerei. 212 pp.
MacDonald, N.	1939	The orchid hunters, 1st ed. A jungle adventure. Farrar & Rinehart, New York (has their colophon). 294 pp., B&W photographs. Introduction by Rex Stout (author of Nero Wolfe mysteries).
Macdonald, R.	1994	Cool growing orchids throughout the year. Ross MacDonald Publisher, 12 Bens Place, Wanganui, New Zealand. 144 pp., numerous color photographs.
Macdonald, R.	1999	More flowering orchids throughout the year. Published by the author at 12 Bens Place, Wanganui, New Zealand. 132 pp., numerous color photographs.
Mackenzie, J. S. E. and C. E. Talbot Ponsonby	1911	British orchids, how to tell them one from another, 49 pp., and 36 color plates on white paper glued onto brown pages. Unwin Brothers, Ltd. London.
Madalski, J.	1971, 1977	Atlas flory Polskieji ziem osciennych. Parts 1 and 2 (two volumes).
Maekawa, F.	1971	The wild orchid of Japan in colour. S+S, Tokyo, Japan, 495 pp., 185, 187 color paintings, in Japanese with some English text and English captions.
Magdefrau, K.	1973	Geschichte der Botanik. Gustav Fuscher Verlag, Stuttgart.
Magrou, J.	1943	Titres et travaux scientifiques de Joseph Magrou. Imprimerie Barnéoud, Laval, France. 76 pp., 6 B&W plates.
Magrou, J.	1943	Des orchidées a la pomme de terre. Gallimard Publishers. 203 pp., line drawings and B&W photographs.

(continued)

(continued)

Author(s)	Date	Title and publisher
Maidement, R.	1985	Basic guide to orchid growing.
Mamaev, S. A., M. S. Knyasev, P. V. Kulikov, E. G. Filippov	2004	Orchids of the Urales: systematics, biology, protection. UrO RAN, Ekaterinburg. 123 pp., 18 line drawings and 21 color photographs. (In Russian.)
Mansfeld, R.	1954	Uber die verteilung der merkmalei nnerhalb der Orchidaceae-Monandrae
Maranan, C. H.	1994	ASEAN orchids. The official handbook of the 9th ASEAN Orchid Congress, Manila. No pagination, numerous B&W and color photographs.
Margaret Ilgenfritz Orchids	No date	Cultural instructions. Margaret Ilgenfritz Orchids, Monroe, Michigan. 10 pp., line drawings. Company no longer exists.
Mariat, F.	1951	Contribution à l'etude de la symbiose dans ses rapports avec les facteurs de croissance. Thesis, Faculty of Sciences, University of Paris. 134 pp., B&W photographs, line drawings, graphs.
Mark, B.	1987	Slipper orchids the genus *Paphiopedilum*.
Mark, B.	1992	The color illustration of *Paphiopedilum* species.
Marques, R. P., and J. L. Bocourt Vigil	1987	Iconographia de orquideas Cubanas.
Marsh, J., and K. Greenway	1978	The illuminated language of flowers. Holt, Rinehart and Winston, New York. 78 pp., numerous color line drawings.
Marshall, K.	2004	Orchids, a guide to cultivation. The Crowood Press, Ltd., Ramsbury, Marlborough, Wiltshire, SN8 2HR. Numerous color illustrations, some line drawings.
Martin, L., and L. Rosier	1974	The orchid family. William Morrow & Co., New York, 96 pp., line drawings.
Masaaki, C.	2002	*Phalaenopsis* species. Chiba Masaaki *Phalaenopsis* Species Publishing Society. Sakado, Saitama, Japan. 136 pp.
Masaart, J.	1894	Un botaniste en Malasie. Bull. Soc. Roy. Bot. Belgique 23: 151–319, illustrations. Xerocopy.
Massee, G.	1895	The spot disease of orchids.
Masters, M. T.	1886	On the floral conformation of the genus *Cypripedium*.
Mather, A.	1986	Pale orchid. A novel. Harlequin Books, Toronto. 186 pp.
Mathis, W.	2005	The gardener's guide to growing hardy perennial orchids. The Wild Orchid Company, Doylestown, Pennsylvania. 93 pp., numerous color photographs.

(continued)

(continued)

Author(s)	Date	Title and publisher
Matho, K.	1956	Orchideen der tropen und subtropen, Karl Winter Universitätsverlag, Heidelberg. 182 pp., 64 plates of color plates, 3 line drawings.
Matho, K.	1958	Orchid growing for everyone. Blandford Press, London. 170 pp., 64 color plates.
Matthews, D. (ed.)	1996	The American orchid society celebrates its diamond jubilee. A review of the first 75 years 1921–1996. American Orchid Society, West Palm Beach, FL.
Matuda, E.	1969	Las orquideas del estado de Mexico. Gobierno del Estado de Mexico, Direccion de Agricultura Y Ganaderia, Toluca, Mexico. 58 pp., numerous B&W photographs.
Matuda, E.	1979	Las orquideas del estado de Mexico. Gobierno del Estado de Mexico, Direccion de Agricultura Y GanadeLas orquideas del estado de Mexicoria, Toluca, Mexico. 58 pp., numerous B&W photographs.
Matuzawa, S.	1983	Cattleya. Color books published by Hoikusya, Japan. 160 pp., numerous B&W and color photographs.
Maunier, E.	1932	Les plantes à parfums des colonies Françaises. Institu Colonial, Marseilles, France. 122 pp., has sections on Vanilla.
Mayr, H.	1998	Orchid names and their meanings. A. R. G. Gantner Verlag K.-G., FL 9490, Vaduz, Liechtenstein. 548 pp.
McCook, L.	1998	*Phragmipedium*. Orchid Digest Corporation, 11 pp., 17 color photographs.
McDonald, E.	1998	100 orchids for the American gardener. Workman Publishing, New York. 247 pp., numerous color plates.
McDonald, R.	1994	Cool flowering orchids throughout the year. Ross McDonald, 12 Bens Place, Wanganui, New Zealand.
McDowell, D.	1995	The orchid picture book. Cynpat Lithography, P. O. Box 1201, Franklin Park, IL 60131.
McKinley, M.	2005	Complete guide to orchids. Ortho Books/ Meredith Books, Des Moines, IA. 224 pp., numerous color photographs.
McLeish, N. R. Pearce, B. R. Adams, and J. S. Briggs	1995	Native orchids of Belize. A. A. Balkema, Rotterdam, The Netherlands/Brookfield, VT, UA. 278 pp., color painting frontispiece, 120 line drawings plates, 237 color photographs.
McQueen, J., and B. McQueen	1992	Miniature orchids. Timber Press, Portland, OR. 192 pp., numerous color and B&W photographs.

(continued)

(continued)

Author(s)	Date	Title and publisher
McQueen, J., and B. McQueen	1993	Orchids of Brazil.
McVaugh, R.	1985	Flora Novo-Galiciana. Volume 16. Orchidaceae. The University of Michigan Press, Ann Arbor. 367 pp., one color plate, numerous line drawings, two maps.
Measures, R. I.	1894	Cypripediums, 3rd ed. No publisher listed. 63 pp. The book measures 11.5 × 6.8 cm. Printing is along the longer dimension.
Meeuse, B. J. D.	1961	The story of pollination. The Ronald Press, New York. 243 pp. several color plates, numerous line drawings and B&W photographs.
Mehra, P. N.	1983	Cytology of orchids of Khasi and Jaintia hills. Published by the author at the Himachal Agricultural University, Palampur (Solan Campus), 118 pp., 337 line drawing and B&W illustrations+frontispiece.
Mehra, P. N.	1989	Ecology, structural adaptations and cultural aspects of some Himalayan orchids. Published by the author at the Department of Botany, Panjab University, Chandigarh, India. 58 pp., numerous line and B&W picture illustrations.
Mehra, P. N., and S. K. Kashyap	1989	Cytology of orchids of north-west Himalayas. Professor P. N. Mehra at Pramodh P. Kapur, Raj Bandhu Industrial Co., New Delhi. 108 pp., numerous B&W photographs and micro-graphs.
Meinecke, E. P.	1894	Beitrage zur anatomie der luftwurzein der orchideen.
Menezes, L. C.	1987	*Cattleya labiata* Lindl.
Menezes, L. C.	1994	*Cattleya warneri*.
Menezes, L. C.	1995	*Laelia purpurata*.
Menezes, L. C.	2000	Orchids, genus *Cyrtopodium*, Brasilian species. Bilingual, English-Portuguese. Edições IBAMA, Brasília. 208 pp., numerous color and B&W photographs, line drawings and paintings..
Menezes, L. C.	2002	*Cattleya labiata autumnalis*. 252 pp., numerous color photographs and paintings. IBAMA, Brasilia.
Menezes, L. C.	2000	Genus *Cyrtopodium* – Brazilian species.
Menezes, L. C.	2004	Orchids-Brazilian central plateau.
Meniere, O., and E. Meniere	1855	Note sur la collection d'orchidees exotiques.
Menzies, D.	1991	Orchids.
Mergner, H.	1992	Orchideenkunde.
Metilla, S. S.	No date	Primer on orchid culture.
Meyer, A.	1886	Ueber die knollen der einheimischen orchideen.

(continued)

(continued)

Author(s)	Date	Title and publisher
Meyer, K. H., and K. Beyer	1966	Orchideen in wort und bild. Verlag BEA-Bücherdienst Kurt Bosshard Dietikon., 255 pp., numerous full page color photographs.
Meyer, N., and B. J. Kaplan	1978	Black orchid. A novel. Bantam Books. 404 pp.
Miheev, V. A.	1993	Orhidei. Bilingual Russian-English. Publishing Center "Rossia Molodia," Moscow. 144 pp., color photograohs.
Millar, A.	1978	Orchids of Papua New Guinea, 1st ed. Australian National University Press, Canberra. 101 pp., numerous color photographs and a few line drawings.
Millar, A.	1978	Orchids of Papua New Guinea, an introduction. University of Washington Press, Seattle and Australian National University Press, Canberra. 101 pp., 200+ color photographs.
Millar, A.	1999	Orchids of Papua New Guinea, 2nd ed. Timber Press, Portland, OR. 118 pp., numerous color photographs and a few line drawings.
Millar, A. (ed.)	No date	Orchids of Papua New Guinea. Text & drawings. The Orchid Society of Papua New Guinea. No pagination, 18 full page line drawings.
Miller, D., R. Warren, and I. M. Miler	1994	Orchids of the high mountain Atlantic rainforest in southeastern Brazil. Salamandra Consultoria Editorial SA, Rio de Janeiro, Brazil. 182 pp., numerous line drawings and color photographs.
Miller, J. (ed.)	1999	The *Odontoglossum* compedium. Articles from the *Odontoglossum*Alliance newsletter March 1988–November 1998, 100–200 pp., numerous line drawings. *Odontoglossum* Alliance, P. O. Box 38, Westport Point, MA 02791, USA.
Millican, A.	1891	Travels and adventures of an orchid hunter. Cassell & Company, Ltd., London. xv+222 pp., color frontispiece, B&W paintings and photographs, line drawings.
Min, A.	2004	Empress Orchid. A novel. Houghton Mifflin Co., Boston, MA. 336 pp.
Miner, H. S.	1885	Orchids, the royal family of plants. John Slark, London. 90 pp., 24 color plates.
Ministerie van Landbouw	1955	Collecties orchideeen in cultuur.
Mioulane, P.	1981	Les Orchidees. Dargaud Editeur, Paris.96 pp., line drawings and color photographs.
Mishkin, B. A.	1953	Flora of Khibini Mountains. Moscow-Leningrad. Izdatelstvo Akademii Nauk SSSR. 112 p. (In Russian.)

(continued)

(continued)

Author(s)	Date	Title and publisher
Miranda, F.	1996	Orchids from the Brazilian Amazon.
Miura, J.	1983	Mini *Cattleya, Sophronitis, Sophronitera, Laelia*. Kazuyuki Ikeda Publisher, Tokyo. 144 pp., numerous B&W and color photographs. (In Japanese.)
Miura, J.	1987	*Cattleya intermedia*. Japan Publications Trading Co., Ltd. 128 pp., numerous B&W and color photographs. (In Japanese.)
Moënne-Loccoz, P.	2003	Orchidées sauvages des Ardennes, des Savoies, du Var, du Vercors er d'Andalousie
Mohring, R., and S. Faller	1987	Zauberwelt der orchideen.
Moir, W. W. G., and M. A. Moir	1980	Breeding variegata oncidiums. Harold L. Lyon Arboretum and the University of Hawaii, Honolulu, HI. 122 pp., color and B&W photographs.
Moir, W. W. G., and M. A. Moir	1982	Creating Oncidiinae intergenetics. Harold Lyon Arboretum and The University Press of Hawaii. 95 pp., some color photographs.
Moir, W. W. G., and M. A. Moir	1982	Laeliinae intergenetics. Harold Lyon Arboretum and The University Press of Hawaii. 95 pp., some color photographs.
Monkhouse, S.	1969	Orchids for all. John Carroll Publishing Pty., Ltd. Adelaide, South Australia. 64 pp., line drawings, B&W and color illustrations.
Montes de Oca	1963	Colibries y orquideas de Mexico.
Montes de Oca	1963	Hummingbirds and orchids of Mexico.
Moore, L. B., and E. Edgar	1970	Flora of New Zealand, Volume II. Government Printer, Wellington. 354 pp., line drawings, maps.
Moore, S. Le M.	1879	On a monandrous *Cypripedium*.
Moore, T.	1857	Illustrations of orchidaceous plants. Willis and Sotheran, London. 100 color plates.
More de Retana, D. E.	1992–1993	Orchids of Costa Rica, parts 2 and 3. Icones Plantarum Tropicarum Fasciles 15 and 16. Plates 1401–1600. Marie Selby Botanical Garden, Sarasota, FL.
Morel, C.	1855	Culture des orchidees. DUSACO, Libraire Agricole de la Maison Rustique.
Morel, C.	1856	De kultur der orchdeën. Translated from the French by R. Witte. A. W. Sythoff. Leyden, The Netherlands. 136 pp.
Morris, B.	1970	The epiphytic orchids of Malawi. The Society of Malawi. Numerous B&W line drawings and some color illustrations. 136 pp.
Morris, F., and E. A. Ames	1929	Our wild orchids. Charles Scribner's sons. New York. 464 pp. xxxi+464 pp., 127 and 3 color illustrations.

(continued)

(continued)

Author(s)	Date	Title and publisher
Morrison, G.	1988	The orchid grower's manual. Kangaroo Press, Kenthurst, NSW 2156. 247 pp., numerous line drawings, few color and B&W photographs.
Morrison, G. C.	1986	Growing orchids in Australia and New Zealand. Kangaroo Press, Knehurst 2154, Australia. 80 pp., line drawings, B&W and color photographs.
Morrison, G., and M. A. Webb.	1991	Essentials of orchid growing. Kangaroo Press Pty., Ltd., Kenthurst, N. S. W., Australia. 157 pp., 66 color photographs.
Morton, J. K.	1961	West African lilies and orchids. Longmans, Green and Co., Ltd. 71 pp., 80 species illustrated with color paintings.
Mossberg, B., and S. Nilsson	1977	Nordens orkideer.
Mossberg, B., and S. Nilsson	1987	Orkidéer. Europas vildväxande arter. Wahlström and Widstrand. 253 pp., numerous color drawings and maps.
Motes, M.	1997	Vandas.
Motes, M.	2006	Orchid territory.
Moulen, F.	1958	Orchids in Australia. 148 pp., numerous color photographs, Austral-Edita Pty., Ltd., Sydney.
Muller, W.	1900	Orchideen abbildungen der in Deutschland und den angrenzenden gebieten.
Muller, W., and F. Kraenzlin	1904	Orchideen abbildungen der in deutschland und den angrenzenden gebieten.
Mullins, B. and M. Martin	1980	Australian orchids. Angus & Robertson, London. 34 pp., 60+ color photographs.
Mullins, B. G., P. Lindsay, and R. Kerr	No date	Australian native orchids, No. 2 in a series. Horwitz International, Inc., Australia. 34 pp., 60 color plates.
Musee Internationale de la Parfumerie	No date	Vanilles et orchidees.
Mutel, A.	1840, 1842	Mémoire sur plusieurs orchidées nouvelles ou peu connues parts 1 (pp. 1–28) and 2 (pp. 1–35) plus 5 plates. Mem. Soc. Mus. Hist. Nat. Strasbourg. The author is described as "captain or artillery."
Möller, A.	1920	Fritz Müller: werke, briefe, und leben. Verlag von Gustav Fischer, Jena. 163 pp., illustrations.
Müller, H.	1883	The fertilization of flowers. Translated and edited by D'Arcy Thompson with a preface by Charles Dawin. Has a section of orchids. Macmillan and Co., London. 669 pp., 186 line drawn figures.

(continued)

(continued)

Author(s)	Date	Title and publisher
Nagano, Dr. [Y.] and Mrs.	No date	Miniature orchids in Japan and China. Nagano Orchids, Tokyo, Japan. No pagination, B&W and color illustrations.
Nagano, Y.	No date (prior) to 1957	Oriental miniature orchids. Kashima Shoten, Tokyo. 95 pp., numerous B&W photographs.
Nair, H. (ed.)	1990	Proceedings of the international conference and exhibition on orchids and ornamental plants held in conjunction with the eighth ASEAN orchid congress, Kuala Lumpur, Malaysia, August 22–28. Orchid Society of Malaysia, P. O. Box 138, Petaling Jaya, Malaysia. 279 pp., B&W photographs.
Nair, H., and J. Arditti (eds.)	2005	Proceedings of the 17th World Orchid Conference, Shah Alam, Malaysia. Natural History Publications (Borneo). 428 pp.
Nakao, I., M. Kameyama, H. Takahashi, and H. Noguchi (eds.)	1983	Native orchids of Brasil. Published for the Associação Orquidófila de São Paulo, São Paulo, Brasil by Editora Grãfica Topan Press Ltda., São Paulo, Brasil. 368 pp., numerous color photographs.
Napier, E.	1985	Report of the international centenary orchid conference, 1985, London. Royal Horticultural Society. 155 pp., color photographs. In slip case with a facsimile of the report of the 1885 conference. See also Anonymous, 1886.
Nash, N., and I. La Croix (consultants) and D. Banks (writer)	2005	Flora's orchids. Timber Press, Portland, OR. 368 pp., about 1,500 color photographs.
Natusch, S.	1968	A bunch of wild orchids. Pegasus Press, Christchurch, New Zealand. 24 pp., line drawings and 4 color paintings.
Nayar, M. P.	1984	Key works to the taxonomy of flowering plants of India. Volume 4. Botanical Survey of India, Department of the Environment, Howrah, India. 268 pp.
Nazarov, E. G.	1985	Orchids. Moscow, Kolos. (In Russian.)
Neal, J.	1994	Growing *Phalaenopsis* at home. Earth Production Pty., Ltd., Chatswoodm NSW, Australia. 57 pp., line drawings and color photographs.
Nebel, G., and F. X. Brunner	1958	Orchideen. C. Berteksmann Verlag, Gütersloh, Germany. 49 pp., 24 color painting.
Neil, J.	ca 1884	Rays from the realms of nature or Parables of plant life. Cassell, Petter, Galpin & Co. London. 148 pp., numerous line drawings, colored frontispiece. This is a silly plant bestiary which tries to couple plants to religion. It has a section on orchids.

(continued)

(continued)

Author(s)	Date	Title and publisher
Nelson, C.	1983	John Lyons and his orchid manual. Boethius Press, Kilkenny, Ireland. 79 pp., unpaginated appendix I (23 pp.), appendix II consists of un numbered pages (8 +4) and 96 pp., 2 portraits, some line drawings. Limited to 585 copies
Nelson, E.	1962	Gestaltwandel und artbildung erörtert am beispioel der orchidaceen Europas unde der mittelmerländer. Text and figures. 50 colored figures, 8 monochromes, 8 maps. Published by Erich Nelson, Switzerland.
Nelson, E.	1931	Die orchideen Deutschlands und der angrenzenden gebiete.
Nelson, E.	1968	Monographie und ikonographie der orchidaceengattungen *Serapias, Aceras, Loroglossum, Barlia.*Text and figures. 30 colored figures, 6monochromes, 8 maps. Published by Erich Nelson in Switzerland.
Nevskii, S. A.	1935	Orchidaceae Lindl. pp. 589–730. In: Flora of the USSR. Vol. 4. Academy of Science Press. Plates with line drawings. (In Russian.)
Nelson, E.	1976	Monographie und ikonographie der orchideengattung *Dactylorhiza* with atlas.
NHK	1994	Orchids of Japan.
Nicholls, W. H.	1969	Orchids of Australia. Thomas Nelson (Australia) Ltd. Sydney. 141 text page, 476 colored plates of paintings.
Nickson, J.	ca 1840	Orchids (in manuscript).
Nicolai, G.	1954	Orchideen.
Nicolai, W.	1931	Orchideen.
Nicolai, W.	1939	Orchideen.
Nielsen, R.	1895	Om tropiske orchideer. 113 pp., 16 color paintings. Gyldendalske Boghandels Forlag, Copenhagen.
Nikitina, E. V.	1951	Orchidaceae. pp. 134–142. In: Flora of Kirghizia. Vol. 3. Frunze. (In Russian.)
Niles, G. G.	1904	Bog-trotting for orchids. G. P. Putnam's Sons, New York. 310 pp., numerous B&W and color plates.
Nilson, Ö., and L.-Å. Gustafsson	1976	Project Linné repporterar. Svensk Botanisk Tidskrift volumes 70–76 in one issue. Inconsistent pagination. Maps, line drawings, B&W and color photographs. (In Swedish.)
Nilsson, S.	1979	Orchids of northern Europe. Penguin Books, Ltd., Harmondsworth, Middlesex, UK. 146 pp., n58 color paintings by Bo Mossberg.
Nir, M.	2000	Orchidaceae Antillanae. DAG Media Publishing, Inc., New York. 453 pp., 88 color photographs, 87 line drawings.

(continued)

(continued)

Author(s)	Date	Title and publisher
Noble, M.	1953	You can grow orchids. Published by the author in Jacksonville, FL. 144 pp., numerous line drawings and B&W photographs. Some color.
Noble, M.	1964	You can grow orchids. Published by the author in Jacksonville, FL. 152 pp., numerous line drawings and B&W photographs. Some color.
Noble, M.	1968	You can grow *Cattleya* orchids, revised 2nd ed. McQuerry Orchid Books, 5700 West Salerno Road, Jacksonville, FL 32244–2354..
Noble, M.	1994	You can grow *Phalaenopsis* orchids, revised 2nd ed. McQuerry Orchid Books, 5700 West Salerno Road, Jacksonville, FL 32244–2354. 102 pp., line drawings and B&W and color photographs.
Northen, R. T.	1950	Home orchid growing, 1st ed. D. van Nostrand Co., Inc. 320 pp., numerous B&W and a few color photographs.
Northen, R. T.	1952	Home orchid growing, 2nd ed. D. van Nostrand Co., Inc. 320 pp., numerous B&W and a few color photographs.
Northen, R. T.	1955	Orchids as house plants, 1st ed. D. van Nostrand Company, Inc., Princeton, New Jersey. 122 pp., numerous line drawings.
Northen, R. T.	1970	Home orchid growing, 3rd ed. D. van Nostrand Co., Inc. 374 pp., numerous B&W and a few color photographs, some line drawings.
Northen, R. T.	1976	Orchids as house plants. Dover Publications, New York. 148 pp., 63 figures consisting of line drawings and photographs.
Northen, R. T.	1980	Miniature orchids. Van Nostrand Reinhold Co., New York. 189 pp., 187 plus additional unnumbered B&W photographs.
Northen, R. T.	1990	Home orchid growing. Prentice-Hall, New York. 376 pp., numerous B&W and color photographs.
Northen, R. T.	1996	Miniature orchids and how to grow them. 1991 pp., numerous B&W photographs. Dover Publications, Inc., New York.
Oakeley, H.	No date	An annotated checklist of the genus *Anguloa*. Orchid Digest Corporation. Redlands, California. 32 pp., numerous color photographs.
Oakeley, H. F.	1993	*Lycaste* species, the essential guide. Published by the author at Beckenham, BR3 1NR, Kent, England. 34 pp., numerous color photographs.

(continued)

(continued)

Author(s)	Date	Title and publisher
Obara, E.	1930	A brief history of orchids. A xerocopy of a book in Japanese with a private unpublished English translation. Attached to it is a copy of a book in Chinese.
O'Brien, J.	1898	Hybrid orchids.
O'Brien, J.	1911 (?)	Orchids. J. C. &E. C. Jack, London. 114 pp., 8 color photographic plates.
O'Brien, J.	ca 1916	Orchids. Frederick A. Stokes Co., New York. 114 pp., 8 color plates.
O'Byrne, P.	1994	Lowland orchids of Papua New Guinea. National Parks Board, Singapore. 584 pp., Numerous line drawings and color photographs.
Odakura, M., and S. Naemura	1982	Wild orchids. Fujinseikatsusha Co., 226 pp., numerous color photographs. (In Japanese.)
Odijk, G. Th.	1944	Hoe kweek ik orchideeën voor huis en kas, 2nd ed. Drukkerij Torenlaan (N. V. De Provinciale Drentsche en Asser Courant) – Dir. H. Clewits - Asen. 96 pp., numerous B&W photographs.
Odijk, G. Th.	1968 or earlier	Orchideeën, may be 1st ed, N. V. W. J. Thieme. 190 pp., line drawings and B&W photographs.
Odijk, G. Th.	No date	Orchideeën, not 1st ed., probably 2nd ed. Joh. Enschedé en Zonen R. V., Haarlem. 121 pp., line drawings and B&W photographs.
Odjik, O. T.	1936	Hoe kweek ik orchdeeën voor huis en kas. Joh. Enschede en Zonen, Haarlem, The Netherlands. 96 pp., numerous B&W photographs.
Odijk, O. T.	1944	Hoe kweek ik orchdeeën voor huis en kas, 2nd ed or printing. Drukkerij Torenlaan, Assen, The Netherlands. 96 pp., numerous B&W photographs.
Oertle, C. F.	1987	Index periodicarum Orchidacearum 1975–1985. European Orchid Commission/Tipografia Poncioni, Lausanne, Switzerland. 615 pp.
Ogura, Y.	1953	Anatomy and morphology of the subterranean organs in some Orchidaceae.
Ohlsson, A.	1951	Svenska orchideer.
Ohlsson, A.	1967	Svenska orchideer.
Okada, H.	2002	*Dendrobium* species.
Okada, H., T. Hirota, and M. Wanaka	2001	*Cattleya* species. Sodo Publishing Co., Ltd., Tokyo. 128 pp., numerous maps and color photographs.
Okami, Y.	1964	Orchids, their kinds and cultivation. Seibundo Shinkosha, Tokyo. 436 pp., 26 color pictures, numerous B&W pictures.

(continued)

(continued)

Author(s)	Date	Title and publisher
Okita, Y.	1976	Oriental orchids. Color books published by Hoikusya, Japan. 160 pp., numerous B&W and color photographs.
Okita, Y.	1982	Wild orchids. Color books published by Hoikusya, Japan. 160 pp., numerous B&W and color photographs.
Oliver, D.	1892–1894	Hooker's Icones Plantarum, Vol. XXI–XXII, Indian Orchideae, Vol I–II. Dulau & Co., London. There is also a 1972 reprint by Wheldon and Wesley, Codicote, Herts, UK, Verlag J. Cramer, D-3301 Lehre, Germany and Stechrt-Hafner Service Agency, New York. Text and line drawing plates
Ong, P. Y.	1968/69	Physiological studies on pod development and seed germination of Vanda Miss Joaquim. Honors Thessis supervised by Prof. P. N. Avadhani, University of Singapore. 84 pp., graphs, pictures.
Oplt, J., and J. Kaplická	1970	Orchids. Spring Books, The Hamlyn Publishing Group, Ltd., London. 140 pp., 56 color paintings.
Orchid Club of Southern Australia	1987	Modern orchid growing.
Orchid Council of New Zealand	1998	Orchids 98.
Orchid Society of South East Asia	1993	Orchid growing in the tropics. Times Editions, Singapore.
Orchid Society of South East Asia	1993	Orchids, commemorating the golden anniversary of the Orchid Society of South East Asia. Published by the society.
Orchid Society of the Republic of China	1984–1987	*Cattleya*, superior parentages.
Orchidophilen Club	1905	Wenken voor de kultuur van orchideeen.
Orchidophilen Club	1906	Wenken voor de kultuur van orchideeen.
Orchidophilen Club	1907	Wenken voor de kultuur van orchideeen.
Orlean, S	2000	The orchid thief. A novel even if the author does not admit it. Balantine Books, New York. 297 pp.
Ortiz V., P., A. Martinez M., and G. Misas U.	1982	Orquídeas ornamentales de Colombia-Ornamental orchids of Colombia. Carlos Valencia Editores, Bogota. 172 pp., numerous color photographs. Text in English and Spanish.

(continued)

(continued)

Author(s)	Date	Title and publisher
Ortiz y Valdivieso, P.	2000	La Real expedicion botanica del Nuevo Reino de Granada, volume XI. Microspermae: Orchidaceae, V. Ediciones Cultura Hispanica, Madrid. 98 pp., 54 color and 1 monochrome plates, maps, B&W illustrations. Book measures 54.7 × 37 cm. See also Perez-Albelaez et al., 1954; Schweinfurth et al., 1963, 1969; Fernandez Perez, 1985, 1995
Osorio, L. F.	1941	Colmobian orchids. Félix de Bedout e Hijos, Editors, Medellin, Colombia. 106 pp. (text and figures are printed only on one side of each page and only these pages are numbered), numerous B&W and color (appear hand painted) illustrations.
Ospina H, M. (ed.)	1974	Proceedings of the 7th world orchid conference, April 19–27, 1972. 7a Conferencia Mundial de Orquideologia, 1974, Medellin. 356 pp., maps, line drawings, B&W photographs, color plates, and accompanied by several brochures and magazines.
Ospina Hernandez, M.	1958	Orquideas Colombianas-Colombian orchids. Publicaciones Technicas Ltda., Bogota. 307 pp., Spanish and English, 10 color plates, 109 B&W photographs.
Ospina Hernández, M.	1996	Orchids and ecology in Colombia. To the rescue of paradise. R. Ospina, 511 Flamevine Lane, Vero Beach, FL. 249 pp., numerous color photographs, a few line drawings.
Ospina Hernández, M., and R. L. Dressler	1974	Orquídeas de las Américas. Litografia Arco, Apartado Aéreo 80221, Bogotá, Colombia. 496 pp., 191 color photographs.
Ottman, V.	1922	Der orchideenjager.
Oudemens, C. A. G. A.	1861	Luftwurzeln der orchideen.
Oversteegen, P. M.	1996	Icones Pleurothallidinarum adjunct on Lepanthes 2 volumes.
Oversteegen, P. M.	2000–2001	*Masdevallia* in color with addenda (adjunct to Icones Plurothallidinarum).
Oversteegen, P., and W. Driessen	2005	*Pleurothallis* in kleur
Pabst, G. F. J., and F. Dungs	1975–1977	Orchidaceae Brasilienses. Brücke-Verlag Kurt Schmersow, Hildesheim, Germany. Volume I: 408 pp., numerous B&W and color photographs and drawings. Volume II: 418 pp., numerous B&W and color photographs and drawings.

(continued)

(continued)

Author(s)	Date	Title and publisher
Paireepairit, P.	1978	Textbook of orchids for beginners. Artoin Publishers, Bangkok. 445 pp., numeros B&W and color photographs.
Pantu, Z. C.	1915	Orchidaceele din Romania.
Parham, B. E. V.	1972	Plants of Samoa. Department of Scientific and Industrial Research, Wellington, New Zealand. 161 pp., no illustrations.
Pase, C. P., and R. R. Johnson	1968	Flora and vegetation of the Sierra Ancha experimental forest, Arizona. Research Paper RM-41, USDA Forest Service, Rocky Mountain Forest and Range Experiment Station, Fort Collins, CO.
Pathak, P., R. N. Sehgal, N.Shekhar, M. Sharma, and A. Sood (eds.)	2001	Orchids: Science and commerce. A tribute to Professor Suraj P. Vij. Bishen Singh Mahendra Pal Singh, Dehra Dun 248001, India. 521 pp., line drawings, B&W and color photographs.
Paul, M.	No date	Orchideeen in kleur.
Paul. M.	1963	Orchideeen.
Paul, M.	1964	Orchids, care and growth. Universe Books, Inc., New York. 135 pp., numerous line drawings and color photographs.
Paul, M.	1977	Orchideeen, zelf kweken en verzorgen in kamer, kas en tuin.
Paull, R. C.	1987	Orchid growing in the kingdom of the coconut palm. Our Orchids, 94500 Overseas Highway, Tavernier, FL 33070. 24 pp., 7 line drawings, typewritten.
Paull, R. C.	1992	Orchid growing in the kingdom of the coconut palm. R. C. Paul, 139 Sunset Gardens Dr., Tavernier, FL. 22 pp., typewritten
Peitz, E.	No date	Gestalt- und farb- abwandlungen an orchideen inbesondere by *Ophrys fuciflora*. No publisher. No pagination, color and B&W photographs.
Pelloe, E. H.	1930	West Australian orchids. Emily H. Pelloe, Perth, Western Australia. 77 pp., several color plates of composite paintings, numerous line drawings.
Perez B., A.	1942	The orchids of Costa Rica. San Jose, Costa Rica. Pages 1–30 are in Spanish and 31–52 are in English. B&W photographs.
Perez-Albelaez, E. et al.	1954	La Real expedicion botanica del Nuevo Reino de Granada, Volume I. Ediciones Cultura Hispanica, Madrid. 144 pp., approx. 60 color plates, maps, B&W illustrations. Book measures 54.7 × 37 cm. Schweinfurth et al., 1963, 1969; Fernandez Perez, 1985, 1995; and Ortiz y Valdivieso, 2000 are part of the same series.

(continued)

(continued)

Author(s)	Date	Title and publisher
Perez-Vera, F.	2003	Les orchidees de Cote d'Ivoire.
Perko, M. L.	2004	Die orchideen kärtners.
Perrier de la Bathie, H.	1930	Catalogue de plantes de Madagascar, Orchidaceae. Imprimerie G. Pitoto & Cie, Tananarive, Madagascar. 60 pp..
Perrier de la Bathie, H.	1939	Orchidées de la région Malagache. Memoires du Museum National d'Histoire Naturelle, Vol. 10, Fascicule 5, pp. 237–298.
Perrier de la Bathie, H.	1941	Flore de Madagascar, 49e famile. Orchidées, Vol. II. Imprimerie Officielle, Tananarive, Madagascar. 387 pp., 80 plates of line drawings.
Perrier de la Bathie, H.	1981	Flora of Madagascar (vascular plants), 49th family – Orchids, Volus. I and II. Translated, revised and published by Steven D. Beckman. 621 Palm Avenue, Lodi, CA 95240.
Pertuy, J. and M.	1984	Les orchdées dans l'Ecole de Nancy. La Societe Française d'orchidophile, Paris. 86 pp., numerous B&W and color photographs.
Peterson, R.	1975	Orchid culture under lights. Indoor Light Gardening Society of America, 128 West 58th Street, New York, NY 10019. 11 pp., 15 B&W photographs.
Petrie, W.	1981	Guide to orchids of North America. Hancock House, Blaine, Washington. 128 pp., numerous color photographs and maps.
Pfitzer, E.	1882	Grundzüge einer vergelichenden morpholgie der orchideen. Carl Winter's Universitätbuchhandlung, Heidelberg. 192 pp., one color plate, numerous black and white illustrations.
Pfitzer, E.	1886	Morphologische studien über die orchideenblüthe. Festschrift der Feier des Fünfhudertjährigen Bestehens der Ruperto-Carola Dargebracht von dem Naturhistorisch-Medicinischen Verein zu Heidelberg. B. Naturhistorischer theil, pp. 1–139.
Pfitzer, E.	1903	Das pflanzenreich, IV.50 Orchidaceae-Pleonandrae, oart 12. Reprint 1965 by Verlag von H. R. Engelmann (J. Cramer) Weinheim/ Begrstrasse. 132 pp., 157 line drawing figures in 41 plates.
Pfitzer, E.	1887	Entwurf einer naturlichen anordnung der orchideen. Carl winter's Universitätsbuchhan dlung, Heidelberg. 108 pp.
Pfitzer, E. and Fr. Kränzlin	1907	Das pflanzenreich, IV.50. II. B. 7. Orchidaceae-Monandrae-Coelogynae, part 32. Reprint 1957 by Verlag von H. R. Engelmann (J. Cramer) Weinheim/Begrstrasse. 169 pp., 294 line drawing figures in 54 plates.

(continued)

(continued)

Author(s)	Date	Title and publisher
Phang, V. P. E.	1983	A list of orchid hybrids of Singapore and Malaysia, 1960–1980.
Piers, F.	1951	A book of east African orchids, 1st ed. Patwa Publications, Nairobi. 112 pp., numerous B&W photographs and a few line drawings.
Piers, F.	1959	Orchids of East Africa. Published by the author in Nairobi, Kenya. This copy is number 176 of only 200 published. 148 pp., B&W photographs, line drawings, map, color frontispiece. Mimeograph. Photographs badly printed. The first edition is very rare.
Piers, F.	1959	Orchids of East Africa, 1st ed. Nairobi. Limited edition of 200 mimeographed copies.
Piers, F.	1968	Orchids of East Africa, 2nd ed. Verlag von J. Cramer, Lehre, Germany. 304 pp., 11 color photographs, numerous B&W photographs. There is also a reprint from India. See also Piers, 1951, 1959.
Pinske, J.	1984	Orchideen für zu hause. BLV Verlagsgesellschaft, Munich. 127 pp., B&W and color photographs map and line drawings.
Pinske, J.	1986	The Macmillan book of orchids. Collier Books, Macmillan Publishing Company, New York. 126 pp., numerous color photographs.
Planchon, J.-E.	1858	Hortus Donnatensis. Catalogue des plantes cultivées dans les serre de S. Ex. Le prince Anatole de Démidoff a San Donato Près Florence. Imprimerie de W. Remquet et Co., Paris. 112 pp.
Plantenwerkgroep	1970–1971	Veld determinatietable voor Orchidaceae.
Platiau, R.	2004	L'orchidée à Mariemont.
Plauszewski, P.	1899	Orchidées et plantes de serres, planches en phototypie. Librairies-Impremeries Réunies, Paris. 19 B&W plates of photographs which include orchids and other tropical plants.
Plaxton, E. H.	1983	Proceedings from (sic) symposium II & lectures, North American terrestrial orchids. Michigan Orchid Society c/o Raymond McCullough, 14800 Harrison Ave., Livonia, MI 48154. 143 pp., line drawings, maps, B&W and color photographs.
Plimpton, P. A.	1979	Oakes Ames, Jottings of a Harvard botanist. Botanical Museum of Harvard University, Cambridge, MA. 4–5 pp., 40 B&W photographs, 22 drawings.
Plimpton, S.	2002	The orchid tourist. American Orchid Society, Delray Beach, FL. 64 pp.

(continued)

(continued)

Author(s)	Date	Title and publisher
Poddubnaya-Arnoldi, V. A.	1964	General embryology of angiosperms. (In Russian.) Nauka Publishing House, Moscow. 482 pp., numerous line drawings, some in color. Has chapters on orchids.
Poddubnaya-Arnoldi, V. A.	1976	Cytoembryology of angiosperms. (In Russian.) Hauka Publishing House, Moscow. 508 pp., numerous line drawings and B&W photographs. Has a good chapter on orchids.
Poddubnaya-Arnoldi, V. A. and V. A. Selezneva	1957	Orhidei I ih kultura (Orchids and their cultivation) Akademia Nauk, SSSR, Moscow. (In Russian.) 176 pp., a few color plates, B&W photographs, numerous line drawings. May be the first book on orchids published in the Soviet Union.
Poliakoff, A. T.	1987	Guide to orchid hybrids: the *Vanda* family.
Pope, V.	1985	*Cymbidium* orchids in your garden, a handbook on Cymbidium orchid culture for beginners in South Central Africa (Zimbabwe). No publisher listed. 21 pp., 9 color photographs.
Porsch, O.	1906	Orchidaceae.
Porto, P. C.	1915	Contribuição para o conhecimento de flora orchidacea de serra do Itatiaya. Jardin Botanico do Rio de Janeiro. 22+ additional pages.
Pottinger, M.	1983	African orchids, a personal view. H G H Publications, Wokingham, Berkshire, UK. 64 pp., B&W and color photographs.
Poucel, J.	1942	A la decouverte des orchidées de France, 3rd ed. Editions Stock, Delemain et Boutelleau, Paris. 222 pp., 29 line drawings, 16 B&W photographs.
Pradhan, U. C.	1976	Indian orchids: Guide to identification and culture. Published by the author at Rishi Road, Kalimpong, West Bengal. India. Vols I and II. 747 pp., 162 B&W and color photographs, numerous line drawings.
Pradhan, U. C., and S. C. Pradhan	1997	100 beautiful Himalayan orchids and how to grow them. Primulaceae Books, Kalimpong, West Bengal, India. 131 pp., 102 color photographs.
Praline, M.	2003	The orchids and gumbo poker club. A novel. Disney Press, New York. 126 pp., line drawings.
Prater, W., and E. Waldvogel	1986	Orchideen fur die Fensterbank.
Prater, W., and E. Waldvogel	1987	Orchideeen voor de vensterbank.
Presser, H.	1995	Die orchideen Mitteleuropas und der Alpen.
Presser, H.	2000	Die orchideen Mitteleuropas und der Alpen

(continued)

(continued)

Author(s)	Date	Title and publisher
Prete, [], [] Tichy and []Tosi.	1983	Le orchidee spontanee della provincia di Grosseto
Price, F. A. E.	1919	Orchid culture for beginners. Printed by the Colombo Apothecaries Co., Ltd., Ceylon. 87 pp., color painting frontispiece, numerous painting in B&W.
Price, F. A. E.	1933	Orchid culture in Ceylon and the east. The Colombo Apothecaries Company Limited, Pronters, Publishers and Booksellers. 123 pp., line drawings B&W photographs, B&W and color paintings. This is actually a second edition, the first having been published in 1918. There is also a recent reprint from India.
Pridgeon, A.	1992	The illustrated encyclopedia of orchids. Timber Press, Portland, OR.
Pridgeon, A. M. (ed.)	1994	Proceedings of the 14th world orchid conference, April 27–May 01, 1993 Glasgow, Scotland. Her Majesty's Stationary Office, Glasgow. Only 750 printed. 428 pp., maps, line drawings, B&W and color photographs.
Pridgeon, A. M., P. J. Cribb, M. W. Chase, and F. N. Rasmussen (eds.)	1999	Genera Orchidacearum. Vol. I. General introduction, Apostasioideae, Cypripedioideae. Oxford University Press, Oxford, UK, 197 pp., numerous line drawings, maps and B&W and color photographs.
Pridgeon, A. M., P. J. Cribb, M. W. Chase, and F. N. Rasmussen (eds.)	2001	Genera Orchidacearum. Vol. II. Orchidoideae (Part 1). Oxford University Press, Oxford, UK, 416 pp., numerous line drawings, maps and 120 color photographs.
Pridgeon, A. M., P. J. Cribb, M. W. Chase, and F. N. Rasmussen (eds.)	2003	Genera Orchidacearum. Vol. III. Orchidoideae (Part 2), Vanilloideae. Oxford University Press, Oxford, UK, 360 pp., numerous line drawings, maps and 105 color photographs.
Pridgeon, A. M., P. J. Cribb, M. W. Chase, and F. N. Rasmussen (eds.)	2005	Genera Orchidacearum. Volume 4. Epidendroideae (Part One). Oxford University Press, Oxford, UK.
Pridgeon, A. M., and L. L. Tillman	1990	Handbook on orchid pests and diseases. American Orchid Society, Delray Beach, FL., 108 pp., numerous B&W and color photographs.
Pritchard, H. W.	1989	Modern methods in orchid conservation. Cambridge University Press, Cambridge. 173 pp., few line drawings and graphs.
Procházka, F.	1980	Nase orchideje. Pardubice, Prague. 296 pp., several maps, line drawings and B&W photographs.

(continued)

(continued)

Author(s)	Date	Title and publisher
Procházka, F., and V. Velísek	1983	Orchideje nasi prirody. Cekoslovenska Akademie Ved, Prague. 279 pp., line drawings, color plates.
Prutsch, J., and R. Schill	2000	Die ontongenese der narbe bei den orchideen.
Pucci, A.	1891	Les *Cypripedium* et genres affines. Imprimerie Éditirice L. Niccolai, Florence. 179 pp.
Pucci, A.	1891	Les *Cypripedium*. Imprimerie Éditrice L. Niccolai, Florence. 223 pp.
Pugh, C., C, Beardsell, D. Beardsell, D. Pugh, and A. Strampp	1983	A year of orchids. Richard Griffin Publisher, South Melbourne, Victoria, Australia. 208 pp., numerous line drawings and color paintings.
Puidt, E. de	1880	Les orchidees, histore iconographque.
Puig Verdura, J. M.	2000	Orquideas en la gran ciudad. Palahi Arts Gràfiques, Canonge Dorca, Girona, Spain. 139 pp., numerous line drawings and B&W and color photographs.
Pupulin, F.	1998	Manuel Antonio national park orchids. MesoAmerican Press, San José, Costa Rica. 74 pp., 7 pages of different line drawings, 29 pages of text and color photographs 34 pages of full page line drawing botanical illustrations.
Pupulin, F.	2005	Vanishing beauty, native Costa Rican orchids. Editorial de la Universidad de Costa Rica, San Jose, Costa Rica.
Quarles van Ufford, L. J.	1892	Handleiding tot de kennis en het kweeken van orchdeeën. K. A. van der Weide, Apeldoorn, The Netherlands. 138 pp., line drawings and table temperature conversions Réaumur, Centigrade, Farenheit.
Quednow, K. G.	1930	Beitrage zur frage der aufnahme geloster kohlenstoffverbindungen durch orchideen und andere pflanzen. Inaugural-Dissertation.
Quentin, P.	No date	Synopsis des orchidees Europeennes.
Quisumbing, E.	1972	Philippine orchids and their culture.
R. H. S.	1886	Orchids, being the report on the orchid conference held at South Kensington May 1885.
R. H. S.	1900	Orchid awards given by the RHS in 1859–1899.
R. H. S.	1913	Orchid conference, November 6, 1912.
R. H. S.	1995	Manual of orchids.
Raggardee, D., S. Rakpaibulsombat, and S. Puthom	1975	The Royal horticultural show 1975. Published by Deewan Raggardee, 30/71 Suthisarn Road, Soi Inthamara 27, Bangkok. No pagination. Numerous B&W and color photographs.

(continued)

(continued)

Author(s)	Date	Title and publisher
Raggardee, D., S. Rakpaibulso-mbat, and S. Puthom	1978	The Royal horticultural show 1978. Published by Deewan Raggardee, 30/71 Suthosarn Road, Soi Inthamara 27, Bangkok, Thailand. No pagination. Numerous B&W and color photographs.
Rain, P.	1986	Vanilla cookbook. Celestial Arts Publishers. Berkeley, CA. 124 pp., Line drawings, B&W and color photographs.
Raizada, M. B., H. B. Naithani, and H. O. Saxena	1981	Orchids of Mussoorie. Bishen Singh Mahendra Pal Singh, Dehra Dun, India. 100 pp.
Rakova, M. V.	1992	Biology of the rare plant species in Nature Reserve "Kedrovaya Pad". Vladivostok, Dalnauka. 175 pp. Detail descriptions and drawings of 6 orchid species. (In Russian.)
Ramirez, M. T., M. Ramirez and Q. Ramirez	1994	Introducción a las orquídeas. 1. Published by the author at P. O. Box 362017, San Juan PR. 88 pp., numerous line drawings.
Ramirez, M. T., M. Ramirez and Q. Ramirez	1998	Introducción a las orquídeas. 2. Published by the author at P. O. Box 362017, San Juan PR. 88 pp., numerous line drawings.
Rand, E. S.	1876	Orchids. Hurd and Houghton, New York. 476 pp.
Rankin, R.	1987	Orchid wise, a collection of epigrams and aphorisms. Published by the author at P. O. Box 278, Salisbury, South Australia 5108.
Rao, A. N. (ed.)	1982	Tissue culture of economically important plants. Costed and ANBS, Singapore, c/o Department of Botany, National University of Singapore. xiv+307 pp., numerous B&W illustrations.
Rao, A. N. (ed.)	1986	Proceedings of the Fifth ASEAN Orchid Congress, August 1–3, 1984. Published by the Parks and Recreation Department, Ministry of National Development, Singapore. 183 pp., graphs and B&W photographs.
Rao, A. S.	1979	Orchids of India. National Book Trust, India, New Delhi. 104 pp., five plates of line drawings, numerous B&W and color photographs.
Rasbach, K.	1958	Orchideen in Deutschland.
Rasmussen, F.	1981	The Orchidaceous gynostemium. I. Neottioid orchids. Published by the author in xerography, Copenhagen. 156 pp., one table, 77 pp., of line drawings and photographs.
Rasmussen, H. N.	1995	Terrestrial orchids from seed to mycotrophic plant. Cambridge University Press, Cambridge. 444 pp., line drawings, graphs and B&W photographs.

(continued)

(continued)

Author(s)	Date	Title and publisher
Ratcliffe, E.	1977	The enchantment of paphiopedilums. Published by the author at Chilton, Didicot, England. 101 pp., line drawings.
Raulerson, L., and A. F. Rinehart	1992	Ferns and orchids of the Mariana Islands. Lynn Raulerson and Agnes Reinhart, P. O. Box 428, Agana, Guam. 138 pp., numerous color photographs.
Raven, C. E.	1986	John Ray: Naturalist. Cambridge University Press, Cambridge. 506 pp., one B&W plate.
Raynaud, C.	1985	Les orchidées du Maroc. With preface by Professor Bertil Kullenberg. Société Française d'Orchidophilie, Paris. 120 pp., 27 plates of line drawings, color photographs and color paintings, maps.
Reeve, T. M., and P. J. B. Woods	1989	A revision of *Dendrobium* section Oxyglossum (Orchidaceae). Notes from the Royal Botanic Gardens Edinburgh, Vol. 46, No. 2, pp. 161–305, 19 maps, 37 line drawing figures, 20 plates of color photographs.
Reichenbach, H. G.	1851	Die orchideen der Deutschen flora.
Reichenbach, H. G.	1852	De pollinis Orchidearum genesi structrura et de Orchideis in Artem systema redigendis. Inaugural-Dissertation.
Reichenbach, H. G.	ca 1857	Orchidographische beitrage.
Reichenbach, H. G.	1866	Beiträge zu einer orchideenkunde Central-Amerika's. Druck von Th. G. Meissner, Hamburg. 112 pp., 10 B&W plates of paintings and line drawings. Stafleu & Cowan 8896.
Reichenbach, H. G.	1869	Refugium botanicum. Vol II, Part I, 24 color plates By Fitch. al. John van Voorst, Publisher, London.
Reichenbach, H. G., f.	1869	Beiträge zur orchideenkunde. Druck von E. Blochmann & Sohn, Dresden. Seems to a reprint from vol. 35 of the Austrain Academy of Science. 19 pp., 6 plates of B&W drawings.
Reichenbach, H. G., fil.	1869	Beiträge zur orchideenkunde I (Ein monströses Selenipedium caudatum Rchb. Fil., verglaichen mit Uropedium lindeni Lindl.), II (Ueber Aganisia Lindl.), and III (Ueber Saundersia Rchb. fil). Druck von E. Blochmann & Sohn, Dresden. 19 pp., and B&W drawings plates.
Reichenbach, H. G. f.	1873	Enumeration of the orchids collected by the Rev. E. C. Parish in the neighbourhood of Moulmein, with descriptions of the new species. Transactions of the Linnean Society of London 30: 133–155, plates (line drawings) 27–32.
Reichenbach, H. G.	1874	Enumeration of Parish's orchids.

(continued)

(continued)

Author(s)	Date	Title and publisher
Reichenbach, H. G. f.	1878, 1881	Otia botanica Hamburgensia. Theodor. Theophil. Meisneri, Printer, Hamburg. Parts 1 (1–68) and 2 (pp. 69–1119).
Reichenbach, H. G.	1884	Ueber das system der orchideen.
Reichenbach, H. G.	1884	System der orchideen. Bulletin of the International Congress of Botany and Horticulture, St. Petersburg, pp. 39–48.
Reichenbach, H. G. f.	1892	Xenia Orchidaceae. Three volumes. Published by F. A. Brockhaus.
Reichenbach, H. G.	1969–1978	Refugium botanicum. Vol II. Reprint 1980 by Earl. M. Coleman, Publisher, Conklin Hill Rd., Stanfordville, NY 12581. No pagination, plates (not in color) 73–144.
Reinhard [], [] Golz, and [] Wildemuth	1991	Die Orchideen der Schweiz und angrenzender gebiete.
Reinikka, M. A.	1972	A history of the orchid. University of Miami Press. Coral Gables, FL. 316. pp., numerous illustrations.
Reinikka, M. A.	1995	A history of the orchid. University of Miami Press. Coral Gables, FL. 316. pp., numerous illustrations. With nomenclatural update by Gustavo Romero.
Rendle, A. B.	1899	Catalogue of the African plants collected by Dr. Friedrich Welwitsch in 1853–61. Volume II, Part I, Monocotyledons and gymnosperms. 269 pp..
Renz, J.	1978	Orchidaceae in Flora des Iranischen Hochlandes und der umrahmenden gebirge. Akademische Druck-u7. Verlagsanstalt, Graz, Austria. 148 pp., 56 color and 16 B&W plates.
Renz, J.	1984	Orchidaceae in Flora of Pakistan. Department of Botany, University of Karachi, Pakistan. 63 pp., 6 color photographs in two plates and 10 line drawing plates.
Rentoul, J. N.	1980	Growing orchids, cymbidiums and slippers. Lothian Publishing Co., Ltd., Melbourne, Victoria, Australia. 172 pp., color and B&W illustrations.
Rentoul, J. N.	1982	Growing orchids. Book two, the cattleyas and other epiphytes. Lothian Publishing Co., Ltd., Melbourne, Victoria, Australia. 218 pp., color and B&W illustrations.
Rentoul, J. N.	1982	Growing orchids. Book three, vandas, dendrobiums and others. Lothian Publishing Co., Ltd., Melbourne, Victoria, Australia. 218 pp., color and B&W illustrations.
Rentoul, J. N.	1985	Growing orchids. Book four, the Australasian families. Timber Press, Portland, OR. 190 pp., color and B&W illustrations.

(continued)

(continued)

Author(s)	Date	Title and publisher
Rentoul, J. N.	1987	The specialist orchid grower.
Rentoul, J. N.	1989	Expanding your orchid collection.
Rentoul, J. N.	1991	Growing orchids, the hybrid story. Timber Press, Portland, OR. 200 pp., numerous color photographs.
Reusch, G.	1963	Orchid corsages & leis. Schaeffer's Tropical Gardens. 33 pp., line drawings and B&W photographs.
Reychler, L.	1928	Die mutation bei den orchideen, durch kreuzungen mit *Cattleya*-mutante.
Reychler, L.	1928	La mutation chezles orchidees.
Reychler, L.	1928	Mutatie bij orchideeen
Reychler, L.	1928	Mutation with orchids, Freaks, Phenomena or Telegony (164 pp., 48 B&W photographs) bound together with an addendum. (61 pp., 25 B&W photographs). Goemaere, Printer to the King, Brussels. There may also be a 1930 version.
Rhodehamel, W. A.	1991	A *Masdevallia* cultural guide. Williams Ames Rhodehamel, 8440 West 82nd Street, Indianapolis, IN 46278, USA.
Richard, A.	1828	Monographie de orchidées de Iles de France et de Bourbon. Mémoires de las Société d'Histoire Naturelle de Paris Volume 4: 1–74, 11 plates of line drawings.
Richard, E. H.	1960	Whys and wherefores of orchid culture. Orchid Digest Corporation. No pagination. Some color and many B&W photographs.
Richard, L. C.	1818	De orchideis Europaeis annotationes. Museum National d'Histoire Naturelle 4: 23–61, plate 5. This is similar or the same as L. C. Richard 1917.
Richard, L. C.	1917	De Orchideis Europaeis annotationes. Ex Typographia A. Belin. 42 pp., one plate of line drawings.
Richards, H., R. Wootton, and R. Datodi	1988	Cultivation of Australian native orchids, 2nd ed. Australian Native Orchid Society Victorian Group, Inc. 96 pp., color photographs and line drawings.
Richter, W.	1958	… die schönsten aber sind orchideen. Neuman Verlag, Radebeul, Germany. 280 pp., numerous color photographs.
Richter, W.	1965	The orchid world, translated and revised by E. Launert, edited by P. F. Hunt. Studio Vista and E. P. Dutton, UK. 292 pp., 100 color plates.
Richter, W.	1971	Orchideen pfelegen, vermehren, züchten. Verlag J. Neumann-Neudamm, Basel. 212 pp., numerous color and B&W photographs and line drawings.

(continued)

(continued)

Author(s)	Date	Title and publisher
Richter, W.	1972	Orchid care, a guide to cultivation and breeding. Studio Vista, London. 212 pp., numerous line drawings and 16 B&W and 16 color photographs. There is also a 1977 printing.
Richter, W.	1987	Das orchideen-jahr. Landbuch-Verlag GmbH, Hannover. 247 pp., a few line drawings, numerous color photographs.
Ridley, H. N.	1907	Materials for a flora of the Malay peninsula, part I. Methodist Publishing House, Singapore. 233 pp.
Ridley, H. N.	1924	The flora of the Malay peninsula, vol IV. Monocotyledons. L. Reeve & Co., London. There is also a 1967 reprint by A. Asher & Co., Amsterdam and L. Reeve & Co., Ashford, Kent, UK. 384 pp., 209 line drawings.
Ridley, H. N.	1895–1897	An enumeration of all Orchideae hitherto recorded from Borneo. Journal of the Linnean Society of London, Botany, Vol. 31, pp. 361–306, a few line drawings. See also Coomans de Ruiter, 1935, Chan, Lamb, Shim and Wood, 1991, Vermeulen, 1991, Wood, 1997.
Riehl, M.	1958	Orchids. Andre Deutsch/Ariel Press, London. 59 pp., 16 color plates.
Rijen, W., van	1979	Orchideeen kweken als hobby.
Riley, J. J., and D. P. Banks	202	Orchids of Australia.
Rittershausen, B&W.	1989	Orchids as house plants. Ward Lock Limited, London. 128 pp., numerous line drawings and color photographs.
Rittershausen, B., and Rittershausen, W.	1970	Popular orchids. Published by Rittershausen, Kingsteignton, Newton Abbot, Devon, UK. Numerous line drawings and B&W photographs. There is also a 1976 edition
Rittershausen, B., and W, Rittershausen	1980	Orchids as indoor plants. Blandford Press, Poole, Dorset, UK.
Rittershausen, B., and W. Rittershausen	1976	Popular orchids. Published by Rittershausen, Kingsteignton, Newton Abbot, Devon, UK. Numerous line drawings and B&W photographs.
Rittershausen, B., and W. Rittershausen	1979	All about orchids.
Rittershausen, B., and W. Rittershausen	1979	Orchids in color. Blandford Press, Poole, Dorset, UK. 192 pp., numerous color photographs.
Rittershausen, B., and W. Rittershausen	1990	Orchids in color. Blandford Press, Poole, Dorset, UK. 192 pp., numerous color photographs. Paperback edition.

(continued)

(continued)

Author(s)	Date	Title and publisher
Rittershausen, B.&W, and A. Cooper	1985	Orchid growing illustrated. Blandford Press, London. 159 pp., numerous B&W and color photographs.
Rittershausen, B., and W.	2004	Growing orchids, successful gardening indoors and out. Southwater, an imprint of Anness Publishing, Ltd., London. 256 pp., numerous color and some B&W photographs.
Rittershausen, B., and W. Rittershausen	2004	The practical encyclopedia of orchids. Lorenz Books, Anness Publishing, Ltd., Hermes House, London.
Rittershausen, P. R. C	1953	Success with orchids. Smithmark Publishers, New York. 127 pp., numerous color photographs.
Rittershausen, P. R. C.	1956	Successful orchid culture.
Rittershausen, W.	1983	Orchideeen zelf kweken.
Rittershausen, W.	1989	Successful indoor gardening with exotic orchids.
Rittershausen, W.	1989	Exotic orchids. HP Books, Price Stern Sloan, Los Angeles. 96 pp., numerous line drawings and color photographs.
Rittershausen, W.	1997	Success with orchids.
Rittershausen, W. (ed.)	1982	Growing your own orchids. Salamander Books, Ltd., London. 160 pp., numerous line drawings and color photographs.
Rittershausen, W. (ed.)	1993	100 years of orchids.
Rittershausen, W., and B. Rittershausen	1999	Orchids, a splendid obsession. Soma Books, 555 De Haro, Str. San Fraancisco. 224 pp., numerous color photographs.
Rittershausen, W., and B. Rittershausen	1999	Orchideeen.
Rittershausen, W., and B. Rittershausen	2001	Orchids, the complete grower's guide. Garden Art Press, Woodbridge, England. 222 pp., 292 color photographs.
Rittershausen, W., and B. Rittershausen	2004	Orchids for every home. The Readers Digest Association, Pleasantville, NY. 192 pp., numerous color photographs.
Rittershausen, W., and E. Crichton	1997	Success with orchids. Smithmark Publishers, New York. 127 pp., numerous color photographs.
Rittershausen, W., G. Oakey, D. Oakey, and N. Sutherland	1993	Growing & displaying orchids, a step-by-step guide. Smithmark Publishers, Inc., New York. 124 pp., numerous color photographs.
Roberts, J. A., C. R. Beale, J. C. Benseler, H. N. McGough, and D. C. Zappi	1995	Cites orchids checklist, Vol 1. Royal Botanic Gardens, Kew. 136 pp.

(continued)

(continued)

Author(s)	Date	Title and publisher
Robinson, R. (ed.)	1983	The 8th Australian orchid conference proceedings. 144 pp., line drawings, maps, B&W and color photographs.
Rodigas, []	1986	Une visite a l'etablissement de l'horticulture internationals.
Rodrigues, J. B.	1996	Iconographie des orchidees du Bresil. 2 volumes
Rodríguez Caballero, R. L., D. E. Mora, M. E. Barahona, and N. H. Williams	1986	Géneros de orquídeas de Costa Rica. Editorial Universidad de Costa Rica.
Rodway, G.	1967	Starting with orchids. 104 pp., line drawings, B&W and color photographs.
Rogers, R. S.	1911	An introduction to the study of South Australian orchids. Government Printer, Adelaide, South Australia. 63 pp., 1 color painting, numerous B&W photographs.
Rogers, R. S.	1922	Orchidaceae in Flora of South Australia edited by J. M. Black. R. E. E. Rogers, Government Printer, Adelaide. 154 pp., 34 text figures, an 9 plates, all of line drawings.
Rogers, R. S.	1933	Some developments in orchidology.
Rolando, I. M.	1998	Orchids, Machu Picchu Pueblo Hotel, field guide. Inkaterra Publishing House, Lima, Peru. 108 pp., 2 maps, numerous color photographs. Bilingual, English-Spanish.
Rolfe, R. A.	1896	Natural hybrid orchids.
Rolfe, R. A., and C. C. Hurst	1909	The orchid stud book. Frank Leslie & co., Lawn Crescent, Kew. 328 pp., numerous B&W photographs.
Rolke, L.	1993	Das praktische orchideen-buch.
Romanes, G. J.	1896	Darwin and after Darwin. Open Court Publishing House. 460 pp., numerous line drawings.
Romero, G, A.	1998	Venezuela, orchid paradise. Armitano Editores, Caracas. 204 pp., 186 color plates.
Romero, G. A. and G. Carnevali Fernández-Concha	2000	Orchids of Venezuela, an illustrated field guide. Armitano Editores, C. A., Caracas. Actually a second edition of Dunsterville and Garay, 1979. 3 volumes. XIX+1156 pages, 1120 line drawings plates by Mr. G. C. K. Dunstervile and others.
Rosenberg, J. (ed.)	2005	An introduction to orchids, 3rd ed. South Florida Orchid Society. 124 pp., numerous color photographs.
Ross-Craig, S.	1971	Drawings of British plants. Part XXVIII. Hydrocharitaceae, Orchidaceae. G. Bell & Sons, Ltd., London. No pagination. 53 orchid plates.
Rosvall, S. and B. Petterson	1951	Gotlands orchideer.

(continued)

(continued)

Author(s)	Date	Title and publisher
Rotherham [], and [] Cady.	1988	Australian orchids in colour.
Roubik, D. W., and P. E. Hanson	2004	Abejas de orquideas de la América tropical-biologia y guía de campo.
Royal Botanic Gardens	1904	Hand list of orchids cultivated by the Royal Botanic Gardens.
Royal Botanic Gardens, Kew	1896	Hand list of orchids cultivated by the Royal Botanic Gardens.
Royal Horticultural Society	1972	Sander's list of orchid hybrids, addendum 1961–1970. lxiv+632 pp. The Royal Horticultural Society, London.
Royal Horticultural Society	1977	Sander's list of orchid hybrids, addendum 1971–1975. lxvii+460 pp. The Royal Horticultural Society, London.
Royal Horticultural Society	1981	Sander's list of orchid hybrids, addendum 1976–1980. lxxxi+567 pp. The Royal Horticultural Society, London.
Royal Horticultural Society	1986	Sander's list of orchid hybrids, addendum 1981–1985. ciii+802 pp. The Royal Horticultural Society, London.
Royal Horticultural Society	1991	Sander's list of orchid hybrids, addendum 1986–1990. lxxxv+779 pp. The Royal Horticultural Society, London.
Royal Horticultural Society	1996	Sander's list of orchid hybrids, addendum 1991–1995. xcvi+1070 pp. The Royal Horticultural Society, London.
Royal Horticultural Society	1999	Sander's list of orchid hybrids, addendum 1996–1998. lxxxviii+731 pp. The Royal Horticultural Society, London.
Royal Horticultural Society	2005	Sander's list of orchid hybrids, 3 year addendum 2002–2004. Royal Horticultural Society.
Royen, P., van	1979	The orchids of the high mountains of New Guinea.
Rugiero, L.	No date	I tesori del Salento, Le orchidee. Alberto Santoro Editore, Lecce. No pagination. Numerous color photographs.
Rupp, H. M. R.	1930	Guide to the orchids of New South Wales. Angus & Robertson, Ltd., Sydney. 157 pp., numerous B&W photographs.
Rupp, H. M. R.	1943	The orchids of New South Wales. National Herbarium, Sydney. 152 pp., numerous line drawings. There is also a 1969 reprint by the National Herbarium, Sydney with supplement by D. J. McGillivray. 177 pp.
Ruppert, J.	1938	Die orchideen des Saarlandes.
Ruschi, A.	1986	Orquideas do Estando do Espirito Santo.
Russell, G.	1980	Anselliana, The proceedings of the Ansellia mini-symposium of August 31, 1980. Transvaal Orchid Society, P. O. Box 2678, Johannesburg, South Africa. No pagination, illustrations.

(continued)

(continued)

Author(s)	Date	Title and publisher
Russenberger, H.	1972	Orchideen des Randens.
Sadovsý, O.	1965	Orchideen in eigenen gerten. BLV Bayerisher Landwitschaftsverlag, Munich. 159 pp., 151 B&W and some color photographs.
Sagarik, R.	1974–1975 (?)	Trends of orchid hybridization in Thailand. Bangkok Flowers Centre Co., Ltd., No. 1/15–16, Petchkasem Road, Bangkae Market, Dhonburi, Bangkok, Thailand. No pagination. 32 color photographs.
Sagarik, R.	1978	Thailand, the world of orchids
Sagarik, R.	No date	Orchids and Thai life.
Saito, K., and R. Tanaka (eds.)	1987	Proceedings of the 12th world orchid conference, March 18–25, 1987, Tokyo. Sevretariat, 12th World Orchid Conference, c/o Odakyu Electric Railway Co, Ltd., 1–8–3 Nishi-Shinjuku, Shinjuku-ku, Tokyo. 296+144 pp., maps, line drawings, B&W and color photographs.
Sakdi Sri Nursery	1966	Orchids catalog, 35 Sucharit 1, Lane, Dusit, Rama V Street, G. P. O. Box 150, Bangkok, Thailand. 60 pp., 11 color and numerous B&W photographs.
Salvado, S.	1996	Orquideas portfolio.
Salyers, J., M. L. Porlick, and B. Brain (eds.)	1971	South Florida Orchid Society Culture Notes. South Florida Orchid Society, 1900 S. W. Avenue, Miami, FL 33129. 86 pp., some line drawings.
Sánchez, C. G.	1930	Las mejores orquideas de Mexico. Bartolome Trucco, Editor. 89 pp., 31 figures, mostly B&W photographs and a few color paintings.
Sander F.	1947	Sanders' complete list of orchid hybrids. Sanders (St. Albans) Ltd., UK. 308 pp.
Sander, ?	1907?	Sanders' orchid guide. Sander & Sons, St. Albans. 256 pp.
Sander, []	1950	Korte wenken voor het kweken van orchideeen.
Sander, D.	1956	Orchids and their cultivation. Further revised edition. Blandford Press, London. 135 pp., fold our chart, 21 colored plates, B&W photographs.
Sander, D.	1969	Orchids and their cultivation, 7th ed. Blandford Press, London. 167 pp., line drawings and colot photographs.
Sander, D.	1970	Orchideen und orchideen-pflege, 2nd ed. Brücke Verlag Kurt Schmersow. 204 pp., numerous color and B&W photographs.
Sander, D.	1979	Orchids and their cultivation, revised edition.
Sander, D. F., and W. J. Wreford	1961	David Sander's one-table list of orchid hybrids, 1946–1960, Vol. 1, xx, 309–578 pp. David Sander's Orchids, Ltd., Selsfield, East Grinstead, Susssex, UK.

(continued)

(continued)

Author(s)	Date	Title and publisher
Sander, D. F., and W. J. Wreford	1961	David Sander's one-table list of orchid hybrids, 1946–1960, Vol. I1, xxii, 579–921, 8 pp. David Sander's Orchids, Ltd., Selsfield, East Grinstead, Sussex, UK.
Sander, D., and E. Cooper	No date	Orchids and their cultivation.
Sander, F.	1901	Sander's orchid guide.
Sander, F.	1906	Sander's orchid guide.
Sander, F.	1888–1894	Reichenbachia. F. Sander & Co., Orchid Growers & Importers. St. Albans. A four volume facsimile edition.
Sanders	1927	Sanders' orchid guide. Sanders, St. Albans, UK. 451 pp.
Sanders	1888–1894	Reichenbachia. F. Sander & Co., Orchid Growers & Importers. St. Albans.
Sander's	1936	Orchids and their cultivation.
Sander's	1951	Orchids and their cultivation. American edition. Sander's Royal Orchid Growers, St. Albans, Herts, UK. 75 pp., 9 color photographs, color plan of greenhouse, 20 BW photographs+ foldout chart of the orchid family.
Sander's	1951	Popular orchid growing.
Sanders St. Albans, Ltd.	1946–1954	Sander's list of orchid hybrids, 3 volumes (1946–1948, 1949–1951, 1952–1954. Sanders (St. Albans) Ltd., Royal Orchid Nurseries, St. Albans, UK.
Sanderson, F. R., and T. A. Yong	1972	Orchids of Singapore. Primary Production Department, Ministry of National Development, Singapore. 15 pp., 9color pictures.
Sandhack, H. U.	1931	Orchideen. Verlagsbuchhandlung Paul Parey, Berlin. 66 pp., 24 line drawing and B&W photographs.
Santapu, H., and Z. Kapadia	1966	The orchids of Bombay. Government of India Press, Calcutta. 239+vi pp., 54 line drawing plates.
Sasayama, S., and Y. Nagano	1961	*Cymbidium kanran*, native of Japan. 148 pp., numerous B&W photographs.
Sasayama, S.	1932	Orchids enumerated. (In Japanese).
Sato. M.	Date ?	Japanese orchids. 328 pp., numerous B&W photographs.
Sauleda, R. P., L. A. Sandow, and C. J. Ochipa	1997	An introduction to orchids. South Florida Orchid Society. 104 pp., numerous color photographs.
Saurenman, D. and V.	1965	*Cymbidium* hybrids and awards. Cymbidium Society of America.
Schelpe, E. A. C. L. E.	1966	An introduction to the South African orchids. Macdonald & Co. (Publishers), London. 109 pp., 47 line drawing figures, 64 color Plates.

(continued)

(continued)

Author(s)	Date	Title and publisher
Schelpe, E. A. C. L. E.	1981	Orchid growing for amateurs. Edson-Clyde Press (Pty) Ltd., South Africa. 68 pp., line drawings, color photographs.
Schelpe, S., and J. Stewart	1990	Dendrobiums, an introduction to the species in cultivation. Orchid Sundries, Ltd., New Gate Farm, Stour Provost, Gillingham, Dorset SB* 5LT, UK. 115 pp., numerous color photographs.
Schlechter, R.	1898	Monographie der Disperideae.
Schlechter, R.	1900	Monographie der Podochilinae.
Schlechter, R.	1904	Pflanzengeographische gliederung der insel Neu-Caledonien. Inaugural-Dissertation.
Schlechter, R.	1911	Contributions to knowledge of the flora of Sumatra. Translated and edited by H. J. Katz and J. T. Simmons from Engler's Beiblatt zu den Botanishen Jahrbüchern No. 104, pp. 1–61. Published and distributed by the Australian Orchid Foundation, 107 Roberts Street, Essendon 3040, Australia. 61+VI typewritten pages.
Schlechter, R.	1915	Die Orchiudeen ihre Beschreibung, Kultur und Züchtung, 1st ed. Verlagbuchhandlung Paul Parey, Berlin. VIII+836 pp., 12 color and 242 B&W plates.
Schlechter, R.	1916	Orchidaceae Perrierianae.
Schlechter, R.	1918	Kritische Aufzählung der bisher aus Zentral-Amerika bekanntgewordenen Orhidaceen. Beihefte der Botanische Centralblatt 36, Abt. II, No. 3: 321–520.
Schlechter, R.	1918	Neurordnung der Angraekoiden Orchidaceen.
Schlechter, R.	1918	Orchidaceen von Zentralamerika.
Schlechter, R.	1919	Orchideologiae Sino-Japonicae prodromu.
Schlechter, R.	1919	Die Orchideenfloren der südamerikanischen Kordillerenstaaten. I. Venezuela. Repertorium Specierum Novarum Regni Vegetabilis. Dahlem bei Berlin, Beihefte. Vol. VI. 100 pp. There is also a 1974 reprint by Otto Koeltz Antoquariat, Koeningstein, Germany.
Schlechter, R.	1919	X. Orchidaceae novae et criticae. Decase LVIII. Additamenta ad Orchideologiam Papuanum. Repertorium Speciarum Novarum Regni Vegetabilis, Vol. 16, pp. 42–47.
Schlechter, R.	1919	XIX. Orchidaceae novae et criticae. Decase LIX-LXIII. Additamenta ad Orchideologiam Papuanum II. Repertorium Speciarum Novarum Regni Vegetabilis, Vol. 18, pp. 103–131.

(continued)

(continued)

Author(s)	Date	Title and publisher
Schlechter, R.	1919	XXXIX. Orchidaceae novae et criticae. Decase LIV. Additamenta ad Orchideologiam Papuanum III. Repertorium Speciarum Novarum Regni Vegetabilis, Vol. 18, pp. 214–219.
Schlechter, R.	1920	Die Orchideenfloren der südamerika-nischen Kordillerenstaaten. II. Colombia. Repertorium Specierum Novarum Regni Vegetabilis. Dahlem bei Berlin, Beihefte, Vol. VII. 301 pp. There is also a 1974 reprint by Otto Koeltz Antoquariat, Koeningstein, Germany. Part 1 was published in 1919.
Schlechter, R.	1921	Die Orchidaceen von Mikronesien.
Schlechter, R.	1921	Die Orchideenfloren der südamerikanischen Kordillerenstaaten. III. Ecuador. Repertorium Specierum Novarum Regni Vegetabilis. Dahlem bei Berlin, Beihefte. Vol. VIII. 172 pp. There us also a 1974 repirnt by Otto Koeltz Antoquariat, Koeningstein, Germany.
Schlechter, R.	1921	Die Orchideenfloren der südamerikanischen Kordillerenstaaten. IV. Peru. Repertorium Specierum Novarum Regni Vegetabilis. Dahlem bei Berlin, Beihefte. Vol. IX. 182 pp. There is also a 1977 reprint by Otto Koeltz Antoquariat, Koeningstein, Germany.
Schlechter, R.	1921	Revision der gattugen Schizochilus Sond und Brachycorythis.
Schlechter, R.	1921	XXXIX. Orchidaceae novae et criticae. Decase LXXI–LXIII. Additamenta ad Orchideologiam Papuanum IV. Repertorium Speciarum Novarum Regni Vegetabilis, Vol. 17, pp. 366–382.
Schlechter, R.	1922	Beitrage zur Orchideenkunde von Zentralamerika. Part I.
Schlechter, R.	1922	Die Orchideenfloren der südamerikanischen Kordillerenstaaten. V. Bolivia. Repertorium Specierum Novarum Regni Vegetabilis. Dahlem bei Berlin, Beihefte. Vol. X. 80 pp. There is also a 1977 repirnt by Otto Koeltz Antoquariat, Koeningstein, Germany.
Schlechter, R.	1923	Beiträge zur Orchideenkunde von Zentralamerika. II. Additamenta ad Orchideologiam Costaricensem. Repertorium Specierum Novarum Regni Vegetabilis. Dahlem bei Berlin, Beihefte. Vol. XIX. 307 pp. There is also a 1973 repirnt by Otto Koeltz Antoquariat, Koeningstein, Germany.

(continued)

(continued)

Author(s)	Date	Title and publisher
Schlechter, R.	1924	Beiträge zur Orchideenkunde von Colombia. Repertorium Specierum Novarum Regni Vegetabilis. Dahlem bei Berlin, Beihefte. Vol. XXVII, 183 pp. There is also a 1980 reprint by Otto Koeltz Antoquariat, Koeningstein, Germany.
Schlechter, R.	1925	Die Orchideenflora von Rio Grande do Sul. Repertorium Specierum Novarum Regni Vegetabilis. Dahlem bei Berlin, Beihefte. Vol. XXXV, 108 pp. There is als a 1980 repirnt by Otto Koeltz Antoquariat, Koeningstein, Germany.
Schlechter, R.	1925	Orchidaceae Perrierianae. Ein Beitrag zur Orchideenkunde del Insel Madagascar. Repertorium Specierum Novarum Regni Vegetabilis. Dahlem bei Berlin, Beihefte. Vol. XXXIII. 331 pp.
Schlechter, R.	1925	Orchidaceae Perrierianae. Ein Beitrag zur Orchideenkunde del Insel Madagascar. Repertorium Specierum Novarum Regni Vegetabilis. Dahlem bei Berlin, Beihefte. Vol. XXXIII. 331 pp. There is also a 1980 reprint by Otto Koeltz Antoquariat, Koeningstein, Germany.
Schlechter, R.	1925	Beitrage zur Orchideenkunde des Amazonas-Gebietes.
Schlechter, R.	1927	Die Orchideen.
Schlechter, R.	1929	Figuren-Atlas zu den Orchideenfloren der südamerikanischen Kordillerenstaaten. Repertorium Specierum Novarum Regni Vegetabilis. Dahlem bei Berlin, Beihefte. Vol. LVII. 142 plates of line drawings, 8 pp. There is also a 1973 repirnt by Otto Koeltz Antoquariat, Koeningstein, Germany.
Schlechter, R.	1932	Blutenanalysen neuer orchideen.
Schlechter, R.	1911–1914	Die Orchidaceen von Deutsch-Neu-Guinea. Repertorium Specierum Novarum Regni Vegetabilis. Dahlem bei Berlin, Beihefte, Band I, 1079 pp., There is also a 1874 repirnt by Otto Koeltz Antoquariat, Koeningstein, Germany.
Schlechter, R.	1922, 1923	Beiträge zur Orchideenkunde von Zentralamerika. I and II. Orchidaceae Powellianae Panamensis and Additamenta ad Orchideologiam Costaricensem. Repertorium Specierum Novarum Regni Vegetabilis. Dahlem bei Berlin, Beihefte. Vol. XVII and XIX. 95 and 307 pp. There is also a 1973 repirnt by Otto Koeltz Antoquariat, Koeningstein, Germany.

(continued)

(continued)

Author(s)	Date	Title and publisher
Schlechter, R.	1923–1928	Figuren-Atlas zu den Orchidaceen von Deutsch-Neu-Guinea. Repertorium Specierum Novarum Regni Vegetabilis. Dahlem bei Berlin, Beihefte, Band XXI, 1079 pp., There is also a 1973 repirnt by Otto Koeltz Antoquariat, Koeningstein, Germany.
Schlechter, R.	1930–1934, 1932	Blütenanalysen neuer Orchideen. Repertorium Specierum Novarum Regni Vegetabilis. Dahlem bei Berlin, Bande I–IV, 81 plates of line drawings and. Part III Afrikanische und madagassische Orchideen. Beihefte, Vol. LXVIII There is also a 1973 repirnt by Otto Koeltz Antoquariat, Koeningstein, German.
Schlechter, R.	1966	The Orchidaceae of the Celebes (1911) and the Orchidaceae of the Island of Celebes (1925). Translation by the Australian Orchid Foundation, 107 Roberts Street, Essendon, Victoria, Australia. Translators: H. J. Katz and J. T. Simmons.
Schlechter, R.	1982	The Orchidaceae of German New Guinea, text and figure atlas. Translation of Die Orchidaceen von Deutsch-Neu-Guinea by R. S. Rogers, H. J. Katz, and J. T. Simmons. The Australian Orchid Foundation, 107 Roberts Street, Essendon, Victoria, Australia. 1180 pp., numerous line drawings, foldout map.
Schlechter, R.	1986	Attempt at a natural new classification of the African Angrecoid orchids. Translated from the German in Beihefte zur Botanisches Centrablatt Vol. 36, Section 11 (1911), pp. 62–181 by H. J. Katz and J. T. Simmons and published by the Australian Orchid Foundation, 197 Roberts Street, Essendon, Victoria 3040, Australia. 105 pp.
Schlechter, R.	1986	Contributions to the knowledge of the flora of New Caledonia. Translated from the German in Engler's Botanische Jahrbucher Vol. 39 (1906), pp. 33–91 by H. J. Katz and J. T. Simmons and published by the Australian Orchid Foundation, 197 Roberts Street, Essendon, Victoria 3040, Australia. 64 pp.
Schlechter, R.	1986	Revision of the orchids of German Samoa. Translated from the German in Regni Vegetabilis, Vol 9 (1910–1911), pp. 82–96, 98–112 by H. J. Katz and J. T. Simmons and published by the Australian Orchid Foundation, 197 Roberts Street, Essendon, Victoria 3040, Australia. 34 pp.

(continued)

(continued)

Author(s)	Date	Title and publisher
Schlechter, R.	1986	The orchids of Micronesia (1914) and The orchids of Micronesia (1921). Translated from the German in Beitrage zur flora Mikronesien und Polynesien, and Engler's Botanische Jahrbucher Vol. 56 (1921), pp. 434–501 by H. J. Katz and J. T. Simmons and published by the Australian Orchid Foundation, 197 Roberts Street, Essendon, Victoria 3040, Australia. 66 pp.
Schlechter, R., and F. C. Hoene	1921	Beitraege zur orchideenkunde Brasiliens. Anexos das Memórias do Instituto de Butantan. 48 pp., 11 plates of line drawings. Bilingual Portuguese and German.
Schmeidt, O.	1996	Eestimaa orhideed. Varrak, Tallinn. 144 pp., numerous color photographs.
Schoser, G.	1966	Pfalenzenkultur mit dem Pfalanzenstrahler OSRAM-L->>Fluora<<. OSRAM GmbH, Berlin. 135 pp., illustrations.
Schoser, G.	1993	Orchid growing basics. Sterling Publishing Co., Ltd., New York. 128 pp., numerous color illustrations.
Schoser, G.	1976	Orchideen, eigenart, schnittblumen, topfkultur, pflege. Falken-Verlag, Niederhausen/ Taunus 63 pp., numerous color photographs.
Schoser, G.	1977	Orchideeen, uiterlijk, snijbloemen, potcultuur, verzorging.
Schoser, G.	1979	Orchideen. Falken Verlag, Niedernhausen, Germany, 96 pp., numerous color photographs. There is also a 1982 edition.
Schoser, G. (ed.)	1975	Der Palemengarten, Vol. 39, No. 1, numerous color and B&W photographs.
Schuiteman, A., and E. F. de Vogel	2000	Orchid genera of Thailand, Laos, Cambodia and Vietnam. Lao-English edition. National Herbarium Nederland, Universiteit Leiden branch, The Netherlands, 111+83 pp., 136 color photographs.
Schulze, M.	1894	Die Orchidaceen Deutschlands, Deutsch-Oesterreichs und der Schweiz. Fra Eugen Köhler's Verlag, Gera-Untermhaus, Germany.
Schultes, R. E.	1960	Native orchids of Trinidad and Tobago. Pergamon Press. 275 pp., numerous line drawings and 22 B&W photographic plates.
Schultes, R. E., and A. S. Pease	1963	Generic names of orchids, 1st ed. Academic Press, New York. 331 pp., line drawings and B&W photographs.

(continued)

(continued)

Author(s)	Date	Title and publisher
Schuster, C.	1931–1943	Orchidacearum iconum index. Feddes Repertorium Specierum Novarum Regni Vegetabilis, Berlin-Dahlem. There is also a 1974 reprint by Otto Koeltz Science Publishers Koenig stein, Germany.
Schweinfurth, C.	1967	Orchid of the Guayana Highland. Memoirs of the New York Botanical Gardens 14 (3) 69–214. each.
Schweinfurth, C.	1958–1961	Orchids of Peru. Fieldiana Botany, Volume 30, Nos. 1–4. 1005 pp., 194 line drawings.
Schweinfurth, C., A. Fernandez Perez, and L. A Garay	1969	La Real expedicion botanica del Nuevo Reino de Granada, volume VIII. Microspermae: Orchidaceae, II. Ediciones Cultura Hispanica, Madrid. 62 pp., approx. 50 color plates, maps, B&W illustrations. On the WWW in 2005 for $300. Book measures 54.7 × 37 cm. See also Perez-Albelaez et al., 1954; Fernandez Perez, 1985, 1995; Ortiz y Valdivieso, 2000.
Schweinfurth, C., A. Fernandez Perez, and R. E. Schultes	1963	La Real expedicion botanica del Nuevo Reino de Granada, volume VII. Microspermae: Orchidaceae, I. Ediciones Cultura Hispanica, Madrid. 64 pp., approx. 50 color plates, maps, B&W illustrations. On the WWW in 2005 for $300. Book measures 54.7 × 37 cm. See also Perez-Albelaez, 1954; Fernandez Perez, 1985, 1995; Ortiz y Valdivieso, 2000.
Schötz, F.	1966	Botanischer garten München. Munich Botanical Garden. 96 pp., several orchid illustrations.
Scott, W.	1906	The florists' manual. Florists' Publishing Company, Chicago, IL. 255 pp., numerous B&W photographs. Has a section on orchids.
Scrugli, A.	1990	Orchidee spontanee della Sardegna. Edizioni Della Torre, Via Contivecchi, Cagliari, Italy. 208 pp., numerous line drawings and color photographs.
Seaton, P., and M. Ramsay	2005	Growing orchids from seed. Royal Botanic Gardens, Kew. 83 pp., numerous color line drawings and photographs.
Seeland, H.	1947	Unser frauenschuh. August Lax, Verlagsbuchhandlung, Hildesheim. 16 pp., 1 line drawing. $25
Segerbäck, L. B.	1983	Orchids of Nigeria.
Segerbäck, L. B.	1992	Orchids of Malaya. A. A. Balkema, Rotterdam, The Netherlands /Brookfield, VT, UA. 168 pp., 226 B&W and 26 color photographs.
Seibels, II, G.	1980	Handbook on orchid photography. American Orchid Society. 83 pp., numerous B&W and color photographs.

(continued)

(continued)

Author(s)	Date	Title and publisher
Seidenfaden, G.	1965–2001	(1) Contribution to the orchid flora of Thailand XIII (1997). (2) Orchid genera in Thailand XIV. Fifty-nine vandoid genera (1988). (3) Orchid genera in Thailand VI. Neottiodeae. (4) Contribution to the orchid genera of Thailand V (ca 1975). (5) The description of Epidendrorum of J. G. König 1791. (6) Orchids in the Flora of the Palnu Hill.
Seidenfaden, G., and C. M. Arora	1982	An Enumeration of the orchids of Northwestern Himalaya.
Seidenfaden, G. and T. Smitinand	1959–1965	The orchids of Thailand. A preliminary check-list.
Seidenfaden, G., and J. J. Wood	1992	The orchids of peninsular Malaysia and Singapore. Olsen & Olsen, 1992. 779 pp., 48 color plates, 316 plates of line drawings, one map.
Selezneva, V. A.	1965	Tropitcheskie I subtropitcheskie orhidei. Nauka Publishing House, Moscow. (In Russian.) 173 pp., numerous line drawings and B&W photographs.
Senghas, K.	1993	Orchids, plants of extremes, contrasts, and superlatives. Paul Parey Scientific Publishers, Berlin. 182 pp., 528 color and 17 B&W illustrations, 83 plates.
Senghas, K. (ed.)	1976	Proceedings of the 8th world orchid conference, April 10–17, 1975. Deutsche Orchideen Gesellschaft e. V. 555 pp., 197 B&W photographs, maps, line drawings, 55 color photographs in 8 plates.
Sessler, G. J.	1978	Orchids & how to grow them. Prentice-Hall, Englewood Cliffs, NJ. 370 pp., B&W and color photographs.
Sezik, E. E.	1984	Orkidelerimiz Türkiye'nin orkideleri. Sandoz Kültür Yayinlari No. 6. Department of Pharmacognosy, Hacettepe University, Ankara, Turkey. 166 pp., numerous maps, B&W and color photographs. (In Turkish.)
Sex, S., and B. Sayers	2004	Ireland's wild orchids
Seybold, S.	1986	Die orchideen von Leonardt Fuchs.
Sguazzin, F., and R. Glerean	1985	Orchidee d'Italia. Carlo Lorenzini Editore, Udine, Italy. 254 pp., numerous line drawings and color paintings.
Sharma, J.	1996	Orchids of India: Commercialization and conservation.
Sharp. W. W.	1970	Orchids of Australia.
Sheehan, M. R.	2002	Orchid nonsense coloring book. American Orchid Society, Del Ray Beach. No pagination. Line drawings.

(continued)

(continued)

Author(s)	Date	Title and publisher
Sheehan, R. M.	1984	A portfolio of Florida orchids. Four plates by this well know orchid artist given to speakers are the 11th World Orchid Conference in Miami, 1984. Plates are magnificent and measure 63.2 × 42.6 cm.
Sheehan, T., and M. Sheehan	1979	Orchid genera illustrated. Van Nostrand Reinhold Company, New York. 207 pp., over 50 drawing, 61 color paintings.
Sheehan, T., and M. Sheehan	1994	An illustrated survey of orchid genera. Timber Press, Portland, OR. 412 pp., numerous line drawings, color painting and maps.
Sheehan, T. J.	2001	Ultimate orchid. The Smithsonian Institution, Washington, DC and The American Orchid Society, Delray Beach FL. 146 pp., numerous color and a few B&W photographs.
Shelag-Sosonka, U. R. (ed.)	1996	Red Data Book of Ukraina. Plants. Kiev, Ukr. Encyclopedia. 595 p. (In Ukrainian.)
Sheviak, C. J.	1974	An introduction to the ecology of Illinois Orchidaceae. Illinois State Museum Scientific Papers XIV, Springfield, Illinois. 89 pp., numerous maps.
Shevtchuk, O. N., A. M. Grodsinskii, T. L. Andrienko, G. P. Kushnir, A. N. Lavrentieva, T. K. Maiko, V. G. Sobko, and T. M. Cherevchenko (eds.)	1983	Ohrana I kultivirovanie orhidei. Abstracts of the All-Union Symposium, Kiev, May 1983. Naukova Dumka, Kiev. 112 pp. (In Russian.)
Shirmer, L., and G. Shirmer	1953	How to raise orchids at home. Tropical Flowerland, Los Angeles, CA. 264 pp., two pages of color photographs, numerous B&W photographs. This book is described as a course. It seems to have inspired the book by Snell and Snell, 1963 which is also on this list.
Shobin, D.	1982	The seeding. A novel. The Linden Press/Simon and Schuster. 223 pp.
Shuttleworth, F. S., H. S. Zim, and G. W. Dillon, revised by A. M. Pridgeon	1989	Orchids, revised edition. A Golden Guide. Golden Press, New York. 160 pp., numerous color illustrations.
Shuttleworth, F. S., H. S. Zim, and G. W. Dillon	1970	Orchids. A Golden Nature Guide. Golden Press, New York. 160 pp., numerous color illustrations.

(continued)

(continued)

Author(s)	Date	Title and publisher
Siegerist, E. S.	2001	Bulbophyllums and their allies. A grower's guide. Timber Press, Portland, OR. 251 pp., 77 color photographs.
Silva, W.	1966	Cultivo de orchideas no Brasil, 64 pp., 5 color paintings of pests and the damage they cause and 1 colot picture and many B&W photographs, in Portuguese.
Silva, W.	1972	Cultivo de orquídeas no Brasil, 2nd revised ed. Livraria Nonel, S. A., São Paolo, Brasil. 98 pp., numerous B&W photographs, 6 color plates.
Silva, J., Y. Sagawa, B. Warne, E. Warne, O. Kirsch, M. Warne, R. Tokunaga, and T. Green	1998	Beginner's handbook for orchid growing in Hawaii. Pacific Orchid Society of Hawaii, P. O. Box 1091, Honolulu, HI 96808. 42 pp., numerous B&W photographs and line drawings.
Silvae, J., and M. Silva	1998	Orchideas natives da Amazonia Brasileiro.
Silverberg, R.	1974	The science fiction bestiary, A collection of stories. Laurel Leag Library, Dell Books, New York. 251 pp.
Simon, H.	1975	The private lives of orchids. J. B. Lippincott Co., Philadelphia. 160 pp., numerous color paintings and several maps.
Singapore Trade Development Board	No date	Singapore orchids.
Singh, K. G. (ed.)	1980	Proceedings of the Third ASEAN Orchid Congress, August 22–26, 1980. Published by the Ministry of Agriculture, Malaysia. 226 pp., some graphs.
Skelsey, A.	1978	Orchids. Time-Life Books, Alexandria, Virginia. 160 pp., line drawings, color diagrams, color paintings and color photographs.
Smirnova, E. S.	1990	Morphology of the shoot systems of orchids. Moscow: Nauka. 208 pp., 41 line drawings. (In Russian.)
Slattery, F.	1995	The blooming years.
Sluys, M., van der, and J. G. Artable	1982	Orquideas de Navarra.
Smith, J. J.	1905–1939	Die orchideen von Java.
Smith, J. J.	1905	Die orchideen von Ambon.
Smith, J. J.	1917	An Interpretation of Rumphius's Herbarium Ambonense-Orchidaceae.
Smith, J. J.	1927	Orchidaceae.

(continued)

(continued)

Author(s)	Date	Title and publisher
Smith, J. J.	1934	New Papuasian orchids. A translation of the text of Neue orchideen Papuasiensis from Beitrage Zur Flora von Papuasian XX in Engler's Botanisches Jahrbuch LXVI, pp. 161–215. Edited and translated by D. F. Blaxell, H. J. Katz and J. Simmons. Published and distributed by the Australian Orchid Foundation, 107 Roberts Street, Essendon 3040, Australia. 61+VI typewritten pages.
Smith, J. J.	1934	Artificial key to the orchid genera of the Netherlands Indies, together with those of New Guinea, the Malay Peninsula and the Philippines.
Smith, J. J.	1909–1934	Nova Guinea-Orchidaceae. 7 volumes.
Smith, J. J.	ca 1938	Icones orchidearum Malesianum.
Smith, N.	1991	*Cymbidium* species. Published by the author at 89 Playford Avenue, Whyala, South Australia, no pagination, 10 pp.
Smith, W. R.	1993	Orchids of Minnesota. University of Minnesota Press, Minneapolis. 173 pp., numerous maps, line drawings and color photographs.
Snell, P., and D. Snell	1963	How to grow orchids at home, volumes I (217 pp., 2 pages of color photographs and numerous B&W photographs) and II (191 pp., numerous B&W photographs). Publishes as a course by World of Orchids, 4546 Hollywood Blvd., Los Angeles, CA 90027. These books seem to have been patterned on the book by Shirmer and Shirmer, 1953 which is also on this list.
Sobko, V. G.	1989	Orhidei Ukraini. Academy of Science Ukrainian RSR, Central Republic Botanical Garden. Naukova Dumka Publishing House, Kiev. 192 pp., Line drawings and color photographs. (In Russian.)
Societe Neo Caledonienne d'Orchidophile	No date	Orchidees indigenes de Nouvelle-Caledonie
Soediono	1984	Perimpunan anggrek Indonesia.
Soediono, N.	No date	Indonesia orchid species. 5 page mimeograph.
Soeryowinoto S. M., and M. Soeryowinoto	1977	Perbanyakan vegetatif pada anggrek. Penerbitan Yayasan Kanisius. Yogyakarta, Indonesia. 93 pp., numerous line drawings and color photographs.
Solander, D. C.	1769	Parts of Plantae Otaheitenses.
Solereder, H., and F. J. Meyer	1930	Systematic anatomy of the monocotyledons. Vol. VI, Microspermae. Translated from German by the Israel Program for Scientific Translations, Jerusalem 1969. 159 pp., line drawings.

(continued)

(continued)

Author(s)	Date	Title and publisher
Solereder, H., and F. J. Meyer	1930	Systematische anatomie der monokotyledonen. Heft VI, Scitamineae-Microspermae. Verlag von Gebruder Borntraeger, Berlin. 212 pp.
Somsak, R.	1992	Thai orchid species.
Soo, R.	1929	Revision der Orchideae-Ophrydineae von Ostasien und dem Himalaya.
Soo, R., von	1932	Die pflanzenareale-die orchideen Europas und des Mediterrangebietes.
South African Tourist Corporation	No date	Orchids of the veld.
Sowerby, J.	1790	English botany or, colored figures of British plants. Published by the author in London.
Soysa, E.	1943	Orchid culture in Ceylon. The Caxton Press, Colombo, Ceylon. Some copies include a pamphlet: "An Introduction to orchid culture," 25 pp., numerous illustrations. 153+xxiii pp.
Soysa, E.	No date	The orchids of Ceylon. Text and a folio of painting.
Sparzek, J. F.	1981	Orchid species source book. Published by the author. 90 pp.
Spatzek, J. F.	1984	Orchid species source book. No publisher 135 pp.
Sprengel, C. K.	1793	Das entdeckte geheimnis der natur im bau und in der befruchtung der blumen. There is also a 1972 reprint by Verlag J. Cramer, Lehre, Germany; Whekdon & Wesley, Ltd., Codicote, Herts, UK; Stecher-Hafner Service Agency, Inc., New York 448 pp.
Springer, A.	1953	Orchideen im zimmer und im garten.
Sprunger, S. (ed.)	1986	Orchids from Curtis's Botanical Magazine. Cambridge University Press. Cambridge, UK.
Sprunger, S., P. J. W. Cribb, and W. T. Stearn	1991	Orchids from the Botanical Register. Two volumes, text and illustrations. Birkhäuser Verlag, Basel.
Squire, D.	2006	Tuinspecialist orchideeën
Szczawinski, A. F.	1959	The orchids of British Columbia, third printing. British Columbia Provincial Museum, Handbook No. 16, Victoria, BC, Canada. 124 pp., B&W illustrations, maps.
Szczawinski, A. F.	1975	The orchids of British Columbia, third printing. British Columbia Provincial Museum, Victoria, Canada. 124 pp., B&W photographs, maps.
Szlachetchenko, L.	2001	Flore du Cameroun. Volume 35. Orchidees 2
Szmant, H. H., and J. Wemple	1976	First symposium on the scientific aspects of orchids, October 25, 1974. Chemistry Department, University of Detroit, Detroit, MI 48221. 164 pp., numerous line drawings, some photographs.

(continued)

(continued)

Author(s)	Date	Title and publisher
St. George, I.	1999	The nature guide to New Zealand native orchids. Godwi – Random House New Zealand, Auckland. 176 pp., numerous color photographs and maps.
Standley, P. C.	1925	Orchid collecting in Central America.
Stanford University	1982	The orchid observed.
Starosta, P., and M. Paul	1998	Orchids. Evergreen/Benedikt Taschen Verlag GmbH, Köln, Germany, 128 pp., numerous color photographs.
Stavrovskaya, L. A.	1998	Cypripedium calceolus. Minsk. Centr "Konkordia". 112 pp. (In Russian.)
Stearn, W. T.	1968	Humboldt, Bonpland, Kunth and tropical American Botany. Verlag von J. Craemer, Lehre, Germany. 159 pp.
Stearn, W. T.	1999	John Lindley 1799–1865, gardener – botanist and pioneer orchidologist. The Antique Collector's Club, Ltd., Woodbridge, Suffolk, UK in association with the The Royal Horticultural Society, UK. 232 pp., numerous B&W and color illustrations.
Stebbins, T. E., Jr.	1999	Martin Johnson Heade. Museum of Fine Arts, Boston and Yale University Press, New Haven, CT. 198 pp., numerous color plates of paintings including some of orchids.
Steenis, C. G. G. J., van	1967	Index to enumeration of the Orchidaceae of Sumatra by J. J. Smith
Stehle, H.	1939	Les Orchidales des Antilles Françaises. Bulletin Agricole de Martinique, Vol. 8, Nos. 2 and 3. One map, some line drawings.
Stein, B.	1892	Stein's orchideenbuch. Reprint 1980 by Brücke-Verlag, Hildesheim, Germany. 604 pp., 184 line drawings in the text.
Stein, B.	1892	Stein's orchideenbuch. Verlag von Paul Parey. 604 pp., 184 line drawings in the text.
Stenzel, K. G. W.	1902	Abweichende Bluten Heimischer Orchideen.
Stewart, J.	1991	Orchids. Timber Press, Portland, OR. 124 pp., numerous color illustrations.
Stewart, J.	1993	Miscellaneous genera. Part Four of Thesaurus Woolwardiae. Missouri Botanical Garden in association with the The Royal Botanic Gardens, Kew. No pagination, 15 color plates. Book measures 42 × 30 cm. See also Chase, 1993; Cribb, 1993; Wood, 1993.
Stewart, J.	1988	Orchids.
Stewart, J.	2000	Orchids, revised edition. Timber Press, Portland, OR. 124 pp., line drawings, color paintings, numerous color photographs.

(continued)

(continued)

Author(s)	Date	Title and publisher
Stewart, J. (ed.)	1992	Orchids at Kew. HMSO, London. Numerous B&W, color photographs and color paintings.
Stewart, J., and B. Campbell	1970	Orchids of tropical Africa. W. H. Allen, London. 117 pp., 45 color plates, 3 maps.
Stewart, J., and B. Campbell	1996	Orchids of Kenya. Timber Press, Portland, OR. 176 pp., numerous color photographs.
Stewart, J., and M. Griffiths	1995	Manual of orchids. Timber Press, Portland, OR. 388 pp., numerous line drawings.
Stewart, J., and E. F. Hennessy	1981	Orchids of Africa, a select review. MacMillan South Africa Pty., Ltd., Johannesburg. 157 pp., 50 color plates.
Stewart, J., H. P. Linder, E. A. Schelpe, and A. V. Hall	1982	Wild orchids of Southern Africa. Macmillan South Africa (Publishers) Pty., Ltd., Johannesburg, South Africa. 307 pp., numerous color photographs, a few line drawings.
Stewart, J., and C. N. (K.) van der Merwe (eds.)	1982	Proceedings of the 10th world orchid conference, September, 11–17, Durban South Africa, South African Orchid Council, P. O. Box 2678, Johannesburg. 324 pp., maps, line drawings, color photographs.
Stewart, J., and W. T. Stearn	1993	The orchid paintings of Franz Bauer. Timber Press, Portland, OR. 160 pp., 63 color paintings.
Steyermak, J. A.	No date	Flora del Auyantepui. Acta Botanica Venezuelica Vol. 2, Nos. 5–8:5–370. Maps, line drawings, B&W photographs.
Stolze, L. R., and G. Esser	1972	Orchideen fur Ihre fensterbank.
Stout, R.	1966	Black orchids. A novel. Pyramid Books, New York. 190 pp.
Strasburger, E.	1884	Neue untersuchingen über dem befruchtungsvorgang als grundlage für theorie der zeugung. Verlag von Gustav Fischer, Jena, Germany. 175 pp., 93 illustrations.
Strauss, M. S.	1976	The physiology of pollination induced phenomena in the Orchidaceae. Ph.D. dissertation, University of California, Irvine.
Su, H.-J.	1985	Native orchids of Taiwan, 3rd ed. Published by Harvest Farm Magazine, Taipei, Taiwan. 276 pp., line drawings and color photographs. Bilingual, Chinese and English.
Summerhayes, V. S.	1951	Wild orchids of Britain. Collins, London. 366 pp., 61 color and 39 B&W color photographs, 19 text figures and 43 maps.
Summerhayes, V. S.	1968	Flora of west tropical Africa. Orchidaceae (Part 1). 236 pp., numerous line drawings.
Summerhayes, V. S.	1968	Wild orchids of Britain, 2nd ed. Collins, London. xviii+366 pp., 61 color and 39 B&W photographs, 19 line drawings and 41 maps.

(continued)

(continued)

Author(s)	Date	Title and publisher
Summerhayes, V. S., and P. J. Cribb	1968, 1984	Flora of Tropical East Africa. Orchidaceae. Two parts.
Sunderman, H.	1970	Europäische und mediterrane Orchideen. Brücke-Verlag Kurt Schmersow, Hannover. 224 pp., numerous B&W and color photographs.
Sunderman, H.	1975	Europäische und mediterrane orchideen. Eine bestimmungsflora. Brücke-Verlag Kurt Schmersow, Hildesheim. 243 pp., numerous B&W and color photographs, line drawings and maps.
Sunset Books	1977	How to grow orchids.
Suringar, W. F. R.	1881	Monstrositeiten van *Cypripedium venustum*.
Suringar, W. F. R.	1884	Monstrositeiten van *Cypripedium insigne*.
Suringar, W. F. R.	1888	Catalogue des Orchidees.
Sweet, H.	1980	The genus *Phalaenopsis*. The Orchid Digest, Inc. 128 pp., numerous line drawings, 11 B&W photographs, 82 color photographs and 4 maps.
Swinson, A.	1970	Frederick Sander: The orchid king. Hodder and Stoughton, London. 252 pp., numerous color and B&W photographs.
Synge, P. M., L. Roper, and J. Plat (eds.)	1960	Proceedings of the third world orchid conference, May 30–June 22, 1960, London. Royal Horticultural Society, London. 443 pp., numerous line drawings, B&W photographs and a few color plates.
Syreishchikov, D. P.	1906	Illustrative flora of Moscow Province. Part 1. Moscow – descriptions and line drawings of orchids found in Central Russia. (In Russian.)
Zachos, E.	2002	Orchid growing for wimps. For Wimps Series, Sterling Publishing Co., New York. 128 pp., numerous color photographs.
Zákrej, J.	1980	Orchidey. Príroda, Bratislava, Chechoslovakia. 197 pp., numerous B&W and color photographs.
Zayas, C. V.	1981	Primeros pasos en el cultivo de orquideas. Published by the author, San Juan, Puerto Rico. 63 pp., numerous line drawings and B&W.
Zelenko, H., and M. Chase	1997	The pictorial encyclopedia of *Oncidium*. ZAI Publications, New York City. 164 pp., numerous color paintings by H. Zelenko, B. D. Zelenko and J. Warshaw.
Ziegenspech, H.	1928	21. Families. Orchidaceae, pp. 1–62 in Lebensgeschichte der Blütenpflanzen Mitteleuropas, Band 1, 4 Abteilung, Bogen 1–4. pp. 1–64, 29 figures of line drawings.
Zimmer, K.	1977	Die labormässige vegetative vermehrung der orchideen. Reprint of pp. 158–192 in F. G. Brieger, R. Maatsch and K. Senghas (eds.), Schlechter, Die Orchideen, 3rd ed.
Zimmermann, A., and R. Dougoud	1959	Orchidees exotiques.

(continued)

(continued)

Author(s)	Date	Title and publisher
Zimmermann, A., and R. Dougoud	1961	Tropische orchideen. Kummerly & Fey, Geographischer Verlag, Bern, Switzerland. 331 pp., line drawings, B&W and color photographs.
Zimmermann, W.	1934	Orchidaceae novae Brasilienses I–VII. Bibliotheca Botanica No. 109. 20 pp., 4 B&W plates, 12 line drawings.
Tahourdin, C. B.	1925	Native orchids of Britain. H. B. Grubb, Ltd., Printers, Croydon, UK. xiv+114 pp., 56 B&W photographic plates.
Taiwan Paphiopedilum Society	1997	*Paphiopedilum* in Taiwan.
Taiwan Paphiopedilum Society	2003	*Paphiopedilum* in Taiwan vol III.
Takabayashi, N.	1978	Miniature orchids. Color books published by Hoikusya, Japan. 160 pp., numerous B&W and color photographs.
Tan, H., and C. S. Hew	1993	A guide to the orchids of Singapore. Published by the Singapore Science Centre, Singapore. 160 pp., numerous color photographs.
Tanaka, R., and H. H. Kamemoto	1963	Tabulation of chromosome numbers of orchids. Parts I and II. Japan Orchid Society, Kobe, Japan. 46 and 11 pp., respectively.
Tanaka, R., and H. H. Kamemoto	1972	A complete tabulation of chromosome numbers in Orchidaceae. Reprint from *The Orchids, Culture and Breeding,* pages 667–773. Seibundo-Shinkosha Co., Ltd. Tokyo.
Tang, H. H. T.	No date	The newest orchid cultivation methods. 281 pp., line drawings, B&W and color photographs. (In Chinese.)
Tatarenko, I. V.	1996	Orhidei Rosyi: jisnenie formi, biologia, voprosi ohranyi (Orchids of Russia: Life forms, biology, strategy of preservation.). Argus Publishing House, Moscow. 208 pp., 45 tables, 71 line drawings. (In Russian.)
Taylor, P.	1979	Orchids. The Color Nature Library/Crescent Books, New York. No pagination, very little text, numerous color photographs.
Teo, C. K. H.	1981	Tropical orchid hybrids. FEP International Sendirian Berhad, Petaling Jaya, Selangor, Malaysia. 160 pp., numerous B&W and color photographs.
Teo, C. K. H.	1985	Native orchids of peninsula Malaysia. Times Books International, Singapore. 119 pp., numerous color photographs.

(continued)

(continued)

Author(s)	Date	Title and publisher
Teo, C. K. H. (ed.)	1980	Collection of articles, Third ASEAN Orchid Congress. Published by the Ministry of Agriculture, Malaysia. 156 pp., B&W and color.
Teoh, E. S.	1978	Orchids, commemorating the golden anniversary of the Orchid Society of South East Asia. Published by the society. 78 numbered pages and about 30 unpaginated pages of advertising. B&W and color photographs. Soft cover edition.
Teoh, E. S.	1980	Asian orchids. Times Books International, Singapore. 287 pp., numerous color photographs. A few line drawings.
Teoh, E. S.	1980	Orchids of Asia. Times Books International, Singapore. 317 pp., numerous color photographs.
Teoh, E. S.	1982	A joy forever. *Vanda* Miss Joaquim, Singaopore's National Flower, 1st ed. Times Books International, Singapore. 72 pp., numerous color and some B&W illustrations.
Teoh, E. S.	1982	A joy forever, *Vanda* Miss Joaquim, Singapore's National Flower, 2nd ed. Times Books International. 80 pp., numerous color photographs.
Teoh, E. S.	2005	Orchids of Asia. Times Editions/Marshall Cavendish, Singapore. 367 pp., numerous color photographs.
Teoh, E. S.	ca 1978–1980s	Orchids, commemorating the golden anniversary of the Orchid Society of South East Asia. Published by the society. 78 numbered pages and about 30 unpaginated pages of advertising. B&W and color photographs. Hard cover edition.
Tergit, G.	1960	Orchids.
The Orchid Society of South East Asia	1978	Proceedings of the symposium on orchidology.
Thielens, A.	1875	Les orchidées de la Belgique et du grand-duché de Luxembourg. Bull. Soc. Ror. Bot. Belgique 12: 26–103.
Thiselton-Dyer, W. T.	1898	Flora of tropical Africa. Orchis on pp. 12–292. L. Reeve & Co., Ltd., Brook, Ashford, Kent, U. K. 595 pp.
Thiselton-Dyer, W. T. (ed.)	1923	Flora Capensis, volume V which includes the Orchidaceae. 332 pp.
Thomale, H.	1954	Die orchideen, 1st ed. Eugen Ulmer Verlag, Stuttgart. 189 pp., color frontispiece, 114 B&W photographs.
Thomale, H.	1957	Die orchideen, 2nd ed. Eugen Ulmer Verlag, Stuttgart. 211 pp., 4 color photographs, 117 B&W photographs.

(continued)

(continued)

Author(s)	Date	Title and publisher
Thompson, P. A.	1980	Orchids from seeds.
Thonglor, []	No date	Orchideen in Thailand.
Thorne, A., and P. Cribb	1984	Orchids of the Solomon Islands and Bougainville, a preliminary report. Mimeograph, Royal Botanic Gardens Kew. 33 pp.
Tibbs, M., and R. Bilton	1990	Growing classic orchids. Sterling Publishing Co., Inc., New York. 96 pp., numerous color photographs.
Tibbs, M., and R. Bilton	1990	Orchids, an illustrated identifier and guide to cultivation.
Tibbs, M., and R. Bilton	1999	Growing classic orchids. Sterling Publishing Co., Inc., New York. 96 pp., numerous color photographs.
Tihonova, M. N.	1971	Krasivotsbetytchie orhidei v orangereyah botanitcheskogo instituta AN SSSR. Nauka Publishing Houise Nauka, Leningrad (St. Petersburg). 17 pp. (In Russian.)
Timbal-Lagrave, E.	1887	Memoires sur quelques hybrides de la familie des orchidees.
Timby, S.	1982–1983	The orchid observed: Five centuries of botanical illustration. This is a catalog of a book exhibition at Stanford university.
Ting, W. P., and T. H. Ho	1970	Diseases and pests of orchids. Ministry of Agriculture and Lands Malaysia. 19 pp., 8 B&W photographs, 15 color pictures.
Tinsley, B.	1983	Singapore green, a history and guide to the botanic gardens. Times Books International. 139 pp., foldout map, color and historical B&W photographs.
Tinsley, B.	1989	Visions of delight, the Singapore botanic gardens through the ages. Singapore Botanic Gardens, Cluny Road, Singapore. 96 pp., many color photographs, a few historical B&W pictures.
Tixier, P.	1966	Flore et végétation orophiles de l'Asie tropicale. Socété d'Édition d'Enseignement Supérieur (SEDES), Paris. 240 pp., 30 figures.
Tlustak, [], and [] Jongepierova-Hlobilova	1990	Orchideje bilych karpat.
Tohonova, M. N.	1971	Krasivotsvetushtie orhidei v oranzheryax botanitcheskovo instituta AN SSSR. Nauka Publishing House, Leningrad Department, Leningrad. 16 pp., 7 B&W photographs. (In Russian.)
Tomlinson, P.	1992	Gardening with the experts: Orchids.
Tomlinson, P. C.	1987	Paphiopedilums, a cultural guide. Wellington Orchid Society, Wellington, New Zealand. 144 pp., color and B&W photographs.
Tomlinson, P. C.	No date	Cymbidiums, a cultural guide. Wellington Orchid Society, Wellington, New Zealand. 48 pp., color and B&W photographs.

(continued)

(continued)

Author(s)	Date	Title and publisher
Tomlinson, P. C.	No date	Cymbidums. A cultural guide, 2nd ed. Wellington Orchid Society, Wellington, New Zealand. 48 pp. numerous line drawings and B&W and color photographs.
Tomlinson, P. C.	No date	*Lycaste* and *Anguloa,* a cultural guide. Wellington Orchid Society, Wellington, New Zealand. 54 pp., color and B&W photographs.
Tomlinson, P. C.	No date	*Oncidium,* a cultural guide. Wellington Orchid Society, Wellington, New Zealand. 100 pp., color and B&W photographs.
Tomlinson, P. C. (ed.)	No date	Odonts. A cultural handbook. Wellington Orchid Society, New Zealand. 12 pp., 10 line drawings, 1 map, 1 B&W photograph.
Torikata, H.	1968	Seed formation and sterile culture of the orchids. Seibundo Shinkosya. 285 pp., numerous B&W photographs. (In Japanese.)
Torikata, H.	1976	Seed formation and sterile culture of the orchids, enlarged edition. (In Japanese.) Seibundo Shinkosya. 324 pp., numerous B&W photographs. (In Japanese.)
Toussaint, P. and E. Klopenstein	1987	Orchidaceae Belgicae with atlas
Traherne, M. J.	1869	Ueber *Ophrys insectifera.*
Treub, M.	1879	Notes sur l'embryogènie de quelques orchidees. Royal Netherlands Academy of Sciences at Johannes Müller, Amsterdam. 50 pp., 8 plates of line drawings.
Troyanovski, (or perhaps Troyanovsky), I.	1913	Kultura orhidei (Culture of orchids). Published by the Moscow Society of Orchid Lovers. 161 pp., 36 illustrations, some of them color paintings. (In Russian.) Published before the communist revolution.
Troyanovski, I.	1912	Orhidei. Published in Russia before the revolution. 161 pages. One color and several black and white illustrations.
Tsitisin, N. V.	1959	Embryological research on angiosperms. Works of the Main Botanical Garden Volume VI, Academy of Sciences, USSR. Collection of articles including some on orchids. (In Russian.)
Tsitsin, N. V. (ed.)	1961	Studies on the biochemical evolution of plants. Works of the Main Botanical Garden Volume VIII, Academy of Sciences, USSR. Collection of articles including some on orchids. (In Russian.)
Tsukamoto, Y., S. Sugiyama, Y. Sakanishi, M. Wakisaka, and S. Hori	1969	Coloured illustrations of roses and orchids. Hoikusha Publishing Co., Ltd., Osaka, Japan. 182 pp., 144 color and 67 B&W photographs of orchids.

(continued)

(continued)

Author(s)	Date	Title and publisher
Tullock, J.	2005	Growing hardy orchids. Timber Press, Portland, OR. 244 pp., map, numerous color photographs.
Turner Ettlinger, D. M.	1976	British & Irish orchids a field guide. The Macmillan Press, Ltd., London. 141 pp., 51 maps, 51 B&W photographs, 12 plates of color photographs.
Turrill., W. B.	1963	Joseph Dalton Hooker, botanist, explorer, and administrator. Thomas Nelson and Sons, Ltd. 228 pp., 25 B&W plates, 3 maps.
Täckholm, V., and M. Drar	1969	Flora of Egypt, Vol IV. Cairo University Press. pp. 432.
Uittien, H.	1939	De volksnamen unser orchideeen.
Upton, W. T.	1969	Growing orchids. Rigby, Ltd., Australia. 64 pp., 8 color pages.
Upton, W. T.	1972	Growing orchids. Rigby, Ltd., Australia. 64 pp., 8 color pages.
Upton, W. T.	1989	*Dendrobium* orchids of Australia. Houghton Mifflin Australia. 237 pp., numerous line drawings and color photographs.
Upton, W. T.	1992	Sarcochilus orchids of Australia. Double U Orchids, P. O. Box 215, West Gosford, NSW 2250, Australia. Numerous color illustrations. 119 pp.
Vacherot, M.	1954	Les orchidées. J. B Baillière et fils Éditeurs, Paris. 268 pp., 84 B&W photographs.
Vacherot, M.	1957	Charme et diversité des orchidées. Éditions J. B. Baillière et Fils, Paris. 68 pp., numerous B&W and few color illustrations.
Vacherot, M.	1973	Les orchidees.
Vaddhanaphuti, N.	1992	A field guide to the wild orchids of Thailand.
Vaddhanaphuti, N.	1997	A field guide to the wild orchids of Thailand. Silkworm Books, Chiang Mai, Thailand. 160 pp., numerous color photographs.
Vaddhanaphuti, N.	2005	A field guide to the wild orchids of Thailand, 4th ed. Silkworm Books, Chiang Mai, Thailand. 272 pp., numerous color photographs.
Waes, J., van	1984	In vitro studies van de kiemings – fysiologie van westeuropese orchideeen. Thesis.
Vahrameeva, M. G., L. V. Denisova, S. V. Nikimina, and S. K. Samsonov	1991	Orhidei nashei stranyi. Nauka, Moscow. 224 pp., 128 line drawings, numerous color photographs. (In Russian.)
Valdivieso, P. O.	1976	Orchideas de Colombia, 1st ed. Universidad Javeriana, Bogota. 233 pp., numerous line drawings and color photographs.
Vales, [], and L. Vales	1988	Orchidees de Tunisie.

(continued)

(continued)

Author(s)	Date	Title and publisher
Walford, J. F., and P. F. Hunt	1972	A book of orchid paintings by James H. Walford. J. Walford Orchid Books, Steep Park, Crowborough, Sussex, UK, Distributed by the Medici Society, Ltd., London. 104 pp., 24 plates. The book measure 48.5 × 34.5 cm. A total of 525 copies were published, 490 of these were signed by the artist.
Válka Alves, R. J.	1991	Field guide to the orchids of the Serra de São José M. G. Brazil. Tropicaleaf, Brazil. 148 pp., numerous line drawings and color photographs.
Wallace, A. R.	1869	The Malay Archipelago. Reprint 1987 by Graham Brash (Pte) Ltd., Singapore. 515 pp. + 3 pp. of maps and numerous line figures.
Wallace, E. J. R.	1951	The orchids of Maine. University of Maine Bulletin, Vol. 53, No. 12, University of Maine Studies, Second Series, No. 65. 80 pp., 9 plates of line drawings. Related works are: Bean, R. C. 1922. Some Maine orchids. The Maine Naturalist, Journal of the Knox Academy of Arts and Sciences. Vol. 2, No. 1, pp. 79–81. Fassett, N. C. 1922. Orchids collected at Ocean Point, in the town of Boothbay, Lincoln County, Maine. The Maine Naturalist, Journal of the Knox Academy of Arts and Sciences. Vol. 2, No. 1, pp. 81. Nylander, O. O. 1922. Life history of Calypso. The Maine Naturalist, Journal of the Knox Academy of Arts and Sciences. Vol. 2, No. 1, pp. 82–83. Nye, H. A. Orchids of Fairfield and neighboring towns. The Maine Naturalist, Journal of the Knox Academy of Arts and Sciences. Vol. 2, No. 1, pp. 144–146. Perkins, A. E. 1922. Orchids in Berwick, York County. The Maine Naturalist, Journal of the Knox Academy of Arts and Sciences. Vol. 2, No. 1, pp. 146.
Valmayaor, H. L., and D. Baldovino	1984	Orchidiana Philippiniana, Vols I and II. Eugenio Lopez Foundation, Manila.
Valmayor, H. L.	No date	Hybridization with Philippine orchids. 9 page mimeograph.
Valmayor, H. L. (ed.)	1981	The complete writings of Dr. Eduardo A. Quisumbing on Philippine orchids. Eugenio López Foundation, Inc. Metro Manila, Philippines. Volume I (IV+534 pp., numerous B&W and color illustrations), Vol II (258 pp., numerous B&W and color illustrations).
Walters, S. M. et al. (eds.)	1984	The European garden flora, Vol. 2. Juncaceae to Orchidaceae. Cambridge University Press, Cambridge. 318 pp., some line drawings.

(continued)

(continued)

Author(s)	Date	Title and publisher
van Bodegom, J.	1985	Some orchids of New Guinea. Translated from Einige orchideeën van west Nieuw Guinea by G. Nieuwenhoven and published by the Australian Orchid Foundation, 197 Roberts Street, Essendon, Victoria 3040, Australia.183+XXI pp.
van der Cingel, N. A.	1995	An atlas of orchid pollination. European orchids. A. A. Balkema, Rotterdam, The Netherlands. 175 pp., numerous color photographs and line drawings.
van der Cingel, N. A.	2001	An atlas of orchid pollination. America, Africa, Asia and Australia. A. A. Balkema, Rotterdam, The Netherlands. 296 pp., numerous color photographs and line drawings.
van der Pijl, L. and C. H. Dodson	1966	Orchid flowers, their pollination and evolution. University of Miami Press, Coral Gables, FL. 214 pp., 109 line drawings and B&W photographs, 12 multi picture color photographs.
van der Star, J. A.	1932	Het kweeken van orchideeën door den liefhebber. N. V. Haagsche Drukkerij en Uitgevers Maatschapij z. o. Buitensingel, Hork Pletterijkade en Spui, Den Haag. 31 pp., 6 B&W photographs.
van Royen, P.	1979	The alpine flora of New Guinea, volume 2, Cupressaceae to Poaceae. J. Cramer, Vaduz, Germany. 1232 pp., 392 line drawing figures, 112 B&W photographs.
van Royen, P.	1983	The genus *Corybas* (Orchidaceae) in its eastern area. J. Cramer, FL-9400 Vaduz, 179 pp., 49 line drawings.
Wan, M.	2005	Deadly slipper. A novel. Doubleday, Random House, New York. 301 pp.
Wanigatunga, L. S.	No date	How I grow my dendrobiums. No publisher. 9 pp., two line drawings.
Ward, D. B.	1968	Checklist of the vascular flora of Florida. Part I. Bulletin 726 Technical, Institute of Food and Agricultural Sciences, Agricultural Experiment Stations, University of Florida, Gainesville, FL.
Warming, E.	1883–1884	Orchideae, parts 29, 30, pp. 841–862 + plates IV–XI of line drawings, some partially colored in Symbolae ad floram Brasiliae centralis Cogniaux.

(continued)

(continued)

Author(s)	Date	Title and publisher
Warner, G. C., and D. Sigerson	1940	Garden Handbook for vitamin B1. Published by the authors, Pasadena, CA. 79 pp. 10 photographs. The story is that Warner was a technician for James Bonner at CalTech, faked vitamin B1 experimental results to make the vitamin look good, quit his job and used research papers which were based on the faked data to publicize his vitamin B1 sales company. It not known if this story is true. Some people use this vitamin on orchids.
Warner, R.	1862–1865	Select Orchidaceous plants. There is also a 1975 reprint 5 of volume 1 (40 colored plates), reduced in size (9 × 12.24 in.) by Rare Editions, Ltd.
Warner, R., and B. Williams	1862	Select Orchidaceous plants. Reprint 1970 of 10 plates by Kingsmead Reprints, Bath, UK. Reprint measures 43.7 × 32 cm.
Warner, R., and B. S. Williams	1888	The orchid album, colored figures and descriptions of new, rare and beautiful orchidaceous plants. Colored figures by John Nugent Fitch. Volumes I–XI. B. S. Williams at the Victoria and Paradise Nurseries, Upper Holloway, UK.
Warren, R.	1994	Orchids of the high mountain Atlantic rain forest in southeastern Brazil.
Vásquez Ch. R., and C. H. Dodson	1982	Orchids of Bolivia. Icones Plantarum Tropicarum fascicles 6, Marie Selby Botanical Garden, Sarasota, FL. Line drawings plates 501–600 and the same number of maps.
Vásquez Ch. R., and P. L. Ibisch.	2000	Orquideas de Bolivia/Orchids of Bolivia. Spanish/English. Editorial F. A. N., Santa Cruz de la Sierra, Bolivia. 555 pp., numerous line drawings, maps, graphs, color photographs.
Vásquez, R., and P. L. Ibisch	2004	Orchídeas de Bolivia, Vols I–II
Watanabe, M.	ca 1980s	Quality stream of Cattleya. Jusekisha, Japan. 248 pp., numerous color and some B&W photographs, maps. (In Japanese.)
Waters, V. H, and C. C. Waters	1973	A survey of the slipper orchids. Carolina Press, Printing and Packaging, Inc., Shelby, NC 28150. 145 pp., numerous line drawings and B&W and color photographs.
Watkins, J. V.	1948	ABC of orchid growing. Ziff-Davis Publishing Co., Chicago, IL. 134 pp., numerous B&W photographs.

(continued)

(continued)

Author(s)	Date	Title and publisher
Watkins, J. V.	1949	ABC of orchid growing, revised edition. Ziff-Davis Publishing Co., Chicago, IL. 134 pp., numerous B&W photographs.
Watkins, J. V.	1956	ABC of orchid growing, 3rd ed. Prentice-Hall, Englewood Cliffs, NJ. 190 pp., numerous B&W photographs.
Watson, D	1971	Simple orchid culture. Circular 452, University of Hawaii, Cooperative Extension Service. 21 pp., line drawings.
Watson, J. (ed.)	1998	Growing orchids. American Orchid Society. Delray Beach, FL. 108 pp., numerous color photographs.
Watson, J. (ed.)	2002	Orchid pests and diseases. American Orchid Society, Delray Beach, FL, numerous B&W and color photographs.
Watson, J. B.	2004–2006	Orchid source directory. American Orchid Society, Delray Beach, FL.
Watson, S. W.	1969	The Australian house and garden book of orchids.
Watson, W., and H. J. Chapman	1903	Orchids: their culture and management. L. Upcott Gill, London. 559 pp., numerous color plates and line drawings.
Watson, W., and W. Bean	1893 (?)	Orchids: Their culture and management. L. Upcott Gill, London. 554 pp., numerous color plates and line drawings.
Watsona, U., and Y. Bina	1892	Orhidei yhod za nimi I ih soderzhaniv. Published by Sad I Ogorod, Moscow. 434 pp., spiral bound in three volumes. Translated into Russian by P. E. Volkensteina.
Webb, A., and M.	1993	Orchids for beginners. Graphic World Pty., Ltd., Lewisham, NSW, Australia. 47 pp., numerous color and B&W photographs and illustrations.
Webber, G.	1978	Anyone for orchids? Schiffer, Ltd., Box E. Exton, PA 19341. 234 pp., numerous line drawings and B&W photographs and some color photographs.
Webster, A., and A. Gloudon	1977	Caribbean orchid growing. Webdon Publishing Co., Ltd., Kingston, Jamaica. 88 pp., B&W photographs.
Webster, P.	1992	The orchid genus book. Published by the author. Confusing pagination.
Weinberg, H. G., and J. Leinfelder	1967	Handbuch der Kohlendioxid-Düngung durch verbrennen von Propan in gewächshäusern. BP Benzin und Petroleum AG, Hamburg. 54+additional pages plus reprint of carbon dioxide fertilization.

(continued)

(continued)

Author(s)	Date	Title and publisher
Weinert, M.	1989	Bibliographie über vermehrung, kultur and mykorrhiza terrestrischer orchideen (inbesondere der europäischen arten) 1804–1988. Bibliography on propagation, culture and mykorrhiza of terrestrial orchids (especially of European species) 1804–1988. Koeltz Scientific Books D-6240 Königstein, Germany. 225 pp.
Veitch, J.	1887–1894	Manual of orchidaceous plants.
Velísek, V.	1981	Flora Olomouc, Orchideje. Vidala Tisková, Prague. No pagination. Numerous color photographs.
Wellington (New Zealand) Orchid Society	No date	Cymbidiums, a cultural guide.
Wellington (New Zealand) Orchid Society	No date	*Lycaste* and *Anguloa,* a cultural guide.
Wellington (New Zealand) Orchid Society	No date	Odonts. A cultural handbook.
Wellington (New Zealand) Orchid Society	No date	Paphiopedilums, a cultural guide.
Wells, T. C. E., and J. H. Willems (eds.)	1991	Population ecology of terrestrial orchids. SPB Academic Publishing bv., The Hague. 189 pp., numerous line drawings, graphs and tables.
Wendelberger, E.	1969	Das kleine orchideenbuch, 3rd ed. Umschau Verlag, Frankfurt/Main, Germany and Pinguin Verlag, Innsbruck/Tirol, Austrial. 89 pp., color paintings, line drawings.
Werkhoven, M. C. M.	1986	Orchideeën van Suriname/Orchids of Suriname. Bilingual Ditch/English. Vaco Uitgevermaatschppij, Paramaribo, Suriname. 256 pp., 180 color photographs and numerous maps.
Vermeulen, J. J.	1991	Orchids of Borneo, Vol 2. The Sabah Society and Royal Botanic Gardens, Kew. 342 pp., numerous line drawings in 100 figures and color photographs in 19 plates, a few maps. See also Ridley, 1895–1897, Coomans de Ruiter, 1935, Chan, Lamb, Shim and Wood, 1991, Wood, 1997.
Vermeulen, J. J.	1993	A taxonomic revision of *Bulbophyllum* sections Adelopetalum, Lepanthanthe, Macrouris, Pelma, Peltopus, and Uncifera (Orchidaceae). Inaugural-Dissertation

(continued)

(continued)

Author(s)	Date	Title and publisher
Vermeulen, P.	1947	Studies on Dactylorchids. Drukkerij Fa Schotanus & Jens, Utrecht. 181 pp., line drawing plates A–P, 8 photographic plates.
Vermeulen, P.	1958	Orchidaceae. In: T. Weevers, B. H. Danser, and J. Hermans (eds.), Flora Neerlandica, Koninklijke Nederlandse Botanische Vereniging, Amsterdam. 127 pp.
Wheeler, W. O	No date	Orchid culture. American Orchid Society. 16 pp., B&W photographs and line drawings.
White, E. A.	1927	American orchid culture, 1st ed. A. T. deLa Mare, Inc., New York. 227 pp., numerous color painting plates and B&W photographs.
White, E. A.	1939	American orchid culture. 2nd ed. A. T. De La Mare Company, Inc., New York. Numerous B&W and some color illustrations.
White, E. A.	1942	American orchid culture, 3rd ed. A. T. De La Mare Company, Inc., New York. Numerous B&W and some color illustrations. 276 pp.
White, E. A.	1947	American orchid culture, 3rd ed., fourth printing. A. T. De La Mare Company, Inc., New York. Numerous B&W and some color illustrations. 276 pp.,
White, J.	1996	Taylor's guide to orchids. Houghton Mifflin C., Boston. 385 pp., numerous color photographs, some line drawings.
White, J. H.	1907	On the polystely in roots of Orchidaceae.
White, K., and B. Sharma	2000	Wild orchid Nepal. White Lotus Press, Ltd., Bangkok. 307 pp., numerous color photographs.
White, W. H.	1902	The book of orchids, 1st ed.. John Lane, the Bodley Head, London. 118 pp., 11 B&W plates.
White, W. H.	1911	The book of orchids, 2nd ed. John Lane, the Bodley Head, London. 100 pp., 15 B&W plates.
Whiting, R. E., and P. M. Catling	1986	Orchids of Ontario, The CanaColl Foundation, Ottawa. 168 pp., 36 B&W illustrations, 60 maps.
Wiard, L. A.	1987	An introduction to the orchids of Mexico. Comstock Publishing Associates/Cornell University Press, Ithaca, NY. 239 pp., color photographs, maps.
Wickham, L., J. Hopley and S. Hopley	1957	Orchid for everybody. Robert M. McBride Co., Inc. New York. 63 pp., one color and 32 (very poor) B&W photographs.

(continued)

(continued)

Author(s)	Date	Title and publisher
Vij, S. P.	1986	Biology, conservation, and culture of orchids. Affiliated East-West Press Private Limited, New Delhi. 492 pp.
Vij, S. P. (ed.)	1985	National Seminar on the biology, conservation & culture of orchids. Orchid Society of India, Botany Department, Panjab University, Chandigarh, India. 56 pp.
Vij, S. P. (ed.)	1988	National Seminar on current research trends in Indian orchids. Orchid Society of India, Botany Department, Panjab University, Chandigarh, India. 41 pp.
Williams, B (ed.)	1979	All about orchids. Orchid Review, Ltd. 75 pp., B&W and color photographs.
Williams, B. S.	1852	The orchid grower's manual, 1st ed. Chapman & Hall, London. 108 pp., one color painting.
Williams, B. S.	1862	The orchid grower's manual, 2nd ed. Chapman & Hall, London. 160 pp., one color painting.
Williams, B. S.	1868	The orchid grower's manual, 3rd ed. Victoria and Paradise Nursey, London. 251 pp., No color paintings.
Williams, B. S.	1871	The orchid grower's manual, 4th ed. Victoria and Paradise Nursery, London. 300 pp., no color painting, numerous line drawings.
Williams, B. S.	1877	The orchid grower's manual, 5th ed. Victoria and Paradise Nursery, London. 336 pp., one color painting, numerous line drawings.
Williams, B. S.	1885	The orchid grower's manual, 6th ed. Victoria and Paradise Nursery, London. 660 pp., no color painting, numerous line drawings.
Williams, B. S.	1852	Orchid grower's manual. Chapman & Hall, London. 108 pp.
Williams, B. S., and H. Williams	1894	The orchid grower's manual, 7th ed. Victoria and Paradise Nurseries, London. 797 pp., numerous line drawings. Reprint 1971 by Wheldon and Wesley, Codicote/Herts, UK, Hafner Publishing Co., New York and J. Cramer, Weinheim, Germany.
Williams, B., P. Dumbleton, R. Bilton, W. Rittershausen, D. Stead, P. Phillips, K. Andrew, and A. Greatwood	1980	Orchids for everyone. Salamander Books, London. 208 pp., numerous line drawings and color photographs.
Williams, J. G., A. E. Williams, and N. Arlott	1979	Elseviers orchideengids.

(continued)

(continued)

Author(s)	Date	Title and publisher
Williams, J. G., A. E. Williams, and N. Arlott	1983	A field guide to orchids of North America. Universe Books, New York. 143 pp., 229 species illustrated in color paintings.
Williams, J. O., A. F. Williams, and N. Arlott	1978	A field guide to the orchids of Britain and Europe with North Africa and the Middle East. Collins Publishing House. London. 176 pp., illustrations in color of 245 species.
Williams, L. O.	1942	Contributions towards a flora of Panama. Orchidaceae. *Annals of the Missouri Botanical Garden* Vol. 29, No. 4, pp. 336–350, 5 plates of line drawings.
Williams, L. O.	1956	An enumeration of the Orchidaceae of Central America, British Honduras and Panama.
Williams, L. O.	1951	Orchidaceae of Mexico.
Williams, L. O., P. H. Allen, and R. L. Dressler	1980	Orchids of Panama with a Checklist of the orchids of Panama. Missouri Botanical Garden, St. Louis. Facsimile edition with ddendum.590+XXVI, 213 line drawing plates.
Williams, T. R.	1993	Developing the orchid orchard. A private report. 20 pages, color photographs.
Williamson, G.	1977	The orchids of south central Africa. J. M. Dent & Sons, Ltd., London. 237 pp., 120 line drawing figures, 185 color plates.
Willing, B., and E. Willing	1977	Bibliographie über die orchideen Europas und der Mittlemeerländer 1744–1976. Willdenowia Beiheft 11., 325 pp.
Willoughby, A. C.	1950	Orchids and how to grow them. Oxford University Press, New York. 135 pp., color plate frontispiece, 16 pages of B&W photographs.
Wilshaw, H. S.	1982	Orchid stamp collecting, volume I 1935–1981. Published by the author in Britain. 56 pp., numerous color photographs of postage stamps.
Wilson, E. H.	1927	Plant hunting. The Stratford Company, Publishers, Boston, MA. 276 pp., numerous B&W photographs.
Wilson, K. S.	1994	Descriptive terminology for the orchid judge. Published by the author. 66 pp.
Wilson, M.	1957	A Bibliography of South African Orchids.
Winkler, M.	1978	A casual stroll through Siam's orchid world. Thai Watana Panich Press Co., Ltd., Bangkok, Thailand. 96 pp., numerous color paintings,. Book measures 37 × 24.5 cm.
Winkler, M.	1985	Forays of a self styled orchid stalker. D. D. Books, Bangkok, Thailand. 16 color paintings, line drawings and text.
Winterringer, G. S.	1967	Wild orchids of Illinois. Illinois State Museum. 130 pp., numerous line drawings and B&W photographs, one color photograph.

(continued)

(continued)

Author(s)	Date	Title and publisher
Vishnyakov, V. A.	1914	Orhidei. (In Russian.) Published by A. F. Devriena, St. Petersburg before the communist revolution. 84 pp., 32 line drawing and B&W photographs.
Wisler, G. C.	1989	How to control orchid viruses. Maupin House Publishers, Gainesville, FL. 120 pp., 24 figures.
Withner, C. L.	1981	Le Cattleye. Edizioni Lativa. 83 pp., numerous color photographs.
Withner, C. L.	1988	The cattleyas and their relatives. Volume I. The cattleyas. Timber Press, Portland, OR. 146 pp., line drawings and 77 color pictures.
Withner, C. L.	1990	The cattleyas and their relatives. Volume II. The laelias. Timber Press, Portland, OR. 154 pp., line drawings and 93 B&W and color pictures.
Withner, C. L.	1993	The cattleyas and their relatives. Volume III. Schomburgkia, Sophronitis and other South American Genera. Timber Press, Portland, OR. 154 pp., line drawing, 17 line drawing plates and color pictures.
Withner, C. L.	1996	The cattleyas and their relatives. Volume IV. The Bahamian and Caribbean species. Timber Press, Portland, OR. 154 pp., line drawing, 17 line drawing plates and 83 color pictures.
Withner, C. L.	1998	The cattleyas and their relatives. Volume V. Brassavola, Encyclia and other general of México and Central America. Timber Press, Portland, OR. 198 pp., line drawing, several line drawing plates and 95 color pictures.
Withner, C. L.	2000	The cattleyas and their relatives. Volume VI. The South American Encyclia species. Timber Press, Portland, OR. 194 pp., line drawing, several drawing plates and 72 color pictures.
Withner, C. L. (ed.)	1959	The orchids, a scientific survey. Ronald Press Co., New York. 648 pp., line drawings and B&W photographs.
Withner, C. L. (ed.)	1974	The orchids, scientific studies. John Wiley & Sons, New York. 604 pp., line drawings and B&W photographs.
Withner, C. L., and P. A. Harding	2004	The cattleyas and their relatives. The debatable epidendrums. Timber Press, Portland, OR. 300 pp., line drawing, several drawing plates and 98 color pictures.
Withner, C., and J. H. G. Vella	No date	A book of orchids. No pagination. Numerous color paintings.
Witte, H.	1862	Enumeration des orchidees.
Vivarelli, W., and U. Fusini	1995	Orchidee spontanee dell'appennino bolognese. Savena Setta Sambro Editore, Monzuno, Italy. 64 pp., 30 color photographs.

(continued)

(continued)

Author(s)	Date	Title and publisher
Vlčko, J., D. Ditě, and M. Kolnik	2003	Vstavačovite Slovenska. Orchids of Slovakia. ZO SZOPK orchidea, Zvolen. 120 pp., color photographs.
Wodrich, K. H. K.	1997	Growing South African indigenous orchids. A. A. Balkema, Rotterdam, The Netherlands/ Brookfield, VT, USA. 253 pp., Frontispiece: color map of South Africa, numerous B&W photographs. 120.5.
Vogel, E. de	1969	Monograph of the tribe Apostasieae (Orchidaceae).
Vogel, E., de, A. Schuiteman, N. Feleus and A. Vogel	1998	Hortus Botanicus Leiden catalog. Part I. Orchidaceae.
Vogel, S.	1959	Organographie der Blüten kapländischer Ophrydeen. I. Disinae und Satyriinae. Akademie der Wissenschaften unde der Literatur in Mainz. Volume for the year 1959, No. 6, pp. 267–401, 74 line drawings.
Vogel, S.	1959	Organographie der Blüten kapländischer Ophrydeen. I. Disperidinae. Akademie der Wissenschaften unde der Literatur in Mainz. Volume for the year 1959, No. 7, pp. 405–532, 65 line drawings.
Vogel, S.	1990	The role of scent glands in pollination. Smithsonian Institution Libraries and The National Science Foundation, Washington, DC., 203 pp., 50 figures (line drawings and photographs).
Vogelpoel, L.	No date	*Disa uniflora*, its propagation and cultivation. Disa Orchid Society of South Africa. c/o Disa Distributors, P. O. Box 19, Bergvliet, Cape Town or Vogelpoel, 7 Sunnybrae Rd., Rodenbosch, South Africa.
Vogt, A.	1973	Probleme des artenschutzes in Baden-Wurttemberg. 96 pp., maps, B&W photographs, color illustrations.
Wolff, H.	1926	Zur physiologie des wurzelpilzes von *Neottia nidus avis* Rich. und einigen grunen orchideen. Inaugural-Dissertation
von Sachs, J.	1890	History of botany, translated H. E. F. Garnsy and revised by I. B. Balfour. First printing of the first edition. Has sections on orchid scientists.
von Sachs, J.	1906	History of botany, translated H. E. F. Garnsy and revised by I. B. Balfour Second impression. The Clarendon Press, Oxford. 568 pp. Has sections on orchid scientists.
Wood, H. P.	2006	The dendrobiums. A. R. G. Gantner Verlag, K. G., Ruggell, Lichtenstein.

(continued)

(continued)

Author(s)	Date	Title and publisher
Wood, J. J.	1993	*Dendrobium*. Part Two of Thesaurus Woolwardiae. Missouri Botanical Garden in association with the The Royal Botanic Gardens, Kew. No pagination, 15 color plates. Book measures 42 × 30 cm. See also Cribb, 1993; Chase 1993; Stewart, 1993.
Wood, J. J.	1997	Orchids of Borneo, Vol 3. The Sabah Society and Royal Botanic Gardens, Kew. 299 pp., numerous line drawings in 100 figures and color photographs in 26 plates, a few maps. See also Ridley, 1895–1897, Coomans de Ruiter, 1935, Chan, Lamb, Shim and Wood, 1991, Vermeulen, 1991.
Wood, J. J., R. S. Beaman, and J. H. Beaman	1993	The plants of Mount Kinabalu. Royal Botanic Garden, Kew. 411 pp., 84 color plates, 57 line drawing plates.
Wood, J. J., and P. Cribb	1994	A checklist of the orchids of Borneo.
Woods, S.	1998	Orchid beach. A novel. Harper Paperbacks. New York. 406 pp.
Woods, S.	2003	Blood orchid. A novel. Signet Books. New York. 357 pp.
Woodson, R. E., R. W. Schery, et al.	1946	Flora of Panama, Orchidaceae. Stechert Hafner reprint 1965 of Annals of the Missouri Botanical Garden, Vol. 33, Nos. 1 and 4, and Vol. 36, Nos. 1 and 2. Numerous line drawings.
Woodward, C. H. (ed.)	1967	Handbook on orchids, Plants and Gardens vol. 23, No. 2. 80 pp., numerous B&W and color photographs.
Woolcock, D. T., and C. E. Woolcock	1984	Australian terrestrial orchids. Thomas Nelson Australia, Melbourne, Victoria, Australia. 154 pp., 72 plates of color paintings each containing more than one species and also numerous line drawings.
Woolward, F.	1896	The genus *Masdevallia*. Issued by the Marquess of Lothian, K. T., Newbattle Abbey. No pagination. 87 color plates and 61 woodcuts. Facsimile edition, reduced in size, no date and no publisher.
Woolward, F. H. (with notes by F. C. Lehmann)	1890–1896	The genus *Masdevallia*. Issued by the Marquess of Lothian London. Bound volume (46 × 33 cm, one double-page chromolithographed map, 87 (1 double-page) handcolored d plates and 61 wood engravings in the text. Edition was limited to 250 copies.
Wright, N. P.	1958	Orquideas de Mexico. Editorial Fournier, S. A., Mexico. No page numbers, 23 pp., 40 color plates, each accompanied by text.
Wu, Y.-S.	1991	Chinese *Cymbidium*. Published in China by the China Forestry Publishing House.

(continued)

(continued)

Author(s)	Date	Title and publisher
Vuylsteke, [], and [] Vuylsteke	1995	Fine fleur van de belgische sierteelt.
Wylie, P.	1958	Orchids for home and garden.
Yam, T. W.	1995	Orchids of the Singapore Botanic Gardens. National Parks Board. 152 pp., more that 150 color photographs, a few B&W and some line drawings.
Yam, T. W.	2007	Orchids of the Singapore Botanic Gardens. 3rd ed. National Parks Board. 204 pp., more that 200 color photographs, a few B&W and some line drawings.
Yamamoto, J.	No ate	Yamamoto dendrobiums.
Yanes, L. H.	No date	Orquídeas para aficionados. Editora Grafos. 176 pp., numerous B&W and color illustrations. Not on the WWW in 2005.
Yang, Z., Q. Zhang, and Z. Feng	1998	Orchids. This book is about the native orchids of China. Kunming Institute of Botany, Chinese Academy of Sceinecs, Yunnan Academy of Forestry. Published by the China Esperanto Press.200 pp., numerous color plates.
Yen, T.-K.	1964	Icones Cymbidiorum Amoyensis. (In Chinese.) The Committee of Science and Technology of Amoy, Amoy, Fukien, China. No pagination. Pages measure 37.5 cm by 28.2 cm. Numerous B&W photographs and color paintings. According to the late Prof. Loo Shih Wei of the Shanghai Plant Physiology Institute many copies of this book were destroyed during the cultural revolution because orchids were considered a decadent capitalist flower. The author died or was killed.
Yeoh, B. C. (ed.)	1960	Proceedings of the fourth world orchid conference, October 3–11, 1963, Singapore. Straits Times Press (M), Ltd. Singapore. 301 pp., maps, line drawings, B&W photographs, color plates.
Ying, S.-S.	1977	Coloured illustrations of indigenous orchids of Taiwan, Volume one. Department of Forestry, National Taiwan University, Taipei, Taiwan. vi+ 565 pp., 127 line drawings and 158 color photographs.
Ying, S.-S.	1977	Coloured illustrations of indigenous orchids of Taiwan, Volume one. Department of Forestry, National Taiwan University, Taipei, Taiwan. xvi+ 565 pp., 85 line drawings and 42 color plates.
Ying, S.-S.	1990	Coloured illustrations of indigenous orchids of Taiwan, Volume two. Department of Forestry, National Taiwan University, Taipei, Taiwan. xvi+ 784 pp., 103 line drawings and 115 color photographs.

(continued)

(continued)

Author(s)	Date	Title and publisher
Yocom, E. S.	1974	The *Paphiopedilum* parade, a century of award. Published by the author in Naples, FL. 129 pp., line drawing frontispiece.
Young, G.	1975	Orchids in line. Vol. 1. *Paphiopedilum* species. Radlcliffe Orchids, Ltd.

Index of Persons

Page numbers in *italics* denote figures, in ***bold italics*** denote tables and in **bold** denote reference lists. Those underlined refer to the appendix.

A

Aana, T., 13
Aaron, 142, 143
Abbink, T.E.M., 259, **270**
Ackerman, J.D., xiii, 160, **180**, **182**, 187, **212**, 281
Agadjanyan, M., 269, **277**
Agassiz, L., 223
Agranovsky, A.A., 266, **270**
Ahlquist, P., 268, **270**, **271**, **273**
Ajjikuttira, P.A., xvi, xviii, xix, 260, 268, **270**
Akerblom, L., 269, **273**
Ali, A., 260, **270**
Ali, M., *155*
Alonso, E., 256, **270**
Alphonso, A.G., 2, 4, 5, 9, 11, 12, **34**
Alsagoff, S.Y., 9, 12, 14, *15*, 16, 19, 21, 24, **34**, **35**, 320
Amaral, M.C.E., 161, 164, 165, 169, 171–176, 178, **181–183**
Ames, O., 142, **156**, 282–283, 379
Anandalakshmi, R., 262, **270**
Anjos, J.R., 262, **270**
Aranda, M.A., 260, **272**
Arditti, J., vii, xv, xvi, xvii, xviii, xix, xxi, xxv, 2, 22, **34**, **35**, 37, **39**, **41**, 42, 45, *51*, *52*, *54*, *55*, *58*, *68–72*, *97*, *99*, *102*, *103*, *108*, *116*, *117*, *123*, *125*, 131–133, 134, 136, 141–156, **156**, **157**, **212**, **215**, **229**, 233–247, **247**, **248**, 263, **272**, 279–424
Arditti, M., 263, **272**
Argos, P., 258, **273**
Armour, S.L., 266, **272**
Atabekova, T.I., 263, 266, 268, **276**
Atabekov, J.G., 265, 266, **270**, 274, 275, **276**
Atencio, F.A., 263, **272**
Atwood, J., xiii
Avadhani, P.N., xvii, xix–xx, 37, *39*, **40**, 45, 57, 83, 98, 102–104, 109, **131**, **133**, 350, 375
Avery-Peck, A.J., 154, 156, **156**
Avila-Rincom, M.J., 256, **270**
Axelsson, M., *150*
Ayllon, M.A., 260, **270**

B

Baker, K., 258, **271**
Bao, Y., 259, **270**, **275**
Barajas, D., 263, **272**
Barnes, S.H., 190, **213**
Barnet, P., 220, **228**
Barret, R.L., xiii, 173–176, **180**
Barry, K., 258, **270**
Bateman, J., 221, 293, 333, 334, 336
Baulcombe, D.C., 259, 265, **270**, **274**, **277**
Beachy, R.N., 269, **271**, **272**
Beattie, A.J., 209, **212**, **213**
Beck, D.L., 260, **277**
Beck, G., 161, 163, 164, 169, **180**, 235, 242, **247**
Bee Yeok Wong, 21
Belanger, H., 267, **275**
Bellani, L., 186, 202, 203, **213**
Benzing, D.H., 177, **180**
Berger, P.H., 262, **276**
Bergon, P., 143, **156**, 303
Bicalho, H.D., 164, 296

Biebreicher, C., 260, **271**
Binnendijk, S., 243
Bittrich, V., 175, 176, **182**
Blackman, P., 142, **156**
Blackmore, S., 190, **213**, **214**, 297
Bleykasten-Grosshans, C., 265, 266, 268, **273**
Blume, C.L., 243, **247**, 297, 312
Boccard, F., 259, **274**
Bok Choon, Yeoh., 4, 5, *8*, 9, 32, **35**, 423
Bol, J.F., 259, **270**
Bond, W.J., 160, **181**
Bonney, J., xxii
Borba, E.L., 206, 208, **212**
Boshuizen, R.S., 269, **271**
Bouriquet, G., **235**, 244, **247**, 298
Bousalem, M., 260, **274**
Bouzoubaa, S., 265, 266, 268, **273**
Braga, P.I.S., 169, 173, **180**
Brakke, M.K., 262, 265, **272**
Brannigan, M., 263, **273**
Brewer, R., 11
Brieger, F.G., 164, 300, 406
Brigneti, G., 259, 262, **270**, **274**
Broome, C.R., 187, **215**
Brown, R.C., 187, 200, **212**
Brueghel, J. the Younger, 220, 221, **230**
Bubrick, P., 268, 269, **271**
Buchmann, S.L., 161, 174, 175, **180**
Buchner, R., 194, **212**
Burgeff, H., 3, **149**, **156**, 301, 302
Burkill, H.M., 5
Burns-Balogh, P., 194, 201, **212**
Busse, W., 243–245, **247**
Butler, P.J.G., 256, **276**
Buttrose, M.S., 170, **180**
Byrd, J.D., *153*

C
Caillebotte, G., 226, 230
Calder, D.M., 187, 190, **213**, **214**
Calvery, L.A., 262, **270**
Calzoni, G.L., 205, **215**
Cameron, K.N., 173–176, **180**
Camus, A., *143*, 144, **156**, 303
Camus, E.G., *143*, 144, 302, 303
Cane, J.H., 174, **180**
Canizares, M.C., 260, **270**
Carlsward, B.S., 177, **183**
Carrington, J.C., 262, 266, **273–276**
Carter, S.A., 259, **270**, **271**, **275**
Casal, J.L., 269, **271**
Casper, R., 265, 266, 268, **275**
Castresana, C., 256, **270**
Catling, P.M., 203, **212**, 417
Cattley, W., 221
Chan, Y.S., 254, **271**
Chang, C.A., 260, **270**
Chapman, S.N., 259, **270**, **274**
Charoenvit, Y., 269, **276**
Chase, M.W., vii, xiii, 173–176, **180**, **249**, 304, 381
Chatin, A., 142, 152, 154, **157**
Cheah, K.C., 27
Chen, H.H., vii
Chesselet, P., 174, **182**
Chew, W., 33, **34**, 320
Chia, T.F., vii, 254, 258, 261, **271**
Chien, D., 45, **133**
Childers, N.F., 234–236, 239, **247**, **248**
Chng, C.G., 254, 256, **271**, **277**
Chong, C.C., 18
Chow, Y.N., vii, xvii, xx, 37
Cibes, H.R., 234–236, **247**, **248**
Clementi, G., 245
Clifford, S.C., 187, 206, **212**
Clusius, C., 38, **39**, 236, **238**
Cocucci, A.A., 171, 174–176, **183**, 195, 201, 206, **212**
Cohn, A., 204, **213**
Cole, J.S., 263, **274**
Collison, M.E., 202, **214**
Colinson, P., 38, *39*
Cooper, B., 265, **271**
Corner, E.J.H., 4, 12
Correll, D.S., 236, 239, **248**, **283**, 308
Correvon, D., **107**
Cotton, J., 194, **228**, 231
Cox, P.T., 258, **271**
Crépin, F., 239, **248**
Cruden, R.W., 207, **212**
Culver, J.N., 259, **272**

D
Dafni, A., 156, **157**, 160, **180**, 205, 206, 208, **212**, 217, 311
Dale, J.L., 260, **276**
Dalpethado, M.E., 2, **34**
Dalsgaard, K., 269, **271**
Danesch, E., 143, 144, **157**, 311, 312
Danesch, O., 143, 144, **157**, 311, 312
Dannenbaum, C., 195, 206, 210, **212**
Daoud, H.S., 176, **182**
Darlington, C.D., 193, **212**
Darwin, C., 38, 40, *41*, **131**, 135, 221, 223, 225, **229**, 312
Dathe, S., 177, **180**

Davies, J., 226, **228**
Davies, K.L., xvi, xvii, xx, 159, 161, 163–174, 176, 177, **180–183**, **212**
Davis, G.L., 197, 200, **212**
Dawson, W., 260, 268, 269, **275**, **277**
de Candolle, A., 243, **248**
Deewan Raggardee, 19, 382, 383
de Faria, A.D., 175, 176, **182**, **183**
de Jager, C.P., 263, 265–268, **276**
de Jong, G.A., xxiii
De Jong, W., 268, **271**
de Kochko, A., 263, **272**
de la Cruz, A., 256, **270**
DeLaTorre, J.C., 260, **273**
de Longpré, P., 226
d'Enghien, P., 239
Delteil, A., **234**, 238, 239, **240**, 242–244, 246, **248**
Demski, J.W., 262, **274**
Demuth, C., 227, 231
Deom, C.M., 266, **272**, **275**
Derrick, P.M., **271**
Desjardins, P.R., 268, 269, **271**
de Brito, Toscano, A.V.L., 161, 175, 177, **184**
De Visiani, R., 239, 243, 245, 246, **248**
De Vriese, W.H., **234**, 236, 243, **248**
Diaz-Ruiz, J., 256, **272**
Dietrich, H., 177, **180**, 315
Di Franco, A., 260, **274**
Distel, A., 226, **228**
Dod, D.D., 176, **181**
Dodds, J.A., 265, 266, **271**
Dodson, C.H., xii, 160, 161, 165, 167, 169–174, 176, **181**, **183**, **184**, 316, 317, 320, 328 413, 414
Dolcher, T., 102, **131**, **132**, 137
Dolja, V.V., 265, 266, 268, **274**
Domingo, E., 259, 260, **271**, **276**
Donald, R.G., 266, **271**, **274**
Dondon, R.C., 161, 172, 173, 178, **181**
Don, R.H., 258, **271**
Donson, J., 268, 269, **271**, **275**
Drake, S.A., 221
Drar, M., 143, 144, **157**, 317, 411
Dressler, R., xii 159, 176, 177, 191, 195, 233
Ducasse, D.A., 265, **275**
Duncan, D.R., 171
Dunn, A.S., xv, xvii, xx–xxi, 141
Dunst, ***167***
Dunsterville, G.C.K., 318, 389
Durand, T., 174

E
Ecott, T., 234, 236, **238**, **242**, 244–245, **248**, 319
Ede, A., 5, *31*, 32, 319
Ede, J., 13, *14*, 31, 32, **34**, 319
Edmond, A., **238**, 244, 245
Edwards, D., xx
Edwards, M.C., 266, **271**, **274**
Edwards, T.J., 187, 195, 208, **213**
Eickwort, G.C., 174, **180**
Eigen, M., 260, **271**
El-Gadi, A., 143, *144*, **157**, 346
Elliott, J., 5, 15, 33, **34**, 320
Endress, P.K., 160, 173, 174, **181**, **182**, 190, 206, **213**
Eng Soon, Teoh, xxiv, 2–35, 408
Erhardt, M., 265, 266, 268, **273**
Ernst, R., 45, **133**, 320
Esau, K., 164, 170, **181**
Eun, A.J.C., 258, **271**, 320
Evans, R.K., 256, **271**

F
Faegri, K., ***193***, 203, **212**, **213**
Fallarme, V., 269, **276**
Fargette, D., 260, **274**
Farnham, E., 228, **229**
Fauquet, C.M., 263, **272**, **274**
Fedorkin, O.N., 267, **271**, **272**, **274**
Felix, J., 142, 154, **157**
Ferry, R.J., 226, **229**, 322
Finetti Sialer, M.M., 260, **274**
Firmage, D., 205, 206, **212**, 217
Fisher, D., ***192***, **213**
Fitchen, J., 269, **272**
Fitting, H., xv, xvii, 37–131, **132**, **134–137**
Fitting, S., *124*
Fitzgeral, M.A., 190, **213**
Flach, A., 161, 172, 173, 178, **181**, **183**
Flax, M., ***193***, **213**
Flemish, 220
Fletcher, J.C., 223
Fleysh, N., 267, **275**
Flores, R., 260, **270**
Focard, K., 244, **248**
Fondong, V.N., 263, **272**, **274**
Footitt, S., 204, **213**
Forster, R.L.S., 260, **277**
Foshay, E., 225, **229**
Fraile, A., 260, **272**
Franchi, G.G., 187, 188, 196, 202, 203, 205–208, **213**, **214**, 216, 217

Francki, R.I.B., 252–254, **272**
Frankie, W., 174, **184**
French, R., 268, **270**, **274**
Freudestein, J.V., 173–176, **180**, 187, 197, 203, **213**
Frymire, G.P., 171, **181**
Fuentes, A.L., 265, **272**
Füller, F., 143, 144, **157**, 324, 325
Fulton, E., 225, **229**

G
Gallitelli, D., 260, **274**
Gandawijaja, D.A., xix, **70**
Garay, L.A., 165, 166, 169, 171, 174, 177
Garcia-Arenal, F., 260, **271**, **272**, **274**, **275**
Garcia-Luqus, I., 256, **270**
Garnsey, S.M., 260, **270**
Garside, S., 174, **181**
Ge, X., 262, **270**, **274**
George, A., 4, 5, 11, 12, 16, 324
George, I., 16, 414
Gerdts, W., 221, 227, **229**
Gerlach, G., vii, 131, 208, **213**, **327**
Geske, S.M., 262, **273**
Ghabrial, S.A., 262, **270**
Ghani, A.K.A., 142, 146, *148*, **156**
Giesman-Cookmeyer, D., 265, **272**
Gill, L.K., 269, **276**
Godfery, H.M., *143*, 143, 144, 146, **157**, 328
Godfery, M.J., 143, 144, 146, **157**, 328
Goh, C.J., 261, **271**, **273**, **277**
Goldbach, R.W., 263, 265–268, **276**
Goldberg, K.B., 262, 265, **272**
Gonçalves, R., 175, 176, **183**
Gonzalez-Jara, P., 263, **272**
Goodman, R.M., 262, 263, **272**
Gopinath, K., 269, **272**
Gorbalenya, A.E., 256, 258, **272**, **275**
Goregaoker, S.P., 259, **272**
Goss, G.J., 171, **181**
Graaf, 243
Granoff, A., 254, **277**
Grantham, G.L., 268, 269, **271**
Grdzelishvili, V.Z., 265, 266, 268, **275**
Gregg, K.B., 165, **181**, 194, 203, **213**
Gregg. A., 161, 174, 178, **181**, **183**, **212**
Guerri, J., 260, **270**
Guessan, P.N., 260, **274**
Guilford, P.J., 265, **277**
Guillaumin, A., 326, 331
Guilley, H., 265, 266, 268, **273**
Gumprecht, R., 165, **181**
Gunther, R.T., **153**, **157**, 331
Gustafson, G.B., 266, **272**
Guy, P.L., 260, **277**

H
Habili, N., 256, 258, **272**
Hadley, G., 263, **272**
Haeckel, P., 257, **276**
Hagsater, E., 234, **248**, 332
Halbritter, H.M., 187, **213**
Halbsguth, W., 42, 44, 45, **132**
Haley, B.E., 256, **271**
Hall, G.H., 221, **230**
Hall, J., 220, **229**
Hamamoto, H., 259, 269, **272**, **273**
Hamilton, R.I., 253, 265, **272**, **273**
Hamilton, W.D.O., 269, **271**
Hanau, R., 266, **272**
Harder, L.D., 195, 208, **214**
Harding, R.M., 260, **276**
Hardy, W.B., ***115***
Harris, F.C.L., 209, **213**
Hart, J.H., 235, **248**
Hashida, E., 259, 269, **272**
Hayek, A., 143, 144, **157**, 338
Hays, W.J., 221, 222, 225, **229**, **230**
Heade, M.J., xxii, 222, 223, 226, 228, **228**, **229**, 230, 231, 404
Hee Seng Ang, 21
Hein, M.B., 269, **272**
Heller, A., 169
Henslow, G., 142, 149, *153*, **157**, 339
Hertz, J.H., 149, 154, **157**
Heslop-Harrison, J., 102, **132**, ***191***, ***192***, 198, **213**, **217**, 245
Heslop-Harrison, Y., ***191***, ***192***, **213**, **217**
Hesse, M., 187, 188, 190, 194, 198, 201, 203, *204*, 208, 209, **212–215**
Hew, C.S., vii, xvii, xviii xix, xxi, 2, **34**, 37, **133**, 279–424
Hicks, E., 222
Hilf, M.E., 268, 269, **271**
Himes, M., ***193***, **213**
Hirashima, K., 259, **273**
Hoffmann, K., 262, **273**
Hofmeister, W., 187, **213**
Holland, J.J., 259, 260, **273**, **276**
Holttum, R.E., xv, xx, 3–5, 11, 12, 17–19, 21, 23, 27, 33, **34**, 341–342
Hon Lum, 27
Hoon Siang Tan, 7
Hooper, C., 269, **277**

Hossain, A.B.M.E., 143, 144, **157**, 346
Hough, T., 209, **212**
Hu, J.S., 253, **270**, **273**
Hu, W.W., 258, 264, **273**
Hunt, A.G., 262, **276**
Hunt, E., *54*, *68*
Hunter, B.G., 266, **274**
Hunter, R., 266, **272**

I
Ichihashi, S., vii, 343, 344, 351
Idris, S., xix
Inoue, K., 160, **181**, 344
Ishii, M., 253, **274**
Ishikawa, M., **276**
Iversen, J., ***193***, 212
Ivri, Y., 160, **180**

J
Jackson, A.O., 266, 269, **271**, **272**, **274**, **275**
Jacquet, P., 220, **229**, 236, **248**, 345
Janse, J.M., 161, 163, 165, 169, **181**, 236, **248**
Jarlfors, U., 262, **270**
Jastrow, M., 156, **157**
Jensen, D.D., 252, **273**
Jensen, W., 191, 192, 195, 201, **212**
Ji, L.H., 259, 262, **270**
Jiew Hoe John Tan, 7
Jiménez, R., 234, **248**, 332
Joaquim, A., 2, 3
Jobson, S., 209, **212**
Joelson, T., 269, **273**
Johansen, D.A., ***191***, 213
Johnson, H., 2, **34**,
Johnson, S.D., 160, **181**, 187, 195, 208, **213**
Jonard, G., 265, 266, 268, **273**
Jones, D.L., 169, 171, **181**, **182**, 306, 347
Jones, T.D., 269, **271**
Joshi, R.L., 268, **273**
Joshi, V., 268, **273**
Joyce, L.E., 256, **276**
Judd, W.S., 177, **183**
Juttner, G., 265, 266, 268, **274**

K
Kado, C.I., 252, **273**
Kamada, H., 259, 269, **273**
Kamstrup, S., 269, **271**
Kao, M.C.C., 254, 256, **271**
Kaplan, I.B., 263, 265–268, **276**
Karasev, A.V., 265, 266, 268, **270**, **274**
Karesova, R., 260, **276**
Kasschau, K.D., 262, 266, **273**, **274**
Kay, Q.O.N., 176, **182**
Kelberlau, S., 225, 229
Kearney, C.M., 268, 269, **271**
Keller, G., 143, 144, **157**, 349
Keng Cheong Cheah, 27
Keng Hoe Koh, 16
Keyes, D.D., 222, **229**
Khan, M.A., 262, **274**
Kiah Huat Quek, 9–10, *10*, 34
Kim Hong Lee, 5, *6*, 13, 29, 31, 32
Kjaer, K., 131
Kjellsson, G., 169, **182**
Knox, R.B., 187, 190, **212–214**
Kobayashi, H., vii
Kocyan, A., 160, **182**, 190, 206, **213**
Koehler, S., 164, 165, 171, 172, 173, **181**, **183**
Koh, K.H., 16, 18, **35**
Kok Wah Yong, 33
Koonin, E.V., 258, **275**
Koonin, R.H., 256, **272**
Koprowski, H., 267, 269, **275**, **277**
Koriba, K., 4
Kornmann, P., 116, ***126***, **132**
Kramer, G., 258, **273**, **276**
Kramer, S.R., 258, **276**
Kroner, P., 259, 273
Kubota, K., **276**
Kuehnle, A.R., 261, **270**, **274**, 348
Kull, K., vii
Kull, T., xvi, xviii, xxi, 34, 279–424
Kurzweil, H., xvii, xxi, xxii, 37–133, 134, 174, **182**, 359

L
Lack, A., 160, **182**
La Cour, L.F., 193, **212**
Laibach, F., **102**, 116, *124*, *125*, ***126***, ***130***, **132**, 135, 137, 356
Lai Wah Sum, 13
Lamb, A., 26, **34**, **35**, 304, 387, 416, 422
Lanave, C., 260, **274**
Landwehr, J., 143, 144, **157**, 357
Lane, F.H., xxii, 225
Langeveld, J.P.M., 269, **271**
Lauber, E., 265, 266, 268, **273**
Lawalrée, A., 240, **248**, 292
Lawler, L.J., 38, **132**, 142, 143, 146, 156, **157**, **181**, 236, **248**, **296**, 357

Law Moi Hwa, 33
Lawson, R.H., 263, **273**
Laycock, A., 13
Laycock, J., 5, 9, 13, 31, 33
L'Écluse, C., **39**, 236, **238**
Lee, R., 13, 31–32
Lee, S.C., 258, **273**
Lee Kim Hong, 5–6, *6*, 13, 29, 31, 32
Legras, A., 244, **248**
Lemaire, C., *97*, **247**
Lemmon, B.E., 187, 200, **212**
Leng San Ooi, 27
Leong Fatt Wong, 17–18, 20
Le Pecq, J.B., ***193***, **213**
Lesemann, D-E., 265, 266, 268, **275**
Lewandowsky, D.J., 259, 268, 269, **271**, **272**, **275**
Lewis, G., 27, 28–29, *28*, **34**
Lim, A.L., 269, **273**
Lim, D., 16, 23–24, **34**
Lim, S., 258, **273**
Lim-Ho, C.L., 260, **270**
Lim, S.T., 258, **273**
Lim Chu Kang, 13, 22
Linder, H.P., xxi, 174, **182**, 359, 405
Lindley, J., 163, **182**, 293, 334, 335, 360, 404
Linthorst, H.J.M., 259, **270**
Livingston, S.P., 4
Li, W.X., 259, 262, **270**
Loh, C.S., 260, 268, **270**, **271**, **273**
Lommel, S.A., 265, **272**
Lomonosoff, G.P., 269, **271**, **272**
Longstaff, M., 259, **274**
Lopez, C., 260, **270**
López, M., 234, **248**
Loprasert, S., 261, **276**
Loring, J., 226, **229**
Lucchesi, J., **271**
Lukasheva, L.I., 265, 266, 268, **274**, **275**
Lum, H., 27
Lynes, B.B., 227, **229**

M

Macpherson, K., 161, 172, 173, **182**
Madam Yong, 33
Magli, G., 102, **132**
Magnus, A., 147
Mahmet, A., **155**
Mahtani, P.H., 252, 254, 256–258, **271**, **277**
Mahunu, A., 45, **133**
Mai, G., 116, **126**, **132**
Maiss, E., 265, 266, 268, **275**
Mak, C.O., 13, 17, 20–22, **34**
Makinen, K., **271**, **272**
Mallory, A.C., **270**
Malpica, J.M., **272**
Malyshenko, S.I., 263, 265–268, **270**, **276**
Mapplethorpe, R., 228
Marathe, R., 270
Marchoux, G., 274
Marcks, B.G., 165, **184**
Marcos, J.F., 260, **270**
Mark, 146, 148
Mark, B., 365
Marsaioli, A.J., 175, 176, **181–183**
Martinez-Garcia, B., 263, **272**
Maschmann, E., *116*, ***126***, ***130***, **132**
Matisse, H., 227, **231**
Matsunaga, Y., 259, 269, **272**
Matthew, 146, 148
Matthews, R.E.F., 259, **274**
Mattick, J.S., **271**
Matusiewicz, J., 161, **167**, 172, **182**, **183**
Mazia, D., ***192***, **214**
McConchie, C.A., 187, **213**
McCully, M.E., ***191***, ***193***, **214**
Medema, H.P., 147, 148, **157**
Medina, E.H., 234, 235, **248**
Medvedev, Z.A., 38, **132**
Meeryweather, A., 269, **271**
Meléndez-Ackerman, E.J., 160, **182**
Meléndez, E.J., 160, **180**
Meloen, R.H., **271**
Merits, A., **271**
Meshi, T., **274**, **276**
Meyer, S., 44
Meyerowitz, E.M., *124*, 132
Michener, C.D., 174, **182**
Mickeliunas, L., 161, 175, **182**
Mikheeva, T., 267, **275**
Miller, H., **275**
Miller, J., 131
Miroshnichenko, N.A., **274**
Mise, K., 268, **276**
Miyazaki, S., 160, **184**
Modelska, A., 269, **277**
Moffat, W.N., ***192***, **214**
Mok, C.Y., 22, 23, **34**
Moldenke, A., 142, 146, 148, 149, **157**
Moldenke, H., 142, 146, 148, 149, **157**
Moore, P.D., 202, **214**
Moreno, P., 260, **270**
Morgan, H., 236
Morita, K., 102, 109, 115, **116–118**, **121–123**, **133**
Morozov, S. Yu., 265, 266, 267, 268, **270–272**, **274**, **275**, **277**

Morren, C.F.A., 239–**241**, 243–245, **247–249**
Morris, T.J., **273**
Moses, 142, 143
Moya, A., 260, **274**, **275**
Moyer, J.W., 262, **273**
Mulcahy, D., 207, **214**
Müller, F., 38, 40, 41, 45, 132, 133, 134–137, 370
Müller, R., 117
Musselman, L.J., 142, 147–149, **157**

N
Nagaishi, S., 160, **184**
Nagano, H., 268, **276**
Nair, H., xvi, xix, xxii, **131**, 233–**249**, 371
Nakagawa, N., 259, 269, **272**
Namba, R., 253, **274**
Nan, G.L., 261, **274**
Nang Yip, Au Yong, 25
Narodny, L.H., 236, **249**
Navas-Castillo, J., 260, **270**
Neiland, M.R.M., 160, **182**
Neljubov, D., 57, 131, **133**
Nelson, R.S., 259, **270**, **271**, **275**
Nepi, M., 186, 202, 203, 205, **213**, **214**, **216**, **217**
Neumann, J., *238*, 242, 243, **249**, 386
Newton, G.D., 187, **214**, 313, 387
Neyer, P., 188, **215**
Nilsson, L.A., 160, 174, **181**, **182**, 208, **214**
Nilsson, S., **214**, 370, 372
Noa, E., 4
Noa, J., 4
Nolan, L.C., 235, 236, **249**
Nolde, E., 226, 227, **231**
Novak, B., 225, **229**
Nuernbergk, E.L., 44

O
O'Brien, T.P.N., ***191***, ***193***, **214**
O'Brien, J., 374
O'Byrne, P., 22, 33
Ogawa, T., 259, **274**
Ogwal, S., **274**
Okada, Y., 259, 269, **272–274**, **276**
O'Keeffe, G., xxii, 227, **229**, **231**
Oliva, A., *149*, *150*
Ong, S.H., 5, 32
Ooi, L.S., 27
Otim-Nape, G.W., 263, **274**
Ottaviano, E., 207, **214**
Ow, D.W., 268, **273**
Owens, S.J., 187, 206, **212**
Oxefelt, L., 269, **273**

P
Pabst, G.F. J., 173, 175
Pacini, E., xvi, xxii, 185–212, **213–215**, 216–218
Paek, K-Y., v, **xi**–xii, 261, **277**
Pallas, V., 260, **270**
Pandolfi, T., 187, 190, 195, 197, 198, 200–203, 206, **214**
Pansarin, E.R., 161, **182**
Papf, E., 226, **231**
Parkinson, S., 38, *39*, **133**
Park, W.M., 256, 258, **271**, **275**
Parrella, G., 260, **274**
Pauw, A., 174, **182**
Pearse, A.G., ***192***, **214**
Pearson, M.N., 263, **274**
Perner, H., vii
Petrizk, K., 260, **276**
Petty, I.T., 266, **274**
Pfeffer, W., 42
Pfeiffer, W., 195, **215**
Pfitzner, A.J., 257, **276**
Philibert, P-H., 244, **248**
Phoon, Y.S., 2, 16, 24–25, **34**, **35**
Pinel, A., 260, **274**
Pita, J.S., 263, **272**, **274**
Pohl, F., 174, 175, **182**
Porsch, O., 160, 161, 163, 165, 169, 170, 172–174, **182**, 380
Porta, C., 269, **271**, **272**
Pridgeon, A.M., xiii, xvi, 173, 177, **182**, 221, **229**, 235, 236, **249**, 290, 332, 381, 400
Proctor, H.C., 195, 208, **214**
Proctor, M., 160, 171, **182**
Pruss, G.A., 262, **270**, **274**
Pshennikova, E.S., 266, 268, **276**
Punt, W., 190, **214**
Pyke, G.H., 160, **182**

Q
Qiu, W.P., 265, **275**
Quinn, K., xvi, xviii, xxii, 219–231, **228**, **229**

R
Raffo, A.J., 268, 269, **271**
Rand, E., 221
Rand, E.S., 383
Randles, J.W., 260, **270**

Ranineri, D.M., 261, **275**
Rao, A.L.N., xvi, xviii, xix, xxii–xxiii, 233–247, 266, **271**, **275**, 383
Rasmussen, F.N., 169, **182**, 187, **213**, **249**, 383
Reagan, S., 192, **214**
Reichenbach, H.G., 109, 187, **215**, 384, 385
Reinikka, M.A., 220, 221, **229**, 385
Reinl, S., 269, 276
Reis, M.G., 175, 176, **182**, **183**
Rey, M.E.C., 263, **272**
Richards, K.E., 265, 266, 268, **273**
Richins, R.D., 265, **275**
Ridley, H., 3
Ridley, H.N., 387
Roberts, D.L., 161, **180**, **183**
Robinson, D.J., 267, **275**
Robinson, F., 221, **229**
Rochow, W.F., 262, 265, **275**
Rockefeller, J.D., Jr., **229**
Rodgers, P.B., 269, **271**
Rodriguez-Cerezo, E., 260, **272**, **275**
Rodriguez-Robles, J.A., 160, **180**, **182**
Rolfe, R.A., 236, 249, 389
Romero, G.A., vii, 169, **183**, 385, 389
Ross, A.F., 262, 263, 265, **271**, **272**, **275**
Roth, D.A., 256, **271**
Roubik, D.W., 171, 178, **183**, 390
Roux, K.H., 258, **275**
Rozanov, M.N., 258, **275**
Rungroj, N., 261, **276**
Rupp, H.M.R., 161, 172, 173, **182**, 328, 390
Ryabov, E.V., 267, **275**
Ryu, K.H., 256, 258, 260, **270**, **271**, **275**

S

Saarma, M., 267, **271**
Sacher, R., 268, **270**
Sagarik, R., 9, 348, 391
Salazar, G., 234, 248
Sander, H., 226, 390
Sangare, A., 263, 274
Sappan, Bajuri Bin, 5, 11, 15
Solms-Laubach, H.G. Zu, 42
Savenkov, E.I., 265, 266, 268, **275**, **277**
Sazima, M., 161, **182**
Scheets, K., 263, **275**
Schiemann, J., 265–268, **272**, **274**
Schneider, H., *55*
Schill, R., 187, 188, 190, 194, 195, 197, 198, 202, 206, 208, 210, **212**, **213**, **215**, **217**, 327, 382
Schlag, M., 194, 198, 203, **215**
Schlechter, R., 143, 144, **157**, **183**, 300, 315, 349, 352, 393–397
Schmitz, I., 265, **271**
Scholthof, H.B., 269, **275**
Scholthof, K.B.G., 264, 265, 269, **275**
Schrum, G., 174, **184**
Schulze, M., *107*, *108*, **133**, *143*, 144, **157**, 397
Scott, R., 29
Scott, W., 17, 18, 20, 27, 29–30, **35**, 398
Sears, B.B., 204, **215**, 351
Sek Man Wong, xix, xxiv, 251–**277**
Semir, J., 206, 208, **212**
Senghas, K., 173, **183**, 300, 399, 406
Seoh, M.L., 258, **271**, **275**
Serra, M.T., 256, **270**
Seufertet, C., 194, **215**
Sezik, E.E., 143, 144, *155*, 156, **157**, 399
Shapka, N.A., 266, **270**
Shaw, J.J., 268, 269, **271**
Sheehan, T.J., 253, 277, 400
Shepherd, R.J., 265, **275**
Shintaku, M.H., 259, **275**
Shivprasad, S., 267, 269, **275**, **277**
Shi, X.M., 262, **274–276**
Short, M.N., 260, **275**
Siew Chin Lee, 6
Siew Hong Ong, 5, 32
Silvera, K.I., 174, 176, **183**
Silver, S., 265, **272**
Singer, R.B., xviii, xxiii, 161, 164, 165, 169, 171–176, 178, **181**, **183**, 206, **215**, 279–424
Simon, V., 44
Siu, I.H.P., 187, **215**
Smithson, A., 160, **183**
Smith, T.H., 262, **270**
Soediono, N., 246, 402
Soh, K.K., 18–20, **35**
Solovyev, A.G., 265, 266, 268, **270–272**, **274**, **275**, **277**
Soon, T.E., xv, xxiv, 1–**35**
Soto, M., 234, **248**, 305, 332
Southwick, E.E., 160, **183**
Souvay, C.L., 142, 149, **157**
Speilholz, J., 174, **180**
Speranza, A., 205, 206, **215**
Spitsin, S., 267, **275**
Srifah, P., 261, **276**
Stebbins, T.E., 221–226, **229**, 404
Steiner, K.E., 174, **183**, **184**, **215**
Steinhauer, D.A., 259, 260, **273**, **276**
Stenzel, H., 195, **215**
Steplewski, K., 267, 269, **275**, **277**
Stern, W.L., 177, **183**

Stirton, C.H., 176, **182**
Storer, C., 226, **231**
Stpiczyñska, M., 161, 164, ***167***, ***168***, **169**, 172, 174, 177, **180–183**
Starling, E., 104, ***115***
Strandberg, B., 269, **273**
Strasburger, E., 44, 405
Strugger, ***191***, **212**
Sugaya, N., 142, 152, 154, **157**
Sugii, N., 261, **270**, **274**
Sugiura, N., 160, **184**
Sugiyama, Y., **272**, 410
Summerhayes, V.S., 143, 144, 157, 405, 406
Symons, R.H., 256, 258, **272**
Sweet, H.R., 164, 326

T
Tachibana, S., 269, **273**
Täckholm, V., 143, 144, **257**, 411
Takamatsu, N., 268, 269, **276**
Takemoto, S., 269, **272**
Takhtajan, A., 208, 215
Taliansky, M.E., 263, 265–268, **270**, **275**, **276**
Tamai, A., 268, **276**
Tan Chay Yan, 3, 7, 16, 29, 32
Tan Hoon Siang, 7, 8, **34**
Tan, P., 33, 252, 320
Tan Wee Kiat, 32
Tanaka, I., 203, **215**, 348, 391, 407
Tanakadate, H., 4
Tatarenko, I.V., 407
Taylor, K.M., 269, **272**, 330, 353, 407
Teijsman, J.E., 242, 243, 249
Teirlinck, I., 143, 157
Tenllado, B., 263, 272
Thien, L.B., 165, 184
Thimann, K.V., **102**, 115, 116, *125*, **133**
Thomas, A.L., 13, **41**, **214**
Tim Wing Yam, 33, 37–131
Tjernberg, P.A., 259, **270**
Tokugawa, M., 4
Tomenius, K., 269, **273**
Tran, H., 45, **133**
Tsou, C.H., 164, **184**
Turnbull, C., 209, **212**, **213**
Turner, D.R., 256, **276**
Turner, M.P., 161, 16–165, 166, 167, 169–173, 176, 177, **180**, **181**, **212**
Turpen, T.H., 269, **276**
Tyagi, S., 258, **276**

U
Ulanova, E.F., 266, 268, **276**
Uttenthal, A., 269, **271**

V
Vaewhongs, A.A., 265, **272**
Valentine, B., 253, **273**
Vallius, E., viii
van den Bosch, 243
van der Cingel, N.A., 160, 169, 173–176, **184**, **413**
van der Pijl, L., 160, 161, 165, 169–174, **184**, **213**, **413**
van Gorkom, K.W., 243, **249**
van Kanmen, A., 269, **272**
van Lent, J.W.M., 265, **276**
van Regenmortel, 252, **273**
Vance, V.B., 262, **270**, **274–276**
Vargas, M., **272**
Vaskova, D., 260, **276**
Vela, C., 269, **271**
Verchot, J., 262, **275**
Vermeulen, J.J., 35, 289, 387, **416–418**, **422**
Vinson, S.B., 174, **184**
Vöchting, H., 42
Vogel, S., 161, 169, 174, 176, **184**, 187, **213**, **421**
Voinett, O., 262, **270**
von Bismark, Otto, 38
von Humboldt, A., 223
von Kirchner, O., 161, 172, **184**
Vu, H., 45, **133**

W
Wager, Sir Charles G., 38, *39*
Wainwright, K., 258, **271**
Wai Lai, 19
Wallich, N., 163, 184
Wanitachacorn, R., 260, 276
Watanabe, Y., 259, 271–274, 276
Webb, J.A., 202, 214
Weber, H., 257, 276
Weber, M., 194, 212
Webster, D.E., 260, **277**
Weiner, C., 269, **277**
Wesley, F.R., 174, **180**
Went, F.W., 42, 56, 98, 102, 103, 115, 116, *124*, *125*, ***126***, ***130***, 131, 133
Weston, E., xxii, 227, 229, 231
Whampoa, 2, *3*
Whitehead, V.B., 174, 184
Wiard, L.A., 239, 239, 249, 417

Wicke, B., 256, **270**
Wigman, H.J., 235, 249
Wilcock, C.C., 160, 182
Williams, L.O., 165
Williams, N.H., 176, 187, 212, 214, 215, 389
Williams, W.C., 227
Winters, C., 161, 165, 169, 176, 177, 180, 181
Wisler, G.C., 253, 277, 420
Wither, A., 221
Wollaston, W., 225
Wolter, M., 187, 190, 194, 195, 197, 202, 215
Woodword, 225
Woon, M.H., 260, 270
Wright, N.H., 245, 249
Wu, N., 220, 228

X
Xiu-Lin, Y., xiv
Xu, F., 269, 271

Y
Yam, T.W., vii, xv, xvii, xviii, xxv, 2, 32, 33–35, 38–131, 134–138, 236, 279–424
Yee Peng How, 18, 19
Yan, C.T., 3, 7, 16, 29, 32
Yang, S.H., 261, 277
Yap, K.F., 20, 35
Yelina, N.E., 266, 272, 277
Yeong, C.Y., 258, 273
Yeo, P., 160, 171, 181
Yeoh Bok Choon, 5, 8, 9, 32
Yeung, E.C., v, vii, xiii, xiv, 187, 194, 195, 200, **201**, 215
Yong, M., 33
Yong, N.Y.A., 26, 290
Yoshii, M., 268, 276
Yu, H.H., 257, 259, 277
Yu, S., 274, 276
Yusibov, V., 267, 269, 275, 277

Z
Zavada, M.S., 187, 201, 202, 215
Zee, S.Y., xiv, 187, 215
Zettler, F.W., 252, 253, 277
Zelenina, D.A., 265, 266, 268, 274, 275
Zhang, L., 258, 275
Ziegenspeck, H., 174, 184
Ziegler-Graff, V., 265, 277

Index of Organism Names

Page numbers which refer to illustrations are in **bold face**.

A

Aeranthes arachnites, 161
Aerides falcata, **55**, 138
Aerides lawrenceae, 19
Aerides odorata, 50, **55**, 56, 57, 64, 67, 78, 82, 84–86, 89–92, 105, 106, 109–112, 138, 224
African cassava mosaic virus, 263
Agrobacterium, 261
Alfalfa mosaic virus, 260, 267,
Alpinia hookeriana, 92, **97**, 138, 140
Alpinia latilabris, 138
Amerorchis rotundifolia, 208
Andrena, 174
Anoda cristata, 126, 127, 138, 140
Anoectochilus imitas, 146
Apostasia, 190, 194, 206
Apostasia wallachii, 194
Arachnanthe clarkei, **68**, 138
Arachnanthe hookeriana, 4, 17, 30
Arachnanthe hookeriana var. *alba*., 4
Arachnanthe sulingi, 67, **68**, 77, 89, 90, 105, 109, 110, 112, 138
Arachnis, 9, 14
Arachnis flos-aeries, 4
Arachnis hookeriana, 4, 12
Arachnis hookeriana var *luteola*, 4
Arachnis maingayi, 138
Aranda lowii, 7
Arethusa bulbosa, 165
Armodorum sulingi, 57, **68**, 138
Ascocentrum, 9, 12, 22
Ascocentrum miniatum, 12
Aspidogyne, 165
Avena, 103, **116**, 125–127, 130

B

Baptistonia echinata, 175
Barley stripe mosaic virus, 265
Bean necrotic yellow vein virus, 265
Begonia geogensis, 94, 138, 140
Beta vulgaris, 252, 253
Bismarkia nobilis, 20
Bletia purpurea, 38, 138
Bletia verecunda, 38, **39**, 138
Bletilla striata, 194
Bokchoonara, 9
Bougainvillea, 12
Bougainvillea var. *poultonii*, 12
Brassica campestris, 120, 138
Brassica rapa, 138
Bulbophyllum campanulatum, 175
Bulbophyllum macranthum, 174

C

Calanthe discolor, **116**, 120, 121, 138
Calanthe triplicata, 4, 138
Calanthe veitchii, 120, 138
Calanthe veratrifolia, 65, 67, **69**, 91, 138
Calopogon tuberosus, 165
Calypso bulbosa, 165, 195, 208
Canna, 138
Cassia occidentalis, 252, 253
Catasetum, 138
Catasetum discolor, 204
Cattleya Guarianthe, **103**
Cattleya labiata, 106, 138, 221, 224, 225
Cattleya sp., **197**
Cattleya trianei, **108**, 113, 138
Cephalanthera austiniae, 169

Cephalanthera erecta, 138
Cephalanthera falcata, 138
Ceratandra atrata, 174
Ceratandra bicolor, 174
Ceratandra grandiflora, 174
Chenopodium amaranticolor, 252, 253, 266
Chytroglossa marileoniae, 175
Citrullus lanatus, 149
Citrullus vulgaris, 149
Cleistes divaricata, 165, 194, 203
Cochlioda rosea, 197
Coelogyne, 56, 306, 312, 329
Coelogyne asperata, 49, **54**, 56, 138
Coelogyne massangeana, 126, 127, 138
Coelogyne pandurata, 49, **54**, 56, 138
Coelogyne rochussenii, 22, 49
Coelogyne speciosa, 127
Coelogyne swaniana, **70**, 77, 105, 109, 110, 112, 138
Corycium orobanchoides, 174
Corymbis disticha, 90, 91, 138
Corymborkis veratrifolia, **72**, 138
Cowpea chlorotic mottle virus, 266
Cowpea mosaic virus, 262, 269
Cucumber mosaic virus, 263, 265
Cucumis Dudaim, 149, 152
Cucumis melo, 149, **153**
Cucumis odoratissimus, 152
Cultriformes, **163**, 168
Cymbidium chloranthum, 138
Cymbidium dayanum, 166
Cymbidium finlaysonianum, 48, **54**, 56, 75, 82, 84, 89, 90, 105, 109, 110, 112, 138
Cymbidium goeringii, 138
Cymbidium lowianum, 166
Cymbidium mosaic virus, 252
Cymbidium sanguinolentum, 45, 48, **54**, 56, 138
Cymbidium virens, 109, **117**, 118, 120–122, **123**, 138
Cypripedium acaule, **39**, 138, 194
Cypripedium calceolus, xxi, 174, **189**, 194, 208, 404
Cyrtopodium, 176, 367

D

Dactylorhiza incarnata, 138
Dactylorhiza maculata, 138, 206
Dactylorhiza purpurella, 206
Datura stramonium, 252, 253
Dendrobium, 10, 17, 21, 22, 32, 33, 223, 260
Dendrobium anosmum, **54**, 138
Dendrobium antennatum, 49, 56, 67, 138
Dendrobium crumenatum, 110, 138
Dendrobium delacourii, 165
Dendrobium fimbriatun, 120, 138
Dendrobium insigne, 146
Dendrobium kingianum, 278
Dendrobium macrophyllum, 57, 67, **68**, 138
Dendrobium superbum, 50, 56, 66, **69**, 77, 90, 110, 138
Dendrobium unicum, 166, 169, 171
Dendrobium wardianum, 66, **69**, 138
Deutschen Reichs, 42, 45
Dialictus aff. *creberrimus*, 171
Dimorphis lowii, 12
Disperis bolusiana, 174
Dunstervillea mirabilis, 175

E

Epidendrum ibaguense, xiii, 195
Epidendrum scutella, 195, 201, 204, 206
Epidendrum secundum, 146
Epipactis erecta, **117**, 121, 138
Epipactis falcata, **117**, 120, 121, 138
Epipactis microphylla, 194
Epipactis palustris, 138
Epipactis papillosa, **117**, 121, 138
Epipactis thunbergii, **117**, 121, 138
Eria oblitterata, **163**
Eria paniculata, **163**
Eria pilifera, 166
Eria ridleyi, **163**
Esmeralda clarkei, 138
Eucharis grandiflora, 92, **97**, 138
Eulophia macrostachya, 90, 138
Eulophia pulchra, 138

G

Galeola nudifolia, 146
Gastrodia elata 156
Gastrodia sesamoides, 169, 171,
Gomesa recurva, 176
Gomphrena globosa, 252, 253
Goodyera repens, 121, 138
Grobya amherstiae, 175
Grobya galeata, 176
Groundnut rosette virus, 267
Gymnadenia conopea, **107**, 113
Gymnadenia conopsea, 161, 138, 206
Gymnadenia cucullata, 138

H
Hedychium, 92, 93, **97**, 109, 138
Hevea braziliensis, 3
Hibiscus rosa-sinensis, 94, 96, 138
Hibiscus schizopetalus, 126, 138
Hibiscus syriacus, 126, 138
Hintonella mexicana, 175
Hordeum vulgare, 266
Hyacinthus orientalis, 120, 138, 142

I
Impatiens rodrigesi, 94, 138

K
Khalbetzin, 154, 156

L
Laelia, 224, 225, 252, 253
Laeliocattleya, **149**, 252, 253
Lagerstroemia, 12
Lilium candidum, 142
Liparis latifolia, 45, 48, **54**, 56, 111, 138
Lobus oblongus aromaticus, 236
Loroglopssum hircinum, 195, **200**, 200–203, 206

M
Maize cholorotic virus, 263
Mandragora autumnalis, 149
Mandragora officinarum, 149
Masdevallia glandulosa, 41, 138
Maxillaria acuminata, 166, 172
Maxillaria acutifolia, 171
Maxillaria brasiliensis, 164, 171
Maxillaria buchtienii, 167
Maxillaria camaridii, 165
Maxillaria cf. *notylioglossa*, 172
Maxillaria discolor, 166, 169, 172
Maxillaria divaricata, 172, 173
Maxillaria equitans, 172
Maxillaria friedrichsthalii, 173
Maxillaria grandiflora, 92, **97**, 161, 166, 169
Maxillaria johniana, **163**
Maxillaria lepidota, 161, 167, 169
Maxillaria nasuta, 172
Maxillaria ochroleuca, **163**, 171
Maxillaria parkeri, 167
Maxillaria pseudoreichenheimiana, 167, 173
Maxillaria pulchra, 165
Maxillaria reichenheimiana, 167, 173
Maxillaria rufescens, 167, 170–172
Maxillaria sanderiana, 164, 170
Maxillaria splendens, 167, 169
Maxillaria tenuibulba, **163**, 171, 172
Maxillaria tenuis, 167
Maxillaria villosa, 170, 172
Maxillaria violaceopunctata, 161
Melipona, 171, 234, 235
Mycaranthes, 164–166, 169, 173

N
Narcissus jonquilla, 120, 138
Neotinea maculata, 143
Neottia nidus-avis, 194
Neotinea maculata, **143**, **144**
Neottianthe cucullata, 138
Neuwiedia, 190
Nicotiana benthamiana, 253
Nicotiana glauca, 260

O
Odontoglossum crispum, **108**, 113, 138
Odontoglossum grande, 252–253
Odontoglossum ringspot virus, 252, 269
Oncidium amictum, 175
Oncidium compressicaule, 176
Oncidium cornigerum, 175
Oncidium divaricatum, 138
Oncidium enderianum, 176
Oncidium flexuosum, 66, **69**, 138
Oncidium guianense, 176
Oncidium haitiense, 176
Oncidium hookeri, 175
Oncidium incurvum, 48, **54**, 56, 111, 138
Oncidium kautskyi, 175
Oncidium loefgrenii, 176
Oncidium longicornu, 175
Oncidium osmentii, 176
Oncidium paranaense, 176
Oncidium powellii, 175
Oncidium pubes, 175
Oncidium pumilum, 176
Oncidium quadrilobum, 176
Oncidium sphacelatum, 114, 138
Oncidium sphegiferum, 114, 138
Oncidium trulliferum, 175
Oncidium truncatum, 175
Oncidium welteri, 175
Ophrys insectifera, **189**
Ophrys insectivora, 195

Ophrys sphegodes, **196**
Orchis fusca, **107**, 113, 138
Orchis latifolia, **107**, 113, 114, 138
Orchis maculata, **108**, 113, 138
Orchis mascula, 111, 114, 138
Orchis militaris, 144
Orchis morio, **107**, 111, 113, 138
Orchis purpurea, 138, 142, **143**, 144, 146, **198**, **199**, 201, 204
Orchis pyramidalis, **190**
Orchis sancta, **143**, 151
Ornithocephalus, 175, 177
Ornithocephalus avicula, 175
Ornithocephalus bicornis, 175
Ornithogalum, 156
Ornithophora radicans, 175

P

Pancratum maritimum, 142
Panicum mosaic virus, 264
Paphiopedilum argus, 120, 138
Paphiopedilum barbatum, 12, 114, 138
Paphiopedilum boxallii, 120, 138
Paphiopedilum callosum, 114, 138
Paphiopedilum glaucophyllum, **72**, 91, 138
Paphiopedilum lathamianum, 120, 138
Paphiopedilum linii, 20
Paphiopedilum niveum, 12
Paphiopedilum villosum, 138, **189**, 194
Papillilabium beckleri, 175
Paraphalaenopsis, 9, 14
Paratetrapedia calcarata, 175
Paratetrapedia testacea, 175
Partamona, 165, 171
Phaius amboinensis, **72**, 138
Phaius tankervillae, 204
Phalaenopsis amabilis, 49, **51**, 56, **58**, 64, 73, 74, 82, 89, 90, 96, 100, 105, 106, 109–112, 114, 127, 128, 135, 138
Phalaenopsis bellina, 12
Phalaenopsis cornucervi, 65, 138
Phalaenopsis esmeralda, 49, **54**, 56, 65, 96, 111, 138
Phalaenopsis gigantea, 12
Phalaenopsis pulcherrima, 56, 138
Phalaenopsis regnieriana, 138
Phalaenopsis violacea, 27, 45, 50, **52**, **54**, 56, 64, 80, 82, 84, 90, 93–96, 104–106, 109, 111, 112, 138
Phymatidium delicatulum, 175
Phymatidium falcifolium, 175
Phymatidium tillandsioides, 175
Platanthera bifolia, **107**, 113, 138, 161
Platanthera chlorantha, 161
Platanthera yatabei, 121, 138
Platyrhiza quadricolor, 175
Plebeia, 171
Pleurothallis eumecocaulon, 195
Pogonia ophioglossoides, 165
Pollenhormon, 104, 115, **117**, 126, 129, 131
Polystachia pubescens, 194, 198
Polystachya campyloglossa, **163**, 168
Polystachya concreta, 164
Polystachya cultriformis, **163**
Polystachya flavescens, 171
Polystachya foliosa, **163**
Polystachya maculata, **163**
Ponthieva racemosa, 174
Prunus mume, 120, 138
Pterostylis plumosa, **188**, 190, 197, 198, 201, 202
Pterygodium alatum, 174
Pterygodium caffrum, 174
Pterygodium catholicum, 174
Pterygodium volucris, 174

R

Red clover necrotic mosaic virus, 265
Rediviva gigas, 174
Rediviva peringueyi, 174
Reizung, 56, 57
Renanthera, 14, 263
Renanthera maingayi, 57, 67, 138
Rhyncostylis retusa, 12, 32, 45–47, **53**, 56, 66, 78, 89, 90, 105, 106, 109, 112, 138
Rudolfiella aurantiaca, 169, 176
Rudolfiella picta, 169

S

Salix gracilistyla, 138
Salix thunbergiana, 120, 138
Samanea samaan, 20
Satellite Panicum mosaic virus, 264
Satyrium bicallosum, 174
Satyrium maculatum, 143, 144, 152
Sigmatostalix, 175
Sobralia liliastrum, 169
Spathoglottis, 4, 32, 90
Spathoglottis filuata, 90, 91, 138
Spathoglottis plicata, 4
Spinacia oleracea, 267
Spiranthes australis, **116**, 121, 138
Spiranthes sinensis, 138
Stanhopea, **41**, **70**, 76, 105, 110, 112, 205
Stanhopea insignis, 76, 138

Stanhopea tigrina, 205
Stenoglottis longifolia, 197
Stichorkis latifolia, 138
Sunn hemp mosaic virus, 266

T
Tetragonisca, 171
Tetrapedia, 175, 176
Tetrapedia diversipes, 176
Teuscheria wageneri, 168, 172, 177
Thea japonica, 120, 138
Theatrum Botanicum, 38, 39
Tlilxochitl, 236
Tomato mosaic virus, 266
Tomato spotted wilt virus, 262
Trichocentrum stipitatum, 176
Trichoglottis geminata, 57, 67, **68**, 138
Trigona, **163**, 164, 165, 171, 173
Trigona fulviventris, **163**, 171
Trigonium obtusum, 206
Tulipa, 142

V
Vanda dearei, 4, 6, 12, 25
Vanda hookeriana, 27
Vanda insignis, 67, **69**, 138
Vanda sanderiana, 4
Vanda tricolor, 48, **54**, 56, 65, 67, 87, 112, **125**, 127, 138
Vandaenopsis lissochiloides, 6
Vandopsis, 12, 14
Vandopsis gigantea, 12
Vanila pompona, 244
Vanilla, 138
Vanilla albida, 243
Vanilla aphylla, 243
Vanilla fragrans, 252
Vanilla planifolia, 234, 235, 243
Victoria regia, 2

W
Wheat streak mosaic virus, 263

X
Xylobium cf. *corrugatum*, 167
Xylobium latilabium, 168, 177
Xylobium leontoglossum, 167
Xylobium squalens, 167

Z
Zucchini yellow mosaic virus, 263
Zygopetalum, 252
Zygopetalum mackaii, 114
Zygopetalum mackayi, 104, **108**, 138
Zygopetalum maculatum, 138
Zygopetalum makayi, 87, 112
Zygostates lunata, 175

Subject Index

Page numbers in **bold face** refer to illustrations.

A

A Joy Forever, 22
Acting Director Gardens in 1970, 12
Aleurone grains, 170–172
Aranda Deborah, 4
Aranda Noorah Alsagoff, 15
Aranda Wendy Scott, **30**
Ascda Children's World, 19
Asian women, Singapore and Malaysia, 27
Avena coleoptile assays, 115–117, **125**
Award of Merit by MOS in 1964, 29
Award of Merit from OSSEA, 17
Award of Special Mention, 9
Awards Panel of OSSEA, 24

B

Belle orchid, 221
Blumenprofessor, 42
Bornean orchids, 25
Botanic Gardens, plant propagation, 12

C

Catasetum pollinarium, **41**
Cattleya flower, **102**
Cattleya pollinia, **41**
Cowpea chlorotic mottle virus (CCMV), 266
Christierara Nadeswari Desha, 19
Curator, Botanic Gardens, 1960, 12
Cut flowers
  commercial, 13
  Dendrobium, 10
  major export to Japan, 13
  Vanda, 10
Cymbidium mosaic virus (CymMV)
  genome organization and translational strategy, **255**
  inoculation, mode of transmission, 253
  molecular aspects of, 257
  natural hosts, 252–253
  orchid mosaic virus, 252
  polymerase chain reaction, 258
  population cloning strategy, 257
  potexvirus group, 254
  Taqman® real-time RT-PCR, 258
Cymbidium pollinarium, **41**
Cypripedium calceolus pollen grains, **189**

D

Darwin and orchid pollinia, **41**
Dead pollinia, 83, **90–91**
Dudaim
  as leah, banana and melom, **152**
  different interpretations of, **153**
  mandrake, **151**

E

Elaiophores
  hydrocarbon chains, 175–176
  oil secretion, 174–175
  types, 173–174
Encyclopedia Judaica, 154
Epidermal structures, 177
Eric Holttum Gold Medal, 17, 21, 23
European orchids, **107**
Extinct species, 12

F

Fitting's views. *See* Pollination effects
Flora Magica, 143
Floral food-hairs. *See also* Pseudopollen
  evolution, 176–177
  food content and ecology, 170–172
  trichome morphology, 169
Folia Orchidacae, 221
France, vanilla flower, 242–243
Futterhaare, 163

G

Gas chromatography-mass spectrometry, 173
Granite hills, Penang, 12
Groundnut rosette virus (GRV), 267
Guadeloupe, French island, 246

H

Haare, 163
Handbook on the Orchid, 9
Hans Fitting's experiments
  chemical nature, pollina
    Aerides odoratum, **84**, **86–87**
    others, **87–88**
    Phalaenopsis amabilis, **85–86**
  cold water extraction, **82**
  gynostemia and stigmas wounding
    Aerides odoratum, **67**
    Arachnanthe sulingi, **67**
    Calanthe veratrifolia, **67**
    Dendrobium antennatum, **67**
    Dendrobium macro phyllum, **67**
    Dendrobium superbum, **66**
    Dendrobium wardianum, **66**
    Oncidium flexuosum, **66**
    Phalaenopsis amabilis, **59–64**
    Phalaenopsis cornucervi, **65–66**
    Phalaenopsis esmeralda, **65**
    Phalaenopsis violacea, **64–65**
    Renanthera maingayi, **67**
    Rhynchostylis retusa, **66**
    Trichoglottis gemina, **67**
    Vanda insignis, **67**
    Vanda tricolor, **67**
  live and dead pollinia and pollen extracts
    Aerides odoratum, **78–80**
    Arachnanthe sulingi, **77**
    Coelogyne swaniana, **77**
    Cymbidium finlaysonianum, **75–76**
    Dendrobium superbum, **77**
    Phalaenopsis amabilis, **73–75**
    Phalaenopsis cornu-cervi, **80**
    Phalaenopsis violacea, **80–81**
    Rhynchostylis retusa, **78**
    Stanhopea insignis, **76**
    Stanhopea sp., **76**
  live and dead pollinia effects
    Aerides odoratum, **90**
    Calanthe veratrifolia, **91**
    Paphiopedilum glaucophyllum, **91**
    Phalaenopsis amabilis, **90**
  orchid pollination
    Aerides odoratum, **50**
    Coelogyne asperata and *Coelogyne pandurata*, **49**
    Cymbidium finlaysonianum, **48**
    Cymbidium sanguinolentum, **48**
    Dendrobium antennatum, **49**
    Dendrobium superbum, **50**
    Liparis latifolia and *Oncidium incurvum*, **48**
    Phalaenopsis amabilis and *Phalaenopsis esmeralda*, **49**
    Phalaenopsis violacea, **50**
    Rhynchostylis retusa, **46–47**
    Vanda tricolor, **48**
  other plants pollen effects
    A. *Hedychium* sp., **92–94**
    Aerides odoratum, **92**
    Hibiscus rosa-sinensis var. *genuinus* Hochr., **94–96**
    Impatiens rodrigesi, **94**
    Phalaenopsis amabilis, **92**
  ovaries swelling and greening, **105–106**
  pollen inserting location, **100–101**
  substances effects, **89**
Heade's work, orchids
  Cattleya labiata, 224–225
  chronology, 223–224
  Dendrobium stratiotes, 224
  humming bird paintings, 222–223
History of Orchidology in Southeast Asia, 11
Hormones
  Fitting's proposal, **115**
  gynostemium, 104, 109, 115, **118**, **123**
  Pollenhormon, 104, 109
  Wuchsstoff nature, **130**
Horticulture, Kew Botanic Gardens, 12
Hybrid *Bokchoonara* Khaw Bian Huat, 9
Hybrids, Singapore and Malaysia, 9

I

Iconography, 220
Indonesia, *Vanilla planifolia* flowers, 243

J
Japanese orchids
  Cymbidium virens, 109, 115, **118–120**
  experimental orchids, 109, **116**, **117**
  life span, 109, **121**
  Morita's research, 109
  Wuchsstoff nature, 130
Jewels of Medicine, **155**

K
Kandang Kerbau Hospital, 20
Khalbetzin orchids, 154, 156
Korean War, 16

L
Labellar features, food-hairs, **166–168**
Langkawi islands, 12
Large orchid shows, Singapore, 9
Leading hybridizer of orchids, 15
Leuenkephalin peptide, inhibitor, 269
Liege Botanical Gardens, Belgium, 239
Limestone hills, Perlis, 12
Live pollinia, 83, **90–91**
Long Beach, California, 12

M
Masdevallia glandulosa pollinia, **41**
Malayan Orchid Review, 12, 14, 22, 24
Malayan Orchid Society, 5, 8, 11, 12, 27, 28, 30, 31
Mandai Gardens of Singapore Orchids, 31
Mandai Orchid Garden, 5, 13
Maschmann, Ernst experiments, 130
Monad pollen grains
  elastoviscin engulfment, 190, 194
  isolated and aggregated tetrads, 194
  pollenkitt, 190
MOS, 8, 9
Müller' s research, 38-41, **40**, **41**
Museum National d'Histoire Naturelle, Paris, 243
Mustards seeds, 146
Mycaranthes, 165

N
Nectar, 160–161
Neoteny, 208
Non-orchidaceous plants, 83
Nuclear magnetic resonance analysis (NMR), 173

O
Odontoglossum ringspot virus (ORSV)
  genome organization and replication strategy, **256**
  molecular aspects of, 257
  natural hosts, 252–253
  orchid strain, 252
  Taqman® real-time RT-PCR, 258
  tobamovirus group, 254–255
  transmission modes, 253
Oncidium pollinarium, **41**
Ophrys sphegodes pollinium, **196**
Orchid Biology, Reviews and Perspectives I, 22
Orchid craze, 16
Orchid growing, setback, 7
Orchid hybridization, 4
Orchid Hybrids of Singapore, 5
Orchid infecting viruses
  Cymbidium mosaic and *Odontoglossum ringspot*
    Agrobacterium mediation, 261
    Brome mosaic virus, 259
    cell-to-cell movement, 265
    CP genes sequences variability, 259–261
    detection of, 257–258
    economic significance and incidence, 252
    full-length cDNA clones, synthesis of, 257
    hordeivirus γb protein, 266
    in vitro transcripts, 264–265
    mode of transmission, 253
    molecular aspects, 257
    molecular structure and composition, 254–256
    MP and CP gene complementation, 265–268
    natural hosts, 252–253
    plants species, susceptibility, **253**
    potexvirus and tobamovirus, 263
    potyviral HCpro, 266
    quasispecies concept, 260
    RdRp mutations study, 258–259
    RNA accumulation, 264
    synergism, 262–265
    TaqMan® real-time RT-PCR, 258
    transgenic orchids, regeneration of, 261–262
  foreign genes expression, 268–269
  Groundnut rosette virus
    co-inoculation of, 268
    GRV ORF3 protein, 267
    PVX CP-deficient mutant, 267

Orchid infecting viruses (*cont.*)
leuenkephalin peptide, 269
virus-resistant transgenic orchids, development of, 269
Orchid pollen grains
general stains, **193**
histochemical techniques, **191–192**
PDU types
fossil records, volume and vacuolisation, 202
generative cell and paternal inheritance, 203–204
pollen carbohydrate reserves, 202–203
pollen presentation, 203
pollen wall modifications, 201
tapetum and locular fluid, 197–200
tetrad adhesion, 200–201
tetrad types, 200
Orchid pollinia, toxicity, 42
Orchid Shows in Singapore, 25
Orchid Society of South East Asia, 22
Orchid Society of Southeast Asia, 12
Orchid Society of Thailand, 9
Orchidaceae, 187, 190, 200, 201
Orchidomania, 221
Orchids
camera lucida and Darwin's writings, 225–226
Cattleya labiata, 224–225
Cotton's work, 229
degrees of abstraction, 226–227
O'Keeffe's work, 227–228
in hummingbird paintings, 223–224
Western art
cultivation in United States, 221–222
floral symbolism, 221
Martin Johnson paintings, 222–223
orchidomania, 221
Orchids, Bibical and Talmudic literature
dudaim
as leah, banana and melom, **152**
different interpretations, **153**
mandrake, **151**
Egyptian and Aaron's staffs, **145**
embryos, **149**
Jewish writings
Encyclopedia Judaica, 154
Flower with an edible bulb, 156
khalbetzin, 154
Salep, **155**
mustard seeds *vs*. orchids seeds, 146–149
Orchis purpurea Huds
distribution, 142–143
species, **143**
structure, 142
ripe fruits, mature seeds, 147
Orchid Society of Southeast Asia (OSSEA), xxiv, 7, 9, 10, 17, 18, 23, 24, 26, 28, 33
Orchids of Sabah, 26
Orchis purpurea pollinium, **198–200**
Orchis pyramidalis pollinarium, **190**
Ornithology, 222, 225
OSSEA. *See* Orchid Society of Southeast Asia
OSSEA, 50th anniversary, National Stadium, 10
Ovary swelling and greening, 103–104, **105–106**

P

Padua Botanical Garden, 245–246
Paphiopedilum villosum pollen grains, **189**
Partially dehydrated pollen (PDP), 205–206
Partially hydrated pollen (PHP), 205–206
Penang Botanic Gardens, 11
Perlis and Kedah, 12
Phalaenopsis pollinia, **41**
Pollen dispersal units (PDUs)
definition and types, 187–188
developmental arrest and pollen longevity, 204–206
female part adaptations, 206
monad pollen grains
consistency, 195–197
elastoviscin engulfment, 190, 194
isolated and aggregated tetrads, 194
pollenkitt, 190
neoteny and inherent problems, 208–209
orchid pollen grains
fossil records, volume and vacuolisation, 202
general stains, **193**
generative cell and paternal inheritance, 203–204
histochemical techniques, **191–192**
pollen carbohydrate reserves, 202–203
pollen presentation, 203
pollen wall modifications, 201
tapetum and locular fluid, 197–200
tetrad adhesion, 200–201
tetrad types, 200
pollen load and paternity, 207–208
soft pollinium
type A, 194
type C, 195
Pollenhormon related auxin
active principle identity, 115

auxin discovery, 115–116
Avena coleoptile assays, 115–117, **125**
Laibach, Friedrich experiments, 117, **126–129**
Went's *Avena* coleoptile bioassay, 116–117
Wuchsstoff nature, 117, **130**
Pollination effects
baseline experiments, **46–50**
experimental species, **66**, **67**, **69**
hormones
Fitting's proposal, **115**
gynostemium, 104, 109, 115, **118**, **123**
Pollenhormon, 104, 109
Wuchsstoff nature, **130**
inter-specific and inter-generic pollination
effects on pollen, 83, **92–96**
live and dead pollinia, 83, **90–91**
Japanese orchids
Cymbidium virens, 109, 115, **118–120**
experimental orchids, 109, **116**, **117**
life span, 109, **121**
Wuchsstoff nature, 130
killed pollinia and pollen extract, 57, 72, 83
chemical nature, 83, **84–88**
cold water extraction, 72, **82**
effects of various substances, 83, **89**
gynostemia and stigmas wounding, **59–62**, **64**
inter- and intra specific pollination, **72**, **91**
live and dead pollinia and pollen extracts, **70**, **73–82**
steam-killed pollinia, **71**, **72**
non-orchidaceous plants
plants, 83, **97**
pollen effects, 83, **92–96**
ovary swelling and greening, 103–104, **105–106**
pollen insertion location effects
auxin students, 102, **124**
Cattleya flower, 98, **102**
flowers, 98
gynostemium, **100–101**
pollinia quarters, **99**
post-pollination phenomena, 98, **103**
rostellum, 98
Went's view, 102, 103
pollenhormon related auxin
active principle identity, 115
auxin discovery, 115–116
Avena coleoptile assays, 115–117, **125**
Laibach, Friedrich experiments, 117, **126–129**
Went's *Avena* coleoptile bioassay, 116–117
Wuchsstoff nature, 117, **130**
pollination and other applications to stigmas
Aerides odorata and *Aerides falcatum*, **55**
aqueous solutions, 56
auxin students, 56, **124**
baseline experiments, **46–50**
Cymbdium finlaysonianum, 56
Cymbidium sanguinolentum, 45
experimented orchids, **54**
lateral petals and floral aging, **51**
Liparis latifolia, 45
Phalaenopsis flowers, **52**
Rhynchostylis retusa, 45, **53**
stimulus concept, 56
post pollination phenomena
classification, 104
European orchids, 104, **107**
flowering, **110–114**
inducing/affecting factors, 104, **109**
research, 45
Tropical and European orchids, 104, **108**
wounding
effects and exceptions, 57
experimental orchids, **68**
gynostemia and stigmas, **59–67**
Phalaenopsis amabilis gynostemia, **58**
Post pollination phenomena
Cattleya (Guarianthe) flowers, **103**
classification, 104
European orchids, 104, **107**
flowering, **110–114**
inducing/affecting factors, 104, **109**
lateral petals and floral aging, **51**
ovary swelling and greening, **103–104**
Phalaenopsis flowers, **52**
research, 45
Tropical and European orchids, 104, **108**
Process of raising orchids from seed, 6
Protein body, 170
Pseudopollen
definition, 161–163
development, 170
occurence, 165, 169
terminology, 163–165
Pseudostamens. See Pseudopollen

R

Resin-like materials
  bee glue, 173
  secretion, 172
Rostellum, 98

S

Selective clonal propagation, 10
Seletar farm, 22
Seng Heng Nursery, 17
Singapore Botanic Gardens, 2, 4, 5, 11, 12, 17
Singapore Gardening Society, 7, 12
Singapore Guard Regiment, 11
Singapore-Malayan Orchid Hybrids list, 5
Singapore Orchid Shows, 14
Singapore Orchids, 6, 29, 32
Singapore Turf Club, 9
Singapore's orchid circle, 9
Smooth endoplasmic reticulum (SER), 172
Soft pollinium
  type A, 194
  type C, 195
Spathoglottis Chrysops, 4
Spathoglottis Primrose, 4
Stanhopea pollinarium, **41**
Steam-killed pollinia, **71**, **72**
Stigmas
  Aerides odoratum and *Aerides falcatum*, **55**
  aqueous solutions, 56
  auxin students, 56, **124**
  baseline experiments, **46–50**
  Cymbdium finlaysonianum, 56
  Cymbidium sanguinolentum, 45
  experimented orchids, **54**
  lateral petals and floral aging, **51**
  Liparis latifolia, 45
  Phalaenopsis flowers, **52**
  Rhynchostylis retusa, 45, **53**
  stimulus concept, 56
Sun Kee Nursery, 13
Synergism
  potexvirus and tobamovirus, 263
  potyvirus, 262
  PVX- *Potato virus Y* (PVY) synergism, 262
Syonan-to Botanic Gardens. *See* Singapore Botanic Gardens, 4

T

Tenom Orchid Centre, 26
Tercüme-I Cedide Fil-Havasil Müfrede, **155**
Theatrum Botanicum, **39**
Trichomes
  elaiophores
    hydrocarbon chains, 175–176
    oil secretion, 174–175
    types, 173–174
  morphology, 169
  resin-like and waxy material secretion, 172–173
Tulips, 220

V

Vanda Dawn Nishimura, 18, 19, 20, 21
Vanda hybrid, 3
Vanda Miss Joaquim, 2, 4, 9, 32, 350
Vandaenopsis (now renamed Paravanda) hybrids, 6
Vanilla flower
  Badianus Codex, 236–237, **237**
  capsules, **235**
  edible plantation crop, 234
  female euglossine bees, 234
  floral parts, **234**
  flower, **236**
  hand-pollination
    in Belgium, 239–242
    in France, 242–243
    in Guadeloupe, 246
    in Indonesia, 243
    in Padua Botanical Garden, 245–246
    in Reunion Island, Philippines, 244–245
  ice-cream orchids, 234
  in Mexico, 238–239
  little pod, 238
  melipona bees, natural pollinator, 235
  Morren family, 239–241
  persons associated with, **238**
Vdnps. Patience, 6

W

Waxy material secretion, 172–173
WOC, 9, 11
World Orchid Conference (WOC), 8, 9, 12, 17, 26, 28
World War II, 44
Wounding
  effects and exceptions, 57
  experimental orchids, **68**
  gynostemia and stigmas, **59–67**
  Phalaenopsis amabilis gynostemia, **58**

Y

Yellow rain tree (Samanea samaan), 20

Contents of *Orchid Biology, Reviews and Perspectives*, Volumes I-IX

***Orchid Biology, Reviews and Perspectives*, Volume I, 1977**

Edited by Joseph Arditti.
Published by Cornell University Press, Ithaca, New York.

1. A Personal View of Orchids. - Richard E. Holttum.
2. Fossil History of the *Orchidaceae*. - Rudolf Schmid and Marvin J. Schmid.
3. Orchids in Rumphius' *Herbarium Amboinense*. - Hendrik C. D. de Wit.
4. The Distribution and Chemistry of Alkaloids in the *Orchidaceae*. - Michael B. Slaytor.
5. Anthocyanins of the *Orchidaceae*: Distribution, Heredity, Functions, Synthesis, and Localization. - Joseph Arditti and Michael H. Fisch.
6. Vitamin Requirements and Metabolism in Orchids. - Joseph Arditti and Charles R Harrison.
7. Variations in Clonal Propagation. - Thavorn Vajrabhaya.

Appendix: Clonal Propagation of Orchids by Means of Tissue Culture - A Manual. - Joseph Arditti.

***Orchid Biology, Reviews and Perspectives*, Volume II, 1982**

Edited by Joseph Arditti.
Published by Cornell University Press, Ithaca, New York.

1. Auyán-tepui: Reminiscences of an Orchid Search. - G. C. K. Dunsterville and E. Dunsterville.
2. A General Review of the Orchid Flora of China. - Chen Sing-chi and Tang Tsin.
3. Orchid Mycorrhiza. - Geoffrey Hadley.
4. The Biology of Orchids and Euglossine Bees. - Norris H. Williams.
5. Carbon Fixation in Orchids. - Popuri Nageswara Avadhani, Chong Jin Goh, Adisheshappa Nagaraja Rao, and Joseph Arditti.
6. Mineral Nutrition of Orchids. - Hugh A. Poole and Thomas J. Sheehan.

7. Flower Induction and Physiology in Orchids. - Chong Jin Goh, Michael S. Strauss, and Joseph Arditti.

Appendix: Orchid Seed Germination and Seedling Culture - A Manual. - Joseph Arditti, Mark A. Clements, Gertrud Fast, Geoffrey Hadley, Goro Nishimura, and Robert Ernst.

***Orchid Biology, Reviews and Perspectives*, Volume III, 1984**

Edited by Joseph Arditti.
Published by Cornell University Press, Ithaca, New York.

1. Orchids: Their Innocent Past, Their Promising Yet Perilous Future. - Rebecca Tyson Northen.
2. Ethnobotany of the *Orchidaceae*. - Leonard J. Lawler.
3. Orchid Phytoalexins. - Albert Stoessl and Joseph Arditti.
4. Physiology of Germinating Orchid Seeds. - Joseph Arditti and Robert Ernst.
5. Carbohydrates of the *Orchidaceae*. - Robert Ernst and Eloy Rodriguez.
6. Toxic and Allergenic Orchids. - Björn M. Hausen.
7. A Reassessment of the Sectional Limits in the Genus *Cymbidium* Swartz. - Christopher J. Seth and Phillip J. Cribb.

Appendix: Chromosomes in Orchids: Counting and Numbers. - Ryuso Tanaka and Haruyuki Kamemoto.

***Orchid Biology, Reviews and Perspectives*, Volume IV, 1987**

Edited by Joseph Arditti.
Published by Cornell University Press, Ithaca, New York.

1. Orchidology in Southeast Asia: A History. - Arthur George Alphonso.
2. Major Patterns and Processes in Orchid Evolution: A Critical Synthesis. - David H. Benzing.
3. Pollination of *Orchis* and Related Genera: Evolution from Reward to Deception. - Amots Dafni.
4. Orchid Stomata - Structure, Differentiation, Function, and Phylogeny. - Hanne Rasmussen.

5. The Velamen and Exodermis of Orchid Roots. - Alec M. Pridgeon.
6. Development of Pollen and Accessory Structures in Orchids. - Edward C. Yeung.
7. Respiration in Orchids. - Choy Sin Hew.

Appendix: Orchid Diseases - A Compendium. - Geoffrey Hadley, Mastura Arditti, and Joseph Arditti.

***Orchid Biology, Reviews and Perspectives*, Volume V, 1990**

Edited by Joseph Arditti.
Published by Timber Press, Portland, Oregon.

1. Orchids in My Life. - Frits W. Went.
2. The Western Australian Fully Subterranean Orchid. - Kingsley W. Dixon, John S. Pate, and John Kuo.
3. Water Relations in Orchids. - Russell Sinclair.
4. Auto-Pollination in the *Orchidaceae*. - Paul M. Catling.
5. A Review of the Genus *Dactylorhiza*. - Leonid V. Averyanov.
6. Power and Passion: The Orchid in Literature. - Martha W. Hoffman Lewis.
7. Perspectives of Tropical Orchids in Space Research. - T. M. Czerevczenko and I. V. Kosakovskaya.

Appendix: Flowering Months of Orchid Species under Cultivation. - Robert M. Hamilton.

***Orchid Biology, Reviews and Perspectives*, Volume VI, 1994**

Edited by Joseph Arditti.
Published by John Wiley and Sons, New York, New York.

1. Long-Long Misis Bilong Plaua. - Andrée N. Millar.
2. History of Orchids in Europe, from Antiquity to the 17th Century. - Pierre Jacquet.
3. Interactions Between Orchids and Ants. - Rod Peakall.
4. Resupination. - Robert Ernst and Joseph Arditti.
5. Physiology of Orchid Flowers. - Popuri Nageswara Avadhani, Helen Nair, Joseph Arditti, and Choy Sin Hew.

6. Orchid Cut-Flower Production in ASEAN Countries. - Choy Sin Hew.
7. Fly Pollination in the *Orchidaceae*. - Dorte E. Christensen.

Appendix: Orchid Pests - A Compendium. - Sing Kwong Chuo, Robert Ernst, Joseph Arditti, and Choy Sin Hew.

***Orchid Biology, Reviews and Perspectives*, Volume VII, 1997**

Edited by Joseph Arditti and Alec M. Pridgeon.
Published by Kluwer Academic Publishers, Dordrecht, The Netherlands; Boston MA, USA and London, U. K.

1. Brown Danube, Blue Pacific. - Joseph Arditti.
2. Ovule and Megagametophyte Development in Orchids. - Edward C. Yeung and Sandra K. Law.
3. Molecular Biology of Orchids. - Adelheid R. Kuehnle.
4. Fungi from Orchid Mycorrhizas. - Randolph S. Currah, Carla D. Zelmer, Sarah Hambleton, and Katherine A. Richardson.
5. Orchid Production and Research in Japan. - Syoichi Ichihachi
6. Three Orchids Used as Herbal Medicines in China: An Attempt to Reconcile Chinese and Western Pharmacology. - Choy Sin Hew, Joseph Arditti and Wen Shu Lin.
7. Research on Micropropagation of *Cymbidium, nobile*-type *Dendrobium,* and *Phalaenopsis* in Japan. - Syoichi Ichihachi.

Appendix: Orchid Potting Mixtures - An Abridged Historical Review. - Thomas J. Sheehan.

***Orchid Biology, Reviews and Perspectives*, Volume VIII, 2002**

Edited by Tiiu Kull and Joseph Arditti.
Published by Kluwer Academic Publishers, Dordrecht, The Netherlands; Boston MA, USA and London, U. K.

1. Plinius Germanicus and Plinius Indicus: Sixteenth and Seventeenth Century Descriptions and Illustrations of Orchid 'Trash Baskets', Resupination, Seeds, Floral Segments and Flower Senescence in the European Botanical Literature. - U. Wehner, W. Zierau, J. Arditti.

2. Ontogeny of Orchid Flowers. - H. Kurzweil, A. Kocyan.
3. Population Dynamics of North Temperate Orchids. - T. Kull.
4. Development of Orchid Seeds and Seedlings. - T. Vinogradova, E.V. Andronova.
5. Temperate Oriental Cymbidium Species. - Kee-Yoeup Paek, H.N. Murthy.
6. Orchid Embryos. - T. Wing Yam, E.C. Yeung, Xiu-Lin Ye, Sze-Yong Zee, J. Arditti.
7. Orchid Seeds and their Germination: An Historical Account. - T. Wing Yam, H. Nair, Choy Sin Hew, J. Arditti.

Appendix: Orchid Viruses - A Compendium. - Sek Man Wong.

***Orchid Biology, Reviews and Perspectives*, Volume IX, 2007**

Edited by Kenneth Cameron, Joseph Arditti, and Tiiu Kull.
Published by New York Botanical Garden Press, New York, New York.

1. Equatorial Ruminations. - Calaway H. Dodson.
2. Orchid Discoveries by French Botanists Around the World. - Pierre Jacquet.
3. Growth Habits of Temperate Terrestrial Orchids. - Irina V. Tatarenko.
4. Molecular Phylogenetics of Orchidaceae: The First Decade of DNA Sequencing. - Kenneth M. Cameron.
5. Pollination of Orchids by Lepidoptera: Outcrossing by Long Distance Transport. - Nelis A. Ven der Cingel.
6. The Origin of *Vanda* Miss Joaquim. - Joseph Arditti and Choy Sin Hew.
7. A Translation of the Writings of Noël Bernard. - Pierre Jacquet.

Appendix: Time from Pollination to Fruit Ripening, Seed Maturation and Germination. - Tim Wing Yam, Abdul Karim Abdul Ghani, Syoichi Ichihashi, Aung Thame, Adisheshappa Nagaraja Rao, Popuri Nageswara Avadhani, Helen Nair, Chou Sin Hew, Joseph Arditti and Irina Tatarenko.